27 Springer Series in Solid-State Sciences

Edited by Peter Fulde

Springer Series in Solid-State Sciences

Editors: M. Cardona P. Fulde H.-J. Queisser

Volumes 1 - 39 are listed on the back inside cover

S. V. Vonsovsky   Yu. A. Izyumov   E. Z. Kurmaev

Superconductivity of Transition Metals

Their Alloys and Compounds

Translated by E. H. Brandt and A. P. Zavarnitsyn

With 182 Figures

Springer-Verlag Berlin Heidelberg New York 1982

Professor Dr. *Serghey V. Vonsovsky*
Professor Dr. *Yuri A. Izyumov*
Dr. *Ernst Z. Kurmaev*

Ural Science Research Center of the USSR Academy of Sciences, Pervomaiskaya 91
GSP-169 Sverdlovsk, USSR

Translators:

Dr. *Ernst H. Brandt*

Max-Planck-Institut für Metallphysik, Institut für Physik, Heisenbergstraße 1
D-7000 Stuttgart, 80, Fed. Rep. of Germany

A.P. Zavarnitsyn
Ural Science Research Center of the USSR Academy of Sciences, Pervomaiskaya 91
GSP-169 Sverdlovsk, USSR

Series Editors:

Professor Dr. Manuel Cardona
Professor Dr. Peter Fulde
Professor Dr. Hans-Joachim Queisser

Max-Planck-Institut für Festkörperforschung, Heisenbergstraße 1
D-7000 Stuttgart 80, Fed. Rep. of Germany

Title of the original Russian edition:
Svjerchprovodimost' pjerjechodnych mjetallov ich splavov i sojedinjenij.

ISBN-13: 978-3-642-61821-5 e-ISBN-13: 978-3-642-61819-2
DOI: 10.1007/978-3-642-61819-2

Softcover reprint of the hardcover 1st edition 1982

Offset printing: Beltz Offsetdruck, 6944 Hemsbach/Bergstr. Bookbinding: J. Schäffer oHG, 6718 Grünstadt.
2153/3130-5 4 3 2 1 0

Foreword

This book should fill a gap which has existed in the literature on superconductivity. There have been a number of excellent textbooks available on the *phenomenon* of superconductivity, which describe in detail the variety of effects connected with it and the mathematical techniques to deal with them properly. However, until now there has not been a textbook available in English which concentrates on the *material* aspects of superconductivity. This is a major shortcoming since most physicists working in the field of superconductivity are mainly concerned with specific materials and subsequently often need to know more about the interplay of superconductivity and material properties. On the other hand, people working in the field know that a competent and well-written book by S.V. Vonsovsky, Y.A. Izyumov, and E.Z. Kurmaev has been available in Russian. It presents a thorough discussion of superconducting transition-metal alloys and compounds.

This volume is a translation of an updated version of the Russian edition. The translation was done by A.P. Zavarnitsyn of the authors' institution and by Dr. E.H. Brandt of the Max-Planck-Institut für Metallforschung in Stuttgart. The master manuscript was ably typed by Mrs. C. Pendl. I would like to take this opportunity to thank each of them.

I sincerely hope that the book will turn out to be useful to physicists working in the field of superconductivity as well as to nonspecialists and interested graduate students.

Stuttgart, February 1982 *Peter Fulde*

Contents

1. Introduction

In this introductory chapter we present the basic properties of superconductors on an elementary level. Superconductivity is first reviewed by considering its occurence in pure elements, in alloys and in chemical compounds. Therefore, we shall summarize the essential results of the theoretical concepts, deferring details to the two subsequent chapters.

1.1 The Distribution of Superconductors Among the Elements of the Periodic Table, Their Alloys and Compounds

1.1.1 Superconductivity in Pure Elements

A large number of superconducting elements have been discovered in recent years. The traditional point of view that superconductivity is a rare and anomalous property should therefore be regarded as obsolete. If we turn our attention to the distribution of superconductors in the periodic table (Fig. 1.1 and Table 1.1) it will be readily seen that most of the metallic elements are superconductors at low temperatures [1.1-4]. The family of new superconducting elements has become broader. On the one hand, this is due to the production of sufficient amounts of very pure substances free from impurities (for example, superconductivity has been discovered in lutecium [1.2] and americium [1.4]) and due to a closer approach to the absolute zero in low-temperature measurements (T_c of W has been found to be 0.0154 K). On the other hand, this results from the performance of measurements under high-pressure conditions [Ref.1.1, Table 1.2] where many of the nontransition elements (Si, P, Ge, As, Se, Sb, Te, Bi) and a number of metals (Y, Cs, Ba)

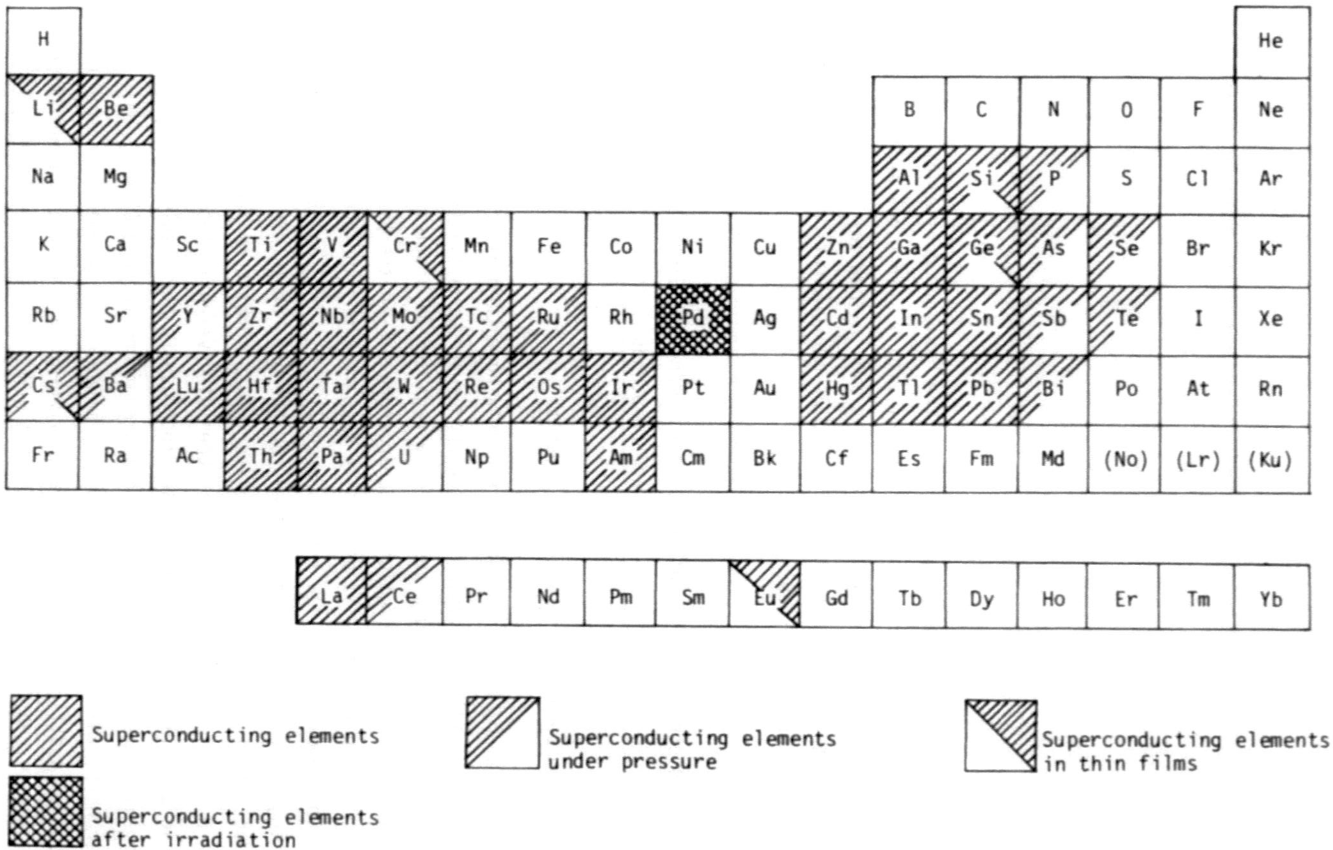

Fig. 1.1. The superconducting elements of the periodic table

Table 1.1. Transition temperature T_c and critical field H_0 for pure elements

Element	T_c [K]	H_0 [Oe]	Element	T_c [K]	H_0 [Oe]
Al	1.175 ± 0.002	104.9 ± 0.3	Nb	9.25 ± 0.02	2060 ± 50
Am	0.6 - 1.0		Os	0.66 ± 0.03	70
Be	0.026		Pa	1.4	
Cd	0.517 ± 0.002	28 ± 1	Pb	7.196 ± 0.006	803 ± 1
Ga	1.083 ± 0.001	58.3 ± 0.2	Re	1.697 ± 0.006	200 ± 5
Ga(β)	5.9 - 6.2	560	Ru	0.49 ± 0.015	69 ± 2
Ga(γ)	7	950	Sn	3.722 ± 0.001	305 ± 2
Ga(Δ)	7.85	815	Ta	4.47 ± 0.04	829 ± 6
Hf	0.128	12.7	Tc	7.8 ± 0.1	1410
Hg(α)	4.154 ± 0.001	411 ± 2	Th	1.38 ± 0.02	160 ± 3
Hg(β)	3.949	339	Ti	0.40 ± 0.04	56
In	3.408 ± 0.001	281.5 ± 2	Tl	2.38 ± 0.02	178 ± 2
Ir	0.1125 ± 0.001	16 ± 5	V	5.40 ± 0.05	1408
La(α)	4.88 ± 0.02	800 ± 10	W	0.0154±0.0005	1.15 ± 0.03
La(β)	6.00 ± 0.1	1600 - 1096	Zn	0.85 ± 0.01	54 ± 0.3
Lu	0.1 ± 0.03	350 ± 50	Zr	0.61 ± 0.15	47
Mo	0.915 ± 0.005	96 ± 3	Zr(ω)	0.65, 0.95	

have been found to exhibit superconducting properties. Several new superconductors, for instance, Li, Cr, Si, Eu have been discovered in thin films[1] [1.8,10,11], including Pd films irradiated by α-particles [1.12,13].

In MATTHIAS' [1.3] opinion, the present situation has taken such a turn that rather than be concerned with the question of why some elements become superconductors one should go into the problem of why all the elements of the periodic table are not superconductors.

To give an answer to this question let us see what elements may not yet be classed as superconducting ones. Primarily, these are metals possessing atomic magnetic order. Here belong transition metals with incomplete d and f shells (Co, Fe, Ni, Gd, etc.). This indicates that superconductivity and atomic magnetism are mutually exclusive phenomena.

1 Subsequent measurements [1.6,7] have not confirmed that superconductivity exists in Mg, Ce, Pr, Nd, Yb [1.5,8,9].

Table 1.2. Superconductivity of pure elements at high pressure

Element	T_c [K]	P [kbar]	Element	T_c [K]	P [kbar]
Al	1.98 - 0.075	0 - 62	La	∿5.5 - 12.9	0 - 210
As	0.31 - 0.50	220 - 140	Lu	0.022 - 1.0	45 - 190
	0.20 - 0.25	∿140 - 100	P	5.8	170
Ba II	∿1 - 1.8	∿55 - 85	Pb II	3.55	160
Ba III	1.8 - 5	∿85 - 144	Re II	2.3	"Plastic compression"
Ba IV	4.5 - 5.4	144 - 190	Sb (prepared: 120 kbar, held below 77 K)	2.6 - 2.7	
Bi II	3.9	25 - 27	Sb III	3.55 - 3.40	85 - 150
Bi III	6.55 - 7.25	28 - 38	Se II	6.75, 6.95	∿130
Bi IV	7.0, 8.7 - 6.0	43, 43 - 62	Si	6.7 - 7.1	120 - 130
Bi V	6.7, 8.3	48 - 80	Sn II	5.2 - 4.85	125 - 160
Bi VI	8.55	90, 92 - 101	Sn III	5.30	113
Bi VII	8.2	30	Te II	2.4 - 5.1	38 - 55
Ce (α)	0.020 - 0.045	20 - 35	Te III	4.1 - 4.2	∿53 - 62
Ce (α')	1.9 - 1.3	45 - 125	Te IV	4.72 - 4.0	63 - 80
Cs V	∿1.5	>125	Te(?)	3.3 - 2.8	100 - 260
Ga II	6.38	≥35	Tl (cubic form)	1.45	35
Ga II'	7.5	≥35 then P removed	Tl (hexagonal form)	1.95	35
Ge	5.35	115	U	2.4 - 0.4	10 - 85
			Y	2.3 - 1.7 - - 2.5	110 - 125 - - 160
			Zr(ω form, metastable	1 - 1.7	60 - 130

Note: II - VI correspond to different structural modifications resulting from polymorphous transformations of elements under pressure

A comparatively small number of superconductors have been revealed in alkali, alkali-earth and noble metals, in rare earths and in actinides. It would, however, be indiscreet here to predict that the above elements are completely "indisposed" to superconductivity. An example indicative of this is the discovery of superconductivity under pressure in Cs and Ba or in thin films of Li, Cr, Pd and Eu. It is quite probable that superconductivity may subsequently be detected in other elements of this group.

The discovery of superconductivity in dilute solid solutions of Gd, Al, In, Sn, and in noble metals (Cu, Ag, Au) has been reported in [1.14]. When investigating the dependence of T_c on the composition in these alloys, extrapolated values of T_c for pure elements were obtained: $T_c(Au) = 2\cdot10^{-4}$ K, $T_c(Ag) = 4\cdot10^{-9}$ K, $T_c(Cu) = 6\cdot10^{-10}$ K.

The superconducting transition temperature T_c of pure elements varies over a very wide range. Under normal conditions (at atmospheric pressure, in bulk material) the highest $T_c = 7.19$ K for nontransition elements belongs to Pb. For transition metals the highest T_c have been registered in Nb ($T_c = 9.25$ K) and Tc ($T_c = 7.8$ K). In thin films of a number of transition and nontransition elements it has been possible to attain a substantial increase of T_c, for example, in W from 0.0154 K to 4.1-5.5 K [1.1,10], in Re from 1.7 K to 6.7 K [1.15], in Be from 0.026 K to 9.75 K [1.1]. An increase in the value of T_c has also been observed in a number of cases under application of high pressure, for example, in La from 5.5 K to 12.9 K at 200 kbar [1.16].

Thus, the T_c of pure elements under normal conditions does not exceed 10 K. By employing special methods of specimen preparation (in thin films of a number of elements) or under high-pressure conditions it becomes possible in special cases to obtain T_c = 10 to 12 K. However, no considerable measure of success has been achieved in this direction as yet. A much larger effect, as will be seen below, has been obtained in binary and ternary systems.

1.1.2 Superconductivity in Alloys

In this subsection we make a cursory inspection of how superconductivity manifests itself in alloys. When elements are alloyed with one another, random alloys such as solid solutions and ordered alloys such as intermetallic compounds, as well as various chemical compounds, are liable to be produced.

As a rule, disordered solid solutions form when we alloy metals which are neighbors in the periodic table. It is among these materials that the superconductors now extensively used in the production of wire for superconducting solenoids were discovered. These are Nb-Ti and Nb-Zr alloys whose T_c is $\sim$ 10 K. Practically all combinations yielding solid solutions have been tried by now. It has been found that only Mo-Tc and Mo-Re alloys have $T_c \sim 11$ to 14 K and, consequently, may compete with Nb-Ti and Nb-Zr alloys.

The superconducting properties have been studied for several hundreds of intermetallic compounds belonging to twenty different structural types (Table 1.3) [1.1, 17-19]. The intermetallic compounds under consideration are listed in the table on the opposite page specifying the characteristics of each structural type with their maximum superconducting transition temperature.

As is evident from the table, the occurrence of superconductors with $T_c > 10$ K is by no means a rare phenomenon in intermetallic compounds. It is worth noting that for each structural type Table 1.3 gives only one representative with a maximum T_c, although within each type of structure there may also exist other compounds with $T_c > 10$ K. It stands to reason that among the intermetallic compounds belonging to the same structural type there may be some that have an appreciable spread in the values of T_c and a number of representatives exhibit no superconductivity at all. This list of the structural types of the intermetallides in which superconductivity has been found to exist just reflects the present situation. It should be continuously updated as new types of intermetallic compounds are studied.

It is noteworthy that the symmetry of intermetallic compounds is highly varied. As can be seen from Table 1.3, and also from Table 1.4, superconductivity has been detected in structures of all crystal systems, and the cubic crystal system is in no way extraordinary or particularly favorable to the occurence of superconductivity, as was otherwise thought earlier.

In Table 1.3 the intermetallides are arranged in a rather arbitrary order, but the compounds whose structural type has received the most attention are among those placed at the top: A-15, C-15, C-14, B-2, A-12, σ phase. The most prominent and interesting of them are undoubtedly the compounds with A-15 structure. Their intensive study began 25 years ago with HARDY and HULM's [1.20] discovery of superconductivity in V_3Si, which has $T_c = 17$ K and possesses this crystal structure. Later, about a dozen compounds with $T_c > 15$ K were revealed in this type of structure: Nb_3Sn (18 K), Nb_3Al (18.8 K), Nb_3Ga (20.3 K), Nb_3Si (19 K), Mo_3Re (15 K) and others. Finally, the compound Nb_3Ge with a superconducting transition temperature exceeding 23 K, which is specified in Table 1.3, was synthesized in 1974 [1.21]. This discovery is of crucial importance since it enables us to use liquid hydrogen instead of liquid helium as a cooling agent [1.22]. Such high values of T_c have naturally caught the attention of many researchers and the physical properties of the compounds

Table 1.3. Superconductivity in alloys

Type of structure	Symmetry	Compound	T_c [K]
	Solid solutions		
A-2	cubic	$Nb_{0.75}Zr_{0.25}$	11.0
A-2	cubic	$Nb_{0.75}Ti_{0.25}$	10.0
	Intermetallides		
A-15	cubic	Nb_3Ge	23.2
C-15	cubic	$(Hf_{0.5}Zr_{0.5})V_2$	10.1
C-14	hexagonal	$ZrRe_2$	6.4
B-2	cubic	VRu	5.0
A-12	cubic	$NbTc_3$	10.5
$D\text{-}8_b$	tetragonal	$Mo_{0.38}Re_{0.62}$	14.6
C-16	tetragonal	$RhZr_2$	11.1
$L\text{-}1_2$	cubic	$NbRu_3$	15-16
$B\text{-}8_1$	hexagonal	BiNi	4.25
B-20	cubic	AuBe	2.64
B-31	rhombic	GeIr	4.70
C-1	cubic	$Ga_{0.7}Pt_{0.3}$	2.9
C_c	tetragonal	Ge_2Y	3.8
$D\text{-}1_c$	rhombic	$AuSn_4$	2.38
$D\text{-}2_d$	hexagonal	Au_5Ba	0.7
$D\text{-}8_8$	hexagonal	Pb_3Zr_5	4.60
$E\text{-}9_3$	cubic	$RhZr_3$	11.0
$L\text{-}1_0$	tetragonal	NaBi	2.25
C-40	hexagonal	$NbGe_2$	16.0

Note: For each type of structure there are specified representatives with maximum T_c

with A-15 structure have been studied closely. The results of these investigations have stimulated quests for superconductivity in other classes of intermetallic compounds. However, many of these compounds are produced by complicated reactions. Thus, their physical properties very often depend on conditions of synthesis. For a long time this field was at a metallurgical stage of evolution when attention was centered on the development of methods for synthesizing pure compounds with a given stoichiometry. However, nowadays effective methods of producing such compounds (including single crystals) have been devised and one can notice clearly defined transition from the metallurgical stage to a stage of systematic physical research.

1.1.3 Superconductivity in Chemical Compounds

A still greater variety of forms in which superconductivity manifests itself is observed in chemical compounds. In contrast to the intermetallides considered above, the chemical compounds are expected to possess a considerable share of ionic and covalent bonds along with the metallic type of bond. The crystal structures of these compounds are formed with the participation of both metallic and nonmetallic atoms or with the participation of nonmetallic atoms only. A typical example of such systems represent compounds with B-1 structure (NaCl). An extraordinarily large number of compounds involving hydrides, carbides, nitrides, oxides of transition metals and their solid solutions crystallize in the B-1 structure. The superconducting transition temperature for individual compounds with B-1 structure exceeds 17 K (Table 1.4 [1.1,17,18,21,23-29]). When considering this type of compounds we encounter a new phenomenon, namely: high T_c may occur in compounds whose components have low T_c or are not superconductors at all. For example, WC has T_c = 10.0 K, ZrN and MoC have T_c = 10.7 K and 14.3 K, respectively, whereas the pure metals W, Zr and Mo have T_c < 1 K, and carbon and nitrogen possess no superconductivity whatsoever.

This situation is typical not only of B-1 structure compounds but also of other types of chemical compounds, for example, sulphides. For the compound $Li_{1.1}TiS_2$ (structural type Ti_3S_4) T_c is about 13 K [1.29], but its constituents Li and S are not superconductors and Ti has T_c = 0.40 K. Moreover, neither of the binary compounds Li-S and Ti-S exhibits superconductivity.

Synthesis of an ever-increasing number of chemical compounds leads to the detection of superconductivity in classes of compounds which have never been regarded as promising. For example, a spinel-type fcc structure was long considered to be unfavorable for superconductivity. Of the 200 known spinel compounds, only three have been found to be superconductive, with T_c not exceeding 5 K. Nevertheless, a spinel structure compound $LiTi_2O_4$ with $T_c \sim 13.7$ K was synthesized in 1973 [1.27].

Analysis of the data given in Table 1.4 shows that, just as in the intermetallides, in this class of chemical compounds we often encounter superconductors with T_c > 10 K. It must be said in addition to this that high T_c

Table 1.4. Superconductivity in chemical compounds

Type of structure	Symmetry	Compound	T_c [K]
	Chemical compounds		
B-1	cubic	NbN	17.3
$D\text{-}5_c$	cubic	$(Y_{0.7}Th_{0.3})_2C_{3.1}$	17.0
Ti_3S_4	hexagonal	$Li_{0.1-0.3}Ti_{1.1}S_2$	10 - 13
Mo_3Se_4	rhombic	$Pb_{0.92}Mo_6S_7$	15.2
$H\text{-}I_1$	cubic	$LiTi_2O_4$	13.7
	hexagonal	MoN	13 - 14.8
C-2	cubic	$Rh_{0.53}Se_{0.47}$	6.0
C-6	trigonal	PdTe	1.53
$D\text{-}0_e$	tetragonal	Mo_3P	5.31
C-32	hexagonal	$B_5Mo_{1.69}Zr_{0.31}$	11.2
	tetragonal	$Lu_{0.75}Th_{0.25}Rh_4B_4$	11.9
	hexagonal	Nb_3S_4	4.0
C-49	orthorhombic	$ZrGe_2$	8.0
	Semiconductors		
B-1	cubic	SnTe	0.02 - 1.1
B-1	cubic	GeTe	0.06 - 0.41
$D\text{-}7_3$	cubic	La_2Se_3	1.0
$E\text{-}2_1$	cubic	$SrTiO_3$	0.03 - 0.35
B-1	cubic	InTe	1.0 - 3.5
	Layered compounds		
C-27	hexagonal	$NbSe_2$	7.0
	hexagonal	NbS_2	5.4
	hexagonal	TaS_2	0.8
	hexagonal	TaS_2 (triton-13)	5.0
	hexagonal	TaS_2 (3-ethylpyridine)$_{0.29}$	4.50
	hexagonal	$C_{8.8}K$	0.125
	Polymers		
	monoclinic	(SN)x	0.26

Note: For each type of structure there are specified representatives with maximum T_c

in these compounds is not infrequently accompanied by high critical fields. Recently it has been reported [1.28] that in the $PbMo_{5.1}S_6$ sulphide (T_c = 14.4 K) the second critical field H_{c2} reaches a record value of 510 kOe at 4.2 K.

Until now we have been considering chemical compounds with metallic conductivity in the normal state. However, in 1964 superconductivity was found to exist in compounds possessing conductivity of a semiconductor, i.e., a low current carrier concentration [1.30]. In these compounds the T_c are comparatively low and, as a rule, do not exceed 1 K. Nevertheless, they merit considerable attention primarily on a theoretical plane, and also in conncetion with the fact that they combine superconductivity with a number of physical properties which are absent in metallic systems (for example, with ferroelectricity). The best known compounds of this type are SnTe, InTe, GeTe, La_2Se_3 and $SrTiO_3$ (a well-known ferro-electric).

For the study of superconductivity, the chemical compounds are particularly interesting because of their wide variety of structural types, including structures with pronounced anisotropy. From this point of view, the so-called layered compounds (mainly transition-metal dichalcogenides such as MeX_2, Me being a transition metal of group IV, V or VI and X standing for S, Se or Te) form an interesting family of superconductors [1.31]. Potassium-intercalated graphite (the compound $C_{8.8}K$, T_c = 0.125 [1.32]) also belongs here. A structural peculiarity of these compounds is the presence of two-dimensional X-Me-X layers (or sandwiches) bound to each other by weak van der Waals forces. It has been possible very often to produce intercalated sandwich compounds such as $AMeX_2$, A being an atom or a molecule (mostly of an organic type) occupying the intermediate sites between adjacent layers. The superconductivity in these compounds exhibits pronounced layered anisotropy. The T_c (from 3 to 7 K) may be shifted by pressure or by intercalation with organic molecules [1.31]. Irrespective of their possible applications, such anisotropic structures are of great theoretical interest since they are examples of quasi-two-dimensional systems [1.33]. Quite recently quasi-one-dimensional superconducting systems have been synthesized, too. Only one representative of this type, a polymer $(SN)_x$ of monoclinic structure with T_c = 0.26 K, is known so far. Sulphure and nitrogen atoms form families of quasi-one-dimensional chains - S - N - S - N - with weak coupling between the chains [1.34]. Apart from the fact that this compound simulates the one-dimensional superconducting system, it is one of the few superconductors which have been constructed without any participation of metallic atoms at all. This discovery apparently marks the beginning of a new stage - the study of superconductivity in organic compounds.

1.1.4 Outlook

The foregoing consideration shows that the manifestation of superconductivity in alloys and chemical compounds has a great variety of forms. Undoubtedly, it is of interest to study in detail each of the above-mentioned types of compounds and to analyze the results to get insight into the causes that lead to the variation of the superconducting characteristics. It would be reasonable to start such analysis with the compounds possessing the highest superconducting parameters. If we again turn to the data quoted in Tables 1.3 and 1.4, it may be noticed that the alloys and compounds based on transition metals have the highest T_c. That is why consideration of their properties is of special practical interest. Our book deals solely with superconductors based on transition elements, including pure metals, their alloys and chemical compounds. Of course, the limited scope of the book does not allow us to describe all the types of superconducting alloys and compounds of transition elements which are enumerated in Tables 1.3 and 1.4. Therefore we shall consider only those of them in which superconductivity is a prominent feature and T_c is high. These compounds which are listed first in Tables 1.3 and 1.4, are: solid solutions of transition metals, intermetallides with A-15, C-15, C-14, B-2 and A-12 structure and σ phases, and also chemical compounds of B-1 structure (NaCl), D-5 structure (sesquicarbides), Ti_3S_4, Mo_3Se_4, $H\text{-}1_1$ structure (chalcogenides). Consideration of these main classes of alloys and chemical compounds makes it possible to reveal the general features peculiar to superconductivity in systems based on transition metals and to map out ways for physical investigation of the new classes of compounds which are as yet insufficiently understood.

The superconductivity of complicated compounds based on transition metals cannot be understood without an analysis of the superconductivity in pure transition metals. We shall therefore continue our treatment with the latter class.

1.2 Survey of Theoretical Concepts

BARDEEN, COOPER and SCHRIEFFER (BCS) [1.35] advanced their microscopic superconductivity theory in 1957. It accounted for the features peculiar to the superconducting state; and the question arose of whether it would be possible to calculate the superconducting parameters of a particular metal on the basis, say, of data characterizing that metal in the normal state. Is it possible to predict whether a given metal will be a superconductor and to calculate the superconducting transition temperature for it?

This question is now particularly important from the practical point of view, in connection with quests for high-temperature superconductivity or, if formulated more moderately, in connection with the problem of improving the major characteristics of superconducting materials. Further, the only parameter which is of principal interest to us is the superconducting transition temperature T_c. If a positive answer to the problem formulated above is possible, then it gives rise to another question: what experimentally measured parameters of the normal phase must one know? It hence follows that the solution of the above problem is determined not only by the level of development of theory but also by capabilities of the experiment.

It is the purpose of this section to review the main concepts of the theory of the superconducting state and to provide the minimum of information about the normal phase which should be known in order to be able to compute the most important parameter of the superconducting state, the T_c. No attempt is made in this review to give an exhaustive analysis of modern superconductivity theory. This is treated in detail in the following two chapters. We wish to enable the reader, without having to study these theoretical chapters, firstly, to gain an insight into the fundamental concepts of modern theory and thus to assess its potentialities in the problem of predicting the properties of a superconductor and, secondly, to give him a key to the experimental evidence presented in the subsequent chapters.

We shall begin with a concise statement of the results of the BCS theory and of what is called the strong coupling method in superconductivity theory, and then try to answer the question which we have posed.

1.2.1 The Fundamentals of the BCS Theory

Since the development of the BCS theory, much progress has been made in our concepts of superconductivity, but the basic ideas advanced by BCS still underlie modern theory.

COOPER [1.36] showed that the ground state of a normal metal is unstable with respect to an arbitrarily weak attraction of electrons on the Fermi surface. As a result of this, it is advantageous for the electrons to condense into bound pairs with zero total momentum and spin. At the same time it was found that the electrons in the vicinity of the Fermi surface in fact attract each other due to an exchange of virtual phonons. The attraction is effective in a k-space layer adjacent to the Fermi surface, whose depth is of the order of the limiting frequency of the phonon spectrum, i.e., of the order of the Debye frequency[2]. From these two results followed the idea about the structure of the ground state of a superconductor. According to the BCS theory, when the superconductor is in the ground state all electrons are bound into Cooper pairs. A measure of the binding is the quantity $\Delta_{\underline{p}}$ determining the gap in the spectrum of one-particle excitations of the superconductor. The energy of these excitations is

$$E_{\underline{p}} = \sqrt{\xi_{\underline{p}}^2 + \Delta_{\underline{p}}^2} \quad , \tag{1.2.1}$$

where $\xi_{\underline{p}}$ is the energy of an electron in the normal phase with momentum $\underline{p}$ measured from the Fermi surface. Thus, for instance, for a quadratic dispersion relation, $\xi_{\underline{p}} = |\underline{p}|^2/2m - E_F$ (E_F is the Fermi energy). The gap is determined from a self-consistency equation of the form

$$\Delta_{\underline{p}} = \sum_{\underline{p}'} V_{\underline{p}\underline{p}'} \frac{\tanh(E_{\underline{p}'}/2T)}{2E_{\underline{p}'}} \Delta_{\underline{p}'} \quad , \tag{1.2.2}$$

where $V_{\underline{p}\underline{p}'}$ is the matrix element for attraction of Cooper pair electrons with momenta $\underline{p}$ and $\underline{p}'$.

The usual approximation in the BCS theory is that $V_{\underline{p}\underline{p}'}$ is assumed to be isotropic and constant in a narrow layer of energies near the Fermi surface. The thickness of the layer is taken to be equal to $\omega_{\mathcal{D}}$, where $\omega_{\mathcal{D}}$ is a limiting (for example, the Debye) frequency of the phonon spectrum, and there is no interaction outside this layer. Thus, the following approximation of the electron-electron attraction matrix element is adopted

$$V_{\underline{p}\underline{p}'} = \begin{cases} V; & |\xi_{\underline{p}}|,\ |\xi_{\underline{p}'}| < \omega_{\mathcal{D}} \\ 0; & |\xi_{\underline{p}}|,\ |\xi_{\underline{p}'}| > \omega_{\mathcal{D}} \end{cases} \quad . \tag{1.2.3}$$

In this approximation, the integral equation (1.2.2) for $\Delta_{\underline{p}}$ becomes transcendental since $\Delta_{\underline{p}}$ ceases to depend on the momentum $\underline{p}$. This equation defines the two well-known relations of BCS theory, viz. the energy gap at zero temperature,

---

[2] Here and henceforth a system of units is used in which the Planck and Boltzmann constants are set equal to unity so that the energy, frequency and temperature are of the same dimension.

$$\Delta = 2\omega_D \exp\left(-\frac{1}{VN(0)}\right) , \tag{1.2.4}$$

and the superconducting transition temperature,

$$T_c = 1.14\omega_D \exp\left(-\frac{1}{VN(0)}\right) , \tag{1.2.5}$$

where N(0) is the density of electron states on the Fermi surface. Both quantities are related by the universal equation

$$2\Delta/T_c = 3.52 , \tag{1.2.6}$$

which is in good agreement with experiment for many superconducting elements.

As is known, the BCS theory has offered an explanation for many experimentally observed phenomena, despite the rough character of the model. Thus, it disregarded two important factors: the Coulomb repulsion of electrons and the retardation effect in the electron-phonon interaction.

1.2.2 The Role of the Coulomb Interaction

Allowance for the Coulomb interaction has been made by BOGOLYUBOV et al. [1.37] who have shown that the superconducting transition temperature should be determined not by (1.2.5) but by the refined relation

$$T_c = 1.14\omega_D \exp\left(-\frac{1}{\lambda - \mu^*}\right) \tag{1.2.7}$$

containing, in addition to the term $\lambda \equiv VN(0)$ describing the electron attraction, a Coulomb pseudopotential μ^*. This is expressed in terms of the matrix element of the Coulomb electron interaction on the Fermi surface as

$$\mu^* = \frac{V_c N(0)}{1 + V_c N(0) \ln(E_F/\omega_D)} . \tag{1.2.8}$$

The logarithmic weakening of the Coulomb potential arises from the difference in the energy scales of interelectron attraction and repulsion, the Debye frequency and the Fermi energy. From a physical point of view, a weakened pseudopotential comes from the fact that the mean distance of the electrons in a Cooper pair, which is characterized by a correlation length $\xi \sim \hbar v_F/\Delta$ (v_F being the velocity of electrons on the Fermi surface), turns out to be much greater than the lattice parameter a. The mean value of the Coulomb interaction of these electrons, thus, proves to be much less than $V_c \sim \ell^2/a$.

The pseudopotential μ^* can be determined experimentally from the isotope effect in superconductors. The BCS theory yields a dependence of T_c on the ion mass M through the Debye frequency which is $\sim 1/\sqrt{M}$, thus

$$\delta T_c/T_c = -0.5\,\delta M/M \quad . \tag{1.2.9}$$

When the Coulomb interaction is taken into account, the quantity T_c defined by (1.2.7) depends on the ion mass not only through the preexponential factor $\sim \omega_D$ but also through the quantity μ^*. The magnitude of the isotope effects is now

$$\delta T_c/T_c = -\alpha\,\delta M/M \tag{1.2.10}$$

with $\alpha = \{1-[\mu^*/(\lambda-\mu^*)]^2\}/2$. Thus, μ^* expresses the isotope effect and may be determined if T_c and the isotopic shift δT_c are measured. For transition metals, for example, μ^* turns out to be small (0.1 - 0.15), whereas the electron-electron attraction constant amounts to several tenths and is close to 1 in a number of cases.

1.2.3 The Effects of Retardation in the Electron-Phonon Interaction

When comparing the BCS theory to experimental data it was immediately noticed that for a number of superconductors there was no quantitative agreement. This primarily applied to lead and mercury. For Pb (T_c = 7.2 K, θ_D = 96 K), the quantity λ defined by (1.2.7) is 0.39, and for Hg (T_c = 4.15, θ_D = 70 K), λ = 0.35. Meanwhile, for a metal such as Al (T_c = 1.17 K, θ_D = 375 K), λ = 0.18. For Hg and Pb the ratio T_c/ω_D proves to be anomalously large compared to other metals, and this leads to anomalously large values of λ. Since λ, as we shall see below, is a measure of the electron-phonon coupling in the metal, superconductors with large λ are called strong-coupling superconductors. The quantitative theory for these superconductors has been constructed not on the basis of the BCS effective Hamiltonian for the electron-electron attraction, but from electron-phonon interaction in higher orders of perturbation theory with frequency-dependent matrix elements, i.e., with due account to the retardation effects.

It turns out in this situation [1.38] that the energy gap Δ is a function of the transition frequency ω. Instead of (1.2.2) there arises an integral equation for $\Delta(\omega)$ with a kernel expressed in terms of electron-phonon interaction matrix elements. The superconducting transition temperature in this case is given by [1.39]

$$T_c \sim \omega_D \exp\left(-\frac{1+\lambda}{\lambda-\mu^*}\right) \qquad (1.2.11)$$

which reduces to (1.2.7) in the case of weak coupling ($\lambda \ll 1$). It turns out that the quantity λ is expressed in terms of the frequency distribution in the phonon spectrum, $F(\omega)$, and through the effective square of the electron-phonon interaction matrix element, $\alpha^2(\omega)$,

$$\lambda = 2\int_0^\infty \frac{\alpha^2(\omega)\,F(\omega)}{\omega}\,d\omega \quad . \qquad (1.2.12)$$

The quantity $\alpha^2(\omega)$ is actually determined from the equation

$$\alpha^2(\omega)F(\omega) = \left\langle \frac{1}{N}\sum_{qj} \frac{|\langle \underline{p}|\underline{e}_j(\underline{q})\nabla U|\underline{p}'\rangle|^2}{2M\omega_j(\underline{q})}\,\delta(\omega-\omega_j(\underline{q}))\right\rangle \quad , \qquad (1.2.13)$$

in which $\omega_j(\underline{q})$ and $\underline{e}_j(\underline{q})$ are the frequency and the polarization vectors of the phonon branch with wave vector $\underline{q} = \underline{p} - \underline{p}'$, and U is the ion potential. The symbol $\langle\ldots\rangle$ in (1.2.13) denotes averaging over the angles of the vectors $\underline{p}$ and $\underline{p}'$ lying on the Fermi surface. This means that superconductivity theory accounting for the electron-phonon interaction, leads to an effective electron attraction λ on the Fermi surface depending on the quantities $\omega_j(\underline{q})$ and $\underline{e}_j(\underline{q})$, which describe the harmonic oscillations of the lattice, on the matrix elements of the electron-ion potential gradient, and on the characteristics of the Fermi surface. The theory of strong-coupling superconductors yields a microscopic definition of the quantity λ (to be more exact, of the quantity V in $\lambda = VN(0)$) which in the BCS theory was a phenomenological parameter.

Equations (1.2.11 - 13) allow the computation of the superconducting transition temperature from the microscopic properties of the normal phase which have been mentioned above. These properties may be obtained either from experimental measurements or by calculation from first principles.

Let us now discuss the accuracy of the equations of the strong-coupling theory. They were introduced in 1960 by Eliashberg generalizing MIGDAL's [1.40] approach in the theory of electron-phonon interaction in normal metals. As shown by ELIASHBERG [1.38], the accuracy of the equations describing superconductors in the strong-coupling case is determined by the parameter $\sim \sqrt{m/M}$, where m is the electron mass and M is the ion mass. Consequently, it may be hoped that if the parameters determining the quantity λ are known exactly, the superconducting transition temperature can be calculated with an accuracy of $\sim 10^{-2}$, which amounts actually to several percent [1.41]. The calculation of T_c performed on the basis of Eliashberg's equations for lead, for example, coincides with the experimental value within several percent [1.41]. Thus, in superconductivity theory there is a method enabling one to calculate very accurately the values of the superconducting transition temperature if the microscopic characteristics of the phonon spectrum and of the electron-phonon interaction on the Fermi surface are known.

The strong-coupling theory permits corrections to a number of expressions following from the BCS theory. For example, the universal equation (1.2.6) [1.42] is replaced by

$$2\Delta/T_c = 3.52\left[1 + C\left(\frac{T_c}{\omega_{\mathcal{D}}}\right)^2 \ln\frac{\omega_{\mathcal{D}}}{T_c}\right] \quad , \qquad (1.2.14)$$

where C is a constant of the order of unity. So, the strong-coupling method leads to characteristic corrections $\sim (T_c/\omega_{\mathcal{D}})^2 \ln(\omega_{\mathcal{D}}/T_c)$, which for a ratio $\omega_{\mathcal{D}}/T_c \sim 10$ amount to several percent and thus appreciably exceed the total inaccuracy of the equations $\sim \sqrt{m/M}$. It is important to note here that many of the transition-metal alloys and compounds with which we are concerned have values of λ of the order of unity and should therefore be treated by the strong-coupling theory.

1.2.4 The Electron-Phonon Coupling Constant

According to the strong-coupling theory (see (1.2.11)), the superconducting transition temperature is uniquely determined by the quantity λ which is commonly called the electron-phonon coupling constant. Its value for a given superconductor can be obtained from a (1.2.11)-type formula, by substituting into the equation the experimentally measured T_c and μ^*. However, there exist effects (pertaining to the normal phase) whose measurements also permit determination of λ.

The point is that a number of equations relating to the thermodynamics and kinetics of the electron gas in the metal are renormalized as a consequence of the electron-phonon interaction. In some cases this renormalization is expressed by the same parameter λ of (1.2.12) which defines the superconducting transition temperature. The electronic part of the heat capacity, which is linear in temperature, is renormalized by the electron-phonon interaction according to [1.43],

$$C_v = \gamma(1+\lambda)\,T \quad . \tag{1.2.15}$$

It is natural that this renormalization is possible only at temperatures below, or of the order of, the Debye temperature θ_D. For $T \gg \omega_D$, most of the electronic states near the Fermi surface which contribute to C_v are not sensitive to the electron-phonon interaction and $C_v = \gamma T$. By comparing the coefficient of the linear term in C_v for low and high temperatures one can find the electron-phonon coupling constant λ.

Since γ is proportional to N(0), Eq.(1.2.15) means that the density of states at the Fermi surface is renormalized,

$$N^*(0) = N(0)(1+\lambda) \quad , \tag{1.2.16}$$

where N*(0) is the density of states being measured, and N(0) is the bare density of states calculated without taking account of the electron-phonon coupling. The last equation may be interpreted as an increase in the cyclotron mass of an electron, i.e.,

$$\frac{m^*}{m} = 1+\lambda \quad . \tag{1.2.17}$$

Therefore the parameter λ is sometimes called the linear enhancement coefficient. The velocity of an electron on the Fermi surface also decreases as

$$v_F^*/v_F = \frac{1}{1+\lambda} \tag{1.2.18}$$

and can be measured using TOMASCH's [1.44] effect.

Thus, there is a sufficiently rich spectrum of possibilities for determining λ from measurements of a number of quantities pertaining to the normal phase of the metal. It is, however, possible to determine λ by measuring a purely superconducting property, viz., the tunneling current across a superconductor-normal metal contact with superconductor and normal metal separated by a thin insulating layer. It is known that the differential characteristics of the tunnel currents J_{sn} and J_{nn} (from a superconducting and a normal metal to a normal metal) are related to the gap function $\Delta(\omega)$ by the equation [1.45]

$$\frac{dJ_{sn}}{dV} \Big/ \frac{dJ_{nn}}{dV} = \mathrm{Re}\left\{\frac{\omega}{\sqrt{\omega^2-\Delta^2(\omega)}}\right\} , \tag{1.2.19}$$

where $\omega = eV$ (V being the potential difference applied to the contact, and e the electron charge). Measurement of the quantities defining the left-hand side of this equation permits calculation of the right-hand side and enables one, with the aid of the dispersion relations, to find the real, $\mathrm{Re}\{\Delta(\omega)\}$, and imaginary, $\mathrm{Im}\{\Delta(\omega)\}$, parts of $\Delta(\omega)$. Substituting the quantities thus determined into the strong-coupling theory Eliashberg's equations yield the function $\alpha^2(\omega)F(\omega)$ on solving them numerically. Then one can calculate λ according to the basic equation (1.2.12).

Thus, we have at our disposal the following techniques for finding the most important parameter determining the superconducting transition temperature, i.e., the electron-phonon coupling constant: a) by measuring a number of effects in the normal phase of the metal; b) by measuring the tunnel effect in the superconductor with the subsequent use of the Eliashberg equations; c) by calculation according to (1.2.12,13) from first principles or by using the phonon frequencies found experimentally from inelastic neutron scattering. Knowing λ one can calculate T_c according to a (1.2.11)-type formula and thus verify the fundamentals of the tight-binding superconductivity theory which has turned out to be entirely self-consistent in the scheme described above. This means that in order to reveal the dependence of T_c on the microcharacteristics of the normal phase of the metal, it is required, in the first place, to investigate the dependence of the electron-phonon coupling constant λ on the same quantities. In Chapter 2, a detailed analysis will be given of λ and its dependence on the structure of the phonon and electron spectra in the metal, and in the subsequent chapters a comparison will be drawn between the empirical data on λ and T_c and the existing theories.

The concepts described above apply to superconducting metals in general and do not take into account the specifics of the transition metals. Therefore we shall now briefly describe the chief features peculiar to the band structure of these metals.

1.2.5 The Peculiarities of the Transition Metals

A quantitative theory of superconductivity in transition metals should take into account the main features peculiar to their electronic structure, as compared to simple metals. At least three must be noted.

Firstly, the density of states on the Fermi surface is very high for the majority of the transition metals, and what prevails in the vicinity of it are states of the d-type, i.e., with orbital moment $\ell = 2$, which tend to be localized near ions to a larger degree than do the s and p-states. This implies that the contribution of d-electrons to the appearance of the superconducting state is substantial if not decisive. Secondly, the Fermi surface of the transition metals is very complicated, both in the symmetry of the states which form it and in topology. Thus, the high anisotropy of the Fermi surface is a factor which must be taken into account in the calculation of the electron-phonon coupling constant.

Finally, in contrast to simple metals for which a weak-potential approximation is often applicable, in transition metals there exists a strong electron-ion potential which, in principle, cannot be described following perturbation theory since a quasi-bound state is likely to be produced in it. This potentia exerts a most substantial effect on the scattering of d-electrons (states with orbital moment $\ell = 2$). The effective potential for them may have a well. Indeed, the potential $V(r)$ describing the radial motion of an electron in the field of a separate ion is composed of two parts, viz., the intrinsic Coulomb potential of attraction and the centrifugal potential of repulsion which is due to the angular momentum

$$V(r) = -\frac{Z(r)e^2}{r} + \ell(\ell+1)\frac{\hbar^2}{2mr^2} \quad .$$

When far from the centre, the effective ion charge $Z(r)$ tends to zero and therefore the centrifugal potential predominates there; but as the centre becomes closer, by virtue of an increase in $Z(r)$ the effective potential $V(r)$ diminishes until the second term in $V(r)$ begins prevailing again due to the higher power of $1/r$. As a result, for large values of the angular momentum $\ell = 2$, there arises the well shown in Fig.1.2. This well turns out to be rather shallow for the transition elements placed in the beginning of the period [1.46], but it becomes deeper if we move along this series, due to the increase of the ion charge.

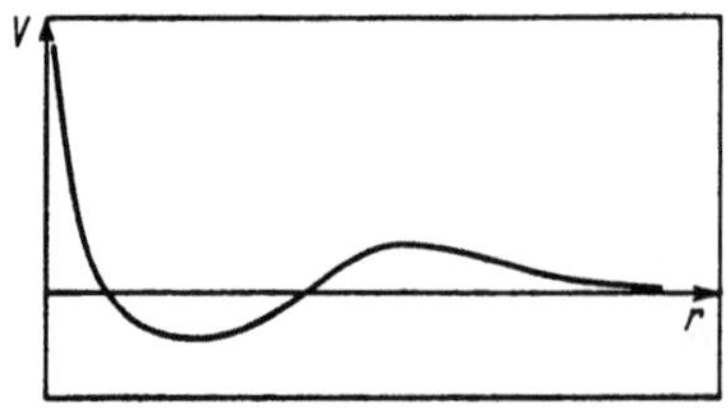

Fig. 1.2. The effective potential for an electron in the d-state

If an electron is scattered by such a potential, a sharp peak corresponding to a virtual, or resonance, state may appear in the scattering cross-section. The resonance width, which depends on the s-d mixing matrix elements, determines the lifetime of the electron near a given ion. In this situation a broadened virtual level rather than a real atomic state corresponds to each lattice. In order to describe the motion of an electron in the lattice, Bloch's functions must be written as linear combinations of states belonging to the resonance levels. The Bloch's functions, thus constructed, simultaneously combine the features of tight and weak binding. The ideology of tight binding here manifests itself in the choice of one-centre scattering wave functions, and the ideology of weak binding in that these functions are not localized but are a superposition of a plane wave and of a divergent spherical wave.

The resonance scattering of the d-state by the isolated ion potential effects that the coherent electron scattering in the crystal forms a narrow width d-band. For many transition metals this resonance lies in the vicinity of the Fermi level and thus determines the motion of the electrons contained in the superconducting condensate. The fact that d-states are produced due to "atomic" resonance states leads (we shall see this in the following) to the concept that the superconductivity in the transition metals is, as it is said, rather an "atomic" property. This should be understood in the sense that important characteristics of the superconductor such as the superconducting transition temperature are determined solely by the local, atomic, parameters of the metal.

We have not yet mentioned the peculiarities of the phonon spectrum. For transition metals, the usual phonon spectrum calculation techniques based on the weakness of the pseudopotential [1.47] are inapplicable due to the abovementioned essential features of the ion potential. Because of this, the calculation of phonon spectra from first principles is not yet possible. This imposes certain restrictions on first principle calculations of characteristics such as the superconducting transition temperature. In the following, in order to calculate T_c for transition metals and compounds based on them, we shall strive to perform first principle calculations of the required parameters of the electronic system or of the electron-ion interaction matrix elements by taking the relevant information on the phonon spectrum from the experiment. In this respect, the transition metals are much more complicated than the normal "pseudopotential" metals.

It must be added here that the consideration of transition metal compounds containing several atoms per unit cell enhances all the difficulties arising in pure metals. This compels us henceforth to utilize a qualitative theory of superconductivity in complicated compounds, proceeding from a quantitative or semiquantitative theory of superconductivity in pure d-metals.

1.2.6 Superconductivity and Magnetism

Since it is the intent of our book to treat superconductivity in transition metals, it is necessary in this brief review of the theory to briefly mention the relationship between superconductivity and magnetism [1.48]. It was noticed long ago that superconductivity is absent in metals having localized magnetic moments. The mechanism of their destructive effect on the superconducting state may be understood within the framework of the BCS theory if one takes into account the exchange interaction of conduction electrons with localized atomic magnetic moments. In fact, in the superconducting state the electrons of the metal are coupled into Cooper pairs which unite states with opposite momentum and spin. If one of the electrons of such a pair is scattered by a paramagnetic atom, there is a large probability that its spin will change orientation due to the exchange interaction [1.49]. As a result the pair will collapse. As shown by ABRIKOSOV and GOR'KOV [1.50], if a superconductor has a paramagnetic ion concentration amounting to several percent, the latter suffices to completely suppress the superconductivity. In a pure metal, the formation of a superconducting state in the presence of atomic magnetic moments is obviously impossible since the 100% concentration of magnetic atoms is much larger than the critical concentration.

In a magnetically ordered transition metal additional mechanisms for the suppression of superconductivity should exist. Owing to the s-d exchange interaction a virtual exchange may occur between electrons via spin waves. In contrast to the phonon exchange this leads to a repulsion of the electrons on the Fermi surface and prevents a superconducting state [1.48,51]. The indirect exchange of electrons through spin waves turns out to be of repulsive character in the case of ferromagnetic and antiferromagnetic ordering. Besides this, a ferromagnetic metal exhibits one more superconductivity suppression mechanism which is due to the exchange magnetization of conduction electrons by spontaneously ordered magnetic moments of incomplete d-shells. This magnetization results in a Zeeman splitting of the electronic states. When the splitting exceeds the binding energy of the electrons in a Cooper pair superconductivity should collapse.

It thus becomes clear why there are no superconductors among the metals with localized magnetic moments. However, some transition metals, for example, palladium, are not superconductors in spite of the fact that localized magnetic moments are absent there. This fact has been attributed to spin density fluctuations by BERK and SCHRIEFFER [1.52]. The point is that Pd is characterized by anomalously high electronic susceptibility, and its other properties also indicate that the state of its electronic system is close to the critical conditions for the occurrence of localized magnetic moments. In this situation spin density fluctuations must play an essential role since the interaction of electrons with these fluctuations should destroy Coulomb pairs due to spin reorientation processes. It follows from theory [1.46,52] that when spin fluctuations are taken into account the superconducting transition temperature is given not by the BCS formula but by a formula of the type

$$Tc \sim \omega_D \exp\left(-\frac{1}{\lambda - \lambda_m}\right) , \qquad (1.2.20)$$

where λ_m is the constant of the coupling of electrons to spin fluctuations. The above suppression mechanisms in highly magnetic transition metals should be borne in mind for the discussion of the properties of a number of alloys and compounds based on them.

Here we conclude our brief survey of the major theoretical concepts about the nature of the superconducting state. All the details of the theory can be found in Chapters 2 and 3. Chapter 2 gives a systematic description of the general theory of strong-coupling superconductors, and Chapter 3 deals with problems associated with the coexistence of superconductivity and magnetism. We recommend to the reader who does not wish to go into the details of the theory to read Sect.2.4, especially the second part. Thereupon he may immediately start reading Chapter 4 which is devoted to experimental data concerning the properties of transition metals and to their analysis on the basis of the existing theories.

2. The Theory of Strong Coupling Superconductors

The theory of superconductors with strong electron-phonon coupling is presented. The mathematical framework is based on the Gor'kov-Nambu matrix formalism. Eliashberg equations are derived for the Green's function of electrons and formulae are found for the superconducting transition temperature. Different representations of the electron-phonon coupling constant are also considered, e.g., that using the Green's function of phonons, the McMillan representation, and the Hopfield representation.

2.1 Perturbation Theory for Green's Functions

2.1.1 The Gor'kov-Nambu Matrix Formalism

The theory of strong-coupling superconductors is inadequately dealt with in most books. To fill this gap, we shall begin with a detailed description of the mathematical apparatus based on the Green's function technique. The first to introduce the Green's function method into superconductivity theory was GOR'KOV [2.1] who demonstrated that the superconducting state should be described by two Green's functions, one of which corresponds to the usual prescription of the Green's function of the normal state of the metal, whereas the other, an anomalous one, allows for the occurrence of a condensate of Cooper pairs in the superconductor. Thus, the mathematical theory of the superconducting state has a 2×2 matrix structure.

Independently, NAMBU [2.2] had the similar idea of describing a superconductor with the aid of two-component field operators corresponding to the electron-hole formalism and leading automatically to the same matrix structure of the theory. The Gor'kov-Nambu matrix formalism is convenient in that the perturbation expansion for the superconductor's matrix Green's function is of the same form as that for the usual Green's function describing a normal metal. In this section we show how this situation arises and formulate rules for constructing perturbation theory series for electrons interacting with phonons and with each other through Coulomb forces.

In the mathematical theory, the occurrence of a superconducting state in the electronic system of the metal is associated with the presence of what is known as the anomalous averages

$$\langle \psi_{\downarrow}(\underline{r})\psi_{\uparrow}(\underline{r})\rangle \quad \text{and} \quad \langle \psi^{+}_{\uparrow}(\underline{r})\psi^{+}_{\downarrow}(\underline{r})\rangle \quad . \tag{2.1.1}$$

Here $\psi^{+}_{\alpha}(\underline{r})$ and $\psi_{\alpha}(\underline{r})$ stand for the operators of creation and annihilation of an electron at point $\underline{r}$ with spin $\alpha = \uparrow,\downarrow$, and <...> is the symbol of statistical averaging with the Hamiltonian

$$\langle\ldots\rangle = \mathrm{Sp}\{e^{-\beta H}\ldots\}/\mathrm{Sp}\{e^{-\beta H}\} \quad , \quad \beta = \frac{1}{kT} \quad . \tag{2.1.2}$$

The total Hamiltonian H of a superconductor should include the electron-phonon interaction giving rise to an effective electron attraction and also include the Coulomb repulsion. If we consider these terms in the Hamiltonian within the framework of perturbation theory, a superconducting state may be obtained provided that we choose correct zero-order functions describing the peculiar "symmetry" of the ground state of the superconductor, i.e., the possibility of the appearance of anomalous averages as in (2.1.1). From a formal standpoint, averages of this type are liable to occur if the Hamiltonian of free electrons of the metal (Bloch electrons) contains sources of Cooper pairs. This means that as a zero-approximation Hamiltonian one must choose the operator [2.3]

$$H_{\Delta} = \int d\underline{r} \left[\sum_{\alpha}\psi^{+}_{\alpha}(\underline{r})H_0(\underline{r})\psi_{\alpha}(\underline{r}) + \Delta(\underline{r})\psi^{+}_{\uparrow}(\underline{r})\psi^{+}_{\downarrow}(\underline{r}) + \Delta^{*}(\underline{r})\psi_{\downarrow}(\underline{r})\psi_{\uparrow}(\underline{r})\right] \tag{2.1.3}$$

where $\Delta(\underline{r})$ is some function of the coordinates with infinitesimal amplitude.

The first term in (2.1.3) is the operator of the energy of free electrons in the metal, which are described by a one-particle Hamiltonian. The two subsequent terms are energy operators associated with the creation and annihilation of singlet electron pairs.

The electronic-field operators satisfy the usual Fermi permutation relations

$$\begin{aligned} \psi_{\alpha}(\underline{r})\psi^{+}_{\beta}(\underline{r}') + \psi^{+}_{\beta}(\underline{r}')\psi_{\alpha}(\underline{r}) &= \delta_{\alpha\beta}\delta(\underline{r}-\underline{r}') \quad , \\ \psi_{\alpha}(\underline{r})\psi_{\beta}(\underline{r}') + \psi_{\beta}(\underline{r}')\psi_{\alpha}(\underline{r}) &= 0 \quad . \end{aligned} \tag{2.1.4}$$

The Hamiltonian (2.1.3) may be formally transformed into a usual operator of free particles in the second-quantization representation by introducing two-component field operators $\Psi(\underline{r})$ and $\Psi^{+}(\underline{r})$ related to the usual field operators by the equations

$$\Psi(\underline{r}) = \begin{pmatrix} \psi_\uparrow(\underline{r}) \\ \psi_\downarrow^+(\underline{r}) \end{pmatrix} , \quad \Psi^+(\underline{r}) = (\psi_\uparrow^+(\underline{r}), \psi_\downarrow(\underline{r})) \tag{2.1.5}$$

(Nambu's representation). In fact, in terms of these operators the Hamiltonian (2.1.3) assumes the simple form

$$H_\Delta = \int d\underline{r}\ \Psi^+(\underline{r})h(\underline{r})\Psi(\underline{r}) \quad , \tag{2.1.6}$$

where according to (2.1.5), Ψ and Ψ^+ are two-component vectors, and $h(\underline{r})$ is a 2×2 matrix

$$h(\underline{r}) = \begin{pmatrix} H_0(\underline{r}) & \Delta(\underline{r}) \\ \Delta^*(\underline{r}) & -H_0(\underline{r}) \end{pmatrix} \quad . \tag{2.1.7}$$

By analogy with the normal state of the electronic system, we now introduce a one-particle temperature Green's function, following the notation adopted in [2.4]:

$$g(x,x') = -\langle T\Psi(x)\overline{\Psi}(x')\rangle \quad . \tag{2.1.8}$$

The operators $\Psi(x)$ and $\overline{\Psi}(x')$ here are Heisenberg field operators $\Psi(\underline{r})$ and $\Psi^+(\underline{r})$ with the total Hamiltonian H:

$$\Psi(x) = e^{H\tau}\Psi(\underline{r})e^{-H\tau} \quad , \quad \overline{\Psi}(x) = e^{H\tau}\Psi^+(\underline{r})e^{-H\tau} \quad , \tag{2.1.9}$$

$x = \{\underline{r},\tau\}$ is a set of the coordinates and of the "thermodynamic time" prescribed in the interval $0 \le \tau \le \beta$, T in (2.1.8) is the symbol of ordering with respect to the variables τ.

It follows from the definition (2.1.8) and from (2.1.5) that the Green's function $g(x,x')$ is a 2×2 matrix

$$g(x,x') = \begin{pmatrix} -\langle T\psi_\uparrow(x)\overline{\psi}_\uparrow(x')\rangle & -\langle T\psi_\uparrow(x)\psi_\downarrow(x')\rangle \\ -\langle T\overline{\psi}_\downarrow(x)\overline{\psi}_\uparrow(x')\rangle & -\langle T\overline{\psi}_\downarrow(x)\psi_\downarrow(x')\rangle \end{pmatrix} \tag{2.1.10}$$

involving, along with a normal (diagonal) function, two anomalous functions which are due to the appearance of anomalous (2.1.1)-type averages in the superconductor. Let us first of all construct an equation of motion for $g(x,x')$ using the well-known representation of the T product in terms of the θ function,

$$g(x,x') = -[\theta(\tau-\tau')\langle\Psi(x)\overline{\Psi}(x')\rangle - \theta(\tau'-\tau)\langle\overline{\Psi}(x')\Psi(x)\rangle]$$

with $\theta(z) = 1$ if $z > 0$ and with $\theta(z) = 0$ if $z < 0$. Differentiating this equality with respect to the variable τ we obtain

$$-\frac{\partial g(x,x')}{\partial\tau} + \left\langle T\frac{\partial\Psi(x)}{\partial\tau}\bar{\Psi}(x')\right\rangle = \delta(x - x') \quad , \tag{2.1.11}$$

where the derivative of the field operator with respect to "time" can be found from the general quantum-mechanical equation

$$\frac{\partial\Psi(x)}{\partial\tau} = [\,H\,,\,\Psi(x)] \quad . \tag{2.1.12}$$

Now write an equation for the Green's function $g_0(x,x')$ which corresponds to the zero-approximation Hamiltonian (2.1.3). Calculating the commutator $\Psi(x)$ with the operator (2.1.6) we get the following matrix equation

$$\left[-\frac{\partial}{\partial\tau} - h(\underline{r})\right]g_0(x,x') = \delta(x - x') \quad . \tag{2.1.13}$$

As is evident from (2.1.7), the Green's function $g_0(x,x')$ is determined by the free-electron Hamiltonian and by the quantity $\Delta(\underline{r})$ characterizing the source of Cooper pairs. If we let $\Delta(\underline{r})$ tend to zero, (2.1.5,6,8) will yield correct zero-approximation functions which should be employed when constructing a perturbation theory for the superconducting state [2.5].

2.1.2 Perturbation Theory

As known, the main interactions essential to the formation of a superconducting state are the electron-phonon and Coulomb interactions. In the second-quantization representation these interactions are written as [2.4]

$$\begin{aligned} H_{\text{int}} &= \int d\underline{r}\, g(\underline{r}) \sum_\alpha \psi^+_\alpha(\underline{r})\,\psi_\alpha(\underline{r})\,\varphi(\underline{r}) \\ &+ \frac{1}{2}\int d\underline{r}_1\, d\underline{r}_2\, U(\underline{r}_1 - \underline{r}_2) \sum_{\alpha\beta} \psi^+_\alpha(\underline{r}_1)\psi^+_\beta(\underline{r}_2)\psi_\beta(\underline{r}_2)\psi_\alpha(\underline{r}_1) \quad , \end{aligned} \tag{2.1.14}$$

where $g(\underline{r})$ is the local electron-phonon interaction energy, $\varphi(\underline{r})$ is the phonon field operator and $U(\underline{r}_1 - \underline{r}_2)$ is the Coulomb potential for the electrons. The structure of that part of the Hamiltonian (2.1.14) which describes the electron-phonon interaction will be discussed in terms of a momentum representation in more detail below.

It can easily be verified that in the representation of Nambu's operators (2.1.5) the interaction Hamiltonian (2.1.14) becomes

$$H_{int} = \int d\underline{r}\, g(\underline{r})\, [\Psi^{+}(\underline{r})\tau_3\Psi(\underline{r})]\, \varphi(\underline{r})$$
$$+ \frac{1}{2}\int d\underline{r}_1 d\underline{r}_2\, U(\underline{r}_1 - \underline{r}_2)\, [\Psi^{+}(\underline{r}_1)\tau_3\Psi(\underline{r}_1)]\, [\Psi^{+}(\underline{r}_2)\tau_3\Psi(\underline{r}_2)] \quad , \qquad (2.1.15)$$

with τ_3 being one of the Pauli matrices

$$\tau_1 = \begin{pmatrix} 0 & 1 \\ 1 & 0 \end{pmatrix} , \quad \tau_2 = \begin{pmatrix} 0 & i \\ -i & 0 \end{pmatrix} , \quad \tau_3 = \begin{pmatrix} 1 & 0 \\ 0 & -1 \end{pmatrix} , \quad \tau_o = \begin{pmatrix} 1 & 0 \\ 0 & 1 \end{pmatrix} . \qquad (2.1.16)$$

Strictly speaking, the diagonal terms, quadratic with respect to the field operators Ψ, which arise from the Coulomb interaction when we pass from (2.1.14) to (2.1.15) should be included in the one-particle Hamiltonian H_{Δ}.

The zero-approximation Hamiltonian in Nambu's representation may be written as

$$H_{\Delta} = \int d\underline{r}\, \Psi^{+}(\underline{r})\, \tau_3 H_0(\underline{r})\, \Psi(\underline{r}) \quad . \qquad (2.1.17)$$

This formula follows directly from (2.1.6) and (2.1.7) when $\Delta(\underline{r})$ tends to zero.

The calculation of the Green's function of a system of interacting particles is based on the use of the interaction representation with (2.1.8) replaced by

$$g(x,x') = -\left\langle T\Psi(x)\overline{\Psi}(x')\sigma(\beta)\right\rangle / \left\langle \sigma(\beta)\right\rangle . \qquad (2.1.18)$$

Here $\sigma(\beta)$ is the temperature dependent scattering matrix,

$$\sigma(\beta) = \sum_{n=0}^{\infty} \frac{(-1)^n}{n!} \int_0^{\beta} d\tau_1 \ldots \int_0^{\beta} d\tau_n\, T[H_{int}(\tau_1)\ldots H_{int}(\tau_n)] \quad . \qquad (2.1.19)$$

In (2.1.18) the field operators $\Psi(x)$ and $\Psi(x')$ are taken in the interaction representation ((2.1.9) with H_0 instead of H).

Just as in the case of normal Fermi systems, the calculation of Green's functions for a superconductor has reduced to a calculation of the averages of the T products of the field operators in the interaction representation. This can be done on the basis of Wick's theorem generalized so as to take into account the anomalous averages (2.1.1) or the anomalous Green's functions in the matrix (2.1.10). Such anomalous Green's functions occur when the Nambu operators $\Psi(x)$ and $\overline{\Psi}(x')$ are paired. If, for example, the Nambu operators $\Psi(x)$ and $\Psi(x')$ are paired, then there arises a Green's functions matrix of the form

$$\begin{pmatrix} \langle T\psi_\uparrow(x)\psi_\uparrow(x')\rangle & \langle T\psi_\uparrow(x)\overline{\psi}_\downarrow(x')\rangle \\ \langle T\overline{\psi}_\downarrow(x)\psi_\uparrow(x')\rangle & \langle T\overline{\psi}_\downarrow(x)\overline{\psi}_\downarrow(x')\rangle \end{pmatrix} \qquad (2.1.20)$$

which corresponds either to normal averages off-diagonal with respect to spin, or to pairs in the triplet state which are disregarded in the BCS model.

Thus, when calculating the average of a T product of Nambu's operators one should formally proceed as follows: only the "normal-type" pairing of the operators $\Psi(x)$ and $\overline{\Psi}(x')$ must be taken into consideration. Such pairings will lead to Green's functions like (2.1.10). In other words, in terms of the Nambu operators, Wick's theorem for superconductors is of the same form as that for normal metals. This implies that the series of perturbation theory will be represented by the same diagrams as those in a normal metal. The difference will consist solely of the matrix nature of the zero-approximation Green's functions $g_0(x,x')$ and of the matrix nature of the vertices. The structure of the latter is easily obtained from the form of the interaction Hamiltonian. Namely, it follows from (2.1.15) that the vertex of the electron-phonon interaction corresponds to $g(\underline{r})\tau_3$, and the vertex of the Coulomb interaction to τ_3. In this fashion, the perturbation theory diagrams for the normal and superconducting states formally coincide, and for the superconductor the analytic expressions corresponding to them should be written with due regard for the matrix structure of the Green's functions and vertices.

2.1.3 The Momentum Representation

Let us now transform the field operators $\psi_\alpha(\underline{r})$ and $\psi^+_\alpha(\underline{r})$ to the momentum representation

$$\psi_{\underline{p}\alpha} = \frac{1}{\sqrt{V}}\int d\underline{r}\, e^{-i\underline{pr}}\psi_\alpha(\underline{r}), \qquad \psi^+_{\underline{p}\alpha} = \frac{1}{\sqrt{V}}\int d\underline{r}\, e^{i\underline{pr}}\psi^+_\alpha(\underline{r}) \quad .$$

Then the Nambu operators $\Psi(\underline{r})$ and $\Psi^+(\underline{r})$ will, by definition (2.1.5), have the components

$$\Psi_{\underline{p}} = \begin{pmatrix} \psi_{\underline{p}\uparrow} \\ \psi^+_{-\underline{p}\downarrow} \end{pmatrix}, \qquad \Psi^+_{\underline{p}} = (\psi^+_{\underline{p}\uparrow}, \psi_{-\underline{p}\downarrow}). \qquad (2.1.21)$$

From these we construct a Green's function of the form

$$g(\underline{p},\tau-\tau') = -\left\langle T\Psi_{\underline{p}}(\tau)\overline{\Psi}_{\underline{p}}(\tau')\right\rangle \quad . \qquad (2.1.22)$$

The explicit matrix representation of this expression,

$$g(\underline{p},\tau-\tau') = \begin{pmatrix} -\langle T\psi_{\underline{p}\uparrow}(\tau)\bar{\psi}_{\underline{p}\uparrow}(\tau')\rangle & -\langle T\psi_{\underline{p}\uparrow}(\tau)\psi_{-\underline{p}\downarrow}(\tau')\rangle \\ -\langle T\psi_{-\underline{p}\downarrow}(\tau)\bar{\psi}_{\underline{p}\uparrow}(\tau')\rangle & -\langle T\bar{\psi}_{\underline{p}\downarrow}(\tau)\psi_{\underline{p}\downarrow}(\tau')\rangle \end{pmatrix} \tag{2.1.23}$$

indicates that the Nambu formalism picks out only those correlations which correspond to Cooper pairs with zero total momentum.

In the momentum representation, the interaction Hamiltonian (2.1.14) may be written as a sum of two terms

$$H_{int} = H_{int}^{c} + H_{int}^{e-ph} \quad ,$$

where H_{int}^{c} describes the Coulomb interaction

$$H_{int}^{c} = \frac{1}{2} \sum_{\underline{p}_1+\underline{p}_2=\underline{p}_1'+\underline{p}_2'} (\underline{p}_1\underline{p}_2|V^c|\underline{p}_1'\underline{p}_2')(\Psi_{\underline{p}_1}^{+}\tau_3\Psi_{\underline{p}_1'})(\Psi_{\underline{p}_2}^{+}\tau_3\Psi_{\underline{p}_2'}) \tag{2.1.24}$$

and H_{int}^{e-ph} describes the electron-phonon interaction

$$H_{int}^{e-ph} = \sum_{\underline{p}-\underline{p}'=q} \sum_{j} g_j(\underline{p},\underline{p}')(\Psi_{\underline{p}}^{+}\tau_3\Psi_{\underline{p}'})(b_{\underline{q}j} + b_{-\underline{q}j}^{+}) \quad . \tag{2.1.25}$$

The electron-phonon interaction matrix element $g_j(\underline{p},\underline{p}')$ is expressed in terms of the electron-ion interaction matrix element

$$g_j(\underline{p},\underline{p}') = -\sum_{\kappa}\frac{1}{\sqrt{2NM_\kappa\omega_j(\underline{q})}}\langle \underline{p}|\underline{e}_j(\underline{q},\kappa)\underline{\nabla}U_\kappa(\underline{r})|\underline{p}'\rangle\, e^{i\underline{q}\underline{\rho}_\kappa} \quad , \tag{2.1.26}$$

where $\omega_j(\underline{q})$ and $\underline{e}_j(\underline{q},\kappa)$ stand for the phonon frequency and for the polarization vector of the branch j with momentum $\underline{q} = \underline{p} - \underline{p}'$; U_κ is the potential of an ion at $\underline{\rho}_\kappa$ with the mass M_κ; N is the number of unit cells in the crystal.

Expression (2.1.25) derives from (2.1.15) if one expands $\underline{U}(\underline{r})$ in a power series of the phonon creation $b_{\underline{q}j}^{+}$ and annihilation $b_{\underline{q}j}$ operators:

$$\underline{U}(\underline{r}) = \sum_{\underline{q}jl}\sum_{\kappa}\frac{\underline{e}_j(\underline{q},\kappa)}{\sqrt{2NM_\kappa\omega_j(\underline{q})}}\, e^{i\underline{q}\underline{r}}\,(b_{\underline{q}j} + b_{-\underline{q}j}^{+})\,\delta(\underline{r}-\underline{R}_l-\underline{\rho}_\kappa) \quad , \tag{2.1.27}$$

where $\underline{R}_l$ is the radius vector of the lth cell.

The zero-order Hamiltonian (2.1.17) in the momentum representation is

$$H_o = \sum_{\underline{p}}\xi_{\underline{p}}(\Psi_{\underline{p}}^{+}\tau_3\Psi_{\underline{p}}), \tag{2.1.28}$$

where $\xi_{\underline{p}}$ is the self-energy part of the one-particle Hamiltonian $H_0(\underline{r})$. Here and henceforth the self-energy part is measured from the Fermi level of a normal metal.

So far, we have considered only the coordinate dependence of the operators and of the Green's functions. A complete transition to the momentum representation may be achieved by writing the time dependence of the function $g(x,x')$ as a Fourier series with discrete frequencies so that the complete expansion is of the form

$$g(x,x') = T\sum_n \int \frac{d\underline{p}}{(2\pi)^3} \exp\{i[\underline{p}(\underline{r}-\underline{r}') - \omega_n(\tau-\tau')]\}\, g(\underline{p},\omega_n) \quad ,$$
$$\omega_n = (2n+1)\pi T \tag{2.1.29}$$

($T = 1/\beta$ is the temperature of the system).

Equation (2.1.13) for the zero-order Green's function has the form

$$(i\omega_n\tau_0 - \xi_{\underline{p}}\tau_3)\, g_0(\underline{p},\omega_n) = \tau_0 \quad , \tag{2.1.30}$$

whence it follows that

$$g_0^{-1}(\underline{p},\omega_n) = i\omega_n\tau_0 - \xi_{\underline{p}}\tau_3 \quad . \tag{2.1.31}$$

The exact Green's function corresponding to the complete series in H_{int} may be expressed in terms of the total irreducible self-energy part $\Sigma(\underline{p},\omega_n)$ according to the usual diagram technique

$$g^{-1}(\underline{p},\omega_n) = g_0^{-1}(\underline{p},\omega_n) - \Sigma(\underline{p},\omega_n) \quad . \tag{2.1.32}$$

Each of these quantities written in (2.1.32) is a two-row matrix and may therefore be expanded in Pauli matrices, including the unit matrix τ_0. First of all, write the expansion for Σ in the following standard form

$$\Sigma(\underline{p},\omega_n) = [1 - Z(\underline{p},\omega_n)]\, i\omega_n\tau_0 + \chi(\underline{p},\omega_n)\tau_3 + \varphi(\underline{p},\omega_n)\tau_1 \quad . \tag{2.1.33}$$

This expansion does not involve the τ_2 matrix since its coefficient may always (in the absence of a magnetic field) be set equal to zero by selecting an appropriate phase. By substituting (2.1.33) into (2.1.32), it is not difficult now to obtain an expression for Σ in terms of Z, χ and φ,

$$g(\underline{p},\omega_n) = \frac{Z(\underline{p},\omega_n)i\omega_n\tau_0 + \bar{\varepsilon}(\underline{p},\omega_n)\tau_3 + \varphi(\underline{p},\omega_n)\tau_1}{[Z(\underline{p},\omega_n)i\omega_n]^2 - \varphi^2(\underline{p},\omega_n) - \bar{\varepsilon}^2(\underline{p},\omega_n)} \quad , \tag{2.1.34}$$

where

$$\bar{\varepsilon}(\underline{p},\omega_n) = \xi_{\underline{p}} + \chi(\underline{p},\omega_n) \quad . \tag{2.1.35}$$

The expression for $\Sigma(\underline{p},\omega_n)$ should be found, according to perturbation theory, by summing a certain sequence of diagrams. As already noted above, the rules of the diagram technique remain the same as those for a normal metal. The elements of the diagrams are propagators, interaction lines and vertices. The rules the analytic expressions must obey are illustrated in Fig.2.1.

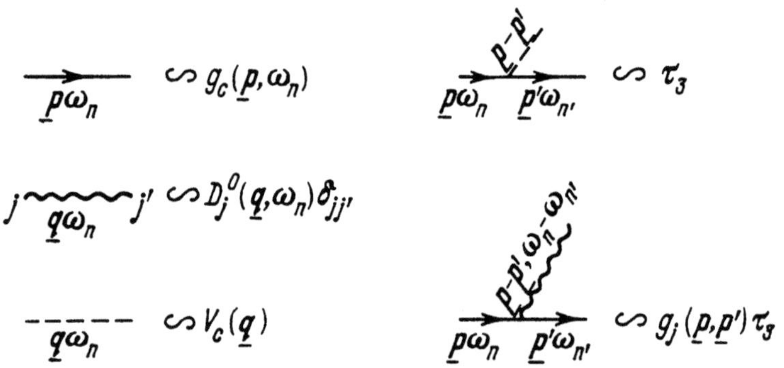

Fig.2.1 Basic diagrams for the electron Green's function of the superconductor

2.1.4 Lower-Order Graphs for Σ

Consider the first-order graphs for Σ (Fig.2.2).

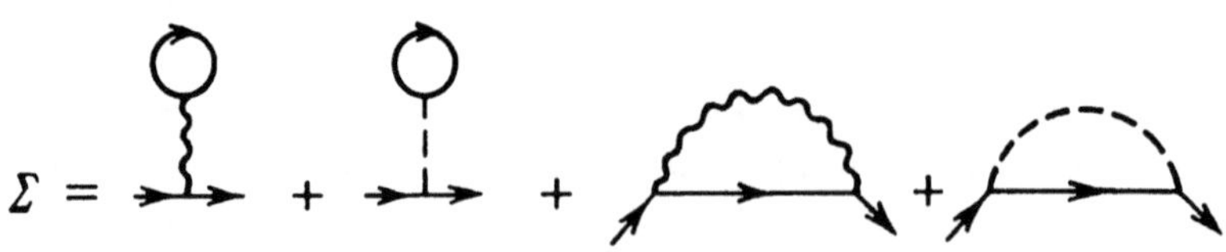

Fig.2.2 First-order graphs for the self-energy part of the Green's function

To the first of them there corresponds the analytic expression

$$= T\sum_m \int \frac{d\underline{p}'}{(2\pi)^3}\,\tau_3\,\mathrm{Sp}\{g_o(\underline{p}',\omega_m)\tau_3\} \times \sum_j g_j(\underline{p},\underline{p})D_j^0(0,0)\,g_j(\underline{p}',\underline{p}') \tag{2.1.36}$$

(the trace, as always, corresponds to the electron loop). Without explicitly writing out this expression, we see that it is proportional to τ_3. From the general expression (2.1.33) we see that the graph (2.1.36) only contributes to $\chi(\underline{p},\omega_n)$, thus leading to corrections in the electron spectrum in normal metals and to a shift of the chemical potential. In other words, it has no effect on the anomalous averages which are due to the terms in Σ, proportional to τ_1. For the same reason, the contribution from the second graph displayed in Fig.2.2, which stems from the Coulomb interaction, is also nonessential for the occurrence of a superconducting state.

Let us now consider the other pair of first-order graphs. We denote their contribution to Σ by $\Sigma_1(\underline{p},\omega_n)$,

$\Sigma_1(\underline{p},\omega_n) =$ 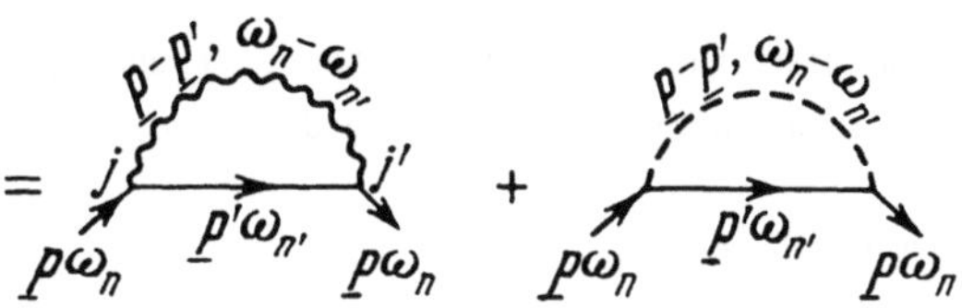 . (2.1.37)

According to the rules introduced

$$\Sigma_1(\underline{p},\omega_n) = -T\sum_{n'}\int \frac{d\underline{p}'}{(2\pi)^3}\,\tau_3 g_o(\underline{p}',\omega_{n'})\tau_3 \times \Big[\sum_j g_j(\underline{p},\underline{p}')D_j^0(\underline{p}-\underline{p}',\omega_n-\omega_{n'})\,g_j(\underline{p}',\underline{p}) + V_c(\underline{p}-\underline{p}')\Big] \quad . \tag{2.1.38}$$

Graphs of higher order may be obtained from (2.1.37) by dressing either the propagators or the vertices, or by combining the two processes. For a normal metal, the electron-phonon interaction was studied by MIGDAL [2.6] who showed that dressing of the vertices leads to corrections to the latter. These corrections are of the order of the square root of the ratio of the Debye frequency ω_D to the Fermi energy E_F, which amounts to $\sqrt{m/M}$ where m and M are the masses of an electron and of an ion, respectively. Hence, it is possible with this accuracy to take for $\Sigma(\underline{p},\omega_n)$ an infinite series of graphs obtained from (2.1.37) by dressing only the propagators, leaving the bare vertices unrenormalized. As shown by ELIASHBERG [2.7], this situation persists in a superconducting metal. For $\Sigma(\underline{p},\omega_n)$ in a superconductor one may therefore use the graphs (2.1.37) but with the thin electron line replaced by a thick one. The latter is determined by a certain self-consistent equation for the self-energy part. It will be observed that, if the expression for $g_0(\underline{p},\omega_n)$ obtained from (2.1.31) is substituted into (2.1.38) then (2.1.38) contains no terms proportional to τ_1, i.e., no effects of the occurrence of anomalous averages. Every set of graphs with elementary Green's functions will not contain similar terms, either. However, as we shall see, the self-consistent equation for $\Sigma(\underline{p},\omega_n)$, which is obtained by thickening the Green's functions and is an infinte series of elementary graphs, will have a solution containing the component $\sim \tau_1$ in the Pauli matrix expansion. In superconductivity theory such an equation has been called the Eliashberg equation (Sect.2.2).

2.2 Eliashberg Equations for the Electron Green's Function

2.2.1 Formulation of the Equations and Summation over Frequencies

In the above approximations the equation for the self-energy part of the electron Green's function of a superconductor should be chosen to be

$$\Sigma(\underline{p},\omega_n) = \Sigma_{ph}(\underline{p},\omega_n) + \Sigma_c(\underline{p},\omega_n)$$

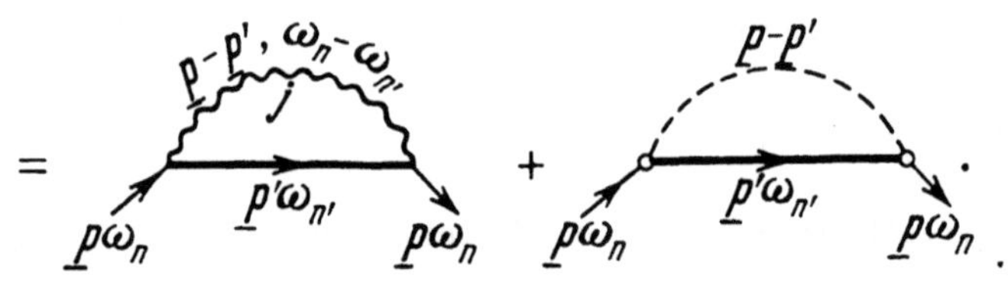

(2.2.1)

The analytic expressions for the phonon Σ_{ph} and Coulomb Σ_c parts, Σ being generalizations of (2.1.38), are

$$\Sigma_{ph}(\underline{p},\omega_n) = -T\sum_{n'}\int\frac{d\underline{p}'}{(2\pi)^3}\,\tau_3 g(\underline{p}',\omega_{n'})\tau_3 \times \sum_j |g_j(\underline{p},\underline{p}')|^2 D_j^0(\underline{p}-\underline{p}',\omega_n-\omega_{n'}) \qquad (2.2.2)$$

$$\Sigma_c(\underline{p},\omega_n) = -T\sum_{n'}\int \frac{d\underline{p}'}{(2\pi)^3}\,\tau_3 g(\underline{p}',\omega_{n'})\tau_3 V_c(\underline{p}-\underline{p}') \quad . \tag{2.2.3}$$

When writing (2.2.2) it was taken into account that the phonon Green's function is diagonal with respect to the phonon branch indices j and use was also made of the equation

$$g_j(\underline{p},\underline{p}') = g_j^*(\underline{p}',\underline{p}) \quad . \tag{2.2.4}$$

By virtue of (2.1.34) and of the property of the Pauli matrices

$$\tau_3\tau_1\tau_3 = -\tau_1 \tag{2.2.5}$$

we obtain the expression

$$\tau_3 g(\underline{p},\omega_n)\tau_3 = \frac{Z(\underline{p},\omega_n)i\omega_n\tau_o + \bar{\varepsilon}(\underline{p},\omega_n)\tau_3 - \varphi(\underline{p},\omega_n)\tau_1}{[Z(\underline{p},\omega_n)i\omega_n]^2 - \varphi^2(\underline{p},\omega_n) - \bar{\varepsilon}^2(\underline{p},\omega_n)} \tag{2.2.6}$$

from which it immediately becomes clear that $\Sigma(\underline{p},\omega_n)$ may contain a term proportional to the τ_1 matrix. As has been stated above, the appearance of this term is attributed to the occurrence of a superconducting state.

To perform the summation over frequencies it is convenient in (2.2.2,3) to use the spectral representation of the electron and phonon Green's functions

$$g(\underline{p},\omega_n) = \frac{1}{2\pi}\int_{-\infty}^{\infty} dz'\,\frac{a(\underline{p},z')}{i\omega_n - z'} \quad , \tag{2.2.7}$$

$$D_j(\underline{p},\omega_n) = \frac{1}{2\pi}\int_{-\infty}^{\infty} dz\,\frac{b_j(\underline{p},z)}{i\omega_n - z} \quad . \tag{2.2.8}$$

Upon substitution of these expressions into (2.2.2,3), the sums over the frequencies are easily calculated

$$T\sum_{n'}\frac{1}{i\omega_{n'} - z'}\cdot\frac{1}{i(\omega_n - \omega_{n'}) - z} = -\frac{1}{2}\,\frac{\tanh\frac{z'}{2T} + \coth\frac{z}{2T}}{i\omega_n - z - z'} \quad , \tag{2.2.9}$$

$$T\sum_{n'}\frac{1}{i\omega_{n'} - z} = -\frac{1}{2}\tanh\frac{z'}{2T} \quad . \tag{2.2.10}$$

Account must be taken of the fact that the spectral density of the temperature Green's function $a(\underline{p},z)$ is related to the retarded Green's function $g(\underline{p},z)$ by

$$a(\underline{p},z) = -2\,\mathrm{Im}\{g(\underline{p},z)\} \quad . \tag{2.2.11}$$

If in expression (2.2.2,3) one performs an analytic continuation from the imaginary axis to the real axis using the substitution

$$i\omega_n \to \omega + i\delta \quad , \tag{2.2.12}$$

one obtains equations for the self-energy part of the retarded Green's function. As a result, in place of (2.2.2,3) we now have

$$\Sigma_{ph}(\underline{p},\omega) = -\int \frac{d\underline{p}'}{(2\pi)^3} \sum_j |g_j(\underline{p},\underline{p}')|^2 \int_{-\infty}^{\infty} \frac{dz}{2\pi} \int_{-\infty}^{\infty} \frac{dz'}{2\pi} b_j(\underline{p}-\underline{p}',z) \times \tau_3 \mathrm{Im}\{g(\underline{p}',z')\}\tau_3 \frac{\tanh\frac{z'}{2T} + \coth\frac{z}{2T}}{\omega - z - z' + i\delta} \quad , \tag{2.2.13}$$

$$\Sigma_c(\underline{p},\omega) = -\int \frac{d\underline{p}'}{(2\pi)^3} V_c(\underline{p}-\underline{p}') \int_{-\infty}^{\infty} \frac{dz'}{2\pi} \tau_3\, \mathrm{Im}\{g(\underline{p}',z')\}\tau_3 \tanh\frac{z'}{2T} \; . \tag{2.2.14}$$

To these expressions we should add (2.2.6) and (2.1.33) which are obtained by analytical continuation with the help of (2.2.12),

$$\tau_3 g(\underline{p},\omega)\tau_3 = \frac{Z(\underline{p},\omega)\omega\tau_0 + \bar{\varepsilon}(\underline{p},\omega)\tau_3 - \varphi(\underline{p},\omega)\tau_1}{[Z(\underline{p},\omega)\omega]^2 - \varphi^2(\underline{p},\omega) - \bar{\varepsilon}^2(\underline{p},\omega)} \quad , \tag{2.2.15}$$

$$\Sigma(\underline{p},\omega) = [1 - Z(\underline{p},\omega)]\omega\tau_0 + \chi(\underline{p},\omega)\tau_3 + \varphi(\underline{p},\omega)\tau_1 \quad . \tag{2.2.16}$$

Expressions (2.2.13-16) form a single matrix equation for the determination of the three quantities $Z(\underline{p},\omega)$, $\varphi(\underline{p},\omega)$ and $\chi(\underline{p},\omega)$. One of them, the quantity $\chi(\underline{p},\omega)$, following from (2.1.35), leads to a shift of the chemical potential, i.e., to an effect which little affects the formation of a superconducting state. Therefore in (2.2.15,16) we put $\chi(\underline{p},\omega) \approx 0$ and then from (2.1.35) we have $\bar{\varepsilon}(\underline{p},\omega) \approx \xi_p$, the free electron energy measured from the Fermi level. Thus, there remains a matrix equation which determines only two quantities, namely $Z(\underline{p},\omega)$ and $\varphi(\underline{p},\omega)$. In order to obtain an explicit form of the equations for Z and φ it proves useful to carry out integration over the momentum $\underline{p}'$, whereas the corresponding integral in (2.2.13,14) should be calculated individually.

2.2.2 Calculation of the Phonon Part $\Sigma_{ph}(\underline{p},\omega)$

Since the electron-phonon interaction essentially changes the electron states only near the Fermi surface in a layer whose width is equal to the maximum frequency ω_D of the phonon spectrum, we shall be concerned with the quantity $\Sigma_{ph}(\underline{p},\omega)$ solely for momenta lying in this interval. Since the denominator of (2.2.15) entering (2.2.13) varies with momentum $\underline{p}'$ at least as $\xi^2_{\underline{p}'}$, the integrand has a sharp maximum near $\xi_{\underline{p}} = 0$, i.e., for $\underline{p}'$ on the Fermi surface. This means that the dependence of $Z(\underline{p},\omega)$ and $\varphi(\underline{p},\omega)$ in (2.2.13) on $|\underline{p}|$ is weak and, accordingly, the integration in (2.2.13) over the momentum modulus may be performed immediately by

$$\int \frac{d\underline{p}'}{(2\pi)^3} \cdots = \int d\xi \int_{S(\xi)} \frac{d^2\underline{p}'}{v_{\underline{p}'}} \tag{2.2.17}$$

where $v_{\underline{p}'}$ is the velocity on the isoenergetic surface $S(\xi)$, and $d^2\underline{p}'$ is an element of the area of this surface. Integration over ξ may be carried out by extending the limits of integration from $-\infty$ to ∞ since the function under the integral sign falls off quickly for large ξ. Thus, in the approximations specified above the integral which occurs in (2.2.13) is of the form

$$\int_{-\infty}^{\infty} \frac{d\xi}{\xi^2 - A} = i\frac{\pi}{\sqrt{A}} \operatorname{sign}\{\sqrt{A}\} \quad . \tag{2.2.18}$$

After that integration has been performed, an integral remains in (2.2.13) over the direction of the momentum lying on the Fermi surface. It is helpful to average the obtained equation with respect to the Fermi surface S_F. To this end we introduce the quantity

$$\Sigma_{ph}(\omega) = \int_{S_F} \frac{d^2\underline{p}}{v_{\underline{p}}} \Sigma_{ph}(\underline{p},\omega) \Big/ \int_{S_F} \frac{d^2\underline{p}}{v_{\underline{p}}} \quad . \tag{2.2.19}$$

Equation (2.2.13) then becomes

$$\Sigma_{ph}(\omega) = \int_{-\infty}^{\infty} \frac{dz'}{2\pi} \int_{-\infty}^{\infty} \frac{dz}{2\pi} \left(\int_{S_F} \frac{d^2\underline{p}}{v_{\underline{p}}} \right)^{-1} \int_{S_F} \frac{d^2\underline{p}}{v_{\underline{p}}} \int_{S_F} \frac{d^2\underline{p}'}{v_{\underline{p}'}} \sum_j |g_j(\underline{p},\underline{p}')|^2 b_j(\underline{p}-\underline{p}')$$

$$\times\ \mathrm{sign}\{z'\} \pi \mathrm{Re} \left\{ \frac{Z(\underline{p}',z')z'\tau_0 - \varphi(\underline{p}',z')\tau_1}{\sqrt{z'^2 Z^2(\underline{p}',z') - \varphi^2(\underline{p}',z')}} \right\} \cdot \frac{\tanh\frac{z'}{2T} + \coth\frac{z}{2T}}{\omega - z - z' + i\delta} \quad . \tag{2.2.20}$$

All the momenta in (2.2.20) lie on the Fermi surface.

In the case when the anisotropy of the Fermi surface is high one should allow for the dependence in $Z(\underline{p}',z')$ and $\varphi(\underline{p}',z')$ on the direction of the vector $\underline{p}'$. If the anisotropy is not very high it may be assumed that the angular dependence of the functions $Z(\underline{p}',z')$ and $\varphi(\underline{p}',z')$ is weaker than that for the quantities $g_j(\underline{p},\underline{p}')$ and $b_j(\underline{p}-\underline{p}',z)$ characterizing the electron-phonon interaction and the phonon spectrum[1]. Then $Z(\underline{p}',z')$ and $\varphi(\underline{p}',z')$ may be replaced by their (2.2.19)-type expressions averaged with respect to the Fermi surface. We denote these expressions by $Z(z')$ and $\varphi(z')$. In this not altogether strict approximation, (2.2.20) becomes an integral over frequencies only,

$$\Sigma_{ph}(\omega) = \int_{-\infty}^{\infty} dz' \int_{-\infty}^{\infty} dz \frac{1}{4\pi} \int_{S_F} \frac{d^2\underline{p}}{v_{\underline{p}}} \int_{S_F} \frac{d^2\underline{p}'}{v_{\underline{p}'}} \sum_j |g_j(\underline{p},\underline{p}')|^2 b_j(\underline{p}-\underline{p}',z) \left(\int_{S_F} \frac{d^2\underline{p}}{v_{\underline{p}}} \right)^{-1}$$

$$\times \frac{\tanh\frac{z'}{2T} + \coth\frac{z}{2T}}{\omega - z - z' + i\delta} \mathrm{sign}\{z'\} \mathrm{Re} \left\{ \frac{z' Z(z')\tau_0 - \varphi(z')\tau_1}{\sqrt{z'^2 Z^2(z') - \varphi^2(z')}} \right\} \quad . \tag{2.2.21}$$

---

[1] ALLEN [2.31] has developed a method of successively allowing for the anisotropy in equations for strong-coupling superconductors. This technique consists of expanding each of the quantities $\Sigma_{ph}(\underline{p},\omega)$ and $g(\underline{p},\omega)$ as a series of functions $F_J(\underline{p})$, i.e., "Fermi surface harmonics" [2.32,33] which constitute a full orthonormalized basis for the functions prescribed on the constant-energy surface. The following equations hold for these functions

$$\sum_{\underline{p}} F_J(\underline{p}) F_{J'}(\underline{p}) \delta(\xi_{\underline{p}} - \xi) = \delta_{JJ'} N(\xi) , \qquad N(\xi) = \sum_{\underline{p}} \delta(\xi_{\underline{p}} - \xi) .$$

They enable the integration over the energy ξ and the angle of the vector $\underline{p}$. As a result, one obtains combined integral equations for the expansion coefficients Σ_J and g_J whose kernels represent a matrix with respect to the indices J and J'. In the approximation that in the sum over J there remains only one term corresponding to the neglect of spatial anisotropy, one obtains (2.2.23,24).

In view of (2.2.16) and allowing for all the approximations made, we have

$$\Sigma_{ph}(\omega) = [1 - Z_{ph}(\omega)]\omega\tau_0 + \varphi_{ph}(\omega)\tau_1 \quad . \tag{2.2.22}$$

Substituting (2.2.22) into (2.2.21) and equating the coefficients of the matrices τ_0 and τ_1 on the left-hand and right-hand sides of the equation so obtained, we find relations for $Z(\omega)$ and $\varphi(\omega)$, i.e.,

$$[1 - Z_{ph}(\omega)]\omega = -\int_{-\infty}^{\infty} dz' \, K_{ph}(z',\omega)\mathrm{Re}\left\{\frac{z'}{\sqrt{z'^2 - \Delta^2(z')}}\right\} \mathrm{sign}\{z'\} \quad , \tag{2.2.23}$$

$$Z_{ph}(\omega)\Delta_{ph}(\omega) = \int_{-\infty}^{\infty} dz' \, K_{ph}(z',\omega)\mathrm{Re}\left\{\frac{\Delta(z')}{\sqrt{z'^2 - \Delta^2(z')}}\right\} \mathrm{sign}\{z'\} \tag{2.2.24}$$

where instead of $\varphi_{ph}(\omega)$ and $\varphi(\omega)$ we have introduced the quantities

$$\Delta_{ph}(\omega) = \varphi_{ph}(\omega)/Z_{ph}(\omega) \quad , \quad \Delta(\omega) = \varphi(\omega)/Z(\omega) \quad . \tag{2.2.25}$$

The quantity $\Delta(\omega)$ is called the gap function since it determines the gap in the spectrum of elementary excitations in a superconductor [2.5].

The kernel of the integral equations (2.2.24,25) has the form

$$K_{ph}(z',\omega) = \int_0^{\infty} dz\alpha^2(z)F(z)\frac{1}{2}\left\{\frac{\tanh\frac{z'}{2T} + \coth\frac{z}{2T}}{z' + z - \omega - i\delta} - \frac{\tanh\frac{z'}{2T} - \coth\frac{z}{2T}}{z' - z - \omega - i\delta}\right\} \quad . \tag{2.2.26}$$

The notation introduced here is

$$\alpha^2(z)F(z) = \int_{S_F}\frac{d^2p}{v_{\underline{p}}}\int_{S_F}\frac{d^2p'}{v_{\underline{p}'}}\sum_j|g_j(\underline{p},\underline{p}')|^2\delta(z - \omega_j(\underline{q}))\Big/\int_{S_F}\frac{d^2p}{v_{\underline{p}}} \tag{2.2.27}$$

When passing from (2.2.21) to (2.2.23,24) we have used the zero-approximation spectral phonon Green's function

$$b_j(\underline{q},z) = 2\pi[\delta(z - \omega_j(\underline{q})) - \delta(z + \omega_j(\underline{q}))] \quad . \tag{2.2.28}$$

Note some general properties of the solution of (2.2.23,24). The kernel (2.2.26) of the electron-phonon interaction possesses the symmetry property

$$K_{ph}(-z',-\omega) = -K^*_{ph}(z',\omega) \quad . \tag{2.2.29}$$

This leads to the symmetry and

$$Z_{ph}(-\omega) = Z^*_{ph}(\omega) \quad , \tag{2.2.30}$$

$$\Delta_{ph}(-\omega) = \Delta^*_{ph}(\omega) \quad . \tag{2.2.31}$$

Generally speaking, such symmetry properties with respect to the variable ω hold for the total self-energy part (involving, for example, the Coulomb interaction). Such equations may be derived immediately from the general equations (2.2.13,14) by using the symmetry for the Green's function $g(\underline{p},\omega)$ relative to the change of sign for ω and $\underline{p}$ simultaneously.

2.2.3 Calculation of the Coulomb Part $\Sigma_c(\underline{p},\omega)$

Let us now transform the Coulomb term (2.2.14). The integral over z' between the limits $-\infty$ and ∞ may be transformed to an integral which is only over the positive values of z'. By substituting expression (2.2.15) into (2.2.14) and by using the parity of the functions $Z(\underline{p}',z')$ and $\varphi(\underline{p}',z')$ one notices that in (2.2.14), the coefficient in front of the matrix τ_0 vanishes. Thus, the Coulomb interaction gives no contribution to the renormalization of $Z(\underline{p},\omega)$. We also neglect the contribution of the Coulomb interaction to $\chi(\underline{p},\omega)$ since, as already mentioned, $\chi(\underline{p},\omega)$ will renormalize only the chemical potential of the system. Consequently, there remains only the equation for the quantity $\varphi_c(\underline{p},\omega)$:

$$\varphi_c(\underline{p},\omega) = \frac{1}{\pi}\int_0^\infty dz' \tanh\frac{z'}{2T} \int \frac{d\underline{p}'}{(2\pi)^3} V_c(\underline{p}-\underline{p}') \, \mathrm{Im}\left\{\frac{\varphi(\underline{p}',z')}{z'^2 Z^2(z') - \varphi^2(\underline{p}',z') - \xi^2_{\underline{p}'}}\right\} \quad . \tag{2.2.32}$$

We note that the quantity $Z(z')$ is momentum-independent because it is determined only by the electron-phonon interaction. It has also turned out that $\varphi_c(\underline{p},\omega)$ does not depend on the frequency, which is the result of neglecting the retardation effects in the Coulomb interaction. We will designate this quantity as $\varphi_c(\underline{p})$.

Equation (2.2.32) is an integral not only over the frequency z' but also over the momenta. Let us average this expression with respect to the isoenergetic surface corresponding to some energy ξ:

$$\int\limits_{S(\xi)} \frac{d^2p}{v_p}\varphi_c(p) \Big/ \int\limits_{S(\xi)} \frac{d^2p}{v_p} = \frac{1}{\pi}\int\limits_0^\infty dz' \tanh\frac{z'}{2T} \int d\xi \left(\int\limits_{S(\xi)} \frac{d^2p}{v_p}\right)^{-1}$$

$$\times \int\limits_{S(\xi)} \frac{d^2p}{v_p} \int\limits_{S(\xi)} \frac{d^2p'}{v_{p'}} V_c(p - p')\,\mathrm{Im}\left\{\frac{\varphi(p',z')}{z'^2Z^2(z') - \varphi^2(p',z') - \xi^2_{p'}}\right\} \quad . \tag{2.2.33}$$

Just as in the case of electron-phonon interaction, we assume that the quantity $\varphi(p',z')$ varies with the direction of the momentum more weakly than the Coulomb interaction matrix element $V_c(p - p')$ does. By virtue of this, (2.2.33) may be rewritten approximately in the form

$$\varphi_c(p) = \frac{1}{\pi}\int\limits_0^\infty dz' \tanh\frac{z'}{2T} \int d\xi N(0)V_c(p,p')\,\mathrm{Im}\left\{\frac{\varphi(p',z')}{z'^2Z^2(z') - \varphi^2(p',z') - \xi^2_{p'}}\right\} \quad . \tag{2.2.34}$$

where

$$N(0)V_c(p,p') = \int\limits_{S_F} \frac{d^2p}{v_p} \int\limits_{S_F} \frac{d^2p'}{v_{p'}} V_c(p - p') \Big/ \int\limits_{S_F} \frac{d^2p}{v_p} \quad . \tag{2.2.35}$$

$V_c(p,p')$ is the quantity $V_c(p,p')$ averaged with respect to the Fermi surface and $N(0) = \int_{S_F} d^2p/v_p$ is the density of electron states on that surface. With an approximation such as (2.2.34), the quantities $\varphi_c(p)$ and $\varphi(p',z')$ are functions of the momentum modulus only. This approximation holds only for weakly anisotropic Fermi surfaces.

In contrast to the electron-phonon interaction which leads to an effective interaction of electrons only near the Fermi surface in a layer whose width is of the order of the Debye frequency, the energy interval of the Coulomb interaction is much larger and is of the order of the Fermi energy E_F. Consequently, the contributions of $\varphi_c(p)$ and $\varphi_{ph}(\omega)$ are essential in different frequency ranges. To explore this problem it is convenient to introduce some frequency ω_c with

$$\omega_D \ll \omega_c \ll E_F \tag{2.2.36}$$

and to split the integral over z' in (2.2.34) into two parts corresponding to integration between 0 and ω_c and between ω_c and ∞.

Over the frequency range $z' > \omega_c$, the electron-phonon interaction is ineffective and therefore $Z(z') \to 1$ and $\varphi(p',z') \to \varphi_c(p')$. Beside this, in view of the evident inequality $T_c \ll \omega_c$, $\tanh z'/2T \to 1$.

As a result, the integral over z' in this region can be calculated exactly

$$\frac{1}{\pi}\int_{\omega_c}^{\infty} dz' \,\mathrm{Im}\left\{\frac{\varphi(p')}{z'^2 - \varphi^2_c(p') - \xi^2_{p'}}\right\}$$

$$= \frac{\varphi(p')}{\pi}\int_{\omega_c}^{\infty} dz' \,\mathrm{Im}\left\{\frac{1}{(z'+i\delta)^2 - E^2_{p'}}\right\} = \frac{\varphi(p')}{2E_{p'}}\,\theta(E_{p'} - \omega_c), \tag{2.2.37}$$

where $E_p = \sqrt{\xi^2_{p'} + \varphi^2_c(p)}$ and $\theta(x)$ is the θ function. In the calculation of the imaginary part in (2.2.37) we have taken into account that all the frequencies involved in the integral equations presented in this section have an infinitesimal imaginary part.

Now (2.2.34) becomes

$$\varphi_c(p) + \int \frac{d\xi_{p'}}{2E_{p'}}\,\theta(E_{p'} - \omega_c)N(0)V_c(p,p')\varphi_c(p') = \int d\xi_p V_c(p,p')F_{p'} \;, \tag{2.2.38}$$

where, for the sake of brevity, the notation

$$F_{p'} = N(0)\int_0^{\omega_c} dz' \,\frac{1}{\pi}\,\mathrm{Im}\left\{\frac{\varphi(p',z')}{z'^2 Z^2(z') - \varphi^2(p',z') - \xi^2_{p'}}\right\}\tanh\frac{z'}{2T} \tag{2.2.39}$$

was introduced. We try a solution of (2.2.38) in the form

$$\varphi_c(p) = \int d\xi_p U_c(p,p')F_{p'} \quad . \tag{2.2.40}$$

For the quantity $U_c(p,p')$ we immediately obtain the integral equation

$$U_c(p,p') + \int \frac{d\xi_{p''}}{E_{p''}}\,\theta(E_{p''} - \omega_c)N(0)\,V_c(p,p'')\,U_c(p'',p') = V_c(p,p') \;. \tag{2.2.41}$$

In its sense, $U_c(p,p')$ is a pseudopotential accounting for the contribution to the effective interaction of electrons near the Fermi surface of multiple Coulomb scattering of electrons excited in the region of higher energies.

Let us now calculate the integral (2.2.40), assuming the solution to integral equation (2.2.41) is known. As follows from the definition (2.2.39), the quantity $F_{p'}$ starts decreasing sharply at energies $\xi_{p'} \gtrsim \omega_c$. This implies that the main contribution to the integral (2.2.40) is made by the values of $\xi_{p'}$ near the Fermi energy within an interval $\sim \omega_c$. Over this interval, the pseudopotential changes little since the scale of its variation is of the order of the Fermi energy. Therefore the quantitiy U_c may be put before the integral sign in (2.2.40) if $p' \approx p_F$. Then for $p \approx p_F$ we get the approximate relationship

$$\varphi_c(p) \approx U_c \int_{-\infty}^{\infty} d\xi_{p'} F_{p'} \tag{2.2.42}$$

with $U_c \equiv U_c(p_F, p_F)$ being the value of the pseudopotential on the Fermi surface. Carrying out the integration over $\xi_{p'}$ we get the following final expression for the Coulomb contribution

$$\varphi_c(p) = -U_c N(0) \int_0^{\omega_c} dz' \tanh\frac{z'}{2T} \mathrm{Re}\left\{\frac{\Delta(z')}{\sqrt{z'^2 - \Delta^2(z')}}\right\} \quad . \tag{2.2.43}$$

To finally solve the problem we need the pseudopotential. As an estimate, we approximate the matrix element of the Coulomb interaction in the form

$$V_c(p,p') = \begin{cases} V_c, & |\xi_p|, |\xi_{p'}| < E_F \\ 0, & |\xi_p|, |\xi_{p'}| > E_F \end{cases} \quad . \tag{2.2.44}$$

The integral equation (2.2.41) then becomes a simple algebraic equation

$$U_c + N(0)V_c \int_{-E_F}^{E_F} \frac{d\xi}{\sqrt{\xi^2 + \varphi_c^2}} \theta(\sqrt{\xi^2 + \varphi_c^2} - \omega_c) U_c = V_c \tag{2.2.45}$$

Since $\varphi_c \sim T_c << \omega_c$, the integration over ξ is actually carried out between the limits ω_c and E_F, which gives $\ln(E_F/\omega_c)$. Thus, we obtain the following expression for the pseudopotential

$$U_c = \frac{V_c}{1 + V_c N(0) \ln(E_F/\omega_c)} \quad , \tag{2.2.46}$$

which is somewhat weaker than the Coulomb potential since $E_F >> \omega_c$. This pseudopotential is often referred to as the "intermediate" pseudopotential since it enters the actual Coulomb pseudopotential μ^* which we have determined in Sect.1.2.

2.2.4 Basic Equations for a Superconductor

Let us now write equations for the quantities $Z(\omega)$ and $\Delta(\omega)$ determining the total self-energy part in a superconductor. According to the results obtained above

$$Z(\omega) = Z_{ph}(\omega), \quad Z(\omega)\Delta(\omega) = Z_{ph}(\omega)\Delta_{ph}(\omega) + \varphi_c \quad , \tag{2.2.47}$$

where $Z_{ph}(\omega)$, $\Delta_{ph}(\omega)$ and φ_c are determined by (2.2.23), (2.2.24) and (2.2.43), respectively. Writing these out in explicit form, we obtain the following combined equations for $Z(\omega)$ and $\Delta(\omega)$:

$$[1 - Z(\omega)]\omega = -\int_{-\infty}^{\infty} dz' K_{ph}(z',\omega) \mathrm{Re}\left\{\frac{z'}{\sqrt{z'^2 - \Delta^2(z')}}\right\} \mathrm{sign}\{z'\} \quad ,$$

$$Z(\omega)\Delta(\omega) = \int_{-\infty}^{\infty} dz' K_{ph}(z',\omega) \mathrm{Re}\left\{\frac{\Delta(z')}{\sqrt{z'^2 - \Delta^2(z')}}\right\} \mathrm{sign}\{z'\} \tag{2.2.48}$$

$$- U_c N(0) \int_0^{\omega_c} dz' \tanh\frac{z'}{2T} \mathrm{Re}\left\{\frac{\Delta(z')}{\sqrt{z'^2 - \Delta^2(z')}}\right\} \quad .$$

It is sometimes convenient to use the equations in a different form where the integration over z' is performed only over the positive region. This form is easily obtained by substituting (2.2.26) for the phonon interaction kernel into (2.2.48). Then the first of these equations becomes

$$[1 - Z(\omega)]\omega = \frac{1}{2}\int_0^{\infty} dz\alpha^2(z)F(z) \int_0^{\infty} dz' \mathrm{Re}\left\{\frac{z'}{\sqrt{z'^2 - \Delta^2(z')}}\right\}$$

$$\times \left[\left(\tanh\frac{z'}{2T} + \coth\frac{z}{2T}\right)\left(\frac{1}{z' + z + \omega + i\delta} - \frac{1}{z' + z - \omega - i\delta}\right)\right.$$

$$\left. - \left(\tanh\frac{z'}{2T} - \coth\frac{z}{2T}\right)\left(\frac{1}{z' - z + \omega + i\delta} - \frac{1}{z' - z - \omega - i\delta}\right)\right] \tag{2.2.49}$$

The second equation has a similar structure

$$Z(\omega)\Delta(\omega) = \frac{1}{2}\int_0^{\infty} dz\alpha^2(z)F(z) \int_0^{\infty} dz' \mathrm{Re}\left\{\frac{\Delta(z')}{\sqrt{z'^2 - \Delta^2(z')}}\right\}$$

$$\times\left[\left(\tanh\frac{z'}{2T}+\coth\frac{z}{2T}\right)\left(\frac{1}{z'+z+\omega+i\delta}+\frac{1}{z'+z-\omega-i\delta}\right)\right.$$

$$\left.-\left(\tanh\frac{z'}{2T}-\coth\frac{z}{2T}\right)\left(\frac{1}{z'-z+\omega+i\delta}+\frac{1}{z'-z-\omega-i\delta}\right)\right]$$

$$-U_cN(0)\int_0^{\omega_c}dz'\tanh\frac{z'}{2T}\,\mathrm{Re}\left\{\frac{\Delta(z')}{\sqrt{z'^2-\Delta^2(z')}}\right\} \quad . \qquad (2.2.50)$$

This is a pair of coupled nonlinear integral equations for the quantities $Z(\omega)$ and $\Delta(\omega)$. The first of these equations describes the renormalization of the electron spectrum due to the electron-phonon interaction, whereas the other describes the occurence of the anomalous part of the Green's function due to the readjustment of the ground state of a normal metal.

Equations of this type (without the Coulomb term in the energy) were first obtained by ELIASHBERG [2.7], and those with the Coulomb term by MOREL and ANDERSON [2.8] (see also [2.9,10,11]). By tradition, (2.2.49,50) are called the Eliashberg equations.

We give one more expression for the Green's function of the superconductor, $g(\underline{p},\omega)$, for the momentum $\underline{p}$ on the Fermi surface. It is given by (2.1.34) in which an analytical continuation $i\omega_n \rightarrow \omega + i\delta$ should be done, i.e.,

$$g(\underline{p},\omega) = \frac{Z(\omega)\omega\tau_0 + \xi_p\tau_3 + \varphi(\omega)\tau_1}{\omega^2 Z^2(\omega) - \varphi^2(\omega) - \xi_{\underline{p}}^2} \quad . \qquad (2.2.51)$$

Instead of $\varphi(\omega)$ one can use $\Delta(\omega) = \varphi(\omega)/Z(\omega)$ as in the Eliashberg equations. Eqs. (2.2.49-51) completely solve the problem of the one-particle electron Green's function of a superconductor.

2.2.5 Dispersion Relations

It is useful to note the dispersion relation between the quantities $Z(\omega)$ and $\Delta(\omega)$ which arises from the dispersion relations for the electron Green's function. From the spectral representation (2.2.7) of the function $g(\underline{p},\omega)$ and from (2.2.11) we get the following general dispersion relation for the retarded Green's function

$$g(\underline{p},\omega) = -\frac{1}{\pi}\int_{-\infty}^{\infty} dz\, \frac{\mathrm{Im}\{g(\underline{p},z)\}}{\omega - z + i\delta} \quad . \qquad (2.2.52)$$

We substitute (2.2.51) for $g(\underline{p},\omega)$ into the left-hand and right-hand sides of this equation and then integrate both sides of (2.2.52) over ξ between $-\infty$ and ∞. The coefficient of the matrix τ_3 disappears because of the oddness of the integrand with respect to ξ, and the coefficients of τ_0 and τ_1 contain a (2.2.18)-type integral. Thus we have

$$i\frac{Z(\omega)\omega\tau_0 - \varphi(\omega)\tau_1}{\sqrt{\omega^2 Z^2(\omega) - \varphi^2(\omega)}}\,\mathrm{sign}\{\omega\}$$

$$= -\frac{1}{\pi}\int_{-\infty}^{\infty}\frac{dz}{\omega - z + i\delta}\,\mathrm{Re}\left\{\frac{zZ(z)\tau_0 - \varphi(z)\tau_1}{\sqrt{z^2 Z^2(z) - \varphi^2(z)}}\right\}\mathrm{sign}\{z\} \quad .$$

Equating the coefficients of the identical matrices on both sides of this equation and passing from $\varphi(\omega)$ to $\Delta(\omega)$, we obtain the desired dispersion relations

$$i\frac{\omega\,\mathrm{sign}\{\omega\}}{\sqrt{\omega^2 - \Delta^2(\omega)}} = -\frac{1}{\pi}\int_{-\infty}^{\infty}\frac{dz}{\omega - z + i\delta}\,\mathrm{Re}\left\{\frac{z}{\sqrt{z^2 - \Delta^2(z)}}\right\}\mathrm{sign}\{z\} \quad , \tag{2.2.53}$$

$$i\frac{\Delta(\omega)\,\mathrm{sign}\{\omega\}}{\sqrt{\omega^2 - \Delta^2(\omega)}} = -\frac{1}{\pi}\int_{-\infty}^{\infty}\frac{dz}{\omega - z + i\delta}\,\mathrm{Re}\left\{\frac{\Delta(z)}{\sqrt{z^2 - \Delta^2(z)}}\right\}\mathrm{sign}\{z\} \quad . \tag{2.2.54}$$

We note that when linearized with respect to $\Delta(\omega)$, the first of these ex-ressions becomes an identity and the second is replaced by

$$i\,\frac{\Delta(\omega)}{\omega} = -\frac{1}{\pi}\int_{-\infty}^{\infty}\frac{dz}{\omega - z + i\delta}\,\mathrm{Re}\left\{\frac{\Delta(z)}{z}\right\} \quad . \tag{2.2.55}$$

This equation will be of use to us later when we analyze the contribution of low-frequency modes to the superconducting transition temperature.

2.3 The Superconducting Transition Temperature

2.3.1 Linearized Eliashberg's Equations

In this section we use Eliashberg's equations (2.2.48) to calculate the temperature of transition to a superconducting state, T_c, defined as the temperature at which a nonzero solution of (2.2.48) just begins to appear. Therefore, to determine T_c it suffices to use the equations linearized with respect to $\Delta(\omega)$:

$$[1 - Z(\omega)]\omega = -\int_{-\infty}^{\infty} dz' K_{ph}(z',\omega) \quad , \tag{2.3.1}$$

$$Z(\omega)\Delta(\omega) = \int_{-\infty}^{\infty} dz' K_{ph}(z',\omega) \mathrm{Re}\left\{\frac{\Delta(z')}{z'}\right\} - U_c N(0) \int_0^{\omega_c} dz' \tanh\frac{z'}{2T} \mathrm{Re}\left\{\frac{\Delta(z')}{z'}\right\} \quad . \tag{2.3.2}$$

The first of these equations yields an explicit expression for $Z(\omega)$ so that there is actually one equation (2.3.2) for $\Delta(\omega)$. It is the condition of solvability of this homogeneous integral equation that determines the superconducting transition temperature which is a functional of the quantity $\alpha^2(z)F(z)$ determining the effective matrix element of electron-phonon interaction $\alpha(z)$ and the density of states of the phonon spectrum $F(z)$. To investigate the dependence of T_c on the characteristics of the superconductor which enter $\alpha^2(z)F(z)$, as well as on the Coulomb interaction, it would be desirable to have an explicit analytic expression for T_c. This, however, cannot be obtained in the general strong-coupling case. We shall therefore consider several approximate solutions to (2.3.2) for the cases of strong, intermediate and weak coupling, depending on the validity of the inequality $\omega_{\mathcal{D}}/T_c \gg 1$.

2.3.2 Weak Coupling

In the limit of weak electron-phonon coupling we should obtain from Eliashberg's equations the basic equations of the BCS theory. This transition may be accomplished by a certain approximation of the interaction kernel $K_{ph}(z',\omega)$. As is evident from (2.2.27), $K_{ph}(z',\omega)$ begins decreasing sharply at the values of the arguments $|z'|$, $|\omega| \gtrsim \omega_{\mathcal{D}}$. We therefore adopt the following approximation: we equate the kernel $K_{ph}(z',\omega)$ to a constant in the interval of variation of the arguments from 0 to the frequency $\omega_{\mathcal{D}}$ and to zero elsewhere. The quantity $K_{ph}(0,0)$ may be written as

$$K_{ph}(0,0) = \frac{\lambda}{2} \tanh\frac{z'}{2T} \quad , \tag{2.3.3}$$

where

$$\lambda = 2 \int_0^\infty dz \frac{\alpha^2(z)F(z)}{z} \tag{2.3.4}$$

is called the electron-phonon coupling constant. Thus, our approximation for the interaction kernel yields the analytic expression

$$K_{ph}(z',\omega) = \frac{\lambda}{2} \tanh\frac{z'}{2T} \, \theta(\omega_D - |z'|)\theta(\omega_D - |\omega|). \tag{2.3.5}$$

It then follows from (2.3.1) that

$$Z(\omega) = 1 \tag{2.3.6}$$

Since the kernel (2.3.5) is factorized, the integral equation (2.3.2) may be solved exactly. Obviously $\Delta(\omega)$ is a real quantity. Using (2.3.5,6) and the evenness of $\Delta(\omega)$ we rewrite (2.3.2) in the form

$$\Delta(\omega) = \lambda \, \theta(\omega_D - |\omega|) \int_0^{\omega_D} \frac{dz'}{z'} \tanh\frac{z'}{2T_c} \Delta(z') - U_cN(0) \int_0^{\omega_c} \frac{dz'}{z'} \tanh\frac{z'}{2T_c} \Delta(z') \quad . \tag{2.3.7}$$

The solution of this equation may be tried in the form

$$\Delta(\omega) = \begin{cases} \Delta_{ph}, & |\omega| < \omega_D \\ \Delta_c, & \omega_D < |\omega| < \omega_c \end{cases} \tag{2.3.8}$$

with Δ_{ph} and Δ_c being some constants. For these we obtain a set of two homogeneous algebraic equations:

$$\begin{aligned} &\left\{1 - [(\lambda - U_cN(0)]\, K\!\left(\frac{\omega_D}{2T_c}\right)\right\} \Delta_{ph} + U_cN(0) \ln\frac{\omega_c}{\omega_D} \Delta_c = 0 \quad , \\ &U_cN(0)K\!\left(\frac{\omega_D}{2T_c}\right) \Delta_{ph} + \left[1 + U_cN(0) \ln\frac{\omega_c}{\omega_D}\right] \Delta_c = 0 \quad , \end{aligned} \tag{2.3.9}$$

where

$$K(y) = \int_0^y dx \frac{\tanh x}{x} \quad . \tag{2.3.10}$$

The condition of solvability of the homogeneous system (2.3.9) yields an equation for T_c. This equation may be written in the form

$$K\left(\frac{\omega_D}{2T_c}\right) = \frac{1}{\lambda - \mu^*} \quad , \tag{2.3.11}$$

where

$$\mu^* = \frac{U_c N(0)}{1 + U_c N(0) \ln \frac{\omega_c}{\omega_D}} = \frac{V_c N(0)}{1 + V_c N(0) \ln \frac{E_F}{\omega_D}} \quad . \tag{2.3.12}$$

When writing the last equality in (2.3.12) we have made use of (2.2.46) for the "intermediate" pseudopotential U_c.

Since $\omega_D \gg T_c$ for all of the known superconductors, the integral (2.3.10) should be calculated for $y \gg 1$. This yields

$$K(y) = \ln\left(\frac{4\gamma}{\pi} y\right) \quad , \tag{2.3.13}$$

where $\ln\gamma = C = 0.577\ldots$ in terms of Euler's constant. Thus, (2.3.11) leads to

$$T_c = \frac{2\gamma}{\pi} \omega_D \exp\left(-\frac{1}{\lambda - \mu^*}\right) \quad . \tag{2.3.14}$$

This completes the BCS result by taking into account the Coulomb interaction of electrons. The parameter μ^* is a pseudopotential representing the effective electron repulsion on the Fermi surface due to the Coulomb interaction. From (2.3.12), which was first obtained by BOGOLYUBOV et al. [2.12], follows that the pseudopotential is significantly reduced because of the large logrithmic term in the denominator.

2.3.3 Intermediate Coupling

The above transition from the strong-coupling equations of the superconductivity theory under consideration has been made in a purely formal way, by approximating the kernel of electron-phonon interaction by (2.3.5). We now consider a more rigorous solution of the linearized equations which gives corrections to the BCS formula for T_c and thus corresponds to the case of intermediate electron-phonon coupling. This will also enable us to obtain a criterion for weak coupling, just as the subsequent case of strong coupling yields a criterion for intermediate coupling.

It is convenient to proceed not from equations such as (2.3.1,2) but from the linearized equations (2.2.49,50) with the explicit kernel of electron-phonon interaction . We shall assume the characteristic frequencies z of the spectral function of phonons F(z), near the main "bulk" of the distribution F(z), to be much larger than the superconducting transition temperature ($z \gg T_c$). This allows us to neglect the thermal phonon distribution function $N(z) = [\exp(z/T_c)-1]^{-1}$ in (2.2.49,50) (the contribution of low-frequency phonons $z \lesssim T_c$ will be considered separately). Since

$$\mathrm{ctanh}\frac{z}{2T} = 1 + 2N(z) \quad ,$$

the approximation specified corresponds to the replacement of ctanh z/2T by unity. The combinations of fractions in (2.2.49,50), proportional to tanh z'/2T, split into two types. In one type finite expressions arise when the arguments ω, z and z' tend to zero. With these arguments it is possible to put $\tanh z'/2T_c = 1$ since $z' \gg T_c$. In the other type a logarithmic singularity arises when $\omega, z \to 0$. To compensate this singularity in the integral over z' it is necessary to retain tanh z'/2T. After the above approximation and linearization of (2.2.49,50) we obtain the following homogeneous system of equations for T_c:

$$Z(\omega)\Delta(\omega) = \int_0^\infty \frac{dz'}{z'} \tanh\frac{z'}{2T_c} \mathrm{Re}\{\Delta(z')\} \times \left[\int_0^\infty dz\alpha^2(z)F(z)K_+(z',z,\omega) - \mu(\omega,z')\right] \quad , \tag{2.3.15}$$

$$[1 - Z(\omega)]\omega = \int_{-\infty}^{\infty} dz' \int_0^\infty dz\alpha^2(z)F(z)f(-z')K_-(z',z,\omega) \quad , \tag{2.3.16}$$

where

$$K_\pm(z',z,\omega) = \frac{1}{z'+z+\omega+i\delta} \pm \frac{1}{z'+z-\omega-i\delta} \quad , \tag{2.3.17}$$

$$\mu(\omega,z') = U_cN(0)\theta(\omega_c - \omega)\theta(\omega_c - z') \quad . \tag{2.3.18}$$

The quantity $f(z)$ in (2.3.16) is a Fermi distribution function $f(z) = [\exp(z/T)+1]^{-1}$ which arises from the identities

$$\tanh(z'/2T) = 1-2f(z') = -1+2f(-z') \quad . \tag{2.3.19}$$

We call (2.3.15,16), which are obtained from the general equations (2.2.49, 50) for a superconductor with neglect of the contribution of thermal phonons, intermediate-coupling equations. Their solution was obtained by KARAKOZOV et al. [2.13] using the iteration technique [2.14] used for the parameter $\lambda/(1+\lambda)$. Here we give only the final result,

$$T_c = \frac{2\gamma}{\pi}\omega_0 \exp\left(-\frac{1+\lambda}{\lambda-\mu^*\left[1+\frac{\lambda A}{1+\lambda}\right]}\right) \tag{2.3.20}$$

Thus, T_c is determined by four parameters: λ, A, ω_0 and μ^*, of which the first three depend on the form of the phonon spectrum $F(z)$: λ through (2.3.4), and A and ω_0 by definition

$$A = -\frac{4}{\lambda^2}\int_0^\infty dz_1\alpha^2(z_1)F(z_1)\int_0^\infty dz_2\alpha^2(z_2)F(z_2)\ln\frac{z_1}{z_2}\cdot\frac{z_2}{z_1}\cdot\frac{1}{z_2^2-z_1^2} \quad , \tag{2.3.21}$$

$$\omega_0 = \omega_{\log}\, e^{-A} \quad , \tag{2.3.22}$$

where

$$\ln\omega_{\log} = \frac{2}{\lambda}\int_0^\infty dz\frac{\alpha^2(z)F(z)}{z}\ln z \tag{2.3.23}$$

is a certain mean logarithmic frequency.

We estimate the quantities $\omega_{\log}$ and A for a phonon-spectrum model with one Einstein frequency Ω_0. In this case the spectral density of the electron-phonon interaction is a δ function

$$\alpha^2(z)F(z) = \frac{\lambda\Omega_0}{2}\delta(z-\Omega_0) \quad . \tag{2.3.24}$$

In this case $A = 1/2$ and (2.3.20) becomes

$$T_c = \frac{\Omega_0}{1{,}43}\exp\left(-\frac{1+\lambda}{\lambda-\mu^*\left[1+0{,}5\frac{\lambda}{1+\lambda}\right]}\right) \quad . \tag{2.3.25}$$

The numerical preexponential factor is

$$\frac{2\gamma}{\pi}\, e^{-1/2} = \frac{1}{1{,}43} \quad . \tag{2.3.26}$$

One expects that in the case of a sufficiently narrow phonon spectrum with one maximum (or two close maxima) (2.3.24) approximately holds, and the transition temperature T_c of such superconductors should be described by a (2.3.25)-type formula where Ω_0 is close to the limiting (Debye) frequency of the phonon spectrum.

Formula (2.3.25) was obtained in [2.13]. Formulae of a similar form for T_c in the case of intermediate coupling were derived in ([2.14-17]). It follows from (2.3.25) that the transition to the weak-coupling theory yielding the BCS formula (2.3.14) for T_c occurs for $\lambda \ll 1$. This condition determines the criterion of weak coupling. In contrast to strong coupling ($\lambda \sim 1$), it is rather difficult to formulate a criterion for intermediate coupling. One must bear in mind those approximations which have been made in the derivation of the basic equations (2.3.15,16) underlying the intermediate-coupling theory.

2.3.4 Strong Coupling

The approximate Eliashberg equations (2.3.15,16) underlying the theory of superconductors with intermediate coupling have been obtained from the more general equations (2.2.9,50) as a result of neglecting the thermal phonons and replacing $\tanh(z'/2T_c)$ by unity in a number of expressions. This corresponds to the neglect of temperature effects when they are of the order of $(T_c/\omega_D)^2$, for example, in $\mathrm{Im}\{Z(\omega)\}$, and to keeping the terms of the order of $(T_c/\omega_D)^2 \ln(\omega_D/T_c)$, for example, in $\mathrm{Re}\{Z(\omega)\}$. Incidentally, the presence of such a temperature dependence of $\mathrm{Re}\{Z(0)\}$ may lead to a deviation of the ratio $2\Delta/T_c$ from the universal value of 3.52 following from the BCS model.

The full Eliashberg equations (2.2.49,50), in which the terms taking into account the thermal phonons and the imaginary parts of the quantities $\Delta(\omega)$ and $Z(\omega)$ are kept, ensure an accuracy that permits retention of the term $(T_c/\omega_D)^2 \ln(\omega_D/T_c)$, along with the terms $\sim (T_c/\omega_D)^2$. In the strong-coupling case (realistically at $T_c/\omega_D \gtrsim 0.1$) when such an accuracy is required it is very difficult to obtain a (2.3.20)-type analytic formula for T_c. A numerical solution of Eliashberg's equations is then required. Calculations of this kind were carried out in 1968 by McMILLAN [2.18], using an ad hoc assumption of the nature of the spectral function $\alpha^2(\omega)F(\omega)$. As $F(\omega)$ he used the niobium phonon spectrum found experimentally from inelastic neutron scattering. This spectrum is rather typical for a number of bcc metals and therefore one may expect the solutions obtained for T_c to be applicable at least to a number of bcc metals. Secondly, McMILLAN assumed in his calculations that the effective matrix element $\alpha^2(\omega)$ of electron-phonon interaction does not depend on frequency. This also follows approximately from a comparison of the results obtained for $\alpha^2(\omega)F(\omega)$ from tunnel effect data with the experimental data for $F(\omega)$ obtained from neutron experiments.

For a given value of parameters α^2 and μ^* it is possible to obtain T_c from the Eliashberg equations. The resulting series of numerical data for T_c as a function of α and μ^* may be written as a function of the two parameters λ and μ^* since λ and α^2 are related by the equation $\lambda = 2\alpha^2 \int_0^\infty (d\omega/\omega)F(\omega)$. McMILLAN assumed that T_c depends on $\alpha^2(\omega)F(\omega)$ only through λ. The analytic function $T_c = T_c(\lambda,\mu^*)$ which spectacularly represents the table of calculated values of T_c at prescribed λ and μ^* is

$$T_c = \frac{\theta_D}{1.45}\exp\left(-\frac{1.04(1+\lambda)}{\lambda-\mu^*(1+0.62\lambda)}\right) \tag{2.3.27}$$

(θ_D stands for the Debye temperature).

This formula for T_c is known as the McMillan formula and has found extensive applications in the analysis of strong-coupling superconductors. Again, one should mention the particular assumptions that underlie its derivation and which do not allow one to regard it as a universal formula for T_c. Firstly, use is made of the form of the phonon spectrum of niobium. Secondly, the effective matrix element of electron-phonon interaction $\alpha^2(\omega)$ is assumed to be independent of frequency. Finally, the functional dependence of T_c on $\alpha^2(\omega)F(\omega)$ is assumed to be determined only by the quantity λ. Nevertheless, McMillan's formula gives a good fit to experimental data for T_c if the quantities λ and μ^* are taken from independent experiments. Such a comparison will be drawn in Chapter 4. As some substantiation of the McMillan formula, we refer to (2.3.25) obtained in [2.13] from the Eliashberg equations in the strong-coupling case. The fit of the expressions in the exponent of (2.3.27) and (2.3.25) is surprisingly good. As for the preexponent, the fit of the numerical factors may be fortuitous since the quantity θ_D involved in the McMillan formula stands for the Debye temperature, whereas the quantity Ω_0 in Karakozov et al.'s formula pertains to the Einstein oscillator model.

In his work [2.18] McMILLAN assumed that the frequency $\langle\omega\rangle$ averaged with respect to the spectrum $F(\omega)$ instead of the Debye temperature θ_D might be inserted into (2.3.27) as a preexponential factor. For niobium the equation $\theta/1.45 = \langle\omega\rangle/1.20$ occurs. If one assumes that the preexponent is determined by $\langle\omega\rangle$ for other superconductors as well, then the second McMillan formula

$$T_c = \frac{\langle\omega\rangle}{1.20}\exp\left(-\frac{1.04(1+\lambda)}{\lambda - \mu^*(1+0.62\lambda)}\right) \tag{2.3.28}$$

should hold in place of (2.3.27).

McMillan's formula (2.3.28) represents analytic results of the numerical solution of the Eliashberg equation for $\lambda \lesssim 1.5$. Recently, ALLEN and DYNES [2.30] performed similar calculations for large values of λ and found a discrepancy in calculations between the exact results and the data obtained on the basis of McMillan's formula already for values of $\lambda \approx 2$. Instead of a maximum in T_c as a function of λ in the vicinity of $\lambda = 2$, as predicted by McMillan, they obtained a monotonic increase of T_c and found that in the ultrastrong-coupling limit ($\lambda \gtrsim 10$) T_c reaches the asymptote (subject to the assumption that $\mu^* = 0.1$):

$$T_c = 0.15\sqrt{\lambda\langle\omega^2\rangle} \quad .$$

However, when $\lambda \lesssim 1.5$, McMillan's formulae (2.3.27,28) are quite correct.

ALLEN and DYNES have also investigated the dependence of T_c on the form of the phonon spectrum, having solved Eliashberg's equations for the Pb and Hg phonon spectra. If in McMillan's formula (2.3.28) one replaces the mean frequency of the phonon spectrum $\langle\omega\rangle$ by the mean logarithmic frequency $\omega_{\log}$ determined by (2.3.23), the λ dependence of T_c remains the same for three substances: Pb, Hg and Nb, as well as for the Einstein phonon spectrum model. This gives an answer to the old question as to whether McMillan's formula obtained involving the use of the phonon spectrum of niobium, a bcc metal, may be applied to metals of a different crystalline structure. As follows from numerical calculations [2.31], the dependence of T_c on the form of the phonon spectrum manifests itself only when $\lambda > 2$.

The McMillan formulae (2.3.27,28) expressing T_c in terms of the electron-phonon coupling constant enable one, in principle, to study the dependence of T_c on the form of the phonon spectrum. However, since a number of particular assumptions are used in the derivation of these formulae, attempts have been made in the literature to obtain other approximate formulae for T_c in the strong-coupling case. One such formula is due to MEDVEDEV et al. [2.19] who have tried to obtain T_c by approximating the solution of Eliashberg's equation (2.3.2) by a (2.3.8)-type step function which reverses sign at a frequency of the order of the limiting frequency of the phonon spectrum. It must be emphasized, however, that numerical solutions of the Eliashberg equation show that the gap function $\Delta(\omega)$ is a much more complicated function for strong-coupling superconductors. As an example, reference could be made to McMillan's calculation for niobium which is presented in Fig.2.3 for $\mathrm{Re}\{\Delta(\omega)\}$ and $\mathrm{Im}\{\Delta(\omega)\}$. It is seen from the figure that in the frequency range $\omega < \omega_D$, $\mathrm{Re}\{\Delta(\omega)\}$ is a highly nonmonotonic function of ω which reflects the complicated structure of the phonon spectrum of niobium.

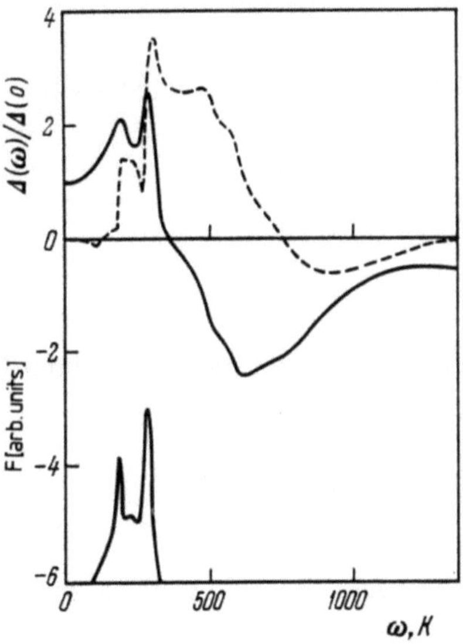

Fig.2.3 Real (solid curve) and imaginary (dotted) parts of the gap function $\Delta(\omega)$ and the phonon density of states $F(\omega)$ for a superconductor with $T_c = 10$ K, $\lambda = 0.91$, $\mu^* = 0.15$ corresponding to Nb [2.18]

2.3.5 Contribution of Low-Frequency Modes

According to the expressions for T_c in the case of weak (2.3.14), intermediate (2.3.25) and strong (2.3.27) coupling, the superconducting transition temperature is determined mainly by the parameter λ which is a functional of the density of states in the phonon spectrum. As follows from (2.3.4), the quantity λ depends on the form of the phonon spectrum. Other conditions being equal, the quantity λ will be larger, the higher the spectral intensity of low frequencies. From this point of view, one might expect higher T_c's in superconductors whose lattice oscillations have a high intensity of low-frequency phonon modes. It must, however, be noted that in the derivation of the above formulae for T_c we have always used the assumption that all the characteristic frequencies ω of the phonon spectrum $F(\omega)$ are much larger than T_c. Therefore, the contribution of low-frequency modes with

$$\omega \lesssim T_c \qquad (2.3.29)$$

to T_c cannot be obtained from (2.3.14,25,27). To investigate this contribution we turn again to the exact linearized Eliashberg equations. The two equations (2.3.1,2) may be replaced by a single equation for the gap function if we rewrite it in the form

$$\Delta(\omega)\left[1+\frac{1}{\omega}A_{ph}(\omega)\right]$$

$$=\int_{-\infty}^{\infty} dz' K_{ph}(z',\omega)\mathrm{Re}\left\{\frac{\Delta(z')}{z'}\right\} - U_c N(0)\int_{0}^{\omega_c} dz' \tanh\frac{z'}{2T_c}\,\mathrm{Re}\left\{\frac{\Delta(z')}{z'}\right\}\ , \qquad (2.3.30)$$

where

$$A_{ph}(\omega) = \int_{-\infty}^{\infty} dz' K_{ph}(z',\omega) \quad . \qquad (2.3.31)$$

The integration here can be carried out exactly if we insert the phonon interaction kernel (2.2.26) into (2.3.31) to obtain

$$A_{ph}(\omega) = \int_0^\infty dz\alpha^2(z)F(z)\mathrm{Re}\left\{\psi\left(\frac{1}{2}+i\frac{\omega+z}{2\pi T_c}\right) - \psi\left(\frac{1}{2}+i\frac{\omega-z}{2\pi T_c}\right)\right\}$$

$$+ i\frac{\pi}{2}\int_0^\infty dz\alpha^2(z)F(z)\left\{\tanh\frac{\omega-z}{2T_c} - \tanh\frac{\omega+z}{2T_c} + 2\mathrm{ctanh}\frac{z}{2T_c}\right\}\ . \qquad (2.3.32)$$

$\psi(z) = d\ln\Gamma(z)/dz$ here is the logarithmic derivative of the Γ function.

To reveal the influence which low frequencies $\omega \lesssim T_c$ exert on T_c we split the function $\alpha^2(z)F(z)$ into two parts,

$$\alpha^2(z)F(z) = \alpha^2(z)F(z)\Big|_{z<\bar{\omega}} + \alpha^2(z)F(z)\Big|_{z>\bar{\omega}}\quad , \qquad (2.3.33)$$

where each part represents it in the interval $z < \bar{\omega}$ or $z > \bar{\omega}$, with $\bar{\omega}$ of the order of T_c.

The phonon interaction kernel (2.2.26) is then represented by two terms

$$K_{ph}(z',\omega) = K^<_{ph}(z',\omega) + K^>_{ph}(z',\omega)\quad , \qquad (2.3.34)$$

where the symbol $K^<_{ph}$ denotes that integration over z in (2.2.26) is taken between 0 and $\bar{\omega}$, and $K^>_{ph}$ stands for integration between $\bar{\omega}$ and ∞. In the expression for $K^<_{ph}(z',\omega)$ the following approximation is possible: we neglect the frequency z in the denominator compared with z' and ω and, as a result, the terms with $\tanh(z'/2T\,)$ cancel out so that we have

$$K^{<}_{ph}(z',\omega) = \frac{\Gamma^{<}_{ph}}{\pi}\frac{1}{z'-\omega-i\delta} \quad , \tag{2.3.35}$$

where

$$\Gamma^{<}_{ph} = \pi\int_0^{\overline{\omega}} dz\alpha^2(z)F(z)\coth\frac{z}{2T_c} \tag{2.3.36}$$

stands for the magnitude of the electron decay on the Fermi surface due to the interaction with low-frequency phonons. Now substitute the decomposition (2.3.34) of the interaction kernel into two constituents with low and high phonon frequencies into the right-hand side of (2.3.30). With the dispersion relation (2.2.55) for the superconductor Green's function, we get

$$\int_{-\infty}^{\infty} dz'K^{<}_{ph}(z',\omega)\mathrm{Re}\left\{\frac{\Delta(z')}{z'}\right\} = i\Gamma^{<}_{ph}\frac{\Delta(\omega)}{\omega} \tag{2.3.37}$$

On the other hand, in view of expression (2.3.32), for the low-frequency contribution of A_{ph} to the main order in the parameter ω/T_c we have $A^{<}_{ph}(\omega) = i\Gamma^{<}_{ph}$. Substituting the last two equations into (2.3.30) we transform it to

$$\Delta(\omega)\left[1+\frac{1}{\omega}\Gamma^{<}_{ph}+\frac{1}{\omega}A^{>}_{ph}(\omega)\right] = i\Gamma^{<}_{ph}\frac{\Delta(\omega)}{\omega}+\int_{-\infty}^{\infty} dz'K^{>}_{ph}(z',\omega)\mathrm{Re}\left\{\frac{\Delta(z')}{z'}\right\} - U_cN(0)$$
$$\cdot\int_0^{\omega_c} dz'\tanh\frac{z'}{2T_c}\,\mathrm{Re}\left\{\frac{\Delta(z')}{z'}\right\} \quad . \tag{2.3.38}$$

We see that the contributions of the low-frequency phonon modes cancel. Thus, the superconducting transition temperature determinable from (2.3.38) does not depend on the low-frequency phonon modes satisfying the condition (2.3.29). This result has been obtained in a phonon-spectrum model with two Einstein-oscillators by KARAKOZOV et al. [2.20] and comes, as is seen from the above conclusion [2.21], from the dispersion relations for the electron Green's function of a superconductor. A similar conclusion as to the independence of T_c on low-frequency modes is drawn by the authors of [2.22] who have shown that the variational derivative $\delta T_c/\delta F(\omega)$ tends to zero when $\omega \to \theta$. This physical result may be regarded as a manifestation of ANDERSON's [2.23] theorem of the independence of T_c on a static perturbation acting on the system of conduction electrons in a superconductor since, for $\omega < T_c$, phonons may be considered as being a static perturbation.

The result of the investigation carried out in this section may be summarized as follows. For strong-coupling superconductors, the superconducting transition temperature is determined chiefly by the electron-phonon coupling parameter λ. Hereafter we shall deal with the various representations for this quantity which enable us to see what microcharacteristics of the metal affect it. Thereby we shall investigate the problem of expressing transition temperatures in terms of normal-phase parameters.

2.4 Calculations of the Electron-Phonon Coupling Constant λ

2.4.1 Green's Function of Phonons

As seen from the previous section, the superconducting transition temperature T_c is determined primarily by the electron-phonon coupling constant λ. For this quantity, which is some integral (2.3.4) over the phonon spectrum of the effective square of the electron-ion interaction matrix element, it is possible to obtain convenient expressions in terms of the Green's function of phonons. In contrast to Sect.2.1, where the phonon Green's function was introduced in a continuum approximation, we shall now define it for a discrete crystal lattice.

We introduce the retarded Green's function [2.24] of phonons

$$\mathcal{D}^{\alpha\beta}(\underline{1}\kappa,\underline{1}'\kappa';t) = -i\theta(t)\left\langle\left[U^{\alpha}_{\underline{1}\kappa}(t),\, U^{\beta}_{\underline{1}'\kappa'}(0)\right]\right\rangle \quad . \tag{2.4.1}$$

All the notations introduced here are standard ones: <...> is the symbol of statistical averaging (2.1.2), [,] denotes a commutator, and $\theta(t)$ is the theta function. The quantity $U^{\alpha}_{\underline{1}\kappa}$ stands for the displacement of an atom κ in the 1th cell in the direction $\alpha = x,y,z$, $U^{\alpha}_{\underline{1}\kappa}(t)$ represents the displacement at the moment of time t

$$U^{\alpha}_{1\kappa}(t) = e^{iHt}U^{\alpha}_{\underline{1}\kappa}e^{-iHt} \quad , \tag{2.4.2}$$

where the Hamiltonian H describes the phonon system to a harmonic approximation

$$H = \sum_{\underline{1}\kappa}\frac{\hat{P}^2_{\underline{1}\kappa}}{2M_\kappa} + \frac{1}{2}\sum_{\underline{1}\kappa,\underline{1}'\kappa'}\Phi^{\alpha\beta}_{\underline{1}\kappa,\underline{1}'\kappa'}U^{\alpha}_{\underline{1}\kappa}U^{\beta}_{\underline{1}'\kappa'} \quad . \tag{2.4.3}$$

$\hat{\underline{P}}_{\underline{l}\kappa}$ here is the operator of the momentum of an atom κ having a mass M_κ in the cell $\underline{l}$; and $\Phi^{\alpha\beta}_{\underline{l}\kappa,\underline{l}'\kappa'}$ is the matrix of force constants.

The equation (2.4.3) for the Hamiltonian eigenvalues is well known. For a perfect lattice it is convenient to write this equation in the form

$$\omega_j^2(\underline{q})e_j^\alpha(\underline{q},\kappa) = \sum_{\beta\kappa'} \frac{e^{-i\underline{q}(\underline{\rho}_\kappa - \underline{\rho}_{\kappa'})}}{\sqrt{M_\kappa M_{\kappa'}}} \Phi^{\alpha\beta}_{\kappa\kappa'}(\underline{q}) e_j^\beta(\underline{q},\kappa') \quad , \qquad (2.4.4)$$

where

$$\Phi^{\alpha\beta}_{\kappa\kappa'}(\underline{q}) = \sum_{\underline{l}-\underline{l}'} e^{-i\underline{q}(\underline{l}-\underline{l}')} \Phi^{\alpha\beta}_{\underline{l}\kappa,\underline{l}'\kappa'} \qquad (2.4.5)$$

is the Fourier component of the matrix of force constants. The matrix depends on the difference of the vectors $\underline{l}-\underline{l}'$ ($\underline{q}$ is a wave vector, N is the number of cells in the crystal). The eigenvectors $e_j^\alpha(\underline{q},\kappa)$ corresponding to the phonon frequency $\omega_j(\underline{q})$ are known as polarization vectors and satisfy the orthogonality and completeness relations,

$$\sum_{\alpha\kappa} e_j^{*\alpha}(\underline{q},\kappa)e_{j'}^\alpha(\underline{q},\kappa) = \delta_{jj'} \quad , \quad \sum_j e_j^{*\alpha}(\underline{q},\kappa)e_j^\beta(\underline{q},\kappa') = \delta_{\alpha\beta}\delta_{\kappa\kappa'} \quad . \qquad (2.4.6)$$

To calculate the Green's function we expand the atomic displacements in a series of the phonon creation ($b^+_{\underline{q}j}$) and annihilation ($b_{\underline{q}j}$) Bose operators

$$U^\alpha_{\underline{l}\kappa} = \sum_{\underline{q}j} \frac{1}{\sqrt{2NM_\kappa\omega_j(\underline{q})}} e_j^\alpha(\underline{q},\kappa) e^{i\underline{q}\underline{R}_{\underline{l}\kappa}} (b_{\underline{q}j} + b^+_{-\underline{q}j}) \quad . \qquad (2.4.7)$$

These expansions arise from the permutation relations for the momentum and displacement operators

$$\left[iP^\alpha_{\underline{l}\kappa}, U^\beta_{\underline{l}'\kappa'}\right] = \delta_{\underline{l}\underline{l}'}\delta_{\kappa\kappa'}\delta_{\alpha\beta} \quad , \quad P^\alpha_{\underline{l}\kappa} = M_\kappa \dot{U}^\alpha_{\underline{l}\kappa} \quad . \qquad (2.4.8)$$

Now, taking into account the diagonal form of the Hamiltonian $H = \sum_{\underline{q}j} \omega_j(\underline{q}) b^+_{\underline{q}j} b_{\underline{q}j}$, we obtain from (2.4.1) the following expression

$$\mathcal{D}^{\alpha\beta}(\underline{l}\kappa,\underline{l}'\kappa';t) = -i\theta(t) \sum_{\underline{q}j} \frac{e_j^\alpha(\underline{q},\kappa)e_j^\beta(\underline{q},\kappa')}{2N\omega_j(\underline{q})\sqrt{M_\kappa M_{\kappa'}}} \cdot e^{i\underline{q}(R_{\underline{l}\kappa} - R_{\underline{l}'\kappa'})} \left(e^{-i\omega_j(\underline{q})t} - e^{i\omega_j(\underline{q})t}\right) \quad . \qquad (2.4.9)$$

With the help of the integral representation for the $\theta(t)$ function

$$\theta(t) = \frac{i}{2\pi}\int\limits_{-\infty}^{+\infty} \frac{e^{-i\kappa t}}{\kappa + i\delta} d\kappa \quad , \tag{2.4.10}$$

we find the final expression for the Fourier component of the Green's function (2.4.9)

$$D^{\alpha\beta}_{\kappa\kappa'}(\underline{q},\omega) = \sum_j \frac{e^{\alpha}_j(\underline{q},\kappa)e^{\beta}_j(\underline{q},\kappa')}{2\omega_j(\underline{q})\sqrt{M_\kappa M_{\kappa'}}} e^{i\underline{q}(\underline{\rho}_\kappa - \underline{\rho}_{\kappa'})}$$

$$\cdot\left(\frac{1}{\omega - \omega_j(\underline{q}) + i\delta} - \frac{1}{\omega + \omega_j(\underline{q}) + i\delta}\right) \tag{2.4.11}$$

($\underline{\rho}_\kappa$ being the radius-vector of the κth atom in the zero cell).

Let us now note some general properties of the phonon Green's function. The following summation rule holds:

$$\int\limits_0^\infty \omega \,\mathrm{Im}\left\{D^{\alpha\beta}(\underline{l}\kappa,\underline{l}'\kappa';\omega)\right\}d\omega = -\frac{\pi}{2M_\kappa}\delta_{\alpha\beta}\delta_{\underline{l}\underline{l}'}\delta_{\kappa\kappa'} \quad , \tag{2.4.12}$$

From this expression follows another useful equation

$$\int\limits_0^\infty \frac{1}{\omega}\mathrm{Im}\left\{D^{\alpha\beta}(\underline{l}\kappa,\underline{l}'\kappa';\omega)\right\}d\omega = -\frac{\pi}{2}\Phi^{-1\,\alpha\beta}_{\underline{l}\kappa,\underline{l}'\kappa'} \quad , \tag{2.4.13}$$

where Φ^{-1} is a matrix reciprocal to the force-constant matrix Φ. Equation (2.4.13) is easily obtained from

$$\omega^2 M_\kappa D^{\alpha\beta}(\underline{l}\kappa,\underline{l}'\kappa';\omega) - \sum_\gamma \sum_{l''\kappa''} \Phi^{\alpha\gamma}_{\underline{l}\kappa,l''\kappa''} D^{\gamma\beta}(\underline{l}''\kappa'',\underline{l}'\kappa';\omega)$$

$$= \delta_{\alpha\beta}\delta_{\underline{l}\underline{l}'}\delta_{\kappa\kappa'} \quad . \tag{2.4.14}$$

2.4.2 Expression of λ in Terms of the Phonon Green's Function

We show now that the electron-phonon coupling constant

$$\lambda = 2\int\limits_0^\infty \frac{d\omega}{\omega}\alpha^2(\omega)F(\omega) \tag{2.4.15}$$

can be expressed in terms of the imaginary part of the phonon Green's function.

Substituting (2.1.26) for the electron-phonon interaction matrix $g_j(\underline{p},\underline{p}')$ into (2.2.27) for $\alpha^2(\omega)F(\omega)$ we get

$$\alpha^2(\omega)F(\omega)$$

$$= \int \frac{d^2p}{v_{\underline{p}}} \int \frac{d^2p'}{v_{\underline{p}'}} \sum_{\kappa\kappa'} \sum_{\alpha\beta} \sum_j \frac{e_j^\alpha(\underline{q},\kappa)e_j^\beta(\underline{q},\kappa')}{2N\omega_j(\underline{q})\sqrt{M_\kappa M_{\kappa'}}} \delta(\omega - \omega_j(\underline{q}))$$

$$\cdot\, e^{i\underline{q}(\underline{\rho}_\kappa - \underline{\rho}_{\kappa'})} \left\langle \underline{p} \left| \frac{\partial U_\kappa}{\partial r_\alpha} \right| \underline{p}' \right\rangle \left\langle \underline{p}' \left| \frac{\partial U_{\kappa'}}{\partial r_\beta} \right| \underline{p} \right\rangle \Big/ \int \frac{d^2p}{v_{\underline{p}}} \tag{2.4.16}$$

(here $\underline{q} = \underline{p} - \underline{p}'$). It follows from (2.4.11) that for $\omega > 0$

$$\mathrm{Im}\left\{\mathcal{D}^{\alpha\beta}_{\kappa\kappa'}(\underline{q},\omega)\right\} = -\pi \sum_j \frac{e_j^\alpha(\underline{q},\kappa)e_j^\beta(\underline{q},\kappa')}{2\omega_j(\underline{q})\sqrt{M_\kappa M_{\kappa'}}} \cos \underline{q}(\underline{\rho}_\kappa - \underline{\rho}_{\kappa'})\delta(\omega - \omega_j(\underline{q})) \quad . \tag{2.4.17}$$

Having compared these two equations we obtain

$$\alpha^2(\omega)F(\omega) = -\frac{1}{\pi} \sum_{\kappa\kappa'} \sum_{\alpha\beta} \int \frac{d^2p}{v_{\underline{p}}} \int \frac{d^2p'}{v_{\underline{p}'}} \left\langle \underline{p} \left| \frac{\partial U_\kappa}{\partial r_\alpha} \right| \underline{p}' \right\rangle \left\langle \underline{p}' \left| \frac{\partial U_{\kappa'}}{\partial r_\beta} \right| \underline{p} \right\rangle$$

$$\cdot\, \mathrm{Im}\left\{\mathcal{D}^{\alpha\beta}_{\kappa\kappa'}(\underline{p} - \underline{p}',\omega)\right\} \Big/ \int \frac{d^2p}{v_{\underline{p}}} \quad . \tag{2.4.18}$$

Now insert (2.4.18) into (2.4.15) for λ. The resulting integral over frequencies is taken by means of the identity (2.4.13) which may be written in the Fourier representation

$$\int_0^\infty \frac{d\omega}{\omega} \mathrm{Im}\left\{\mathcal{D}^{\alpha\beta}_{\kappa\kappa'}(\underline{q},\omega)\right\} = -\frac{\pi}{2} \Phi^{-1\alpha\beta}_{\kappa\kappa'}(\underline{q}) \quad , \tag{2.4.19}$$

where $\Phi^{-1}(\underline{q})$ is the Fourier component of the reciprocal matrix of force constants

$$\Phi^{-1\alpha\beta}_{\kappa\kappa'}(\underline{q}) = \frac{1}{N} \sum_{\underline{l}-\underline{l}'} e^{-i\underline{q}(\underline{l}-\underline{l}')} \Phi^{-1\alpha\beta}_{\underline{l}\kappa,\underline{l}'\kappa'} \quad . \tag{2.4.20}$$

In the final analysis we obtain the following expression for λ

$$\lambda = \int \frac{d^2p}{v_{\underline{p}}} \int \frac{d^2p'}{v_{\underline{p}'}} \sum_{\alpha\beta} \sum_{\kappa\kappa'} \Phi^{-1\alpha\beta}_{\kappa\kappa'}(\underline{p}-\underline{p}')$$

$$\cdot \left\langle \underline{p} \left| \frac{\partial U_\kappa}{\partial r_\alpha} \right| \underline{p}' \right\rangle \left\langle \underline{p}' \left| \frac{\partial U_{\kappa'}}{\partial r_\beta} \right| \underline{p} \right\rangle \Big/ \int \frac{d^2p}{v_{\underline{p}}} \quad . \tag{2.4.21}$$

Thus, λ is determined by the matrix of force constants and by the electron-ion interaction matrix elements. The phonon frequencies and their dependence on the ion masses have fallen out completely. This conclusion was first drawn by MAKSIMOV [2.25], and the generalizations given here for many-atom lattices were obtained by TAYLOR and VASHISHTA [2.26].

It is useful to represent (2.4.21) in one more form. Since, in view of (2.4.14), the phonon Green's function is a matrix in a space of 3sN dimensions (s stands for the number of atoms in a unit cell)

$$\mathcal{D}^{\alpha\beta}(\underline{l}\kappa,\, \underline{l}'\kappa';\omega) = \left(\frac{1}{\omega^2\hat{M} - \Phi}\right)^{\alpha\beta}_{\underline{l}\kappa,\underline{l}'\kappa'} \quad , \tag{2.4.22}$$

the expression for the force constant reciprocal matrix Φ^{-1} which defines λ is a Green's function of phonons at zero frequency. More precisely,

$$\Phi^{-1\alpha\beta}_{\kappa\kappa'}(\underline{q}) = -\mathcal{D}^{\alpha\beta}_{\kappa\kappa'}(\underline{q},0) \quad . \tag{2.4.23}$$

Thus, by virtue of (2.4.21,23), the quantity λ is determined by the Green's function of zero-frequency phonons and by the matrix elements of the electron-ion potential. In Chapter 4, the representation obtained for λ will be used for numerical calculations of the electron-phonon coupling constant.

2.4.3 The McMillan Representation

Expression (2.4.15) for λ may be brought approximately into a form in which the characteristics of the phonon spectrum and of the electron-ion interaction are represented by multiplicative factors. Special consideration will be given to the case of a one-atom lattice with an ion mass M. Then (2.4.16) for the function $\alpha^2(\omega)F(\omega)$ simplifies to

$\alpha^2(\omega)F(\omega)$

$$= \int \frac{d^2\underline{p}}{v_{\underline{p}}} \int \frac{d^2\underline{p}'}{v_{\underline{p}'}} \sum_j \frac{|\langle\underline{p}|\underline{e}_j\underline{\nabla}U|\underline{p}'\rangle|^2}{2NM\omega_j(\underline{q})} \delta(\omega - \omega_j(\underline{q})) \Big/ \int \frac{d^2\underline{p}}{v_{\underline{p}}} \quad . \tag{2.4.24}$$

The first moment of the function α^2F is independent of phonon frequencies,

$$\int_0^\infty \omega\alpha^2(\omega)F(\omega)d\omega = \frac{N(0)\langle\underline{J}^2\rangle}{2M} \quad , \tag{2.4.25}$$

where $\langle\underline{J}^2\rangle$ denotes the averaged value of the square of the electron-ion interaction matrix element with respect to the Fermi surface

$$\langle\underline{J}^2\rangle \equiv \int \frac{d^2\underline{p}}{v_{\underline{p}}} \int \frac{d^2\underline{p}'}{v_{\underline{p}'}} \sum_j |\langle\underline{p}|\underline{e}_j\underline{\nabla}U|\underline{p}'\rangle|^2 \Big/ \int \frac{d^2\underline{p}}{v_{\underline{p}}} \int \frac{d^2\underline{p}'}{v_{\underline{p}'}} \quad . \tag{2.4.26}$$

Now rewrite (2.4.15) for λ with the aid of the following supplementary notation

$$\lambda = 2 \int_0^\infty d\omega\, \omega\alpha^2(\omega)F(\omega)\frac{1}{\langle\omega^2\rangle} \quad , \tag{2.4.27}$$

where

$$\langle\omega^2\rangle \equiv \int_0^\infty d\omega\, \omega\alpha^2(\omega)F(\omega) \Big/ \int_0^\infty d\omega\, \omega^{-1}\alpha^2(\omega)F(\omega) \tag{2.4.28}$$

is some mean-square frequency of the phonon spectrum. If $\alpha^2(\omega)$ is assumed to have a weak frequency dependence, $\langle\omega^2\rangle$ is a characteristic mean frequency of the phonon spectrum since then

$$\langle\omega^2\rangle = \int_0^\infty \omega F(\omega)d\omega \Big/ \int_0^\infty \omega^{-1}F(\omega)d\omega \quad . \tag{2.4.29}$$

Expressing the first moment in (2.4.27) through (2.4.25) we obtain McMillan's representation for the electron-phonon coupling constant

$$\lambda = \frac{N(0)\langle\underline{J}^2\rangle}{M\langle\omega^2\rangle} \quad . \tag{2.4.30}$$

This expression is remarkable in the respect that it explicitly defines the dependence of λ on the form of the phonon spectrum. The mean square of the electron-ion interaction matrix element may be rewritten as a double integral over the entire three-dimensional space of momenta $\underline{p}$ and $\underline{p}'$, using the equation

$$\int_{SF} \frac{d^2p}{v_{\underline{p}}} \ldots = \int \frac{d\underline{p}}{(2\pi)^3}\,\delta(\varepsilon_{\underline{p}} - E_F) \quad . \tag{2.4.31}$$

Also taking into account the fact that $\int d^2\underline{p}/v_{\underline{p}}$ is the density of states N(0) on the Fermi surface, we can write the following equation [2.27] instead of (2.4.26):

$$\langle \underline{J}^2 \rangle = \int \frac{d\underline{p}}{(2\pi)^3} \int \frac{d\underline{p}'}{(2\pi)^3}\,\delta(\varepsilon_{\underline{p}} - E_F)\delta(\varepsilon_{\underline{p}'} - E_F)$$

$$\cdot \sum_j |\langle \underline{p}|\underline{e}_j \nabla U|\underline{p}'\rangle|^2 / N^2(0) \quad . \tag{2.4.32}$$

This turns out to be more convenient in a number of further applications.

Thus, thanks to McMillan's equation (2.4.30), the calculation of λ breaks down into two problems: calculating $\langle \underline{J}^2 \rangle$ and $\langle \omega^2 \rangle$, i.e., some mean quantities with respect to the Fermi surface and with respect to the phonon spectrum.

2.4.4 The Hopfield Representation for a Transition Metal

HOPFIELD's idea of analyizing the electron-phonon interaction through the expansion of wave functions in angular-momentum eigenfunctions proves to be fruitful for d-metals having a wide valence electron s-p band and a narrow d-band in which the Fermi level lies [2.28]. Let us assume, for simplicity, that the atomic potential in a metal is spherically symmetric. Since the magnitude of the angular momentum l and its projection m should be conserved, we can insert a wave function of an electron with a quasimomentum $\underline{p}$ in the vicinity of a given atom

$$\psi_{\underline{p}}(\underline{r}) = \sum_{lm} Y_{lm}(\theta,\varphi) g_{l\underline{p}}(\underline{r}) \quad , \tag{2.4.33}$$

where r, θ and φ are the polar coordinates of a vector $\underline{r}$, and Y_{lm} is a spherical function. It will be assumed that the electron wave function of a metal is of this form near every single atom within the Wigner-Seitz cell and is repeated in other cells with the phase vector $\exp(-i\underline{p}\underline{R}_s)$, R_s being the radius vector of the sth cell.

Consider the electron-ion interaction matrix element in an angular-momentum representation. We have

$$<\underline{p}|\underline{e}\nabla U|\underline{p}'> = \sum_{lm}\sum_{l'm'} \underline{e}\left\langle Y^*_{lm} g^*_{l\underline{p}}|\underline{\nabla} U|Y_{l'm'}g_{l'\underline{p}'}\right\rangle . \tag{2.4.34}$$

Since $\underline{\nabla}U$ transforms as a vector, we can use the selection rules [2.29]. According to these rules, only those matrix elements for which $l = l' \pm 1$ are nonzero in (2.4.34). In a transition metal, the main contribution to the density of states on the Fermi level is made by d-electrons corresponding to $l = 2$. Due to the selection rules, they are connected by electron-phonon interaction with the f and p states corresponding to $l = 3$ or 1. In d-metals, f states lie significantly higher than the Fermi energy, so the transitions between d and p states prove to be essential. This means that only the terms with (l,l') equal to (2,1) and (1,2) should be retained in the sum over l and l'. We consider below the case of a cubic crystal for which the three major crystallographic directions are equivalent. Then the matrix element of $\underline{e}\nabla U$ will be equal to three times the matrix element of $e_z\nabla_z U$. Passing to integration in spherical coordinates ($d\underline{r} = r^2 dr d\underline{\Omega}$) we see that for the matrix element of interest one may write

$$\left\langle Y^*_{2m} g^*_{2\underline{p}}|\underline{e}\nabla U|Y_{1m'}g_{1\underline{p}'}\right\rangle \sim \int_0^{r_{ws}} dr\, r^2 g^*_{2\underline{p}}(r) e_z \nabla_z U(r) g_{1\underline{p}'}(r) . \tag{2.4.35}$$

The proportionality coefficient resulting from the integration over angles is a numerical factor of the order of unity. The integration over the entire space in (2.4.35) reduces to an integral within the Wigner-Seitz cell which we have replaced by a sphere of equal volume with radius r_{ws}.

According to what has been stated above with respect to the selection rules, we now obtain the following equation

$$|<\underline{p}|\underline{e}\nabla U|\underline{p}'>|^2 \sim \left| \int_0^{r_{ws}} dr\, r^2\, g^*_{2\underline{p}}(r) e_z \nabla_z U\, g_{1\underline{p}'}(r)\right|^2 . \tag{2.4.36}$$

We now substitute this expression for the square of the electron-ion interaction matrix element into (2.4.30) for λ, devide and multiply the integrand by the quantity

$$\int_0^{r_{ws}} dr\, r^2\, |g_{2\underline{p}}(r)|^2 \cdot \int_0^{r_{ws}} dr\, r^2\, |g_{1\underline{p}'}(r)|^2 \quad ,$$

then write the expression for λ in the following form

$$\lambda \sim \frac{1}{M\langle\omega^2\rangle N(0)} \int \frac{d\underline{p}}{(2\pi)^3} \int \frac{d^2\underline{p}'}{(2\pi)^3} \delta(\varepsilon_{\underline{p}} - E_F)\delta(\varepsilon_{\underline{p}'} - E_F)$$

$$\cdot \left(\frac{dU}{dz}\right)^2 \int_0^{r_{ws}} dr\, r^2\, |g_{2\underline{p}}(r)|^2 \int_0^{r_{ws}} dr\, r^2\, |g_{1\underline{p}}(r)|^2 \quad , \tag{2.4.37}$$

where

$$\left(\frac{dU}{dz}\right)^2 = \frac{\left| \int_0^{r_{ws}} dr\, r^2\, g^*_{2\underline{p}}(r) e_z \nabla_z U(r) g_{1\underline{p}'}(r) \right|^2}{\int_0^{r_{ws}} dr\, r^2 |g_{2\underline{p}}(r)|^2 \cdot \int_0^{r_{ws}} dr\, r^2\, |g_{1\underline{p}'}(r)|^2} \tag{2.4.38}$$

stands for a quantity which varies slowly with the vectors $\underline{p}$ and $\underline{p}'$ lying on the Fermi surface. Bringing this quantity out of the integrals over $\underline{p}$ and $\underline{p}'$, we write (2.4.37) as

$$\lambda \sim \frac{1}{M\langle\omega^2\rangle} \left(\frac{dU}{dz}\right)^2 \frac{N_d(0)N_p(0)}{N(0)} \quad . \tag{2.4.39}$$

Here we have introduced the quantities

$$N_l(0) = \int \frac{d\underline{p}}{(2\pi)^3} \delta(\varepsilon_{\underline{p}} - E_F) \int_0^{r_{ws}} dr\, r^2\, |g_{l\underline{p}}(r)|^2 \tag{2.4.40}$$

for the values $l = 1$ and 2 which correspond to p and d-type states.

We now show that $N_l(0)$ are the densities of states of an electron on the Fermi surface with a given orbital moment l. In fact, by definition, the density of states on the Fermi surface is

$$N(0) = \int \frac{d\underline{p}}{(2\pi)^3} \delta(\varepsilon_{\underline{p}} - E_F) \quad . \tag{2.4.41}$$

We now transform (2.4.40) using the wave function normalization condition (2.4.33)

$$\int d\underline{r}\, \psi^*_{\underline{p}}(\underline{r})\, \psi_{\underline{p}}(\underline{r}) = 1 \quad . \tag{2.4.42}$$

Using the orthogonality of the spherical functions, (2.4.42) may be brought into the form

$$\sum_{lm} \int_0^{r_{ws}} dr\, r^2\, |g_{l\underline{p}}(r)|^2 = 1 \quad . \tag{2.4.43}$$

Then expression (2.4.41) for the total density of states may be represented as a sum

$$N(0) = \sum_l N_l(0) = N_s(0) + N_p(0) + N_d(0) + \dots \quad , \tag{2.4.44}$$

where

$$N_l(0) = \int \frac{d\underline{p}}{(2\pi)^3} \delta(\varepsilon_{\underline{p}} - E_F) \sum_m \int_0^{r_{ws}} dr\, r^2\, |g_{l\underline{p}}(r)|^2 \tag{2.4.45}$$

is the contribution to the density of states corresponding to a particular value of the angular momentum.

Since the contribution from the d-states on the Fermi surface prevails for the majority of transition metals, we approximately have $N(0) \approx N_d(0)$ so that, in view of (2.4.39), the final expression for λ may now be written as

$$\lambda = \frac{1}{M\langle\omega^2\rangle} \left(\frac{dU}{dz}\right)^2 N_p(0) \quad . \tag{2.4.46}$$

(An unessential numerical factor of the order of unity may be included in the determination of the paramter $(dU/dz)^2$ which may be regarded as phonomenological).

This expression obtained by Hopfield shows that for transition metals the electron-phonon coupling constant is expressed mainly in terms of the atomic constants: the potential gradient dU/dz and the partial density of p-type states on the Fermi surface. Both quantities should vary weakly in a group of metals with the same lattice and valency, in the same way as the mean (quadratic) frequency of the phonon spectrum does. Instead of the total density of states $N(0)$ the partial density $N_p(0)$ enters λ. Since T_c is determined mainly by λ, the result (2.4.46) enabled Hopfield to draw the conclusion that in transition metals the superconducting transition temperature is determined mainly by parameters of atomic (chemical) nature. This problem will be dealt with in more detail in Chapter 4.

3. Superconductivity and Magnetism

The theory of strong-coupling superconductors presented above applies to metals that do not exhibit any specific magnetic properties. To be more exact, the metals concerned are assumed to be either weak diamagnets or Pauli-type paramagnets. In the treatment of superconductivity in transition metals and their compounds one encounters situations when, in addition to the coupling electron-phonon interaction, it is also necessary to take into account the interaction of conduction electrons with the magnetic system of a metal, or, as we shall sometimes refer to it, the spin system. This interaction may arise when the superconducting metal contains impurities of paramagentic atoms conserving the localized magnetic moments in the alloy. The exchange interaction of electrons with paramagnetic impurities is the source of a number of phenomena in such dilute alloys. These phenomena will be considered in this chapter.

It is known that in a number of cases the localized moments on the impurity atoms vanish when a transition-metal impurity is dissolved in the superconducting matrix. However, magnetic moment fluctuations may in this situation be essential to the conditions for the formation of a superconducting state. The effect of localized spin fluctuations on superconductivity is another problem which is treated in this chapter. The interaction of electrons with spin fluctuations may play a significant role also in some transition metals, for example, in Pd which is paramagnetic but, in its electronic parameters, is located in the vicinity of the conditions for the occurrence of ferromagnetism. This situation, too, will be analyzed in detail in the present chapter.

Finally, one more type of problem arises if, in a superconducting metal (for example, in a dilute alloy of a transition element in a superconducting matrix) a ferromagnetic state occurs in the system of localized magnetic moments. Among them is the problem of the simultaneous occurrence of two mutually exclusive phenomena in one volume, viz., superconductivity and ferromagnetism. The mutual antagonism of these phenomena stems from their contradictory nature itself: in a superconductor the electrons tend to group into pairs with antiparallel spins, whereas in a ferromagnet there is a tendency to make all spins parallel. Bringing to light the conditions under which both states may occur simultaneously is a most interesting problem, with which we shall begin our third chapter.

Experimental studies of the above problems have begun quite recently. Therefore we cite the available experimental evidence, which is not ample but has been gathered up and is systematized here, not in separate sections but in the general text as illustrations to the theoretical problems outlined below.

3.1 Conditions for the Occurrence of Superconductivity in a Ferromagnetic Metal

3.1.1 Four-Dimensional Matrix Formalism

If in a superconducting metal there are localized magnetic moments, for example, as impurities of paramagnetic ions, then one should take into consideration the exchange interaction of conduction electrons with electrons of incomplete shells of impurity ions. The Hamiltonian V of this interaction has the following structure [3.1]

$$V(\underline{r}) = \sum_a J(\underline{r} - \underline{R}_a)(\underline{\sigma}\underline{S}_a) \quad , \tag{3.1.1}$$

where $\underline{\sigma}/2$ is the electron spin operator ($\underline{\sigma}$ stands for a Pauli matrix), $\underline{S}_a$ is the spin operator of an ion in a site $\underline{R}_a$, and $J(\underline{r} - \underline{R}_a)$ is the so-called s-d exchange integral.

What corresponds to this Hamiltonian in the second-quantization representation is the operator

$$H_m = \int d\underline{r}\, \psi^+_\alpha(\underline{r})\, V_{\alpha\beta}(\underline{r})\, \psi_\beta(\underline{r}) \tag{3.1.2}$$

(the index m indicating the interaction of an electron with the magnetic system).

Since the operator H_m connects states with different directions of the electron spin, anomalous averages (2.1.1) are likely to occur in such a superconductor. These averages correspond not only to singlet pairing but also to a triplet one, i.e., with a parallel orientation of the spins of both electrons. To take into account these types of anomalous averages, the sources of the corresponding Cooper pairs should be introduced into the Hamiltonian. Thus, in this case one should proceed from an "initiating" Hamiltonian generalized with respect to (2.1.3)

$$H_\Delta = \int d\underline{r} \sum_\alpha \psi_\alpha^+(\underline{r})\, H_0(\underline{r})\, \psi_\alpha(\underline{r})$$

$$- \frac{1}{2} \int d\underline{r} \sum_{\alpha\beta} \left[\psi_\alpha^+(\underline{r})\, \psi_\beta^+(\underline{r})\, \Delta_{\beta\alpha}(\underline{r}) + \psi_\beta(\underline{r})\, \psi_\alpha(\underline{r})\, \Delta_{\beta\alpha}^*(\underline{r})\right] \qquad (3.1.3)$$

in which the quantity $\Delta_{\beta\alpha}$ should in actual calculations tend to zero.

In order to bring the Hamiltonian (3.1.2,3) into the usual form of the energy operator of second-quantized normal systems one should not introduce the two-component field operators (2.1.5) but the four-component operators [3.2,3]:

$$\Psi(\underline{r}) = \begin{pmatrix} \hat{\psi}(\underline{r}) \\ \hat{\psi}^+(\underline{r}) \end{pmatrix} \quad , \quad \Psi^+(\underline{r}) = \left(\hat{\psi}^+(\underline{r}), \hat{\psi}(\underline{r})\right) \quad , \qquad (3.1.4)$$

where $\hat{\psi}(\underline{r})$ and $\hat{\psi}^+(\underline{r})$ are two-component spinors

$$\hat{\psi}(\underline{r}) = \begin{pmatrix} \psi_\uparrow(\underline{r}) \\ \psi_\downarrow(\underline{r}) \end{pmatrix} \quad , \quad \hat{\psi}^+(\underline{r}) = \left(\psi_\uparrow^+(\underline{r}), \psi_\downarrow^+(\underline{r})\right) \quad . \qquad (3.1.5)$$

The operator (3.1.3) in this representation is of the form

$$H_\Delta = \frac{1}{2} \int d\underline{r} \left[\Psi^+(\underline{r}) h(\underline{r}) \Psi(r)\right] \quad , \qquad (3.1.6)$$

where $h(\underline{r})$ is a 4×4 matrix

$$h(\underline{r}) = \begin{pmatrix} \hat{H}_0(\underline{r}) & \hat{\Delta}(\underline{r}) \\ \hat{\Delta}^*(\underline{r}) & -\hat{H}_0^{tr}(\underline{r}) \end{pmatrix} \quad , \qquad (3.1.7)$$

since $\hat{H}_0(\underline{r})$ and $\hat{\Delta}(\underline{r})$ are two-row matrices with respect to the spin indices: $\hat{\Delta} = \{\Delta_{\alpha\beta}\}$, $\hat{H}_0 = \{H_0 \delta_{\alpha\beta}\}$, Δ^* denotes Hermitian conjugation over Δ, and H_0^{tr} is the transposed H_0. It is easy to verify by back transformation that (3.1.6) yields (3.1.3), but with an accuracy to some constant $Sp\{H_0\}/2$. The Hamiltonian, when written in the conventional representation with the field operators $\psi_\alpha(\underline{r})$ and in the four-dimensional formalism, will always differ by such constants.

Similarly, the Hamiltonian H_m will now take on the form

$$H_m = \frac{1}{2} \int d\underline{r} \left[\Psi^+(\underline{r}) v(\underline{r}) \Psi(\underline{r})\right] \qquad (3.1.8)$$

where

$$v(\underline{r}) = \begin{pmatrix} \hat{V}(\underline{r}) & 0 \\ 0 & -\hat{V}^{tr}(\underline{r}) \end{pmatrix} \qquad (3.1.9)$$

is a four-dimensional matrix since $\hat{V}(\underline{r})$, in view of (3.1.1), is a 2×2 matrix with respect to the spin indices of an electron.

Let us now state the Coulomb and electron-phonon interaction Hamiltonian (2.1.14) in terms of the four-dimensional formalism. To this end, let it be noted that there holds the relation

$$(\Psi^{+}T\Psi) = 2\Big(\sum_{\alpha} \psi_{\alpha}^{+}\psi_{\alpha} - 1\Big) \quad , \qquad (3.1.10)$$

where T is a 4×4 matrix of the form

$$T = \begin{pmatrix} \sigma_0 & 0 \\ 0 & -\sigma_0 \end{pmatrix} \qquad (3.1.11)$$

and σ_0 is a 2×2 unit matrix. The matrix T may be represented as a direct product of the matrices σ_0 and τ_3. Thus, (3.1.10) may be rewritten as

$$\sum_{\alpha} \psi_{\alpha}^{+}\psi_{\alpha} = 1 + \frac{1}{2}\Big(\Psi^{+}\sigma_0\tau_3\Psi\Big) \quad . \qquad (3.1.12)$$

As a result, we write the Hamiltonian H_{int} (2.1.14) in its final form

$$H_{int} = \frac{1}{2}\int d\underline{r}\, g(\underline{r})\Big[\Psi^{+}(\underline{r})\sigma_0\tau_3\Psi(\underline{r})\Big]\varphi(\underline{r})$$
$$+ \frac{1}{2^2\cdot 2}\int d\underline{r}_1 d\underline{r}_2 U(\underline{r}_1 - \underline{r}_2)\Big[\Psi^{+}(\underline{r}_1)\sigma_0\tau_3\Psi(\underline{r}_1)\Big]\Big[\Psi^{+}(\underline{r}_2)\sigma_0\tau_3\Psi(\underline{r}_2)\Big] \qquad (3.1.13)$$

[to an accuracy of a constant and of the one-particle terms arising from the unity in (3.1.12)]. The occurrence of 1/2 with each factor of the type $(\Psi^{+}\sigma_0\tau_3\Psi)$ is a new element as compared to the Nambu representation. Below we shall see, however, that the representation (3.1.4) leads to a formulation of the rules for calculating the average of the T produce of the field operators (Wick's theorem) which results in the disappearance of the 1/2 factors in the Hamiltonians (3.1.6,8,13) so that we return to a formula usual for Hamiltonians of a normal metal.

3.1.2 Wick's Theorem

Consider the average of the T product of the field operators where a certain concrete system of pairings is taken:

$$\langle T\Psi(x)\bar{\Psi}(x')H_{int}(\tau_1)H_{int}(\tau_2)\dots H_{int}(\tau_n)\rangle \qquad (3.1.14)$$

which occurs in the nth order of perturbation theory for a one-particle Green's function. Let, for example, H_{int} contain two Fermi operators only, i.e., $H_{int} \sim (\bar{\Psi}\alpha\Psi)$ where α is some four-row matrix.

Every operator may be paired with one of the two operators involved in H_{int}, either with $\bar{\Psi}$ or with Ψ. Thus, to the system of pairing which connects the external operators with H_{int} in (3.1.14), there correspond actually 2^n terms including the average of both types: $\langle T\Psi\bar{\Psi}\rangle$ and $\langle T\Psi\Psi\rangle$. Let us show that all these 2^n terms are equal to one and the same term in which only normal-type pairings $\langle T\Psi\bar{\Psi}\rangle$ are taken.

Denote the components of the four-dimensional operators Ψ and $\bar{\Psi}$ by Ψ_l and $\bar{\Psi}_l$ $(l = 1,2,3,4)$,

$$\Psi = \begin{pmatrix}\Psi_1\\ \Psi_2\\ \Psi_3\\ \Psi_4\end{pmatrix} \qquad \bar{\Psi} = (\bar{\Psi}_1,\bar{\Psi}_2,\bar{\Psi}_3,\bar{\Psi}_4) \quad . \qquad (3.1.15)$$

Take some $H_{int}(\tau_1)$ and consider the involved operators $\bar{\Psi}_l(x)$ and $\Psi_{l'}(x')$ with a definite pair of field operators $A_i(x_o)$ and $A'_{i'}(x'_o)$, where A_i and $A'_{i'}$ may be any of the eight operators Ψ_l and $\bar{\Psi}_l$. All in all there are two systems of pairing which connect H_{int} with the other pair of operators A_i and $A'_{i'}$. The expression below for the average of the T product corresponds to these two systems of pairing shown by brackets from above and below

$$\begin{aligned}&\langle T \dots A_i(x_o) \dots A'_{i'}(x'_o) \dots [\bar{\Psi}_l(x_1)\alpha_{ll'}\Psi_{l'}(x_1)] \dots\rangle\\ &= \langle TA_i(x_o)\bar{\Psi}_l(x_1)\rangle\alpha_{ll'}\langle T\Psi_{l'}(x_1)A'_{i'}(x'_o)\rangle\langle\dots\rangle\\ &- \langle TA_i(x_o)\Psi_{l'}(x_1)\rangle\alpha_{ll'}\langle T\bar{\Psi}_l(x_1)A'_{i'}(x'_o)\rangle\langle\dots\rangle \quad ,\end{aligned} \qquad (3.1.16)$$

where $\langle\dots\rangle$ denotes the contribution by all the other pairings that do not pertain to the operators singled out. The repeated indices 1 and 1' imply summation and therefore 1 in the second term may be replaced by 1' to transform (3.1.16) into

$$\Big[\langle TA_i(x_0)\bar{\Psi}_1(x_1)\rangle\,\alpha_{11'}\,\langle T\Psi_{1'}(x_1)A'_{i'}(x'_0)\rangle - \langle TA_i(x_0)\Psi_1(x_1)\rangle\,\alpha^{tr}_{11'}\,\langle T\bar{\Psi}_{1'}(x_1)A'_{i'}(x'_0)\rangle\Big]\,\langle\dots\rangle\ , \qquad (3.1.17)$$

where α^{tr} is the transposed matrix of α.

Now let us pay attention to the structure of the four-row matrix α characterizing the perturbation. The matrices (3.1.9,11) involved as α matrices in the Hamiltonians (3.1.8,13), which can be regarded as perturbations, have the following structure:

$$\alpha = \begin{pmatrix} \hat{\alpha} & 0 \\ 0 & -\hat{\alpha}^{tr} \end{pmatrix}\ , \qquad (3.1.18)$$

where $\hat{\alpha}$ is a two-row matrix (with respect to spin indices). After this it is not difficult to see that the second term in the braces of expression (3.1.17) is equal, with opposite sign, to the first term. Accordingly, for the average in (3.1.16) we have

$$\begin{aligned}&\langle T\ \dots\ A_i(x_0)\ \dots\ A_{i'}(x'_0)\ \dots\ [\bar{\Psi}_1(x_1)\alpha_{11'}\Psi_{1'}(x_1)]\ \dots\rangle\\ &= 2\,\langle TA_i(x_0)\bar{\Psi}_1(x_1)\rangle\,\alpha_{11'}\,\langle T\Psi_{1'}(x_1)A'_{i'}(x_0)\rangle\,\langle\dots\rangle\\ &= 2\,\langle T\ \dots\ A_i(x_0)\ \dots\ A'_{i'}(x_0)\ \dots\ [\bar{\Psi}_1(x_1)\alpha_{11'}\Psi_{1'}(x_1)]\rangle\ . \qquad (3.1.19)\end{aligned}$$

Thus, the contributions from the pairing system shown by the upper and lower brackets turn out to be equal to each other. This means that one can choose a system which corresponds to quasi-normal-type pairings $\langle T\Psi_1\bar{\Psi}_{1'}\rangle$ only. Considering the other H_{int} in the initial average (3.1.14) we arrive at the conclusion that the general-type average of the T product of the four-component field operators ψ and $\bar{\psi}$:

$$\langle T\Psi(x)\bar{\Psi}(x')[\bar{\Psi}(x_1)\alpha\Psi(x_1)][\bar{\Psi}(x_2)\alpha\Psi(x_2)]\dots[\bar{\Psi}(x_n)\alpha\Psi(x_n)]\rangle$$

is equal to the expression, obtained through a formal use of Wick's theorem which takes into account normal pairings $\langle T\Psi(x)\bar{\Psi}(x')\rangle$ only, multiplied by 2^n If H_{int} has a four-fermion structure $[H_{int} \sim (\bar{\Psi}\alpha\Psi)(\bar{\Psi}\alpha\Psi)]$, then the multiplicative factor is 4^n. These factors 2^n or 4^n will be compensated for by factors $1/2^n$ or $1/4^n$ arising because of the additional factors 1/2 or 1/4 in the Hamiltonians (3.1.8,13) describing the perturbation in the four-dimensional formalism.

Thus, we come to the following conclusion. If one takes the expressions

$$H_m = \int d\underline{r}\left[\Psi^+(\underline{r})v(\underline{r})\Psi(\underline{r})\right] \quad , \tag{3.1.20}$$

$$H_{int} = \int d\underline{r}\, g(\underline{r})\left[\Psi^+(\underline{r})\sigma_0\tau_3\Psi(\underline{r})\right]\varphi(\underline{r}) + \frac{1}{2}\int d\underline{r}_1 d\underline{r}_2 U(\underline{r}_1 - \underline{r}_2)\left[\Psi^+(\underline{r}_1)\sigma_0\tau_3\Psi(\underline{r}_1)\right]\left[\Psi^+(\underline{r}_2)\sigma_0\tau_3\Psi(\underline{r}_2)\right] \tag{3.1.21}$$

as perturbation Hamiltonians, then when expressing the averages of the T product of field operators according to Wick's theorem it suffices to allow only for "normal-type" averages $\langle T\Psi(x)\bar{\Psi}(x)\rangle$. In this manner, in the four-dimensional formalism the perturbation theory for a superconductor is of the character of the perturbation theory for a normal metal with the Green's function being replaced by Green's functions 4×4 matrices

$$g_0(x,x') = -\left\langle T\Psi(x)\bar{\Psi}(x')\right\rangle \quad , \tag{3.1.22}$$

which include anomalous averages, corresponding to a superconductor, as their components. In virtue of what has been said of the quasi-normal structure of the perturbation Hamiltonian (3.1.20,21) and in view of Wick's theorem, the diagram technique for a superconductor in this (four-dimensional) formalism, too, is of the same form as that for a normal metal with substitution of matrix vertices for scalar vertices. In contrast to Nambu's two-component formalism described in Sect.2.1, what will correspond to the electron-phonon interaction vertex and to the Coulomb interaction vertex in this case will be the matrix $\sigma_0\tau_3$.

3.1.3 Perturbation Theory

Next, consider a superconductor with ferromagnetic ordering of localized magnetic moments. The exchange interaction of electrons with localized spins will be treated within the framework of perturbation theory. However, just as in the case of a normal ferromagnetic metal, it is convenient to single out immediately the effects of the self-consistent field in the s-d interaction that lead to the magnetization of conduction electrons. They are contained in the portion of the Hamiltonian (3.1.1) averaged with respect to the spin system

$$V_0(\underline{r}) = \Big[\sum_a J(\underline{r} - \underline{R}_a) \langle S_a^z \rangle\Big]\sigma_3 \equiv J(\underline{r})\sigma_3 \quad , \tag{3.1.23}$$

where $\langle S_a^z \rangle$ is the mean value of the spin projection at site a on the direction of the spontaneous moment. Thus, the zero-approximation Hamiltonian H_0 should, according to (3.1.6-9,23), be written as

$$H_0 = \int d\underline{r}\Big[\Psi^+(\underline{r}) h_0(\underline{r}) \Psi(\underline{r})\Big] \quad , \tag{3.1.24}$$

where

$$h_0(\underline{r}) = H_0(\underline{r})\sigma_0\tau_3 + J(\underline{r})\sigma_3\tau_3 \quad . \tag{3.1.25}$$

The Green's function $g_0(x,x')$ of (3.1.22) corresponding to the zeroth approximation Hamiltonian (3.1.24) obeys the equation of motion

$$\Big[-\frac{\partial}{\partial\tau}\sigma_0\tau_3 - h_0(r)\Big] g_0(x,x') = \delta(x - x') \quad , \tag{3.1.26}$$

from which

$$g_0^{-1}(\underline{p},\omega_n) = i\omega_n\sigma_0\tau_0 - \xi_{\underline{p}}\sigma_0\tau_3 \cdot J\sigma_3\tau_3 \tag{3.1.27}$$

is found in the momentum representation. ξ_p here is the eigenvalue of the one-electron operator $H_0(\underline{r})$, and $J \equiv J(0)$ is the Fourier component of the s-d exchange integral $J(\underline{r})$ [see definition (3.1.23)].

Using the zeroth approximation Green's function, let us consider the electron-phonon and s-d exchange interactions according to perturbation theory. In this problem we disregard the Coulomb interaction since we know its major effect in superconductivity theory. It will be our aim here to compare the roles of the electron-phonon and electron-magnon interaction, involved in the s-d exchange interaction, in the formation of a superconducting state.

The s-d exchange interaction Hamiltonian generates new elements in the perturbation-theory series, viz., spin Green's functions

$$\mathcal{D}_{ab}^{\mu\nu}(\tau,\tau') = -\Big\langle T\Big[S_a^\mu(\tau) - \langle S_a^\mu\rangle\Big]\Big[S_b^\nu(\tau') - \langle S_b^\nu\rangle\Big] \quad , \tag{3.1.28}$$

where $S_a^\mu(\tau)$ is the operator of the μ projection of the localized spin at site a at the moment of "time" τ:

$$S_a^\mu(\tau) = e^{H\tau} S_a^\mu e^{-H\tau} \tag{3.1.29}$$

and H is the total Hamiltonian of the system which also includes the direct exchange interaction H_{ex} of localized spins (if it exists in a metal)

$$H_{ex} = -J \sum_{a \neq b} \underline{S}_a \underline{S}_b \quad . \tag{3.1.30}$$

In the graphs we shall denote the spin Green's function by a dotted line, and the vertex corresponding to the s-d interaction by a small circle

$\backsim D_{ab}^{\mu\nu}(\tau,\tau')$;

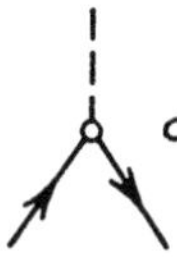

$\backsim J(\underline{r}-\underline{R}_a)\alpha^\mu$, (3.1.31)

where α^μ is a four-row matrix arising from (3.1.9)

$$\alpha^\mu = \begin{pmatrix} \sigma_\mu & 0 \\ 0 & -\sigma_\mu^{tr} \end{pmatrix} \quad . \tag{3.1.32}$$

Let us write the lowest-order contribution to the irreducible self-energy part of the electron due to the s-d interaction. It is represented by the graph

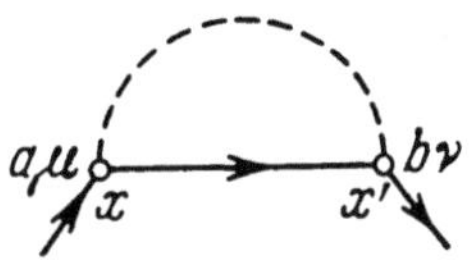

$$= - \sum_{ab} J(\underline{r} - \underline{R}_a) I(\underline{r}'-\underline{R}_b) D_{ab}^{\mu\nu}(\tau,\tau')\alpha^\mu g_0(x,x')\alpha^\nu \tag{3.1.33}$$

The contribution from the first-order graph describes the effects of the self-consistent field in the s-d interaction which have already been included in the zeroth approximation Hamiltonian. Higher-order graphs may be regarded as a dressing of the graph (3.1.33) in two aspects, i.e., the dressing of propagators and of the s-d interaction vertex. As in the description of the electron-phonon interaction, we shall neglect the renormalization of the s-d exchange vertex since the correction associated with it should have the smallness $\sim \sqrt{T_k/E_F}$, where T_k is the Curie temperature of the ferromagnetic metal. Then the graphic structure of the equation for the self-energy part of the electron Green's function of the superconductor will be

$$\Sigma(\underline{p},\omega_n) = \Sigma_{ph}(\underline{p},\omega_n) + \Sigma_m(\underline{p},\omega_n)$$

$=$ 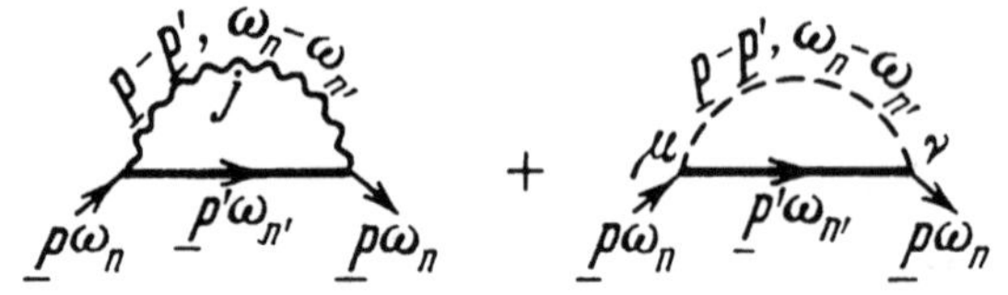

, (3.1.34)

where the analytic expressions

$$\Sigma_{ph}(\underline{p},\omega_n) = -T\sum_{n'}\int\frac{d\underline{p}'}{(2\pi)^3}$$

$$\times\sum_j |g_j(\underline{p},\underline{p}')|^2 \mathcal{D}_j^0(\underline{p}-\underline{p}',\omega_n-\omega_{n'})\sigma_0\tau_3 g(\underline{p}',\omega_{n'})\sigma_0\tau_3 \quad , \tag{3.1.35}$$

$$\Sigma_m(\underline{p},\omega_n) = -T\sum_{n'}\int\frac{d\underline{p}'}{(2\pi)^3}$$

$$\times\ |J(\underline{p}-\underline{p}')|^2 \mathcal{D}^{\mu\nu}(\underline{p}-\underline{p}',\omega_n-\omega_{n'})\alpha^\mu g(\underline{p}',\omega_{n'})\alpha^\nu \tag{3.1.36}$$

correspond to the phonon and magnon contributions. Here $J(\underline{q})$ is the Fourier component of the $J(\underline{r}-\underline{R})$ exchange interaction, and $\mathcal{D}^{\mu\nu}(\underline{q},\omega_n)$ is the Fourier component of the spin Green's function (3.1.28)

$$\mathcal{D}^{\mu\nu}_{ab}(\tau,\tau') = T\sum_n\int\frac{d\underline{p}}{(2\pi)^3}\exp\left[i\underline{p}\left(\underline{R}_a-\underline{R}_b\right)-i\omega_n(\tau-\tau')\right]\mathcal{D}^{\mu\nu}(\underline{p},\omega_n) \quad . \tag{3.1.37}$$

Summation over frequencies in (3.1.35,36) may be performed by using the spectral representation for the Green's functions of electrons, (2.2.7), and phonons, (2.2.8), and the spin Green's function

$$\mathcal{D}^{\mu\nu}(\underline{p},\omega_n) = \frac{1}{2\pi}\int_{-\infty}^{\infty}dz\,\frac{B^{\mu\nu}(\underline{p},z)}{i\omega_n-z} \quad , \tag{3.1.38}$$

where the frequencies ω_n are even numbered (see, for example, [3.4]). As a result, for the analytically continued ($i\omega_n \to \omega + i\delta$) functions we obtain the expressions

$$\Sigma_{ph}(\underline{p},\omega_n) = -\int\frac{d\underline{p}'}{(2\pi)^3}\sum_j |g_j(\underline{p},\underline{p}')|^2\int_{-\infty}^{\infty}\frac{dz}{2\pi}\int_{-\infty}^{\infty}\frac{dz'}{2\pi}$$

$$\times\, b_j(\underline{p}-\underline{p}',z)\sigma_0\tau_3\,\mathrm{Im}\{g(\underline{p}',z')\}\sigma_0\tau_3\,\frac{\tanh\frac{z'}{2T}+\mathrm{ctanh}\frac{z}{2T}}{\omega-z-z'+i\delta}\quad, \tag{3.1.39}$$

$$\Sigma_m(\underline{p},\omega_n) = -\int\frac{d\underline{p}'}{(2\pi)^3}|J(\underline{p}-\underline{p}')|^2\int_{-\infty}^{\infty}\frac{dz}{2\pi}\int_{-\infty}^{\infty}\frac{dz'}{2\pi}$$

$$\times\,\alpha^{\mu}\mathrm{Im}\{g(\underline{p}',z')\}\alpha^{\nu}\,\frac{\tanh\frac{z'}{2T}+\mathrm{ctanh}\frac{z}{2T}}{\omega-z-z'+i\delta}\quad. \tag{3.1.40}$$

The equation for the self-energy part of the retarded electron Green's function

$$\Sigma(\underline{p},\omega) = \Sigma_{ph}(\underline{p},\omega) + \Sigma_m(\underline{p},\omega) \tag{3.1.41}$$

will be of closed form if to (3.1.41), (3.1.39) and (3.1.40) we add the relationship between Σ and g:

$$\Sigma(\underline{p},\omega) = g_0^{-1}(\underline{p},\omega) - g^{-1}(\underline{p},\omega)\quad, \tag{3.1.42}$$

where, in view of (3.1.27),

$$g_0^{-1}(\underline{p},\omega) = \omega\sigma_0\tau_0 - \xi_{\underline{p}}\sigma_0\tau_3 - J\sigma_3\tau_3\quad. \tag{3.1.43}$$

3.1.4 Matrix Structure of Green's Functions

Equation (3.1.41) is a matrix equation for the four-row matrices Σ or g. The matrix structure of its solution is easiest to obtain by using a structure for Σ similar to the procedure performed in Chap.2 when employing the Nambu formalism. An arbitrary 4×4 matrix may be expanded in 16 independent unit "vectors" which are the direct product $\sigma_i\tau_k$ of two systems of Pauli matrices with $i,k=0,1,2,3$ where 0 corresponds to the unit matrix.

According to (3.1.43), g_0^{-1} is expanded only in three matrix components $\sigma_0\tau_0$, $\sigma_0\tau_3$ and $\sigma_3\tau_3$. For a superconductor in a singlet pairing model, the self-energy part of the Green's functions should contain components $\sigma_1\tau_1$ and $\sigma_2\tau_2$. The presence of a spontaneous moment in the system may lead to terms of the type $\sigma_3\tau_0$. Thus, the series expansion of Σ or g^{-1} in matrix components should be tried in the form

$$g^{-1}(\underline{p},\omega_n) = \tilde{\omega}\sigma_0\tau_0 - \tilde{\xi}\sigma_0\tau_3 - \tilde{J}\sigma_3\tau_3 - \Delta\sigma_2\tau_2 - A\sigma_3\tau_3 + B\sigma_1\tau_1 \tag{3.1.44}$$

where six expansion coefficients are functions of the variables $\underline{p}$ and ω_n. In the absence of perturbation, as is seen by comparing (3.1.44) with (3.1.43), $\tilde{\omega}\to\omega$, $\tilde{\xi}\to\xi$, $\tilde{J}\to J$, but Δ, A and B tend to zero.

It is not difficult to invert the matrix g^{-1}:

$$g(\underline{p},\omega) = \left(\frac{\tilde{P}_+ + \tilde{P}_-}{2}\sigma_0\tau_0 + \frac{\tilde{P}_+ - \tilde{P}_-}{2}\sigma_3\tau_3\right) \times\left(\tilde{\omega}\sigma_0\tau_0 + \tilde{\xi}\sigma_0\tau_3 - \tilde{J}\sigma_3\tau_3 + \Delta\sigma_2\tau_2 + A\sigma_3\tau_0 - B\sigma_1\tau_1\right) \quad , \tag{3.1.45}$$

where the notation introduced is as follows

$$\tilde{P}_\pm = \frac{1}{\tilde{\omega}_\pm^2 - \tilde{\varphi}_\pm^2 - \tilde{\xi}_\pm^2} \quad , \tag{3.1.46}$$

$$\tilde{\omega}_\pm = \tilde{\omega} \mp \tilde{J}, \quad \tilde{\varphi}_\pm = \Delta \pm B, \quad \tilde{\xi}_\pm = \tilde{\xi} \pm A \quad . \tag{3.1.47}$$

If we write (3.1.45) as a matrix, we shall make sure that all its components will be expressed in terms of combinations of the quantities $\tilde{P}_\pm$, $\tilde{\omega}_\pm$, $\tilde{\varphi}_\pm$, and $\tilde{\xi}_\pm$:

$$g(\underline{p},\omega) = \begin{pmatrix} \tilde{P}_+(\tilde{\omega}_+ + \tilde{\xi}_+) & 0 & 0 & -\tilde{P}_+\tilde{\varphi}_+ \\ 0 & \tilde{P}_-(\tilde{\omega}_- + \tilde{\xi}_-) & \tilde{P}_-\tilde{\varphi}_- & 0 \\ 0 & \tilde{P}_-\tilde{\varphi}_- & \tilde{P}_-(\tilde{\omega}_- - \tilde{\xi}_-) & 0 \\ -\tilde{P}_+\tilde{\varphi}_+ & 0 & 0 & \tilde{P}_+(\tilde{\omega}_+ - \tilde{\xi}_+) \end{pmatrix} \tag{3.1.48}$$

All the quantities involved on the right-hand side of (3.1.48) depend on $\underline{p}$ and ω. The dependence on momentum near the Fermi surface is assumed to be weak for them, just as in the case when only the electron-phonon interaction is taken into account. As functions of frequency, these quantities obey the symmetry relations

$$\tilde{\omega}_\pm(-\omega) = -\tilde{\omega}^*_\mp(\omega), \quad \tilde{\varphi}_\pm(-\omega) = \tilde{\varphi}^*_\mp(\omega), \quad \tilde{\xi}_\pm(-\omega) = \tilde{\xi}^*_\mp(\omega) \quad , \tag{3.1.49}$$

which are a generalization of (2.2.30) for the quantities involved in Eliashberg's equations.

The subsequent scheme according to which the integral equation (3.1.41) is transformed, coincides completely with the derivation of Eliashberg's equations performed in Sect.2.2. Thus, after carrying out the integration over $\underline{p}'$ in (3.1.39,40) and after averaging the resulting equations with respect to the

Fermi surface, we write the equations for the quantities defining the Green's function (3.1.48) of the ferromagnetic superconductor in the form [3.3,5]:

$$[1 - Z_{\pm}(\omega)]\omega = \pm J - \int_{-\infty}^{\infty} dz' \Big\{ \Big[K_{ph}(z',\omega) + K_m^{zz}(z',\omega) \Big] \times \mathrm{Re}\left\{ \frac{z' \mathrm{sign}(z')}{\sqrt{z'^2 - \Delta_{\pm}^2(z')}} \right\} + K_m^{\mp\pm}(z',\omega)\, \mathrm{Re}\left\{ \frac{z' \mathrm{sign}(z')}{\sqrt{z'^2 - \Delta_{\mp}^2(z')}} \right\} \Big\} \tag{3.1.50}$$

$$Z_{\pm}(\omega)\Delta_{\pm}(\omega) = \int_{-\infty}^{\infty} dz' \Big\{ \Big[K_{ph}(z',\omega) - K_m^{zz}(z',\omega) \Big] \times \mathrm{Re}\left\{ \frac{\Delta_{\pm}(z') \mathrm{sign}(z')}{\sqrt{z'^2 - \Delta_{\pm}^2(z')}} \right\} - K_m^{\mp\pm}(z',\omega)\, \mathrm{Re}\left\{ \frac{\Delta_{\mp}(z') \mathrm{sign}(z')}{\sqrt{z'^2 - \Delta_{\mp}^2(z')}} \right\} \Big\} . \tag{3.1.51}$$

The quantities $\tilde{\varphi}_{\pm}$ and $\tilde{\omega}_{\pm}$ are related to $\Delta_{\pm}$ and $Z_{\pm}$ by the equations

$$\tilde{\omega}_{\pm}(\omega) = Z_{\pm}(\omega)\omega \quad , \quad \tilde{\varphi}_{\pm}(\omega) = Z_{\pm}(\omega)\Delta_{\pm}(\omega) \quad . \tag{3.1.52}$$

In (3.1.50,51) $K_{ph}(z',\omega)$ is the electron-phonon interaction kernel (2.2.26) determined earlier, and $K_m^{\mp\pm}(z',\omega)$ is the electron-"magnon" interaction kernel [3.6]:

$$K_m^{\mp\pm}(z',\omega) = \mp \int_0^{\infty} dz G_m^{\perp}(z) \frac{1}{2} \frac{\tanh \frac{z'}{2T} \mp \mathrm{ctanh} \frac{z}{2T}}{z' + z - \omega - i\delta} \theta(\Omega_m - |z'|) \tag{3.1.53}$$

with

$$G_m^{\perp}(z) = \int \frac{d^2\underline{p}}{v_{\underline{p}}} \int \frac{d^2\underline{p}'}{v_{\underline{p}'}} |J(\underline{p} - \underline{p}')|^2 \frac{1}{2\pi} B^{\perp}(\underline{p} - \underline{p}', z) \Big/ \int \frac{d^2\underline{p}}{v_{\underline{p}}} \quad , \tag{3.1.54}$$

where $B^{\perp}$ corresponds to Green's functions of the type $\langle TS^+S^- \rangle$ and $\langle TS^-S^+ \rangle$ with circular projections of the spin operator $S^{\pm} = S^x \pm iS^y$.
In (3.1.53) Ω_m is a cutoff frequency corresponding to the finite layer near the Fermi surface where the matrix element J of s-d exchange interaction (the s-d exchange integral) is nonzero. We admit for it the following approximation

$$|J(\underline{p} - \underline{p}')|^2 = |J(\underline{p}_F - \underline{p}_F')|^2 \theta(\Omega_m - |\xi_{\underline{p}'}|) \tag{3.1.55}$$

with the help of which the integration over the energy $\xi_{\underline{p}'}$ is performed in (3.1.40). The fact that $J(\underline{p} - \underline{p}')$ is not replaced by a constant (or by a δ function in real space) is essential for the convergence of the integral over z' in the last term of (3.1.50). The divergence in the absence of the cutoff is due to the difference in the electron-magnon (3.1.53) and electron-phonon (2.2.26) interaction kernels. The electron-phonon interaction kernel leads to no divergence when integrating over z' in (3.1.50), as a result of the compensation of two terms in (2.2.26). In the electron-magnon interaction kernel we have only one term, therefore at large z' we have to cut off the integration at some frequency Ω_m which is an additional parameter of the theory. This was first noted in [3.6].

Equations (3.1.50,51) are equations for a ferromagnetic superconductor with strong coupling, both electron-phonon and electron-magnon. We see that in the presence of a spontaneous magnetic moment the superconductivity is characeterized by a set of four related quantities $Z_\pm(\omega)$ and $\Delta_\pm(\omega)$. This is a direct consequence of the splitting of the electron spectrum with respect to spin, since all the differences in the quantities $Z_+(\omega)$, $Z_-(\omega)$ and $\Delta_\pm(\omega)$ are due to the first term on the right-hand side of (3.1.50) which causes the exchange Zeeman electron energy and the difference in the scattering of electrons with spin wave absorption and emission.

In the absence of the spontaneous magnetization the combined equations (3.1.50,51) become

$$Z(\omega)\Delta(\omega) = \int_{-\infty}^{\infty} dz'\left[K_{ph}(z',\omega) - K_m(z',\omega)\right]\mathrm{Re}\left\{\frac{\Delta(z')\,\mathrm{sign}(z')}{\sqrt{z'^2-\Delta^2(z')}}\right\}\ ,$$

$$[1 - Z(\omega)]\omega = -\int_{-\infty}^{\infty} dz'\left[K_{ph}(z',\omega) + K_m(z',\omega)\right]\mathrm{Re}\left\{\frac{z'\,\mathrm{sign}(z')}{\sqrt{z'^2-\Delta^2(z')}}\right\} \qquad (3.1.56)$$

where $K_m(z,\omega)$ is the electron-magnon interaction kernel. It is given by a (3.1.53)-type expression in which instead of $G_m^\perp(z)$ it is necessary to put

$$G_m(z) = \int \frac{d^2p}{v_{p'}}\int\frac{d^2p'}{v_{p'}}\,|J(\underline{p}-\underline{p}')|^2\frac{1}{2\pi}B(\underline{p}-\underline{p}',z)\Big/\int\frac{d^2p}{v_p}\ , \qquad (3.1.57)$$

where $B(\underline{p},z)$ is the spectral density for the Fourier component of the spin Green's function

$$\mathcal{D}_{ab}(\tau,\tau') = -\left\langle T\underline{S}_a(\tau)\underline{S}_b(\tau')\right\rangle\ . \qquad (3.1.58)$$

Finally, when there is no electron-magnon interaction ($K_m(z,\omega) = 0$) Eq.(3.1.56) reduces to Eliashberg's equations (2.2.23,24) for the phonon model of the superconductor.

3.1.5 Superconducting Transition Temperature of a Ferromagnetic Metal

For a discussion of T_c we write the linearized combined equations (3.1.50, 51) with the integration over z' replaced by integration over positive z'. Introducing the standard notation

$$K_{\pm}(z',z,\omega) = \frac{1}{z'+z+\omega+i\delta} \pm \frac{1}{z'+z-\omega-i\delta} \tag{3.1.59}$$

we write the linearized equations (3.1.51) for the gap function in the form [3.5]

$$Z_{\pm}(\omega)\Delta_{\pm}(\omega) + \int_{-\infty}^{\infty} dz' K^{zz}(z',\omega)\,\mathrm{Re}\left\{\frac{\Delta_{\pm}(z')}{z'}\right\} + i\pi \int dz G_m^{\perp}(z)\,\mathrm{ctanh}\frac{z}{2T_c}\,\frac{\Delta_{\mp}(\omega\pm z)}{\omega\pm z}$$

$$- i\frac{\pi}{2}\int dz\alpha^2(z)F(z)\,\mathrm{ctanh}\frac{z}{2T_c}\left[\frac{\Delta_{\pm}(\omega-z)}{\omega-z} + \frac{\Delta_{\pm}(\omega+z)}{\omega+z}\right]$$

$$= \frac{1}{2}\int_0^{\infty} dz' \tanh\frac{z'}{2T_c}\left\{\frac{1}{2}\int dz\alpha^2(z)F(z)\left[K_+(z',z,\omega) + K_+(-z',z,\omega)\right]\right.$$

$$\times\,\mathrm{Re}\left\{\frac{\Delta_+(z')+\Delta_-(z')}{z'}\right\} - \int dz G_m^{\perp}(z)\left[K_+(z',z,\omega)\,\mathrm{Re}\left\{\frac{\Delta_+(z')}{z'}\right\}\right.$$

$$\left.\left. + K_+(-z',z,\omega)\,\mathrm{Re}\,\frac{\Delta_-(z')}{z'}\right]\right\}$$

$$\mp \frac{1}{2}\int_0^{\infty} dz' \tanh\frac{z'}{2T_c}\left\{\frac{1}{2}\int dz\alpha^2(z)F(z)\left[K_-(z',z,\omega) - K_-(-z',z,\omega)\right]\right.$$

$$\times\,\mathrm{Re}\left\{\frac{\Delta_+(z')-\Delta_-(z')}{z'}\right\} + \int dz G_m^{\perp}(z)\left[K_-(z',z,\omega)\,\mathrm{Re}\left\{\frac{\Delta_+(z')}{z'}\right\}\right.$$

$$\left.\left. + K_-(-z',z,\omega)\,\mathrm{Re}\left\{\frac{\Delta_-(z')}{z'}\right\}\right]\right\} \quad . \tag{3.1.60}$$

To this equation we add the linearized equation (3.1.50) for $Z_{\pm}(\omega)$:

$$[1 - Z_{\pm}(\omega)]\omega = \pm J - A_{ph}(\omega) - A_m^{zz}(\omega) - A_m^{\mp\pm}(\omega) \quad , \tag{3.1.61}$$

where

$$A_{ph}(\omega) = \int_{-\infty}^{\infty} K_{ph}(z',\omega)dz' \quad , \quad A_m^{\mu\nu}(\omega) = \int_{-\infty}^{\infty} K_m^{\mu\nu}(z',\omega)dz' \tag{3.1.62}$$

The quantity $A_{ph}(\omega)$ is given by (2.3.32), and for $A_m^{\mp\pm}$ it is not difficult to obtain the expression

$$A_m^{\mp\pm}(\omega) = \pm \int dz\alpha_m^2(z)F_m(z)\left[\ln\frac{\left|\Omega_m^2 - (\omega \pm z)^2\right|}{2\pi T_c}\right.$$

$$\left. - 2\mathrm{Re}\left\{\psi\left(\frac{1}{2} + i\frac{\omega \pm z}{2\pi T_c}\right)\right\} + i\pi\left(\tanh\frac{\omega \pm z}{2T_c} \mp \mathrm{ctanh}\frac{z}{2T_c}\right)\right] \quad , \tag{3.1.63}$$

where $\alpha_m^2(z)F_m(z)$ is defined by

$$\alpha_m^2(z)F_m(z) = \int\frac{d^2p}{v_{\underline{p}}}\int\frac{d^2p'}{v_{\underline{p}'}}|J(\underline{p}-\underline{p}')|^2\delta(z - E_{\underline{q}})\bigg/\int\frac{d^2p}{v_{\underline{p}}} \quad , \tag{3.1.64}$$

which is similar to (2.2.27) for $\alpha^2(z)F(z)$. Equation (3.1.64) defines the effective square of the electron-magnon interaction matrix element and the density of states in the magnon spectrum.

E_q is the energy of a spin wave with momentum $\underline{q} = \underline{p} - \underline{p}'$, which has arisen in (3.1.64) from the substitution of the expression for the spectral density of states of spin Green's functions

$$B^{\mp\pm}(\underline{q},z) = \mp 4\pi\langle S^z\rangle\delta(z \pm E_q) \tag{3.1.65}$$

into (3.1.57).

Further exploration of (3.1.60) is impossible without simplifying assumptions. We shall assume the finite frequency ω_m of the magnon spectrum to be much greater than T_{co}, the superconducting transition temperature in the absence of a ferromagnetic ordering. In this situation ($\omega_D \gg T_{co}$, $\omega_m \gg T_{co}$) both interactions of electrons with phonons and magnons must be considered equally. To discuss (3.1.60) it is convenient to write in symbolic form

$$Z_{\pm}(\omega)\Delta_{\pm}(\omega) - \delta_{ph}^{\pm}(\omega) + \delta_m^{\mp}(\omega) = \Delta(\omega) \mp \delta(\omega) \quad . \tag{3.1.66}$$

The terms denoted by $\delta_{ph}^{\pm}(\omega)$ and $\delta(\omega)$ were discussed in the theory of usual strong-coupling superconductors (Sects.2.2,2.3). It was shown that in the intermediate coupling case the terms $\delta_{ph}^{\pm}(\omega)$ containing cotangents describe the contribution of thermal phonons and may be omitted, along with the terms $\delta(\omega)$ containing the quantity $K_-(z',z,\omega)$ which is small if z' and ω are small compared to the characteristic frequencies in the phonon spectrum. Since the electron-phonon and electron-magnon interactions should be treated to an equal extent, we stay within the framework of intermediate coupling and neglect all the terms $\delta_{ph}^{\pm}$, $\delta^{\mp}$, and δ in (3.1.60). To these approximations, (3.1.60) may be replaced by the following equation for the quantity

$$\Delta(\omega) = \frac{1}{2}\int_0^\infty dz'\tanh\frac{z'}{2T_c}\left\{\frac{1}{2}\int dz\alpha^2(z)F(z)\Big[K_+(z',z,\omega)+K_+(-z',z,\omega)\Big]\right.$$

$$\times\,\mathrm{Re}\left\{\frac{\Delta(z')}{z'Z_+(z')}+\frac{\Delta(z')}{z'Z_-(z')}\right\}-\int dz\alpha_m^2(z)F_m(z)\Big[K_+(z',z,\omega)$$

$$\left.\times\,\mathrm{Re}\left\{\frac{\Delta(z')}{z'Z_+(z')}\right\}+K_+(-z',z,\omega)\mathrm{Re}\left\{\frac{\Delta(z')}{z'Z_-(z')}\right\}\Big]\right\} \quad . \qquad (3.1.67)$$

Proceeding from (2.3.32) and (3.1.63) it may be shown that in the low-frequency limit $\omega\to 0$ they have the asymptote

$$\mathrm{Re}\{A_{ph}(\omega)\} = \lambda\omega \quad , \quad \mathrm{Im}\{A_{ph}(\omega)\} = 0$$
$$\mathrm{Re}\{A^{\mp\pm}(\omega)\} = \mp A(\omega) + \lambda_m\omega \quad , \quad \mathrm{Im}\{A_m^{\mp\pm}(\omega)\} = 0 \quad , \qquad (3.1.68)$$

where

$$\lambda = 2\int\frac{dz}{z}\,\alpha^2(z)F(z) \quad , \quad \lambda_m = 2\int\frac{dz}{z}\,\alpha_m^2(z)F_m(z) \quad . \qquad (3.1.69)$$

Since (3.1.68) in (3.1.67) is approximate, it may be reduced to

$$\Delta(\omega) = \frac{1}{2}\int_0^\infty dz'\tanh\frac{z'}{2T_c}\left\{\Big[L_{ph}(z',\omega)+L_{ph}(-z',\omega)\Big]\frac{z'}{z'^2-\tilde{I}^2}\right.$$

$$\left.-L_m(z',\omega)\frac{1}{z'-\tilde{I}}-L_m(-z',\omega)\frac{1}{z'+\tilde{I}}\right\}\Delta(z') \quad , \qquad (3.1.70)$$

where the notation introduced is

$$L(z',\omega) = \frac{1}{1+\lambda+\lambda_m}\int dz\alpha^2(z)F(z)K_+(z',z,\omega); \quad \tilde{I} = \frac{I+A_m(0)}{1+\lambda+\lambda_m} \quad . \qquad (3.1.71)$$

Since the logarithmic feature prevails in the integrals of (3.1.70) and the kernels $L_{ph}(z',\omega)$ and $L_m(z',\omega)$ behave smoothly at small arguments, we can make the usual approximation to ignore the difference in these kernels at z' and -z'. As a result, we obtain the final equation for the gap function in the region of intermediate coupling

$$\Delta(\omega) = \int_0^\infty dz'\tanh\frac{z'}{2T_c}\,\frac{z'}{z'^2-\tilde{I}^2}\Big[L_{ph}(z',\omega)-L_m(z',\omega)\Big]\Delta(\omega) \quad . \qquad (3.1.72)$$

Using the standard iteration procedure it is possible from this equation to obtain the superconducting transition temperature. It can be found from

$$1 = \chi(0)J(T_c), \qquad (3.1.73)$$

where

$$J(T_c) = \int_0^{\omega_m} dz \tanh\frac{z}{2T_c}\,\frac{z}{z^2-\tilde{I}^2} \approx \frac{1}{2}\ln\frac{|\omega_m^2-\tilde{I}^2|}{(2\pi T_c)^2} - \mathrm{Re}\left\{\psi\left(\frac{1}{2}+i\frac{\tilde{I}}{2\pi T_c}\right)\right\} \quad , \qquad (3.1.74)$$

and $\chi(0)$ is found from the inhomogeneous integral equation for $\chi(\omega)$ which in the first step of iterations (subject to the additional condition $\omega_m \lesssim \omega_D$) leads to

$$\chi(0) = \frac{\lambda^*(1+\alpha)}{1+\lambda+\lambda_m} - \frac{\lambda_m}{1+\lambda+\lambda_m} \quad , \qquad (3.1.75)$$

where λ^* is the electron-phonon coupling constant renormalized due to magnons,

$$\lambda^* = \frac{\lambda}{1-\dfrac{\lambda}{1+\lambda+\lambda_m}\ln\dfrac{\omega_D}{\omega_m}} \qquad (3.1.76)$$

(α is some constant less than unity and proportional to ω_m).

Equation (3.1.73) may be rewritten as

$$\ln\frac{T_c}{\tilde{T}_{co}} = \psi\left(\frac{1}{2}\right) - \mathrm{Re}\left\{\psi\left(\frac{1}{2}+i\frac{\tilde{I}}{2\pi T_c}\right)\right\} \quad , \qquad (3.1.77)$$

where $\tilde{T}_{co}$ stands for the superconducting transition temperature, renormalized due to the electron-magnon interaction, of an intermediate-coupling superconductor

$$\tilde{T}_{co} = \frac{2\gamma}{\pi}\omega_D \exp\left[-\frac{1+\lambda+\lambda_m}{\lambda^*(1+\alpha)-\lambda_m} - \ln\frac{\omega_D}{\omega_m}\right] \quad . \qquad (3.1.78)$$

Equation (3.1.77) takes into account the influence of both effects on T_c, i.e., of the magnetization of conduction electrons and of the interaction via spin waves. If in (3.1.77) we set $\tilde{I} = 0$, then we obtain $T_c = \tilde{T}_{co}$; thus, the quantity $\tilde{T}_{co}$ determines the superconducting transition temperature in a magnetic superconductor when only the second effect, i.e., the interaction of electrons via spin waves is taken into account. The easiest way of ascertaining the structure of (3.1.78) is to consider the case when the width of the phonon and magnon spectra is of the same order of magnitude ($\omega_D \sim \omega_m$), and use the weak coupling limit ($\lambda \ll 1$, $\lambda_m \ll 1$). In this case, in view of the expression for T_{co}, (3.1.78) goes over into the expression

$$\tilde{T}_{co} = \frac{2\gamma}{\pi}\omega_D \exp\left(-\frac{1}{\lambda - \lambda_m}\right) \quad , \tag{3.1.79}$$

where the repulsive nature of the interaction of electrons via spin waves stands out distinctly. In the general case the behaviour of $\tilde{T}_{co}$ as a function of the parameters ω_m and λ_m is shown in Figs.3.1 and 2. On the other hand, if in (3.1.78) we set $\lambda_m = 0$, then (3.1.77) goes into the equation by SARMA [3.7]

$$\ln\frac{T_c}{T_{co}} = \psi\left(\frac{1}{2}\right) - \mathrm{Re}\left\{\psi\left(\frac{1}{2} + i\frac{J}{2\pi T_c}\right)\right\} \quad , \tag{3.1.80}$$

which allows for the paramagnetic effects in superconductors. The dependence of T_c on J which follows from the numerical solution of (3.1.80) is represented in Fig.3.3.

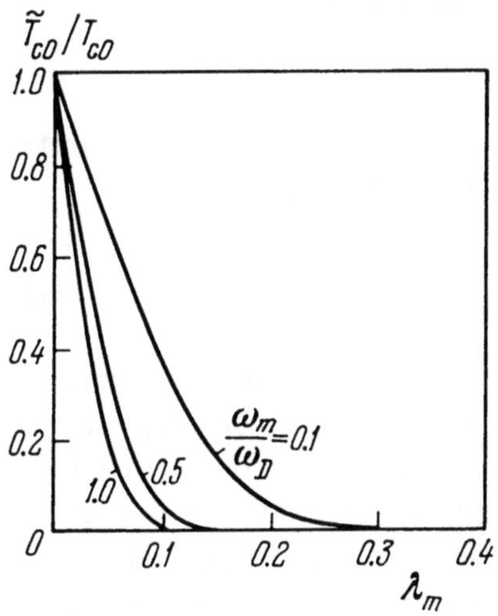

Fig.3.1. Dependence of $\tilde{T}_{co}/T_{co}$ [Eq.(3.1.78) in the limit $\lambda \ll 1$ and $\lambda_m \ll 1$] on the interelectron repulsion parameter λ_m induced by spin waves, for $\lambda = 0.2$

Over the interval of variation of J/Δ_0 from 1/2 to the point where the derivative T_c with respect to J goes to infinity, the curve $T_c(J)$ is a two-valued field function, which indicates that the state is metastable when

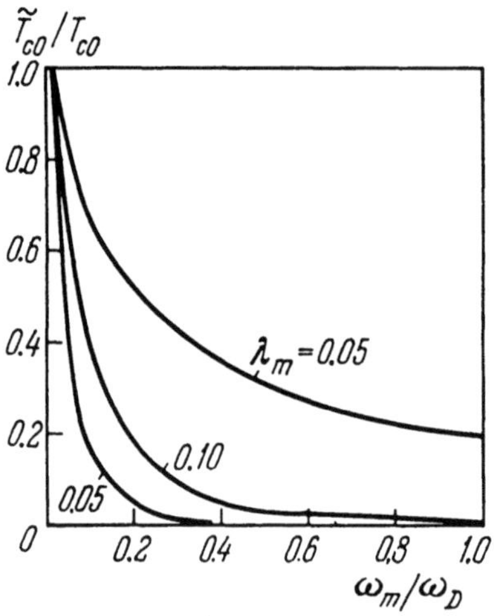

Fig.3.2. Dependence of $\tilde{T}_{c0}/T_{c0}$ on the spin-wave spectrum width ω_m for $\lambda = 0.2$

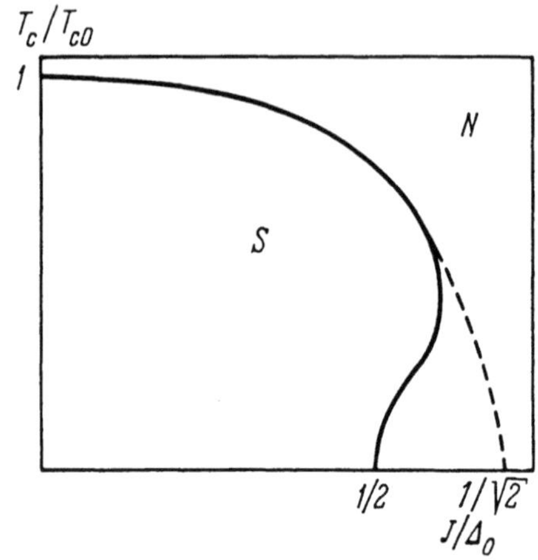

Fig.3.3. Dependence of the superconducting transition temperature in a ferromagnetic superconductor on the magnitude of the exchange magnetization field J/Δ_0 [3.7] ($\Delta_0 = \pi T_{c0}/\gamma$ is the gap at zero temperature

$J/\Delta_0 > 1/2$. In Fig.3.3 the dotted line shows the curve along which the first-order phase transition goes. The curve terminates at the point

$$\frac{J_c}{\Delta_0} = \frac{1}{\sqrt{2}} \quad , \tag{3.1.81}$$

above which superconductivity cannot exist owing to the destructive action of the magnetic (or exchange) field in the electron spin in the Cooper pair. It is natural that in order of magnitude, J_c has turned out to be equal to Δ_0 which is a measure of the Cooper pair coupling energy. The limiting value of the field, J_c, is known as the CHANDRASEKHAR-CLOGSTON [3.8,9] limit.

We have discussed the effect which the interaction of electrons with a magnetically ordered spin system exerts on T_c for the case of weak coupling. In the general case the dependence of T_c on the factors ω_m and λ_m may be determined only from several equations like (3.1.60,61). The results of such calculations by LEE et al. [3.6] are presented in Fig.3.4. As is seen from the figure, such results are in qualitative agreement with the approximation obtained according to (3.1.78). Figure 3.4 demonstrates also the strong influence of the paramagnetic effect on T_c.

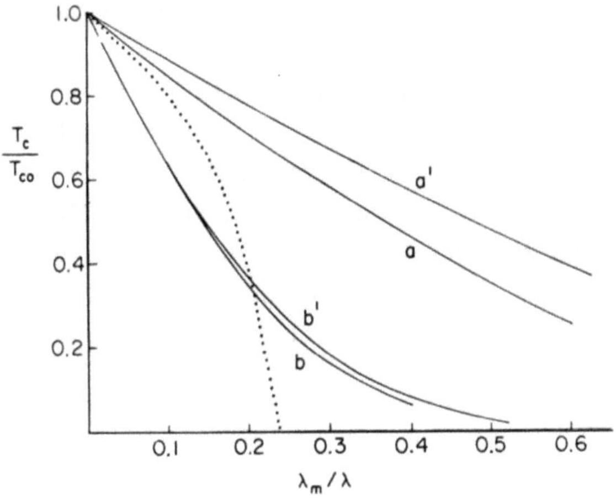

Fig.3.4. T_c/T_{co} as a function of λ_m/λ for $\lambda = 0.5$, $\omega_D = 300$ K [3.6] [Curves a and b result from numerical calculations of equations such as (3.1.60,61) for $\omega_m = 300$ K and $\omega_m = 30$ K, respectively. Curves a' and b' have been obtained from (3.1.78) with the same parameters. For these four curves $\tilde{J} = 0$. The dotted curve is the solution of (3.1.60,61) with $\tilde{J}$ being finite, but $J = 0$, $\omega = 30$ K and $\Omega_m/2\pi T_{co} = 200$]

We see that both effects, i.e., the magnetization of electrons and their repulsion via spin waves, has a destructive effect on the superconducting state. In particular, according to (3.1.81) J_c amounts to several degrees on the temperature scale. At the same time, in ferromagnetic transition metals (Fe, Ni, Co) the value of the magnetizing field at low temperatures is, from (3.1.23), of the order of the s-d exchange integral, which for the transition metals enumerated above amounts to hundreds or thousands of Kelvin degrees. That is why in these metals superconductivity is impossible even in the case when the effective attraction via phonons prevails over the Coulomb repulsion.

In an antiferromagnetic metal, where there is no exchange magnetization of conduction electrons, the mechanism of Cooper pair destruction via spin waves remains, and the problem of whether a superconducting state may occur in it is determined by the magnitude of the indirect interaction through phonons and through spin waves, and also by direct Coulomb repulsion. Superconductivity and ferromagnetism may be expected to coexist, however, in dilute alloys containing paramagnetic impurities which, because of the long-range exchange via conduction electrons, may, at low impurity concentrations, turn out to be magnetically ordered. This problem will be dealt with in the next section.

In conclusion we would like to give a historical background concerning the study of the problems touched upon here. The destructive effect of the exchange magnetization of electrons on superconductivity, and also the interaction with spin waves were first revealed by VONSOVSKY and SVIRSKY [3.10-14]. The same authors have investigated the problem concerning the formation of triplet pairs through the interaction with spin waves. For details we refer the reader to the review by IZYUMOV and SKRYABIN [3.15], and also to original papers [3.16-20].

3.2 Superconductivity in Dilute Ferromagnetic Alloys

3.2.1 The Problem of the Coexistence of Superconductivity and Magnetic Ordering

If superconductivity and ferromagnetism can exist in one volume, such a situation in dilute alloys of transition metals is most likely to occur in a superconducting non-transition matrix. In fact, some dilute alloys such as Cu-Mn become ferromagnetic at a low enough concentration of magnetic impurities due to the indirect exchange interaction of localized magnetic moments via conduction electrons. If the matrix is superconducting, then in the case of high dilution when the ferromagnetic ordering has already occurred, the concentration of magnetic impurities may happen not to exceed the critical concentration at which, according to the theory by ABRIKOSOV and GOR'KOV [3.21] (see below), the superconductivity is completely suppressed. Proceeding from these considerations, one may expect the following phase diagram of coexistence in the temperature-concentration plane. The curve for T_c as a function of the paramagnetic impurity concentration falls continuously, whereas the curve for T_K, the temperature of ferromagnetic ordering, increases. Evidently, it is at the intercept of these curves that the superconducting and ferromagnetic states of the alloy occur simultaneously.

A first experimental investigation of the coexistence problem was made by MATTHIAS et al. [3.22-24] in superconducting rare-earth systems with Gd impurities, namely in $Y_{1-x}Gd_xOs_2$ and $La_{1-x}Gd_x$ alloys. Figure 3.5 presents the phase diagram for $Y_{1-x}Gd_xOs_2$. The experimental points only lie on the curves for the superconducting or for the magnetic transition. It is extraordinarily difficult to ascertain what the magnetic state of an alloy is like below or to the left of the intercept of the T_c and T_K curves (i.e., within the superconducting region) since magnetic measurements in the superconducting phase are not feasible because of the Meissner effect.

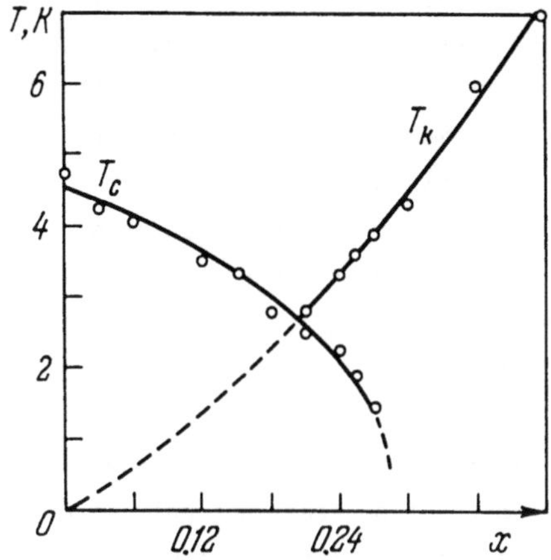

Fig.3.5. The concentration dependence of of the superconducting transition temperature T_c and of the temperature of transition to ferromagnetic state, T_K, in the alloy $Y_{1-x}Gd_xOs_2$ as a function of Gd content

Similar evidence for the possibility of the coexistence of superconductivity and ferromagnetism was obtained when studying $Ce_{1-x}Re_xRu_2$ and $Ce_{1-x}Gd_xRu_2$ [3.25,26].

Apparently, the fact that superconductivity and magnetic ordering coexist was established for the first time on the alloy $La_{1-x}Eu_x$ [3.27] in which a splitting of the Mössbauer line of the nucleus Eu^{151} was observed at very low temperatures. This splitting vanished at a magnetic ordering temperature T_K appreciably lower than T_c. It was impossible in those experiments to determine the type of magnetic ordering: ferro- or antiferromagnetic. Another situation ($T_K > T_c$) was observed in $Ce_{1-x}Tb_xRu_2$ with $x = 0.2$ [3.28] by neutron diffraction. Here the type of ordering turned out to be ferromagnetic.

The theory of superconducting dilute alloys with ferromagnetic ordering of impurity spins was constructed by GOR'KOV and RUSINOV [3.29] on the assumption that the ground state of alloys has a usual ferromagnetic structure. It became clear later that the question of a cooperative behaviour of the spin system of dilute alloys with paramagnetic impurities remains open also for the normal phase, even in the most important point as to whether ferromagnetism can exist in them. Predictions based on results of a self-consistent field theory [3.30] give rise to doubts: The computations [3.31] show that, because of the long-range and oscillating nature of the indirect exchange interaction of spins through conduction electrons, the ground state of the alloy cannot be ferromagnetic. There is a point of view that a peculiar type of magnetic state, known as "spin glass" [3.32], occurs in such alloys at impurity concentrations which are not too low [3.33,34]. Spin glasses are characterized by a high short-range order in the absence of long-range order and, in contrast to a simple paramagnetic state, possess a greater rigidity of the spin system with respect to, for example, external magnetic fields. It is believed that spin glasses are realized in dilute alloys of Mn, Fe and Cr in the matrix of Au, Ag and Cu. As for the rare-earth alloys in which superconductivity and ferromagnetism are expected to coexist, their magnetic state at low concentrations has not been ascertained, but it is quite possible that within certain intervals of paramagnetic impurity concentrations they may belong to spin glasses.

Gor'kov and Rusinov's theory pertains to alloys with ferromagnetic ground state. In a disordered alloy the s-d exchange interaction leads not only to an exchange magnetization of conduction electrons and to a shift of the Fermi surface for states with opposite spins, but also to a decay of the electron states on the Fermi surface. The influence of both effects on superconductivity has been studied thoroughly. The first of them leads to the so-called paramagnetic effect in superconductors which has been described in the previous section, whilst the other, investigated by Abrikosov and Gor'kov in the theory of paramagnetic impurities, leads to gapless superconductivity [3.21].

In a ferromagnetic superconductor both effects should evidently come into play, their interference producing a nonlinear result. Gor'kov and Rusinov have used the so-called static approximation in the description of the spin system, ignoring dynamic effects in the electron-magnon interaction.

3.2.2 Equations Including Dynamic Effects

For a dilute ferromagnetic alloy one should employ the same mathematical scheme as that developed in the previous section, with the only difference being that it is necessary to carry out averaging with respect to impurity configurations of randomly arranged impurities. This averaging technique was initially worked out by ABRIKOSOV and GOR'KOV [3.30] for normal alloys and was then applied by them [3.21] to superconductors containing paramagnetic impurities.

Consider the superconducting Green's function of the four-dimensional formalism. Just as in the previous section, we define it as

$$g(x,x') = -\langle T\Psi(x)\bar{\Psi}(x')\rangle \quad , \tag{3.2.1}$$

where Ψ and $\bar{\Psi}$ are the four-dimensional components (3.1.15) of the field operators. The electron-phonon and Coulomb interaction Hamiltonian in this representation is given by (3.1.21), and the interaction of electrons with an unordered spin system is described by (3.1.20) in which the four-row matrix $v(\underline{r})$ is defined by (3.1.9).

Expanding the Green's function (3.2.1) as a power series in the Hamiltonian $H_m + H_{int}$ and performing term-by-term averaging with respect to the impurity configurations, we get the following expression for the self-energy part [3.35]

$$\Sigma(\underline{p},\omega_n) = \Sigma_{ph}(\underline{p},\omega_n) + \Sigma_{ex}(\underline{p},\omega_n) \quad . \tag{3.2.2}$$

Here $\Sigma_{ph}(\underline{p},\omega_n)$ is the conventional expression (3.1.35) of the theory of strong coupling in superconductivity, and $\Sigma_{ex}(\underline{p},\omega_n)$ is the contribution of the s-d exchange interaction,

$$\Sigma_{ex}(\underline{p},\omega_n) = -cT\sum_{n'}\int \frac{d\underline{p}'}{(2\pi)^3}|J(\underline{p}-\underline{p}')|^2\alpha^\mu g(\underline{p}',\omega_{n'})\alpha^\nu \mathcal{D}^{\mu\nu}(\omega_n-\omega_{n'}) \quad . \tag{3.2.3}$$

This expression corresponds to the same degree of accuracy in the s-d exchange interaction as (3.1.35). In contrast to the latter, it contains not the usual spin Green's function $\mathcal{D}^{\mu\nu}(\underline{p},\omega_n)$ but $\mathcal{D}^{\mu\nu}(\omega_n)$, which is a Fourier component with respect to the variable $\tau-\tau'$ of the one-site Green's function

$$\mathcal{D}^{\mu\nu}(\tau,\tau') = -\left\langle T\left[S_a^\mu(\tau) - \langle S_a^\mu\rangle\right]\left[S_a^\nu(\tau') - \langle S_a^\nu\rangle\right]\right\rangle \tag{3.2.4}$$

describing the spin correlations on one site. The occurrence of this function in (3.2.3) together with an additional factor c (the impurity concentration) results from averaging with respect to impurity configurations. From (3.2.4) we get

$$\mathcal{D}^{\mu\nu}(\omega_n) = \int \frac{d\underline{q}}{(2\pi)^3}\mathcal{D}^{\mu\nu}(\underline{q},\omega_n) \quad , \tag{3.2.5}$$

where $\mathcal{D}^{\mu\nu}(\underline{q},\omega_n)$ is the Fourier component of a (3.1.28)-type spin Green's function averaged with respect to concentration. It is now clear that (3.2.3) differs from the corresponding expression (3.1.36) for a perfect crystal in that it contains an additional integration over momenta which results from averaging with respect to impurity concentrations.

The further program of writing equations for the Green's function (3.2.1) is the same as that in the previous section. It is easy to see that the Green's function is given by the previous expression (3.1.48), but the equations for the quantities $\tilde{\omega}_{\pm}(\omega)$ and $\tilde{\Delta}_{\pm}(\omega)$ become different. We write out only the linearized equations since we are concerned with the superconducting transition temperature only,

$$\tilde{\omega}_{\pm}(\omega) = \omega \mp J + i\frac{\langle S_z^2\rangle}{2\tau_{ex}} + \int_{-\infty}^{\infty} dz'\left[K_{ph}(z',\omega) + K^{\mp\pm}(z',\omega)\right] \quad , \tag{3.2.6}$$

$$\tilde{\Delta}_{\pm}(\omega) = \int_{-\infty}^{\infty} dz'\left[K_{ph}(z',\omega)\,\mathrm{Re}\left\{\frac{\tilde{\Delta}_{\pm}(z')}{\tilde{\omega}_{\pm}(z')}\right\} - K_m^{\mp\pm}(z',\omega)\,\mathrm{Re}\left\{\frac{\tilde{\Delta}_{\mp}(z')}{\tilde{\omega}_{\mp}(z')}\right\}\right]$$
$$- i\frac{\langle S_z^2\rangle}{2\tau_{ex}}\frac{\tilde{\Delta}_{\pm}(\omega)}{\tilde{\omega}_{\pm}(\omega)} \quad . \tag{3.2.7}$$

Here $K_m^{\mp\pm}(z',\omega)$ represents the electron-magnon interaction kernel. As distinct from the corresponding kernel (3.1.53) for an ideal crystal, it has the form

$$K_m^{\mp\pm}(z',\omega) = \pm\frac{\langle S_z\rangle}{2\pi\tau_{ex}}\int\frac{dq}{(2\pi)^3}\frac{1}{2}\frac{\tanh\frac{z'}{2T} - \mathrm{ctanh}\frac{E_q}{2T}}{z' \mp E_q - \omega - i\delta} \quad , \tag{3.2.8}$$

where E_q stands for the energy of a spin wave in a dilute ferromagnetic alloy due to the indirect exchange interaction of localized spins via conduction electrons; τ_{ex} is determined from

$$\frac{1}{\tau_{ex}} = c\frac{1}{2}N(0)\int\frac{d^2p}{v_p}\int\frac{d^2p'}{v_{p'}}|J(p-p')|^2 \Big/ \int\frac{d^2p}{v_p}\int\frac{d^2p'}{v_{p'}} \quad , \tag{3.2.9}$$

in which integration is performed over the Fermi surface. The quantity τ_{ex} is the average time between the collisions of an electron with an impurity which leads to the reorientation of spin.

The considerable difference between (3.2.6,7) and the corresponding linearized equations (3.1.50,51) consists of additional imaginary terms proportional to $1/\tau_{ex}$ which vanish when the impurity concentration tends to zero. It is convenient to rewrite these equations in the following form

$$\tilde{\Delta}'_{\pm}(\omega) = \int_{-\infty}^{\infty} dz'\left[K_{ph}(z',\omega)\,\mathrm{Re}\left\{\frac{\tilde{\Delta}'_{\pm}(z')}{z' \mp J + A^{\mp\pm}(z')}\right\}\right.$$

$$- K_m^{\mp\pm}(z',\omega)\,\mathrm{Re}\left\{\frac{\Delta'_{\mp}(z')}{z' \pm J + A^{\mp\pm}(z')}\right\}\Bigg] \quad , \tag{3.2.10}$$

with

$$A^{\mp\pm}(\omega) = \int_{-\infty}^{\infty} dz'\left[K_{ph}(z',\omega) + K_m^{\mp\pm}(z',\omega)\right] + i\frac{\langle S_z^2\rangle}{\tau_{ex}} \quad . \tag{3.2.11}$$

3.2.3 Quasi-Static Approximation

The integral equation (3.2.10) can be solved only in limiting cases. As one limiting case let us consider the situation when the width of the spin-wave spectrum of an alloy, ω_m, is very small, or more precisely

$$\omega_m \ll T_c \quad . \tag{3.2.12}$$

This case may be considered if the critical concentration at which ferromagnetism occurs in a dilute alloy is small, so that the Curie temperature T_k of the alloy is low enough to be comparable to T_c. When $T_c \sim T_k$ one may expect the coexistence of superconductivity and ferromagnetism.

Inequality (3.2.12) is the condition of quasistaticity in a spin system. In this case the electron-magnon interaction kernel (3.2.8) becomes

$$K_m^{\mp\pm}(z',\omega) = \frac{S(S+1) - \langle S_z^2\rangle}{2\pi\tau_{ex}}\cdot\frac{1}{z' - \omega - i\delta} \quad , \tag{3.2.13}$$

if use is made of the relation

$$\langle S_z\rangle \mathrm{ctanh}\frac{E_{\underline{q}}}{2T} = \frac{1}{2}\left(\langle S_{\underline{q}}^+ S_{\underline{q}}^-\rangle + \langle S_{\underline{q}}^- S_{\underline{q}}^+\rangle\right) \quad , \tag{3.2.14}$$

which follows from the spectral properties of the magnon Green's function. By substituting (3.2.13) into (3.2.10) and integrating over z' with Cauchi's theorem, one gets the following inhomogeneous integral equation for T_c:

$$C_{\pm}(\omega) = \int_{-\infty}^{\infty} dz' K_{ph}(z',\omega)\,\mathrm{Re}\left\{\frac{[z' \pm J + A^{o}(z')]C_{\pm}(z') + ibC_{\mp}(z')}{[z' + A^{o}(z')]^2 - J^2 + b^2}\right\} \quad . \tag{3.2.15}$$

where

$$A^{o}(\omega) = ia + \int_{-\infty}^{\infty} dz' K_{ph}(z',\omega) \quad , \tag{3.2.16}$$

$$a = \frac{S(S+1) + \langle S_z^2 \rangle}{2\tau_{ex}} \quad , \; b = -\frac{S(S+1) - \langle S_z^2 \rangle}{2\tau_{ex}} \quad . \tag{3.2.17}$$

It is useful to note that the function $C_\pm(\omega)$ is related to the function $\tilde{\Delta}'_\pm(\omega)$, which is defined by (3.2.10), by the expression

$$\frac{\tilde{\Delta}'_\pm(\omega)}{\tilde{\omega}_\pm(\omega)} = \frac{[\omega \pm J + A^0(\omega)]C_\pm(\omega) + ibC_\mp(\omega)}{[\omega + A^0(\omega)]^2 - J^2 + b^2} \quad . \tag{3.2.18}$$

If one considers (3.2.15) in the limit of weak electron-phonon coupling, i.e., if one uses the approximation (2.3.5) for the kernel $K_{ph}(z,\omega)$, one obtains the following equation for T_c:

$$1 = \lambda \int_0^{\omega_D} dz' \, \mathrm{Re}\left\{\frac{z' + i(a+b)}{(z' + ia)^2 - J^2 + b^2}\right\} \tanh \frac{z'}{2T_c} \quad . \tag{3.2.19}$$

Integrating (3.2.19) one gets [3.36]:

$$\ln\frac{T_c}{T_{co}} = \psi\left(\frac{1}{2}\right) - \frac{1}{2}\left[\left(1 + \frac{b}{\sqrt{b^2 - J^2}}\right)\psi\left(\frac{1}{2} + \rho_-\right) + \left(1 - \frac{b}{\sqrt{b^2 - J^2}}\right)\psi\left(\frac{1}{2} + \rho_+\right)\right] \quad , \tag{3.2.20}$$

where

$$\rho_\pm = \frac{a \pm \sqrt{b^2 - J^2}}{2\pi T_c} \quad . \tag{3.2.21}$$

T_{co} stands for the superconducting transition temperature of a metal containing no magnetic impurities.

If the scattering parameter $1/\tau_{ex}$ is set equal to zero (i.e., $a = b = 0$), (3.2.20) goes into the Sarma equation (3.1.77) which describes the paramagnetic effect in superconductors due to the electron energy exchange splitting. In the other limiting case when there is no ferromagnetic ordering (i.e., $J = 0$), (3.2.20) goes over into the Abrikosov-Gor'kov equation [3.21] corresponding to the paramagnetic case,

$$\ln\frac{T_c}{T_{co}} = \psi\left(\frac{1}{2}\right) - \psi\left(\frac{1}{2} + \rho\right) \quad , \quad \rho = \frac{S(S+1)}{2\pi T_c \tau_{ex}} \quad . \tag{3.2.22}$$

From this equation one can readily find the critical concentration of paramagnetic impurities at which the superconductivity of an alloy vanishes. This concentration corresponds to a critical value τ_{ex}^p of τ_{ex} given by [3.21]

$$\frac{S(S+1)}{\tau_{ex}^{p}} = \frac{\pi}{2\gamma} T_{co} \quad . \tag{3.2.23}$$

Equation (3.2.20) implicitly defines T_c as a function of the averages $\langle S_z \rangle$ and $\langle S_z^2 \rangle$. Since these quantities themselves are temperature dependent one should write self-consistent equations for them taking into account the effect of the superconducting state on the spin system. Such equations for T_c and $\langle S_z \rangle$ yield the phase diagram shown in Fig.3.6.

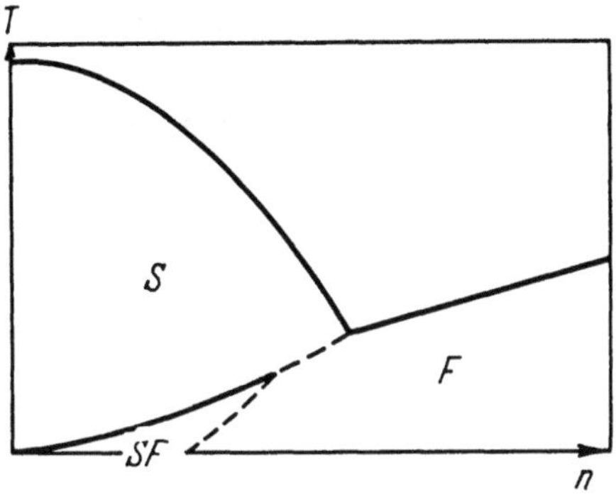

Fig.3.6. Phase diagram for superconducting (S) and ferromagnetic (F) states in a dilute alloy according to [3.29] (without spin-orbit scatterers)

GOR'KOV and RUSINOV [3.29] have also taken into account the spin-orbit electron scattering on impurities which is assumed to arise from other (non-paramagnetic) impurites usually contained in rare-earth materials. In the presence of spin-orbit interactions, the electron spin can flip. This interaction thus leads to the mixing of electron states with different spin projections. As a result, the effect of the exchange splitting on the Cooper pairs decreases and the domain of the coexistence of superconductivity with ferromagnetism widens.

It must be noted, however, that the calculation of the critical impurity concentration at which the superconducting transition temperature tends to zero is not likely to be legitimate in the static approximation considered here since the condition of quasistaticity (3.2.12) in this case (with $T_c \to 0$) turns out to be definitely violated. Therefore, to draw clear-cut boundaries of the coexistence on the straight line $T_c = 0$, the need arises to take into account the dynamic effects in electron-magnon scattering. The various approaches to the problem of dynamic correlations in a spin system may be found in the literature [3.37-43]. Mention should also be made of three reviews dealing with the problem of the coexistence of superconductivity and ferromagnetism [3.44-46].

Following GOR'KOV and RUSINOV, a detailed analysis of phase diagrams and of the conditions for the coexistence of superconductivity and magnetism within the framework of the static approximation has recently been made by BALSEIRO and FALIKOV [3.47]. As opposed to GOR'KOV and RUSINOV, they more accurately took into account the magnetization dependence of the parameter describing the mechanism of destruction of Cooper pairs by the exchange scattering of electrons. It appears as a result that in the metal, where the direct exchange interaction between localized atomic spins is absent, there is no region of coexistence, and the phase transition from the superconducting state to the normal ferromagnetic state is a first-order transition. In the presence of direct exchange interaction there appears to be a small region of coexistence at low temperatures and small concentrations of magnetic atoms of the alloy. The phase diagrams for both cases obtained, as a result of numerical calculation, are shown in Fig.3.7.

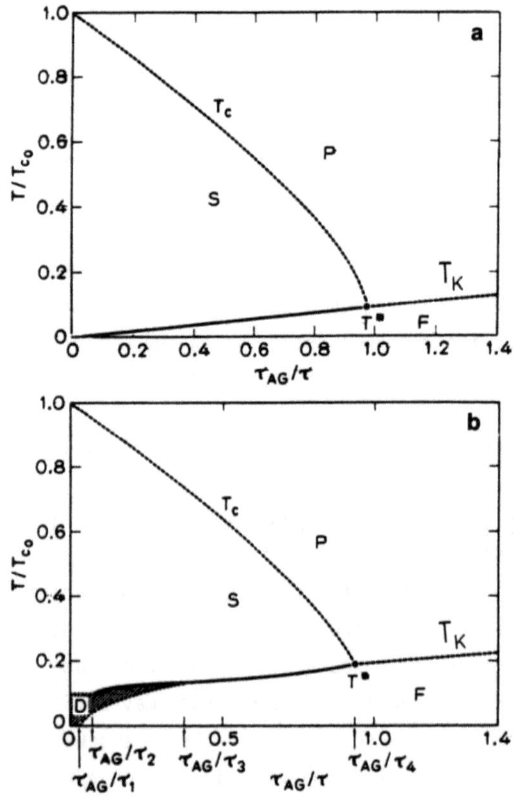

Fig.3.7a and b. Phase diagrams in the temperature-concentration plane, obtained in [3.47] for $T_c = 10$ K, $N(0) = 5 \cdot 10^{-14}$. The solid curves illustrate first-order phase transition lines, the dotted ones are second-order phase transition lines
a) absence of direct exchange
b) direct exchange parameter J_H corresponding to the ratio $J_H S^2/3K = 1$ K

The Gor'kov-Rusinov theory applies to the case where there is no direct exchange between localized spins. The difference in the phase diagrams, seen in Figs.3.7a and 3.5, is due to the fact that, according to GOR'KOV and RUSINOV, a narrow region of coexistence emerges at low T_c and C, whereas it actually occurs only in the presence of direct exchange interaction (Fig.3.7b).

Thus, within the framework of the approximation of a self-consistent field with localized magnetic moments[1], the phase where the superconducting and ferromagnetic order parameters would be non-zero is absent. Among the factors destroying the superconducting state in the magnetically ordered phase which have been considered in the previous section (the paramagnetic effect, repulsion of electrons through their interaction with spin waves, and exchange scattering of electrons with spin flip), the first factor, i.e., the exchange splitting of the conduction electron band, prevails in this situation. We have not discussed the role of one more factor, the fluctuations of the magnetic-order parameter. This phenomenon becomes essential near the magnetic ordering temperature. We shall put off the discussion of this problem until Sect.3.4 where the experimental observations for the coexistence of superconductivity with ferromagnetism and other types of magnetic ordering will be analyzed. Now we pass to a theoretical investigation of the coexistence of superconductivity with helicaltype antiferromagnetic ordering.

3.3 The Coexistence of Superconductivity and Helical Magnetic Ordering

3.3.1 The Anderson-Suhl Hypothesis

Long before the appearance of microscopic theories dealing with problems of the coexistence of superconductivity and ferromagnetism, ANDERSON and SUHL [3.48] proposed that inhomogeneous magnetic order is likely to exist in a superconducting alloy. This magnetic state with ferromagnetic domains of the size of the reciprocal value of some wave vector

---

[1]BALSEIRO and FALIKOV's theory applies both to the case of an alloy with a low concentration of the magnetic component, which has been considered by GOR'KOV and RUSINOV, and to a metal with a regular arrangement of magnetic atoms interacting chiefly via conduction electrons. In both cases the phase diagrams on the T, $1/\tau$ plane coincide. In the case of a disordered alloy the parameter $1/\tau$ may be recalculated for the concentration C (in a dilute alloy $1/\tau \sim C$).

$$Q_0 \sim (p_F \xi_0)^{-1/3} p_F \qquad (3.3.1)$$

($\xi_0 = \hbar v_F / \pi\Delta_0$ being the coherence length, and Δ_0 the gap at a zero temperature) they called cryptoferromagnetic.

This prediction was based on an elementary analysis of the exchange energy of a system of localized spins interacting with each other via conduction electrons

$$H_{ex} = - \sum_{a \neq b} I(\underline{R}_a - \underline{R}_b) \underline{S}_a \underline{S}_b = - \sum_{\underline{q}} I(\underline{q}) |\underline{S}_{\underline{q}}|^2 \quad . \qquad (3.3.2)$$

Here $I(\underline{q})$ is the Fourier component of the indirect exchange interaction integral $I(\underline{R}_a - \underline{R}_b)$. From (3.1.1) it follows that in the second order of perturbation theory $I(\underline{q})$ is proportional to the magnetic susceptibility $\chi(\underline{q})$, i.e.,

$$I(\underline{q}) \sim \chi(\underline{q}) \quad . \qquad (3.3.3)$$

For a normal metal with quadratic electron dispersion relation $\chi_n(\underline{q})$ is a well-known function with a maximum at $\underline{q} = 0$ corresponding to the situation when the ferromagnetic ordering of spins is energetically favourable. It is also known that for a superconductor the electronic susceptibility in a homogeneous field is $\chi_s(0) = 0$, but at large $\underline{q}$ $\chi_s(\underline{q})$ coincides with that for a normal metal. This means that in a superconductor $\chi_s(\underline{q})$ should have a maximum at some finite $\underline{q}$. ANDERSON and SUHL have shown that the position of the maximum in $\chi(\underline{q})$ corresponds to the wave vector Q_0 (3.3.1) (Fig.3.8). They interpreted this maximum as an indication of the energetic alloy favourable existence of inhomogeneous magnetic regions with characteristic size $\sim Q_0^{-1}$ in the superconducting state.

At that time, helical magnetic structures of crystals, which are homogeneous and are specified by some wave vector $\underline{q}$, were not yet known. It might be assumed that the maximum in $\chi_s(\underline{q})$ at $q = Q_0$ corresponds to a helical magnetic structure. Thus the question arises whether a helical magnetic structure may exist in the superconducting state. It would be reasonable to attack this problem using a model for a crystal in which the arrangement of magnetic atoms is regular (the experimental situation described in the next section has to do exactly with such a case). A detailed investigation of this problem has been carried out by BULAEVSKY et al. [3.49], to which we now give our attention.

The major problem here is to obtain equations describing the electronic system of a metal in the simultaneous presence of Cooper pairing and of distortions of the electron spectrum due to the interaction with the helical magnetic order produced by localized atomic spins. Conduction electrons are carriers of indirect exchange interaction and, at the same time, their eigenstates depend on the parameters of the superconducting state in the metal. There arises a system of strongly coupled order parameters characterizing the superconducting state and magnetic ordering. These parameters determine the phase diagram of a system.

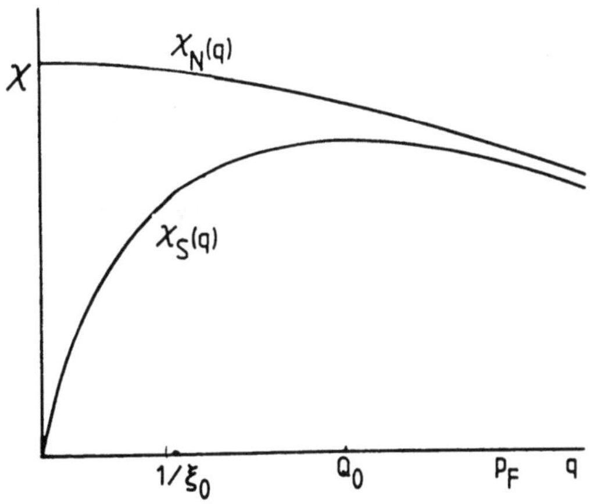

Fig.3.8. The susceptibility of the electron gas in the normal and superconducting states

Now we derive equations describing the interaction of these order parameters.

3.3.2 Equations for the Electron Green's Function in the Weak-Coupling Approximation

In view of the intricacy of taking into account the changes in the electron spectrum of a superconductor due to the interaction of electrons with the helical magnetic order, it is reasonable to reject the strong-coupling method when considering the electron-phonon interaction and to solve the problem in the weak-coupling approximation. It is useful to resort to GOR'KOV's method and, within the framework of the BCS model, to obtain equations for the normal and for the anomalous Green's functions

$$G_{\alpha\beta}(x,x') = -\left\langle T\psi_\alpha(x)\overline{\psi}_\beta(x')\right\rangle \quad , \quad F^+_{\alpha\beta}(x,x') = -\left\langle T\overline{\psi}_\alpha(x)\overline{\psi}_\beta(x')\right\rangle \quad . \tag{3.3.4}$$

In the general case when there is an interaction with some external field these equations are of the form

$$\left.\begin{aligned} &\left[-\frac{\partial}{\partial\tau} - H_o(\underline{r}) - \hat{V}(\underline{r})\right]\hat{G} + \Delta\hat{I}\hat{F}^+ = \hat{I}\delta(x - x') \\ &\left[\frac{\partial}{\partial\tau} - H_o(\underline{r}) - \hat{V}^{tr}(\underline{r})\right]\hat{F}^+ + \Delta^*\hat{I}\hat{G} = 0 \end{aligned}\right\} \quad , \tag{3.3.5}$$

where all the capped quantities are 2×2 matrices with respect to spin indices and

$$\hat{I} = \begin{pmatrix} 0 & 1 \\ -1 & 0 \end{pmatrix} \quad . \tag{3.3.6}$$

The superconducting paramter Δ is found from the self-consistency condition,

$$\hat{F}^+(\underline{r},\underline{r}) = \Delta^*\hat{I} \quad . \tag{3.3.7}$$

After expansion of (3.3.4) in a Fourier series in the variable $\tau - \tau'$, the combined equations (3.3.5) may be rewritten in the form

$$\begin{aligned} &\left\{\left[i\omega_n - H_o(\underline{r})\right]\delta_{\alpha\gamma} - V_{\alpha\gamma}(\underline{r})\right\}G_{\gamma\beta}(\underline{r},\underline{r}';\omega_n) + \Delta I_{\alpha\gamma}F^+_{\gamma\beta}(\underline{r},\underline{r}';\omega_n) = \delta_{\alpha\beta}\delta(\underline{r} - \underline{r}') \\ &\left\{\left[i\omega_n - H_o(\underline{r})\right]\delta_{\alpha\gamma} + V_{\gamma\alpha}(\underline{r})\right\}F^+_{\gamma\beta}(\underline{r},\underline{r}';\omega_n) - \Delta^*I_{\alpha\gamma}G_{\gamma\beta}(\underline{r},\underline{r}';\omega_n) = 0 \end{aligned} \tag{3.3.8}$$

Here $V_{\alpha\gamma}(\underline{r})$ is the s-d interaction matrix element (3.1.1) in the self-consistent field approximation

$$V_{\alpha\beta}(\underline{r}) = \sum_a J(\underline{r} - \underline{R}_a)\left\langle S_a\right\rangle\sigma_{\alpha\beta} \quad , \tag{3.3.9}$$

where $\left\langle S_a\right\rangle$ is the mean value (at a given temperature) of the atomic spin vector at $\underline{R}_a$.

Assume now that in the metal exists helical ordering of magnetic moments with the wave vector $\underline{Q}$ along the z-axis,

$$\left\langle S^x_a\right\rangle = S\sigma\cos(\underline{Q}\underline{R}_a) \quad , \quad \left\langle S^y_a\right\rangle = S\sigma\sin(\underline{Q}\underline{R}_a) \quad , \quad \left\langle S^z_a\right\rangle = 0 \quad , \tag{3.3.10}$$

The matrix (3.3.9) is then

$$\hat{V} = h(\underline{Q})\begin{pmatrix} 0 & \exp(-i\underline{Q}\underline{r}) \\ \exp(i\underline{Q}\underline{r}) & 0 \end{pmatrix} \tag{3.3.11}$$

with

$$h(\underline{Q}) = J(\underline{Q})S\sigma \quad . \tag{3.3.12}$$

Here S is the atomic spin, σ is the magnetic order parameter and $J(\underline{Q})$ is the Fourier component of the s-d exchange integral $J(\underline{r}-\underline{R}_a)$.

Solutions of (3.3.8) with the matrix (3.3.11) should be tried by expanding the functions G and F^+ into a Fourier integral with respect to the variables $\underline{r}$ and $\underline{r}'$:

$$G(\underline{r},\underline{r}';\omega_n) = \int\frac{d\underline{p}}{(2\pi)^3}\int\frac{d\underline{p}'}{(2\pi)^3}\exp[i(\underline{p}\underline{r}-\underline{p}'\underline{r}')]G(\underline{p},\underline{p}';\omega_n) \quad . \tag{3.3.13}$$

It can be easily verified that the eight combined equations (3.3.8) in the Fourier representation split into two independent systems, each containing four equations. Their solution leads to the following anomalous Green's functions [3.49]:

$$F^+_{\uparrow\uparrow}(\underline{p}+\frac{Q}{2},\ \underline{p}';\ \omega_n) = \delta(\underline{p}-\frac{Q}{2}-\underline{p}')\ \frac{2h(i\omega_n+\delta)\Delta^*}{\mathcal{D}(\underline{p},\omega_n)} \quad , \tag{3.3.14}$$

$$F^+_{\downarrow\uparrow}(\underline{p}-\frac{Q}{2},\ \underline{p}';\ \omega_n) = \delta(\underline{p}-\frac{Q}{2}-p')\ \frac{\omega_n^2+\xi^2_{\underline{p}+\frac{Q}{2}}+|\Delta|^2-h^2}{\mathcal{D}(\underline{p},\omega_n)}\Delta^* \quad . \tag{3.3.15}$$

The functions $F^+_{\downarrow\downarrow}$ and $F^+_{\uparrow\downarrow}$ are obtained from $F^+_{\uparrow\uparrow}$ and $F^+_{\downarrow\uparrow}$ by the replacement $\underline{Q}\rightarrow-\underline{Q}$ and by multiplication by -1. The notation introduced is as follows

$$\mathcal{D}(\underline{p},\omega_n) = \left[\omega_n^2+E_1^2(\underline{p})\right]\left[\omega_n^2+E_2^2(\underline{p})\right] \quad , \tag{3.3.16}$$

$$E_\mu^2 = \varepsilon^2+\delta^2+|\Delta|^2+h^2\pm 2\sqrt{\varepsilon^2(\delta^2+h^2)+|\Delta|^2h^2} \quad , \quad (\mu=1,2) \tag{3.3.17}$$

$$\delta = \frac{\xi_{\underline{p}+\frac{Q}{2}}-\xi_{\underline{p}-\frac{Q}{2}}}{2} \quad , \quad \varepsilon = \frac{\xi_{\underline{p}+\frac{Q}{2}}+\xi_{\underline{p}-\frac{Q}{2}}}{2} \tag{3.3.18}$$

(h is an abbreviation for $h(\underline{Q})$).

$E_1(\underline{p})$ and $E_2(\underline{p})$ are the energies of elementary excitations in a superconductor in the presence of helical order. When $h\rightarrow0$ and $Q\rightarrow0$ (3.3.17) goes over into the equation $E_\mu(\underline{p}) = \pm\sqrt{\xi^2+\Delta^2}$ which determines the energy of elementary excitations of a superconductor in the BCS theory. On the other hand, we obtain when $\Delta \to 0$

$$E_\mu(\underline{p}) = \varepsilon \pm \sqrt{\delta^2 + h^2} \quad ,$$

which is the known spectrum of elementary excitations of a normal metal with helical magnetic order [3.50].

The self-consistent equation (3.3.7) determining the parameter Δ is of the form

$$\frac{V}{2} \sum_{\underline{p}} \sum_{\mu} \frac{\partial E_\mu}{\partial |\Delta|^2} \tanh \frac{E_\mu}{2T} = 1 \quad . \tag{3.3.19}$$

3.3.3 Analysis of the Free Energy

The free energy should be a function of four parameters: Δ, σ, Q and T. From the spectrum of elementary excitations (3.3.17) we get the energy

$$F(\Delta,\sigma,Q,T) = F_s(\Delta,T) + F_M(\sigma,Q,T) + F_{int}(\Delta,\sigma,Q,T) \quad , \tag{3.3.20}$$

where F_s and F_M are BCS-type free energies of a superconducting metal and a normal metal with helical ordering of localized spins, respectively; F_{int} gives the variation of free energy due to the simultaneous coexistence of superconductivity and magnetic ordering. We have [3.49]

$$F_s(\Delta,T) = -\frac{1}{2}N(0)\left\{\Delta^2 \ln\frac{e\Delta_0^2}{\Delta^2} + 8T\Delta \int_0^\infty dt \ln\left[1 + \exp\left(-\frac{\Delta}{T}\sqrt{1+t^2}\right)\right]\right\} \quad , \tag{3.3.21}$$

$$F_M(\sigma,Q,T) = -N(0)\, b^2 h_0^2 \sigma^2 \left(1 - a^2 Q^2\right) + T \int_0^\sigma b_s(x)dx \quad , \tag{3.3.22}$$

$$F_{int}(\Delta,\sigma,Q,T) = T \sum_{\underline{p}} \sum_{\mu} \ln \frac{\cosh[E_\mu(0,h,Q)/2T] \cdot \cosh[E_\mu(\Delta,0,0)/2T]}{\cosh[E_\mu(\Delta,h,Q)/2T] \cdot \cosh[E_\mu(0,0,0)/2T]} \tag{3.3.23}$$

The first of these expressions is one of the standard formulae of free energy in the BCS theory. The second constitutes an expansion of the free energy of a helical magnet in powers of Q, the last term in (3.3.22) being a compact representation of the entropy term of a Heisenberg magnet. The first term in (3.3.22) is the mean energy of a magnet in which the exchange interaction is realized via conduction electrons. Here a and b are some constants determined by the Fourier components of the indirect exchange integral in the normal phase of the metal, with wave vectors equal to the reciprocal lattice vectors. Their definition may be found in [3.49]. Here we note only that $b \sim 1$ and $a \sim 1/p_F$. The quantity E_μ in (3.3.23) is defined by (3.3.17). In (3.3.22) $h_0 \equiv h(Q_0)$, and Δ_0 is the gap at zero temperature in the absence of s-d exchange coupling.

We now investigate the conditions for the coexistence of superconductivity and helical ordering, bearing in mind the experimental situation in rare-earth superconducting compounds which is described in detail in the next section. In some compounds of this kind superconductivity vanishes at some temperature $T_{c2} \ll T_c$ as a result of a first-order phase transition, and the normal phase which emerges at $T < T_{c2}$ is ferromagnetic. In the narrow region near T_{c2} in the interval between T_{c2} and T_M we have the region of coexistence of superconductivity and helical ordering. The phase transition to a magnetically ordered state within the superconducting phase is a second-order phase transition.

Analysis of the free energy (3.3.20) shows [3.49] that a phenomenon of this type is possible at certain ratios of parameters. Let us, for example, analyze the transition from the superconducting S phase to the MS phase where the superconducting and magnetic order parameters are both finite. Near the transition point expressions (3.3.21-23) may be expanded into power series in σ. With an accuracy to σ^4 this expansion has the form

$$F(\Delta,\sigma,Q,T) = N(0)h_0^2b^2\left(1-\frac{T}{\theta}+a^2Q^2+\frac{\pi^2\Delta}{2Qv_Fb^2}\right)\sigma^2 + \frac{1}{2}\left[\frac{S+1}{3S}g(S)\frac{T}{\theta} - \frac{\pi^2h_0^2}{16Qv_F\Delta b^2}\right]\sigma^4 - \frac{1}{2}N(0)\Delta^2\ln\frac{e\Delta_0^2}{\Delta^2} \quad . \tag{3.3.24}$$

Here we assumed

$$T \sim \theta \ll T_{c1} \quad , \quad Qv_F \gg h_0 \quad , \tag{3.3.25}$$

θ being the Curie temperature of the magnet when there is no interaction with the superconducting system, and $g(S)$ a numerical constant of the order of 1.

Minimization of the free energy with respect to the paramters Q and Δ leads to the equilibrium values

$$Q_0 = \left(\frac{\pi^2\Delta_0}{4v_Fa^2b^2}\right)^{1/3} \quad , \tag{3.3.26}$$

$$\Delta = \Delta_0 - h_0^2\sigma^2\left(\frac{\pi^2 a}{4v_F\Delta_0^2}\right)^{1/3} \quad ; \tag{3.3.27}$$

the quantity (3.3.26) for the wave vector of the helix here coincides with the estimate of the quantity (3.3.1) obtained by ANDERSON and SUHL. Substitution of (3.3.26,27) into (3.3.24) leads to an expansion in powers of the magnetic order,

$$F(\sigma,T) = -\frac{1}{2}N(0)\Delta_0^2 + \frac{T-T_M}{T_M}\sigma^2 + \frac{1}{2}\frac{S+1}{S}\,g(S)\left(1-\frac{\theta}{\theta^*}\right)\sigma^4 \quad , \tag{3.3.28}$$

where T_M and θ^* are

$$T_M = \theta(1-3a^2Q^2), \qquad \theta^* \sim \Delta_0\left(\frac{\Delta_0}{E_F}\right)^{1/3} \quad . \tag{3.3.29}$$

T_M is the temperature of transition to a magnetically ordered state and practically coincides with θ in view of the smallness of $a^2Q^2T_M$. The sign of the term $\sim\sigma^4$ determines the kind of phase transition to a magnetically ordered state. Thus, everything depends on the ratio of the two temperature parameters θ and θ^*. When $\theta<\theta^*$ the phase transition is of second order.

The temperature T_{c2} of the first-order phase transition from the MS phase to the normal ferromagnetic phase, and the jumps of the order parameters Δ and σ are determined from the equality of the free energies $F(\Delta,\sigma,Q,T)$ and $F(0,\sigma,0,T)$ of these phases. On the assumption that $v_FQ \gg h_0$ and $h_0 \gg \Delta$ one can obtain the following estimate for T_{c2}:

$$\frac{T_M - T_{c2}}{\theta} \sim \left(\frac{\theta}{T_{co}}\ln\frac{E_F}{\Delta_0}\right)^{-1} \quad , \tag{3.3.30}$$

provided that this quantity is much less than 1.

Thus, a simple theory which takes into account the variation of the spectrum of elementary excitations in the system due to the interaction with the magnetically ordered system and which ignores the effects of electron scattering on the magnetic system is capable of predicting the coexistence of superconductivity with helical magnetic ordering. This state is restricted by two temperatures, T_M and T_{c2}, of second-order and first-order phase transitions. These conclusions are essential for explaining the experimental situation in a number of rare-earth superconducting compounds. Let us now have a closer look at the experimental situation.

3.4 Magnetic Ordering in Superconducting Rare-Earth Compounds

Two groups of new superconducting compounds (Chap.7) have been synthesized and investigated recently in which the coexistence of superconductivity and magnetic ordering manifests itself definitely for the first time. Here belong molybdenum chalcogenides ($R\,Mo_6S_8$ and others) and rhodium borides ($R\,Rh_4B_4$) of rare-earth elements (R). In a number of compounds of this type, some phase transition to a magnetically ordered state occurs at low temperatures in the superconducting state region ($T < T_{c1}$). For antiferromagnetic ordering, which has been detected reliably in neutron diffraction experiments, the superconducting state is conserved at temperatures down to zero, but for ferromagnetic ordering it vanishes at some temperature T_{c2} which is somewhat lower than the superconducting transition temperature T_{c1}. Thus, in the narrow temperature interval $T_{c2} < T < T_M$ above the appearance of the ferromagnetic normal state of the metal, coexistence of the superconducting state with some magnetic order is observed. Neutron diffraction analysis shows that in this region a short-range magnetic order occurs with the characteristic wave vector of the fluctuating components of the order parameter. Let us first consider experimental evidence on rare-earth superconductors with a magnetic ordering of the antiferromagnetic and ferromagnetic type.

3.4.1 Superconductivity and Antiferromagnetism

Among the rare-earth molybdenum chalcogenides, substances have been found with an antiferromagnetic order in the superconducting state. The rare-earth molybdenum sulphides ($R_xMo_6S_8$ [3.52]) and selenides ($R_xMo_6Se_8$ [3.53]) with $1.0 < x < 1.2$ possess a rhombohedral crystal structure such as that of a distorted CsCl with complexes Mo_6S_8, Mo_6Se_8 at the vertices and a rare-earth ion at the centre of the cell [3.54]. According to [3.52-56], with the exception of R = Ce, Eu all the other rare-earth chalcogenides are superconducting. Experimental investigation of the superconducting properties of rare-earth molybdenum sulphides [3.52] has shown that it is the 4d-electrons of the molybdenum atoms which are responsible for the superconductivity in sulphides. The indirect exchange interaction of localized magnetic moments of 4f rare-earth ions occurs through the 5d and 6s-electron bands of rare-earth ions which have a large overlap of wave functions with 4f-electrons. The 4d-electrons of the molybdenum atoms also interact with the localized magnetic moments, but this interaction is weak since the overlap of the wave functions of the 4d-electrons of molybdenum with those of the 4f-electrons of localized rare-earth ions is very small. Therefore the contribution of superconducting electrons to the formation of the exchange interaction between rare-earth magnetic moments turns out to be small in these compounds. Although no similar results have been obtained for selenides, the similarity of the electronic properties of selenides and sulphides suggests that the present conclusion will also hold for rare-earth molybdenum selenides.

The first superconducting compound in which antiferromagnetic ordering was assumed to exist was $Gd_{1,2}Mo_6Se_8$ [3.57]. The thermal capacity was found to have a λ-type anomaly at the temperature of 3.5 K, below the temperature of transition to the superconducting state. However, when measuring the magnetic susceptibility of $Gd_{1,2}Mo_6Se_8$ [3.58] at about 3.5 K no traces of magnetic ordering were observed, and at 0.7 K the minimum of the reversible susceptibility was found. As this compound was studied further [3.59], another peak in the thermal capacity was registered in the vicinity of 0.7 K. This peak agrees with the magnetic transition resulting from magnetic susceptibility measurements (Fig.3.9). It was also observed that the major

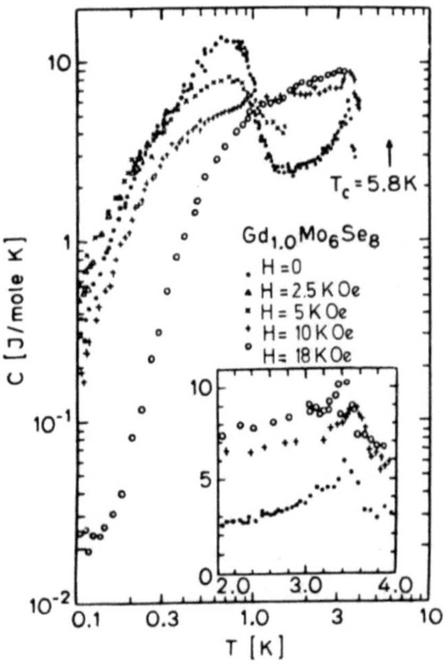

Fig.3.9. The temperature dependence of the heat capacity of $GdMo_6Se_8$ for different values of the magnetic field [3.59]

decrease in the entropy comes about at a temperature below $\sim$ 2 K and is due to a peak in the thermal capacity at 0.75 K. This is indicative of the ordering of Gd^{3+} ions in the superconducting state. In the same investigation, heat capacity measurements with applied magnetic fields were made. The results of these measurements show that the thermal-capacity peak at the temperature of 0.75 K in up to 5 kOe fields decreases and widens. When higher fields are applied, the peak broadens and shifts towards higher temperatures. Such behaviour is typical of an antiferromagnet, therefore these results as well as the presence of a minimum in the reversible susceptibility at $\sim$ 0.7 K are due to the occurrence of antiferromagnetic ordering in the superconducting state of $GdMo_6Se_8$. The nature of the peak in the heat capacity at about 3.5 K in this compound is not clear at present. It should be noted that magnetic measurements in the superconducting state of many rare-earth molybdenum chalcogenides are feasible because these compounds are extreme second-order superconductors for which the magnetic field penetrates almost entirely into the specimen. On the basis of measurements of similar physical quantities, superconductivity and antiferromagnetism were found to coexist in compounds of $R_xMo_6Se_8$ ($x = 1.0$ or 1.2) with R = Tb, Er [3.60].

Investigating the temperature dependence of the upper critical field in $R_{1,2}Mo_6S_8$ with R = Tb,Dy,Er, ISHIKAWA and FISHER [3.61] have found that a magnetic ordering arises in these substances in the superconducting state. The magnetic transition in the superconducting state of $YbMo_6S_6$ has been detected in [3.63].

An attempt has been made recently to study the magnetic structure of $Er_xMo_6Se_8$ [3.62,64] using neutron diffraction technique. It turns out that at present it is impossible to decipher the neutron diffraction pattern of this compound and to determine the type of magnetic structure. Success has been made in the neutron diffraction study of $DyMo_6S_8$. This substance becomes superconducting at $T_c = 2.05$, and at $T = 0.4$ K a magnetic transition has been detected. From neutron diffraction analyses the atoms of Dy have been found to form a simple cubic sublattice with an inappreciable rhombohedral distortion; the magnetic structure consists of (100) planes with magnetic moments $8.77\mu_B$ in magnitude which are parallel and antiparallel to the [111] direction.

A neutron diffraction study of the coexistence of magnetic ordering and superconductivity has also been made recently for the compounds $Dy_{1,2}Mo_6S_8$ and $Tb_{1,2}Mo_6S_8$ [3.65]. Both compounds become superconducting at $T_c = 2.05$ K. At the temperature $T_M = 0.4$ K for $Dy_{1,2}Mo_6S_8$ and at $T_M = 1.0$ K for $Tb_{1,2}Mo_6Se_8$ a long-range antiferromagnetic order arises which does not destroy superconductivity.

3.4.2 Superconductivity and Ferromagnetism

In two compounds, viz., in $ErRh_4B_4$ [3.66] from the group of rare-earth rhodium borides, as well as in $Ho_{1,2}Mo_6S_8$ [3.67], another superconducting transition temperature has been found at which the metal passes to a normal state with ferromagnetic ordering of magnetic moments of rare-earth atoms. Ternary compounds RRh_4B_4 possess a tetragonal crystal structure in which the rare-earth ions are arranged as a body-centred tetragonal lattice [3.68]. The electron structure of $ErRh_4B_4$ has been calculated in [3.69] where it is shown that, again, it is the electrons of the 4d-states of the rhodium atoms (which make the dominant contribution to the density of states on the Fermi level) that appear to be responsible for the superconducting state, and the magnetic ordering of localized rare-earth moments produced by electrons of 4f-states occurs due to the indirect exchange with 4d-electrons of rare-earth atoms.

The superconducting and magnetic properties of rare-earth rhodium borides are highly unusual. Figure 3.10 presents superconducting transition temperatures T_c and magnetic ordering temperatures T_M for various RRh_4B_4 compounds [3.66]. The linear extrapolation of the Curie temperatures for R = Dy,Ho towards Er, which is a superconductor at 8.7 K, enables us to assume that the magnetic moments of Er^{3+} ions may be in the ordered and in the superconducting state simultaneously. However, magnetic susceptibility and resistivity measurements [3.70] have shown that the superconducting state (T_c = 8.7 K) collapses at a temperature below the second superconducting transition temperature,

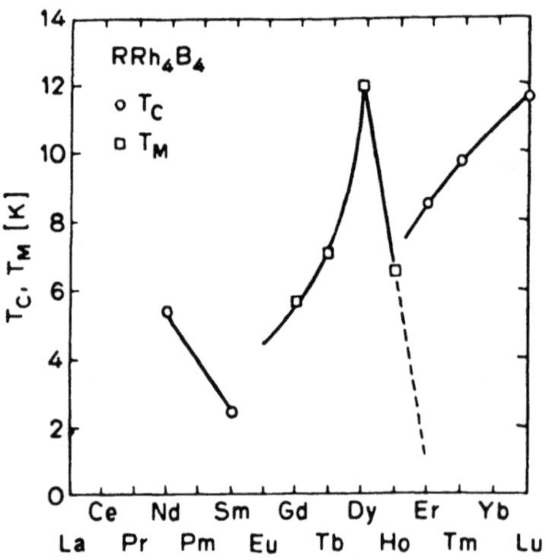

Fig.3.10. The superconducting transition temperature T_c and the magnetic ordering temperature T_M of RRh_4B_4 compounds [3.66]

$T_{c2} = 0.93$ K. At T_{c2} the susceptibility increases in a jump and then drops as the temperature decreases. This behaviour of the susceptibility indicates that the transition from a superconducting state to a normal state at T_{c2} is accompanied by a transition to a magnetically ordered state. Resistivity data are also indicative of a transition to a normal state at T_{c2}, the resistivity below T_{c2} appearing to be much less than that of a normal metal at a temperature above T_c. This fact is due to the presence of small superconducting regions. When the magnetic fields are sufficiently high this superconductivity is suppressed, and the resistivity increases to the value observed in a normal metal at a temperature above T_{c1}. Another interesting feature of the resistivity is the presence of hysteresis phenomena at T_{c2}, which occur when carrying out measurements with decreasing and increasing temperature.

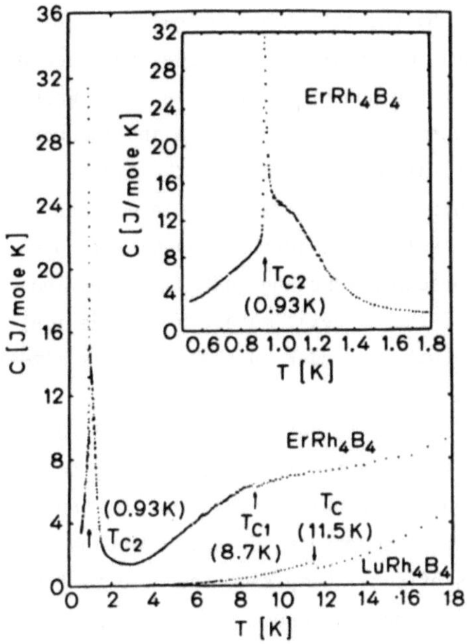

Fig.3.11. The temperature dependence of the heat capacity of $ErRh_4B_4$ [3.71]

In Fig.3.11 a graph is given showing the temperature dependence of the heat capacity of $ErRh_4B_4$ in comparison with the heat capacity of a nonmagnetic superconducting $LuRh_4B_4$ [3.71]. At $T_c = 8.7$ K a jump in thermal capacity takes place which is due to the transition to a superconducting state, and at a lower temperature when $T_{c2} = 0.93$ K a λ-type anomaly has been detected. According to

[3.56,72,73], the magnetic contribution to the heat capacity at low temperatures is due to a Schottky anomaly with a peak in the vicinity of 11 K, in addition to the λ-type anomaly which is associated with the ordering of magnetic Er^{3+} ions.

Early neutron scattering experiments with $ErRh_4B_4$ [3.74] have shown that the destruction of superconductivity at $T_{c2} = 0.93$ K is due to the occurrence of ferromagnetic order of the Er^{3+} ions. The magnetic transition in this case is a phase transition of the second kind, but it is preceded by a strong pretransition scattering at a temperature higher than 1.2 K. The value obtained for the magnitude of the magnetic moment of a Er^{3+} ion is as low as to 5.6 μ_B and differs from the magnetic moment of a free ion, which is 9.58 μ_B. From Mössbauer effect measurements in $ErRh_4B_4$, the value of the magnetic moment turned out to be equal to 8.4 μ_B [3.75]. This difference in the magnitude of the magnetic moment may be accounted for by the fact that in neutron scattering one measures long-range paired spin correlations. One does not succeed in measuring the disordered components of the moment (if these exist), whereas the Mössbauer effect measures the properties of a single ion and thereby detects the total angular momentum. Thus, it may be said that the magnetic moment in $ErRh_4B_4$ below T_M is not yet completely ordered. This incomplete ordering may be due to superconducting fluctuations in the magnetically ordered state. Neutron diffraction in a nonsuperconducting $HoRh_4B_4$ yields about 8.7 μ_B (a value close to the magnitude of the magnetic moment of a free ion) for the magnetic moment. This indicates that the order in the system is almost complete.

Let us now recall that according to resistivity data for $ErRh_4B_4$ [3.70] the superconducting transition is hysteretic and the magnetic ordering determined by neutron scattering is a second-order phase transition. On the other hand, thermal expansion coefficient measurements for this compound [3.76] show that there are no hysteresis effects. Comparing the results for magnetic susceptibility, thermal expansion and heat capacity, it may be assumed that the temperature of magnetic ordering T_M does not coincide with the superconducting transition temperature T_{c2} but is approximately 0.05 K higher than T_{c2}, so that the $ErRh_4B_4$ compound has a narrow region between T_{c2} and T_M for the occurrence of superconductivity and ferromagnetism. To establish this fact, a more detailed measurement of the temperature dependence of the heat capacity (Fig.3.11) was carried out [3.71]. A most interesting feature of this dependence is an acute λ shaped peak 70 mK wide, which is observed at $T = 0.93$ K. At the foot of this peak lies a wider heat capacity maximum originating at ~ 1.5 K. This is in agreement with neutron diffraction results [3.74] where a pretransition scattering starting at ∿ 1.2 K was observed. It is seen from Fig.3.11 that the heat capacity graph is asymmetric in either direction from the peak when T = 0.93 K. From the low-temperature side the specific heat decreases linearly, starting from about 0.90 K. On the other side, the slope of the curve decreases at T = 1.1 K with decreasing temperature. A λ peak arises at a temperature of ∿ 0.98 K. This is the same temperature as that at which the anomaly in the thermal expansion coefficient is observed. According to the conclusion drawn in [3.76], this anomaly arises directly at the magnetic transition. On the other hand, the sharp peak in the specific heat at T = 0.93 K occurs at the same temperature as that at which the transition from a normal state to a superconducting one takes place according to measurements of the magnetic susceptibility with increasing temperature [3.76]. Thus, the turning point of the heat capacity at the temperature of 0.98 K may be associated with the magnetic ordering temperature, whilst the heat capacity peak at 0.93 K occurs at the transition of the superconductor to a normal state at $T = T_{c2}$.

For further study of the superconducting transition at T_{c2}, quasiternary compounds $(Er_{1-x}Ho_x)Rh_4B_4$ [3.77] were considered. The phase diagram of these compounds is given in Fig.3.12. An essential feature of the diagram is the presence of a critical concentration x_c beyond which the superconductivity disappears completely. The temperature interval between T_{c1} and T_{c2} contains the temperature T_M, but it is not clear at the moment whether there is a region of coexistence of superconductivity and magnetic ordering here. In [3.77]

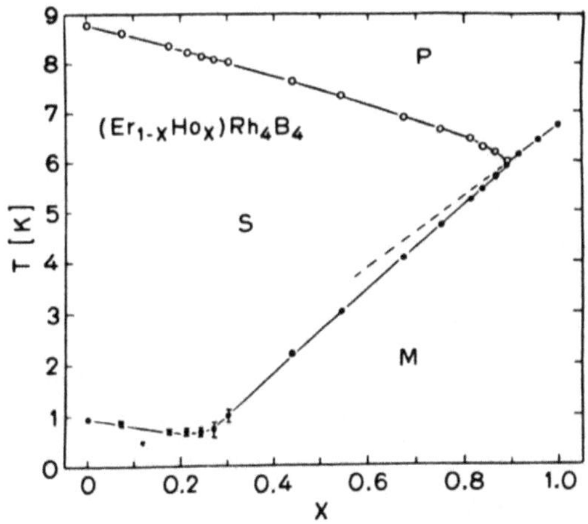

Fig.3.12. The phase diagram for $(Er_{1-x}Ho_x)Rh_4B_4$ [3.77]

heat capacity measurements were made for a number of $(Er_{1-x}Ho_x)Rh_4B_4$ compounds. With $x = 0.813$ being somewhat lower than the critical concentration $x_c = 0.89$, a sharp peak arises in the temperature dependence of the heat capacity on the characteristic stepped increment. The occurrence of an anomaly in the heat capacity at T_{c2} was noted also in $(Lu_{1-x}Ho_x)Rh_4B_4$ [3.72] and $Ho_{1,2}Mo_6S_8$ [3.78] systems. Another important feature of the superconductor-normal metal transition, as pointed out, is the presence of a hysteresis in the resistivity and magnetic susceptibility [3.70]. This means that the transition at T_{c2} is a first-order transition.

In addition to this, to ascertain the nature of the occurrence of a magnetic order in $ErRh_4B_4$ a new investigation of the critical magnetic scattering of neutrons was undertaken [3.79]. Using small-angle neutron scattering to study the fluctuation effects in the vicinity of the superconductor-ferromagnet transition, it was found that magnetization fluctuations exist with a finite wave vector for which the corresponding wavelength is about 100 Å.

We have considered above the properties of a $ErRh_4B_4$ compound. It may now be seen from Fig.3.10 that the extrapolation of the Curie temperature from Gd towards Sm and Nd also admits the possibility that superconductivity and magnetism coexist in RRh_4B_4 compounds with R = Sm, Nd [3.80].

There is one more compound which possesses two superconducting transition temperatures. This is $HoMo_6S_8$ [3.67] with $T_{c1} = 1.2$ K and $T_c = 0.64$ K. Neutron diffraction results for this compound give a simple ferromagnetic structure with magnetic moments aligned with the rhombohedral [111] axis. The value found for the magnetic transition temperature from the temperature dependence of the magnetic intensity is $T_M = 0.67$ K. Measurements of the heat capacity in the compound $Ho_{1,2}Mo_6S_8$ at low temperatures [3.78], just as for $ErRh_4B_4$ [3.77], have shown that at the lower superconducting transition temperature T_{c2} a sharp peak, which may be due to the accurrence of a long-range order of magnetic moments in Ho^{3+}, is superimposed on the heat capacity anomaly. Besides, there also occurs a substantial hysteresis of the magnetic susceptibility at T_{c2} [3.78]. Having analyzed the temperature dependence of the thermal capacity near T_{c2}, WOOLF et al. [3.78] have arrived at the conclusion that in $Ho_{1,2}Mo_6S_8$ there is no region of coexistence of superconductivity and ferromagnetism, whereas in earlier papers [3.67,81] it was assumed to exist.

To conclude this section, we summarize the experimental results on $ErRh_4B_4$. At $T_{c1} = 8.7$ K this compound passes into a superconducting state. At a temperature below $T_M = 0.98$ K a second-order phase transition to some magnetically ordered state takes place (with the conservation of superconductivity) characterized by short-range order with a wave vector of about 100 Å. At $T_{c2} = 0.93$ K

$ErRh_4B_4$ passes from a superconducting state to a normal ferromagnetic state in a phase transition of the first kind. Thus, between T_{c2} and T_M superconductivity coexists with magnetic ordering.

Table 3.1 lists the currently available experimental results on rare-earth compounds in which the magnetic atoms are ordered at a temperature T_M which is lower than the superconducting transition temperature T_{c1}. The experimental facts presented here require an explanation which one might try to give on the basis of the Ginzburg-Landau phenomenological theory.

Table 3.1. Experimental data on coexcistence of superconducting and magnetic order

	Substance	T_{c1} [K]	T_{c2} [K]	T_M [K]	Type of magnetic structure	References
1	$ErRh_4B_4$	8.7	0.93	0.98	FM	[3.70,76,74,71]
2	$HoMo_6S_8$, $Ho_{1.2}Mo_6S_8$	1.2	0.64	0.67	FM	[3.81,67]
3	$ErMo_6Se_8$, $Er_{1.2}Mo_6Se_8$	5.7	1.07	1.07	?	[3.60,62]
4	$DyMo_6S_8$, $Dy_{1.2}Mo_6S_8$	2.05		0.4	AFM	[3.61,64,65]
5	$TbMo_6S_8$, $Tb_{1.2}Mo_6S_8$	2.05		1.0	AFM	[3.61,64,65]
6	$Gd_{1.0}Mo_6Se_8$	5.8		0.75	AFM	[3.59]
	$Gd_{1.2}Mo_6Se_8$	5.5		0.7		[3.57]

3.4.3 Phenomenological Theory of Magnetic Ordering in Superconductors

We first explore the problem concerning the occurrence of magnetic ordering within the superconducting state. We shall describe a magnetic superconductor in the vicinity of phase transition points by a set of two coupled order parameters, viz., by the superconducting pairing parameter $\Delta(\underline{r})$ and the magnetization $M(\underline{r})$. The expression for the free energy of such a system is of the form

$$F = \int d\underline{r}\left[\frac{c}{2}(\nabla M)^2 + \frac{a}{2}M^2 + \frac{b}{4}M^4 + \gamma|\nabla\Delta|^2 + \alpha|\Delta|^2 + \frac{\beta}{2}|\Delta|^4 + \frac{\eta}{2}|\Delta|^2 M^2\right] \quad . \tag{3.4.1}$$

In this expression $a = a_0(T - T_M^0)$, $\alpha = \alpha_0(T - T_c)$, where T_M^0 is the temperature of magnetic ordering, T_c is the superconducting transition temperature in the absence of a coupling of the superconducting and magnetic order parameters, and a_0 and α_0 are constant quantities. Strictly speaking, (3.4.1) only works well provided that the temperatures T_M^0 and T_c are close to each other, i.e., when both order parameters $\Delta(\underline{r})$ and $M(\underline{r})$ are small, but the results usually hold true qualitatively in some temperature interval. It will be seen that the order parameter coupling quantity η should be positive, which reflects the antagonistic nature of the order parameters in the coexistence of superconductivity and magnetic ordering. Namely, at a fixed magnetization $M(\underline{r})$ the superconducting transition temperature T_c decreases, according to the condition $\alpha \to \alpha + \eta M^2/2$, and for a given value of the pairing parameter $\Delta(\underline{r})$ the magnetic ordering temperature T_M decreases in virtue of a similar condition $a \to a + \eta|\Delta|^2/2$. The coefficients b and β are positive quantities, too.

From the condition for the minimum, $\delta F = 0$, we obtain the following Ginzburg-Landau equations for the system of coupled order parameters

$$\left.\begin{aligned} -c\underline{\nabla}^2 M + aM + bM^2 M + \eta|\Delta|^2 M &= 0 \\ -\gamma\underline{\nabla}^2\Delta + \alpha\Delta + \beta|\Delta|^2\Delta + \frac{1}{2}\eta M^2\Delta &= 0 \end{aligned}\right\} \quad . \tag{3.4.2}$$

We first consider the simplest case of a homogeneous system where the order parameters Δ and M are independent of the coordinates. In this case the Ginzburg-Landau equations (3.4.2) reduce to the following expressions

$$\left.\begin{aligned} aM + bM^2 M + \eta|\Delta|^2 M &= 0 \\ \alpha\Delta + \beta|\Delta|^2\Delta + \frac{1}{2}\eta M^2\Delta &= 0 \end{aligned}\right\} \quad . \tag{3.4.3}$$

The combined equations (3.4.3) are capable of four solutions. One solution describes a high-symmetry phase with $M = 0$ and $\Delta = 0$. Two other solutions correspond to a high-symmetry phase with respect to one of the order parameters and to a condensed phase with respect to the other, i.e., correspond to a purely superconducting phase with $M = 0$, $\Delta^2 = -\alpha/\beta$ and to a purely magnetic one with $\Delta = 0$, $M^2 = -a/b$. Finally, there is a solution where both order parameters are nonzero, i.e.,

$$M^2 = \frac{-a\beta + \alpha\eta}{b\beta - \eta^2/2} \quad , \qquad \Delta^2 = \frac{-\alpha b + a\eta/2}{b\beta - \eta^2/2} \quad . \tag{3.4.4}$$

We shall call this solution the MS phase. It is our aim to determine the conditions for the existence of this phase. In order to find the conditions under which the MS phase is stable with respect to the superconducting and magnetic phases it is necessary to compare the values of the free energies in these phases. In the general case, such an analysis for a system with coupled order parameters has been made in [3.82] (see also [3.83]).

The free energy of the MS phase, F_{MS}, is expressed in terms of the free energies of the magnetic phase $F_M = -a^2/4b$, and of the superconducting phase $F_S = -\alpha^2/2\beta$, as follows

$$F_{MS} = F_M - \frac{1}{2b}\frac{(\alpha b - a\eta/2)^2}{b\beta - \eta^2/2} = F_S - \frac{1}{4\beta}\frac{(a\beta - \alpha\eta)^2}{b\beta - \eta^2/2} \quad . \tag{3.4.5}$$

It is easy to see here that the free energy of the MS phase will be less than the free energies of the magnetic and superconducting phases provided that the condition

$$\frac{1}{2}\eta^2 < b\beta \tag{3.4.6}$$

is satisfied. This is the condition for the stability of the MS phase or, in other words, for the simultaneous coexistence of superconductivity and magnetic ordering.

When (3.4.6) holds true, the phase transitions between all the phases are transitions of the second kind. They occur on lines which may be found from the free-energy minimum condition. As a result, the high-symmetry phase appears to be stable when $a > 0$ and $\alpha > 0$. On the lines determined by

$$\alpha\eta = ab \quad , \quad \frac{1}{2}a\eta = \alpha b \quad , \tag{3.4.7}$$

second-order phase transitions take place from the superconducting phase to the intermediate MS phase and from the magnetic phase to the MS phase.

If the inequality (3.4.6) becomes opposite, i.e.,

$$\frac{1}{2}\eta \geq b\beta \quad , \tag{3.4.8}$$

there arises no MS phase, and a first-order phase transition takes place between the magnetic and superconducting phases. This transition occurs on a line [3.83] which may be found from the equality of the free energies of the superconducting and magnetic phases,

$$\sqrt{\beta}a = \sqrt{2b}\alpha \quad . \tag{3.4.9}$$

Thus, depending on the magnitude of the coupling constant η of the order parameters in a homogeneous magnetic superconductor, the MS phase, which describes the phenomenon of coexistence of superconductivity and ferromagnetic ordering, may emerge only in the case of a sufficiently weak coupling η. Then condition (3.4.6) holds true and all the phase transitions being second-order transitions. In the opposite case, when the inequality (3.4.8) is valid, i.e., when the coupling η is strong, the region of coexistence is absent and the transition between the superconducting and magnetic phases will be a transition of the first kind. After having appropriately estimated the values of the quantities involved in the inequality (3.4.6), it was shown in [3.84] that there is no region of coexistence of superconductivity and ferromagnetism in compounds of rare-earth elements, since the inequality $J(0) \gg T_c$ [where $J(0)$ is the Fourier component of the exchange interaction between localized 4f electrons of rare-earth ions and conduction electrons] is satisfied in these substances [the coupling paramter η is expressed in terms of $J(0)$].

Let us now return to the Ginzburg-Landau combined equations (3.4.2). It turns out that among the possible solutions of this system there exists one more simple solution when the superconducting pairing parameter is a quantity independent of the coordinates and the magnetization varies in space according to a simple helix, i.e.,

$$\Delta = \text{const} \quad ; \quad \underline{M} = (M\cos \underline{Q}\underline{r},\ M\sin \underline{Q}\underline{r},\ 0) \quad . \tag{3.4.10}$$

This solution corresponds to the state which Suhl and Anderson have called "cryptoferromagnetic" (Sect.3.3).

After substitution of (3.4.10) into the Ginzburg-Landau equations (3.4.2) we obtain the following equations for the MS phase of interest

$$\left.\begin{aligned} a + cQ^2 + bM^2 + \eta|\Delta|^2 &= 0 \\ \alpha + \beta|\Delta|^2 + \frac{1}{2}\eta M^2 &= 0 \end{aligned}\right\} \quad . \tag{3.4.11}$$

The solution of these combined equations may be represented in a form similar to (3.4.4), except for the replacement of $a \to a + cQ^2$. The expression for the free energy in the "cryptoferromagnetic" state is

$$\begin{aligned} F_{MS} &= -\frac{(a + cQ^2)^2}{4\beta} - \frac{1}{2b}\,\frac{[\alpha b - (a + cQ^2)\eta/2]^2}{b\beta - \eta^2/2} \\ &= -\frac{\alpha^2}{2\beta} - \frac{1}{4\beta}\,\frac{[(a + cQ^2)\beta - \alpha\eta]^2}{b\beta - \eta^2/2} \quad , \end{aligned} \tag{3.4.12}$$

whence it follows that the condition of stability of the intermediate MS phase proves to be the same as (3.4.6). The equilibrium value of the wave vector Q_0 which determines the helical structure may be found from the free-energy minimum of the "cryptoferromagnetic" state with respect to Q.

As a result, we find that for the free energy given by (3.4.1) the quantity Q_0 is equal to zero. The nonzero value of Q_0 can be obtained if we take into account the nonlocal character of the superconducting order parameter [3.85] (Sect.7.5).

It should be borne in mind here that when effecting the exchange interaction of localized magnetic moments the superconducting electrons participate directly in the production of the Anderson-Suhl "cryptoferromagnetic" state. However, the situation seems to be different in compounds such as rare-earth rhodium borides and molybdenum chalcogenides with which we are concerned. We have presented above the results of numerical band-structure calculation for these compounds which have shown that the contribution of superconducting electrons to the exchange interaction of localized magnetic moments may be neglected. This means that the direct relationship between the magnetization and the superconducting order parameter, which is expressed by the term with the coefficient η in (3.4.1), appears to be insignificant. Nevertheless, the magnetization and the superconducting order parameter are interrelated through the electromagnetic field since the magnetic induction of a superconductor interacts with the magnetization of localized magnetic moments. BLOUNT and VARMA [3.86], SUHL [3.87] and MATSUMOTO et al. [3.88] have considered the effects of the interaction between the magnetization, the electromagnetic field (described by the vector potential A) and the superconducting order parameter. Their analysis is based on the use of the Ginzburg-Landau Hamiltonian with the addition of terms which are due to the electromagnetic field

$$F(\Delta,\underline{M},\underline{A}) = \int d\underline{r}\Big[\frac{1}{2}\tilde{c}\,(\underline{\nabla}\underline{M})^2 + \frac{1}{2}\tilde{a}\,\underline{M}^2 + \frac{1}{4}b\,(\underline{M}^2)^2 + \frac{\underline{B}^2}{8\pi} - \underline{B}\,\underline{M} + \gamma|(\underline{\nabla} - ip_0\underline{A})\Delta|^2 + \alpha|\Delta|^2 + \frac{1}{2}\beta|\Delta|^4\Big] \quad , \tag{3.4.13}$$

where the notation

$$p_0 = 2e/\hbar c \quad , \quad \gamma = \hbar^2/2m \quad , \quad \tilde{a} = a + \eta_1|\Delta|^2 \quad , \quad \tilde{c} = c + \eta_2|\Delta|^2 \tag{3.4.14}$$

is introduced. The terms with the coefficients η_1 and η_2 in (3.4.13) describe the coupling between the superconducting and magnetic order parameters which arises because of the s-d exchange interaction of electrons and localized moments and is assumed to be insignificant in the compounds under consideration. To expression (3.4.13) one has to add

$$\underline{B} = \mathrm{rot}\,\underline{A} \quad , \tag{3.4.15}$$

where for the vector potential $\underline{A}(\underline{r})$ it is convenient to choose the Coulomb gauge

$$\underline{\nabla}\underline{A} = 0 \quad . \tag{3.4.16}$$

We shall not explore the Ginzburg-Landau equations which may be obtained by proceeding from the free energy (3.4.13), as was the case in the work by BLOUNT and VARMA [3.86]. We rather integrate the expression for the statistical sum for the functional $F[\Delta,\underline{A},\underline{M}]$, (3.4.13), over the variable $\underline{A}(\underline{r})$ [3.51]. In this manner we shall obtain the expression for the free energy $F[\Delta,\underline{M}]$

$$\exp\left\{-\frac{1}{T}F[\Delta,\underline{M}]\right\} = \int \mathcal{D}\underline{A}\,\exp\left\{-\frac{1}{T}\,F[\Delta,\underline{M},\underline{A}]\right\} \tag{3.4.17}$$

which determines the effective Hamiltonian of a magnetically ordered superconductor with due regard for the fluctuations of the electromagnetic field. Taking into account the explicit form of the functional $F[\Delta,\underline{M},\underline{A}]$, the expression for the free energy $F[\Delta,\underline{M}]$ becomes

$$\begin{aligned} F[\Delta,\underline{M}] &= \int d\underline{r}\left[\frac{1}{2}\tilde{c}(\underline{\nabla}\underline{M})^2 + \frac{1}{2}\tilde{a}\,\underline{M}^2 + \frac{b}{4}(\underline{M}^2)^2 + \alpha|\Delta|^2 + \frac{1}{2}\beta|\Delta|^4\right] \\ &- T\ln\int \mathcal{D}\underline{A}\,\exp\left\{-\frac{1}{T}\int d\underline{r}\left[\gamma|(\underline{\nabla} - ip_0\underline{A})\Delta|^2 + \frac{\underline{B}^2}{8\pi} - \underline{B}\,\underline{M}\right]\right\} \quad . \end{aligned} \tag{3.4.18}$$

For simplicity, we consider the case where the superconducting order parameter does not depend on the coordinates, i.e., $\Delta = \mathrm{const}$. This just means that we are not interested in the fluctuations of the parameter Δ. Then the functional integral over the variable $\underline{A}$ can be calculated, and in the case when the magnetization $\underline{M}_K$, in the momentum representation, satisfies the relation $(\underline{K}\,\underline{M}_K) = 0$ (only the transverse components of the magnetic order parameter are taken into consideration), the integral becomes

$$\begin{aligned} &\int \mathcal{D}\underline{A}\,\exp\left\{-\frac{1}{T}\int d\underline{r}\left[\gamma p_0^2|\Delta|^2\underline{A}^2 + \frac{1}{8\pi}(\mathrm{rot}\,\underline{A})^2 - \underline{M}\cdot\mathrm{rot}\,\underline{A}\right]\right\} \\ &\quad = \exp\left[\frac{V}{(2\pi)^3}\int d\underline{K}\,\ln\frac{8\pi^2 T}{K + K_s^2} + \frac{V}{(2\pi)^3}\int d\underline{K}\,\frac{4\pi K^2}{K^2 + K_s^2}|M_{\underline{K}}|^2\right] \quad , \end{aligned} \tag{3.4.19}$$

where

$$K_s^2 = 8\pi p_0^2|\Delta|^2 \tag{3.4.20}$$

is the reciprocal London depth of penetration of the magnetic field, and V is the volume. Substituting (3.4.19) into (3.4.18) and passing to the momentum representation, we obtain the following expression for $F[\Delta,\underline{M}]$:

$$F[\Delta,\underline{M}] = V\left[\alpha|\Delta|^2 + \frac{1}{2}\beta|\Delta|^4 - \frac{T}{(2\pi)^3}\int d\underline{K}\ \ln\left(\frac{8\pi^2T}{K^2+K_s^2}\right)\right] + \frac{V}{2(2\pi)^3}\int d\underline{K}\chi^{-1}(\underline{K})|M_{\underline{K}}|^2$$
$$+ b\frac{V}{4(2\pi)^4}\int d\underline{K}_1\ldots d\underline{K}_4\,\delta\left(\sum_{i=1}^{4}\underline{K}_i\right)\left(\underline{M}_{\underline{K}_1}\underline{M}_{\underline{K}_2}\right)\left(\underline{M}_{\underline{K}_3}\underline{M}_{\underline{K}_4}\right) \quad . \tag{3.4.21}$$

Here we have introduced the notation for the reversible susceptibility

$$\chi^{-1}(\underline{K}) = \tilde{a} + \tilde{c}K^2 - \frac{4\pi K^2}{K^2+K_s^2} \quad . \tag{3.4.22}$$

It can be readily seen from (3.4.19,21) that the contribution of the electromagnetic field fluctuations leads to two effects. Firstly, if the magnetization is not taken into account, an additional term appears in the free energy of the superconductor. This term, as can be shown, leads to a change of the kind of phase transition. If this additional term is not allowed for, the transition to a superconducting state is a second-order phase transition, and, when this term is taken into account, the transition of second order becomes a first-order transition. This effect was first predicted by HALPERIN et al. [3.89]. In the following treatment we shall discard this contribution which is not essential for our purposes and is described by the first term in the exponent of (3.4.19). Secondly, in the magnetic part of the free energy there occurs an additional term $\sim|\underline{M}_{\underline{K}}|^2$. It turns out that this term plays an essential part, leading to the occurrence of a magnetic structure with a nonzero wave vector in the superconducting state. This can be readily verified by investigating the minimum of $\chi^{-1}(\underline{K})$ with respect to the variable $\underline{K}$. As a result, the value of the wave vector of the magnetic structure will be

$$K_0^2 = \sqrt{\frac{4\pi}{\tilde{c}}}K_s - K_s^2 \quad . \tag{3.4.23}$$

At its minimum $\chi^{-1}(\underline{K})$ assumes the value

$$\chi^{-1}(K_0) = \tilde{a} - 4\pi\left(1 - \sqrt{\frac{\tilde{c}}{4\pi}}K_s\right)^2 \quad . \tag{3.4.24}$$

The temperature at which the magnetic susceptibility $\chi^{-1}(K_0)$ goes to zero is the temperature of transition T_M to a state with a spatially homogeneous pairing parameter and with an oscillating magnetic ordering of the atoms. This temperature is determined by the following expression

$$\frac{\tilde{T}_M - T_M}{\mathcal{T}_M} \approx 4\sqrt{\pi}\frac{K_s\sqrt{\tilde{c}}}{a_0 + 4\pi} \quad , \tag{3.4.25}$$

where

$$\mathcal{T}_M = \left(1 + \frac{4\pi}{a_0}\right)T_M^0 \tag{3.4.26}$$

is the temperature of transition to a magnetic state with $B = 4\pi M$ and the free energy is $F_M = -(a - 4\pi)^2/4b$.

Next, we examine the nature of superconducting phase at temperatures below the transition to a magnetically ordered state. Since the configuration of magnetic moments is most likely polarized circularly according to (3.4.10), with constant M^2, the overall energy appears to be equal to

$$F_{MS} = -\frac{1}{4b}\chi^{-2}(K_0) + \alpha|\Delta|^2 + \frac{1}{2}\beta|\Delta|^4 \quad . \tag{3.4.27}$$

In order to estimate the magnitude of the free energy we insert $|\Delta|^2 = -\alpha/\beta$ and get

$$F_{MS} = -\frac{1}{4b}\chi^{-2}(K_0) - \frac{1}{2\beta}\alpha^2 \quad . \tag{3.4.28}$$

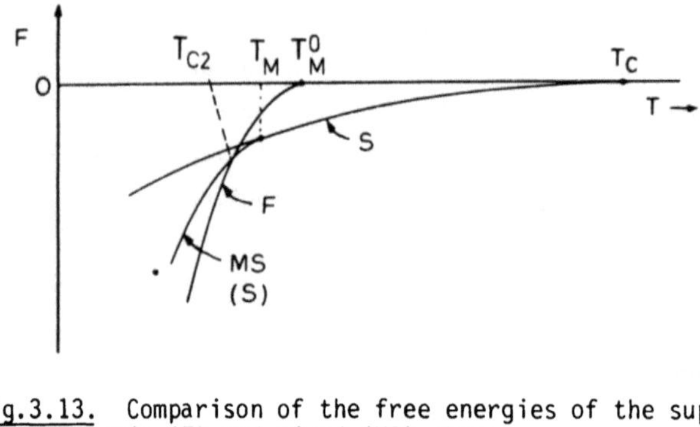

Fig.3.13. Comparison of the free energies of the superconducting (S), ferromagnetic (F) and mixed (MS) phases

Figure 3.13 schematically compares the values of (3.4.28) for the free energies of the MS phase, the purely superconducting state ($F_S = -\alpha^2/2\beta$), and the purely magnetic state ($F_M = -(a - 4\pi)^2/4b$ [3.86]. As is seen from Fig.3.13, when the intermediate phase is not taken into account the transition from a superconducting phase to a magnetic phase is of first order. When the temperatures T_M and $\tilde{T}_M$ are close enough to each other an intermediate helical state with wave vector K_0 is produced in a superconductor. This state arises by a second-order phase transition and then, through a first-order phase transition, transforms into a ferromagnetic normal phase.

In the present treatment we have completely neglected the coupling between the superconducting and magnetic order parameters, which is due to the presence of the s-d exchange interaction. Consideration of such terms will render the picture very complicated. We mention also that in the case of a strong s-d exchange coupling, neglecting the effects of the electromagnetic field, there arises the situation described by ANDERSON and SUHL [3.48], which probably has no bearing on the experimental situation in the substances considered. A self-consistent field theory of ferromagnetic superconductors has also been constructed in [3.90].

The results obtained in [3.48,86] have recently been united by FERREL et. al. [3.91] in their calculation of the reciprocal magnetic susceptibility in magnetic superconductors. In (3.4.22) for the reciprocal susceptibility, which is the coefficient of the quadratic term in the Ginzburg-Landau functional, a term $\sim 1/K$ appears due to the s-d exchange interaction. It is also shown in that paper that BLOUNT and VARMA's use of the local London approximation does not suffice; it is necessary to take into account nonlocal effects which alter the form of the last term in (3.4.22). When renormalized with respect to the fluctuations of the electromagnetic field and the s-d exchange coupling, the magnetic susceptibility has a maximum at a finite value of K. Such a behaviour of the neutron-scattering cross-section in a region above T_{c2} has been found in [3.79,81].

3.4.4 Magnetic Fluctuations and Superconductivity

We have discussed the contribution of electromagnetic field fluctuations in magnetic superconductors. Now we shall study the role of magnetic fluctuations in the formation of the superconducting state and neglect the electromagnetic fluctuations.

Consider the Ornstein-Zernike magnetic fluctuations in magnetic superconductors, the free energy of which is decribed by (3.4.1). It might seem that the last term in this expression describes the interaction of the superconducting order parameter with magnetic fluctuations, but, as first noted by SUHL [3.92], this is not the case.

The point is that near T_M the magnetic fluctuations grow enormously. This must lead to a decrease in the superconducting order parameter, but actually this term does not describe this phenomenon. To see this it suffices to consider the behaviour of the last term near T_M,

$$\int d\underline{r}|\Delta|^2\underline{M}^2 \leq |\Delta|^2_{max} \int d\underline{r}\,\underline{M}^2(\underline{r}) = |\Delta|^2_{max} \sum_{\underline{K}} \underline{M}_{\underline{K}}\underline{M}_{-\underline{K}} \quad . \tag{3.4.29}$$

If we now express the magnetization in the momentum representation $\underline{M}_{\underline{K}}$ in terms of the spins $\underline{S}_a$ on the sites $\underline{R}_a$,

$$\underline{M}_{\underline{K}} = \sum_a \underline{S}_a e^{i\underline{K}\underline{R}_a} \quad , \tag{3.4.30}$$

then the quantity $\sum_{\underline{K}} \underline{M}_{\underline{K}}\underline{M}_{-\underline{K}}$ in (3.4.29) becomes temperature-independent since the following summation rule holds

$$\sum_{\underline{K}} \underline{M}_{\underline{K}}\underline{M}_{-\underline{K}} = \sum_{ab} \underline{S}_a\underline{S}_b \sum_{\underline{K}} \exp[i\underline{K}(\underline{R}_a - \underline{R}_b)] = NS(S+1) \quad . \tag{3.4.31}$$

Since the quantity (3.4.29) yields a variation of the superconducting transition temperature by $\alpha \to \alpha + \eta\left\langle \underline{M}^2 \right\rangle$, it is easily seen from (3.4.31) that its contribution to T_c is temperature independent and describes the electron scattering on paramagnetic impurities which was calculated by ABRIKOSOV and GOR'KOV [3.21]. This simple conclusion about the temperature independence of (3.4.29) does not change in a more rigorous treatment [3.92].

To describe the physical situation near T_M SUHL [3.92] has suggested that the nonlocal nature of the superconducting order parameter should be retained in the phenomenological expression for the free energy, i.e., that $\Delta(\underline{r},\underline{r}')$ should be regarded as dependent on both coordinates. Such nonlocality of the pairing parameter stems particularly from the long-range nature both of the interaction of electrons via phonons and of the interaction via spin fluctuations in coordinate space [3.93]. The pairing parameter $\Delta(\underline{r},\underline{r}')$ corresponds to the Cooper pair wave-function, and $\underline{r}$ and $\underline{r}'$ are the coordinates of the electrons forming a Cooper pair. In the BCS theory, where only the spatially homogeneous case is considered, the nonlocality of the pairing parameter assumes the form

$$\Delta(\underline{r}-\underline{r}') = \frac{1}{(2\pi)^3}\int d\underline{p}\ \Delta_{\underline{p}} \exp[i\underline{p}(\underline{r}-\underline{r}')] \quad . \tag{3.4.32}$$

If we assume that Δ_p is nonzero for momenta within a thin layer δp_F around the Fermi sphere and zero for momenta outside this layer, then (3.4.32) reduces to

$$\Delta(\underline{r}-\underline{r}') = \Delta \int_{(\delta p_F)} \frac{d\underline{p}}{(2\pi)^3} \exp[i\underline{p}(\underline{r}-\underline{r}')] = \Delta \frac{\sin(p_F|\underline{r}-\underline{r}'|)}{p_F|\underline{r}-\underline{r}'|} p_F^2 \delta p_F \quad . \tag{3.4.33}$$

Proceeding from (3.4.33), SUHL [3.92] has written a similar equation for the superconducting pairing parameter in the general form

$$\Delta(\underline{r},\underline{r}') \equiv \Delta\left(\frac{\underline{r}+\underline{r}'}{2}, \underline{r}-\underline{r}'\right) = \Delta\left(\frac{\underline{r}+\underline{r}'}{2}\right) \frac{\sin(p_F|\underline{r}-\underline{r}'|)}{p_F|\underline{r}-\underline{r}'|} \quad , \tag{3.4.34}$$

and for the free-energy term which describes the coupling between the magnetic and superconducting order parameter he has chosen

$$\frac{1}{2}\eta' \int d\underline{r}\, d\underline{r}'\ \underline{M}(\underline{r})\Delta(\underline{r},\underline{r}')\, \underline{M}(\underline{r}')\Delta^*(\underline{r},\underline{r}') \quad . \tag{3.4.35}$$

Upon substitution of (3.4.34) into (3.4.35), the coupling energy of the two order parameters becomes

$$\frac{1}{2}\eta' \int d\underline{r}\, d\underline{r}' \left|\Delta\left(\frac{\underline{r}+\underline{r}'}{2}\right)\right|^2 \underline{M}(\underline{r})\underline{M}(\underline{r}') \frac{\sin^2(p_F|\underline{r}-\underline{r}'|)}{p_F^2|\underline{r}-\underline{r}'|^2} \quad . \tag{3.4.36}$$

This form of the coupling energy allows a consideration of the influence of magnetic fluctuations on superconductivity. Following SUHL, we make a qualitatitve analysis of this problem. The variation of the superconducting transition temperature due to magnetic fluctuations is determined by

$$\alpha(T-T_c) + \frac{\eta'}{2}\sum_{\underline{K}} \langle \underline{M}_{\underline{K}}\underline{M}_{-\underline{K}} \rangle \int d\underline{\rho}\ \frac{\sin^2(p_F\rho)}{p_F^2\rho^2} e^{i\underline{K}\underline{\rho}} = 0 \quad , \tag{3.4.37}$$

where $\underline{\rho} = \underline{r}-\underline{r}'$. Into this expression we substitute

$$\int d\underline{\rho}\ \frac{\sin^2(p_F\rho)}{(p_F\rho)^2} e^{i\underline{K}\underline{\rho}} = \frac{2\pi^2}{Kp_F}\theta(2p_F - K) \tag{3.4.38}$$

and the value of the correlator $\langle \underline{M}_{\underline{K}}\underline{M}_{-\underline{K}} \rangle$ which corresponds to the Ornstein-Zernike approximation (see, for example, [3.94])

$$\left\langle \underline{M}_{\underline{K}} \underline{M}_{-\underline{K}} \right\rangle = \frac{c q_c^2 S(S+1)}{2\pi [cK^2 + a_0(T - T_M)]} \tag{3.4.39}$$

and obtain

$$\sum_{\underline{K}} \left\langle \underline{M}_{\underline{K}} \underline{M}_{-\underline{K}} \right\rangle \int d\underline{\rho} \, \frac{\sin^2(p_F \rho)}{(p_F \rho)^2} e^{i\underline{K}\underline{\rho}} = \frac{S(S+1)}{2p_F^2 q_c} \ln \left[\frac{\min(q_c^2, 4p_F^2) + a_0(T - T_M)}{(a_0/c) \cdot (T - T_M)} \right] \quad . \tag{3.4.40}$$

Here q_c is chosen so that $q_c^3 V = (2\pi)^3 N$. Setting $q_c \sim p_F$ and $\eta' = 2p_F^2 q_c \eta$, we see that the nonlocal nature of $\Delta(\underline{r}, \underline{r}')$ leads to an increase of the effect destroying the superconductivity by paramagnetic impurities by the quantity $\ln\{cq_c^2/[a_0(T - T_M)]\}$ which diverges logarithmically near T_M. It becomes clear from (3.4.37) that superconductivity collapses when the temperature is close enough to T_M.

However, in the derivation of (3.4.40) a number of assumptions have been made which overestimated the effect of magnetic fluctuations. Primarily it is necessary to take into account the effects associated with the finite lifetime of the electrons which lead to the occurrence of the term $a_0(T - T_M)/c + q_0^2$ in the denominator of (3.4.40) instead of $a_0(T - T_M)/c$ where q_0 is a constant having the dimension of a momentum.

Let us now return to SUHL's fundamental idea of accounting for the nonlocal nature of the superconducting pairing parameter to describe the interaction with magnetic fluctuations. As shown in [3.92], microscopic analysis yields a free energy expression which contains a term describing the coupling of the order parameters in the form (3.4.35), the nonlocality of the superconducting order parameter making an essential contribution to just this term; on the other free energy terms the nonlocal character of $\Delta(\underline{r}, \underline{r}')$ has no effect. Thus, the final expression for the free energy accounting for the nonlocality may be written as

$$F = \int d\underline{r} \Big[\frac{1}{2} c(\underline{\nabla} \underline{M})^2 + \frac{1}{2} a \underline{M}^2 + \frac{1}{4} b(\underline{M}^2)^2 + \gamma |\underline{\nabla} \Delta|^2 + \alpha |\Delta|^2 + \frac{1}{2} \beta |\Delta|^4$$
$$+ \frac{1}{2} \eta |\Delta|^2 \int d\underline{r}' \underline{M}(\underline{r}) \underline{M}(\underline{r}') \frac{\sin^2(p_F |\underline{r} - \underline{r}'|)}{p_F^2 |\underline{r} - \underline{r}'|^2} \Big] \quad . \tag{3.4.41}$$

Here the symbol Δ is used to label the superconducting order parameter. In [3.92] the values of the constants involved in (3.4.40) are also given and are expressed through microscopic parameters.

From (3.4.41) it is possible to investigate the conditions for the coexistence of superconductivity and magnetic ordering on the basis of this modified free energy.

Compare (3.4.12) for the free energy of the "cryptoferromagnetic" state (where instead of η we have $g_Q = \eta\pi^2/2p_FQ$) with (3.4.5) for the free energy of the intermediate MS phase of a homogeneous system. The free energy of the intermediate phase turns out to be lower than that of the superconducting phase F_S and, apart from this, less than that of the magnetic phase F_M. The free energy of the "cryptoferromagnetic" state is also less than that of the superconducting phase, but, in contrast to (3.4.5), expression (3.4.12) for F_{MS}^{crypto} contains the term $-acQ^2/2b$ which, at temperatures below the magnetic ordering temperature, increases the magnitude of the free energy of the "cryptoferromagnetic" state. When sufficiently remote from T_M, the free energy F_{MS}^{crypto} may become equal to the free energy of the magnetic state F_M and then a phase transition will take place between these phases. This transition will occur on the line determined by the equation

$$\frac{|a|cQ^2}{2b} - \frac{c^2Q^4}{4b} - \frac{[\alpha b - (a + cQ^2)\eta/2]^2}{2b(b\beta - \eta^2/2)} = 0 \quad . \tag{3.4.42}$$

The magnetic fluctuations in superconductors with a regular arrangement of magnetic atoms have also been considered in [3.93-96] where a microscopic approach is used. According to their results, when scattered on magnetic fluctuations, the electrons forming Cooper pairs weaken the BCS interaction. In a superconductor, on the other hand, the magnetic fluctuations themselves are suppressed by the Meissner effect. Having estimated the weakening of the pairing interaction by magnetic fluctuations and the role of the Meissner effect, MATSUMOTO et al. [3.95] arrived at the conclusion that magnetic fluctuations may not be the cause of the collapse of the superconducting state at the lower superconducting transition temperature in $ErRh_4B_4$-type compounds.

The joint action of the fluctuations of the magnetic-order parameter and of those of the electromagnetic field [whose free energy is written in the form of (3.4.13)] in magnetic superconductors has been considered by BLOUNT and VARMA [3.86]. By extending the method of HALPERIN et al. [3.89] to calculate the contribution to the free energy of magnetization fluctuations and electromagnetic field fluctuations, BLOUNT and VARMA have found that the superconductors under discussion not only have the tendency to decrease the superconductivity due to magnetic fluctuations, but they also suppress the magnetic fluctuations themselves in the superconducting state.

3.4.5 Critical Behaviour of Magnetic Superconductors

We have discussed the problem of the coexistence of superconductivity and magnetic ordering within the framework of Ginzburg and Landau's phenomenological theory whose condition of applicability is, as is well known, the smallness of the order parameter fluctuations. At a temperature close to the transition temperature there is a narrow region Δt known as the "fluctuation region", where the temperature dependence of the thermodynamical quantities differs from the results of self-consistent field theory, and fluctuations play a major role. This region of temperatures is determined by Ginzburg's criterion. For the superconductor, the fluctuation region

$$\Delta t_s = \frac{\beta^2 T_c}{\alpha_0 \gamma^3} \sim 10^{-14} \tag{3.4.43}$$

is extraordinarily small. Therefore the condition for applicability of the Ginzburg-Landau theory $|(T - T_c)/T_c| \gg \Delta t_s$ for superconductors is satisfied practically everywhere and the fluctuations of the superconducting order parameter may be neglected. The fluctuation region of a magnetically ordered system

$$\Delta t_M = \frac{b^2 T_M}{a_0 c^3} \sim 10^{-2} \tag{3.4.44}$$

is much larger than that of a superconductor. The region of applicability of the self-consistent field theory $|(T - T_M)/T_M| \gg \Delta t_M$ narrows, and the fluctuations of the magnetic-order parameter may make an appreciable contribution.

Of particular interest is the coincidence of the superconducting transition and magnetic-ordering temperatures. Such a situation occurs, for example, in quasiternary compounds of the type which has already been mentioned above, i.e., $(Er_{1-x}Ho_x)Rh_4B_4$ [3.77]. From the phase diagram of this compound, presented in Fig.3.12, it is seen that the phase transition lines $T_{c1}(x)$ and $T_{c2}(x)$ intersect at a point x_c. Both the fluctuations of the superconducting-order parameter and the magnetic-order fluctuations become essential at the intercept and it is necessary to consider their action jointly on the basis of the renormalization group technique which is widely used in the theory of critical phenomena.

However, since the fluctuation region of a superconductor is negligible compared to that of a magnetically ordered system, there arises the immediate question of how the fluctuation regions Δt_s and Δt_M change if one takes into account the interaction of the superconducting- and magnetic-order parameters which occurs in magnetic superconductors. If this change is negligible, the effective fluctuation region will turn out to be comparable with the fluctuation region of a superconductor or, to put it differently, the results obtained on the basis of a stricter theory with due regard for the fluctuations of the superconducting- and magnetic-order parameters will be inaccessible to measurements so far. This problem has recently been considered [3.97] for a system of coupled superconducting and magnetic-order parameters which is described by the Ginzburg-Landau functional in the form of (3.4.1). It has been shown that the fluctuation region Δt_S expands largely because of the interaction with the magnetic subsystem and becomes comparable with the fluctuation region Δt_M. Consequently, in magnetic superconductors it is necessary to allow for the the critical fluctuations of the superconducting-order parameter along with critical fluctuations of the magnetic-order parameter when calculating, for example, the critical indices of the thermodynamical quantities characterizing the behaviour of the system. They can probably also be measured experimentally.

Another problem arising with coupled-order parameters is the nature of the interception point itself. This point may be either bicritical (the merging of two second-order phase transition lines and one first-order phase transition line) or tetracritical (the convergence of four second-order phase transition lines), and this fact depends on the number of components of the order parameter, as shown in the theory of critical phenomena of systems with coupled-order parameters [3.83,98]. The effective Ginzburg-Landau Hamiltonian in the form of (3.4.1) (treated in our problem of the coexistence of superconductivity and magnetic ordering) is of the same form as in the general theory [3.83] with a specification of the order parameters applicable to the problem under consideration.

It is known that the superconducting-order parameter has two components, but, with regard to the magnetic-order parameter, we recall that the magnetic moments in $ErRh_4B_4$ are ordered in the basal plane, and therefore the number of components of the magnetic-order parameter is equal to two. Hence the total number of order parameters is equal to four. According to NELSON et al. [3.98], for a total number of order-parameter components equal to four a new critical behaviour, different from the critical behaviour of the isotropic Heisenberg model, is realized in the system of coupled-order parameters and the intercept of the phase transitions lines appears to be tetracritical. The latter circumstance immediately indicates that superconductivity and magnetic ordering have a region of coexistence at temperatures below the intersection point and we note that all the phase transitions will be second-order phase transitions. It is known, however, from experimental data for $ErRh_4B_4$ that the transition from a superconducting phase to a magnetically ordered phase is a first-order transition.

Thus, it has turned out that the simplest form of the free energy (3.4.1), which has formed the basis of this conclusion, does not suffice to explain the experimental phase diagram and the character of the point of intersection.

More appropriate for these purposes is the expression (3.4.13) where account is taken fo the electromagnetic field effects which, as shown above, lead to a first-order transition from the normal to the superconducting state [3.89]. In addition, these lead to a first-order transition, from the superconducting state with a helical arrangement of the spins of magnetic atoms, to the normal state with a ferromagnetic ordering of spins [3.86]. The structure of the phase diagram, which is obtained in mean-field theory for the free energy (3.4.13), correctly describes the relation of the phases in a system in which a transition from the superconducting to the ferromagnetic state occurs. Recently, the character of the point of intersection of phase transition lines for the free energy (3.4.13) has been investigated [3.99] on the basis of the method of Halperin, Lyubensky and Ma. It was shown in [3.99] that this point is a first-order phase transition point but not a tetra- or bicritical point (Fig.3.14). Consequently, this case is a situation more complicated than the previous one with the free energy of the form (3.4.1). To describe the region of the coexistence of superconductivity with helical magnetic order which precedes the transition to the ferromagnetic state, it is not enough to know the character of the intersection point, and therefore one has to involve physical mechanisms responsible for the occurrence of superconductivity, which depend on whether or not the superconducting electrons participate in the indirect exchange of localized magnetic moments. In the first case the cause of the occurrence of a helical phase is the Anderson-Suhl mechanism (the s-d exchange coupling); in the second case it is the Blount-Varma mechanism (electromagnetic field fluctuations).

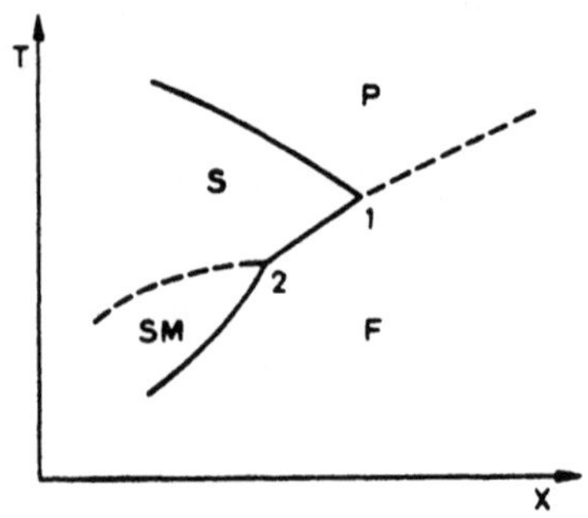

Fig.3.14. The schematic phase diagram of a magnetically ordered superconductor from a self-consistent field approximation [3.99] for the compound $(Er_{1-x}Ho_x)Rh_4B_4$ (The solid line marks the first-order phase transition, the dotted line stands for the second-order phase transition. Points 1 and 2 are no critical points and may, in principle, coincide)

3.5 Magnetic Impurities in Superconductors

3.5.1 The Abrikosov-Gor'kov Theory

In contrast to nonmagnetic impurities which weakly affect the temperature of superconducting transition of the metal, paramagnetic impurities exert a highly suppressing effect on superconductivity. This effect arises from the exchange scattering of one of the electrons of the Cooper pair on the impurity, as a result of which the electron spin is liable to vary and this, in its turn, already means the destruction of the Cooper pair. The theory of superconductors was put forward by ABRIKOSOV and GOR'KOV [3.21] as far back as 1960; it has received tremendous development and stimulated a new trend in experimental research. Since there exist splendid reviews on magnetic impurities in superconductors (see, for example, [3.2]) where detailed presentations of the various aspects of the Abrikosov-Gor'kov theory are given, we shall not elaborate it here and shall restrict our attention to the concentration dependence of the superconducting transition temperature.

It may be found from the known expression [3.21] (see also Sect.3.2)

$$\ln\frac{T_c}{T_{co}} = \psi\left(\frac{1}{2}\right) - \psi\left(\frac{1}{2} + \frac{1}{2\pi T_c \tau_s}\right) \quad , \tag{3.5.1}$$

where τ_s is the mean time between the exchange scatterings of the electron on the impurity. This spin-flip is determined by the expression $\tau_s^{-1} = S(S+1)\tau_{ex}^{-1}$ in which τ_{ex}^{-1} is given by (3.2.9).

For the impurity of a rare-earth element whose atoms conserve the total quantum number J in the metal, the expression of τ_s^{-1} should be replaced by

$$\frac{1}{\tau_s} = \frac{(g-1)^2 J(J+1)}{\tau_{ex}} \quad , \tag{3.5.2}$$

where $(g-1)^2 J(J+1)$ is the De Gennes factor [3.100].

The dependence of T_c on the impurity concentration, which is obtained from numerical solution of the Abrikosov-Gor'kov equation, can be seen in Fig.3.15. The critical concentration at which superconductivity is completely suppressed is determined by the known equation

$$\frac{1}{\tau_s} = \frac{\pi}{2\gamma} T_{co} \quad . \tag{3.5.3}$$

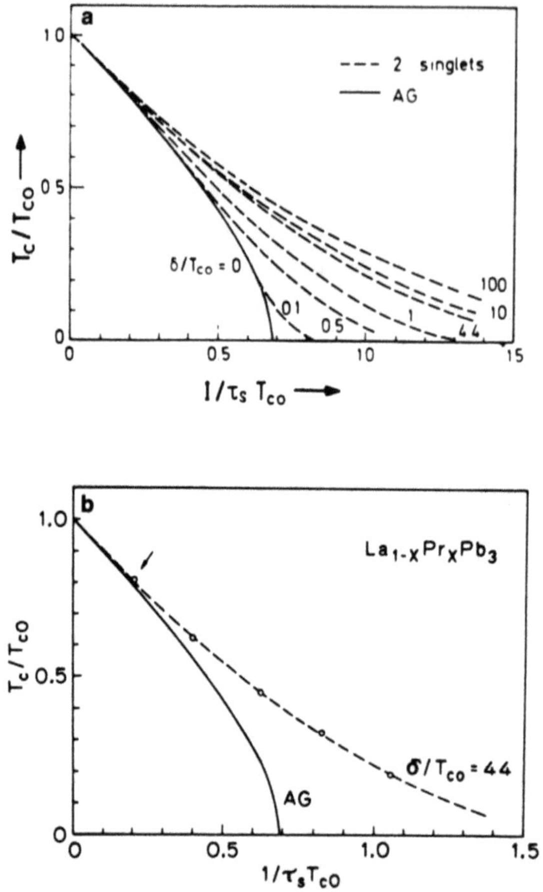

Fig.3.15. T_C as a function of impurity concentration with two singlet levels separated by the energy δ: (a) according to the theory by KELLER and FULDE [3.101], (b) according to experimental data for $La_{1-x}Pr_xPb_3$ [3.102]. AG is from the Abrikosov and Gor'kov theory

We shall not go into these well-known problems but shall discuss one of the latest developments of the Abrikosov-Gor'kov theory: the allowance for the crystalline field acting on magnetic ions [3.103,104].

3.5.2 The Crystalline Field Effects

The destruction effect exerted on Cooper pairs by the exchange scattering on the impurity with the flip of the spin of one of the pair electrons is associated with the possibility of the reorientation of the impurity spin itself.

This process comes about most easily when the paramagnetic spins are free and there is a degeneracy for each impurity ion in the orientation of its spin or of the atomic total angular momentum. If an ion is in a crystalline field and its states are split, the reorientation of its total angular momentum demands a transition between the levels of the crystalline field. The simplest situation occurs in the case of a non-Kramers field with a singlet ground state. If the splitting of the energy levels by the crystalline field is well in excess of the transition temperature, only the virtual excitation of states is possible. This gives rise to the collapse of Cooper pairs. The measure of probability of such a process is the ratio of the effective matrix element

$$\tau_{ij}^{-1} = \frac{(g-1)^2 |\langle i|\underline{J}|j\rangle|^2}{\tau_{ex}} \qquad (3.5.4)$$

to the energy difference corresponding to the states $|i\rangle$ and $|j\rangle$ of the ion in the crystalline field. It is clear that the larger these energy differences the lesser the destructive influence of these impurities on the superconducting temperature. Conversely, when these differences tend to zero the physical picture of the phenomenon must be close to the Abrikosov-Gor'kov theory.

Equations for Green's functions of a superconductor with impurity ions in a crystalline field have been obtained by FULDE et al. [3.101,103]. Their numerical solution leads to results for the superconducting transition temperature which are illustrated in Fig.3.15. These calculations apply to the case when the ground state and the excited state of the ion is a singlet with an energy difference δ. Figure 3.15 shows the dependence of T_c on the dimensionless parameter δ/T_{co}, proportional to the concentration, for several values of the parameter $1/\tau_s T_{co}$. When $\delta \to 0$ the curve for T_c should go over to the relevant curve of the Abrikosov and Gor'kov theory, and deviate from it as δ increases. At finite δ the curvature of the line varies and at large δ a critical concentration is absent. The limit $\delta \to \infty$ corresponds to scattering on a nonmagnetic impurity. To this limit corresponds a smooth and gradual decrease of T_c with increasing concentration. It is significant that at large enough δ (in the figure this already occurs at $\delta/T_c > 4$) the curves are insensitive to δ.

Within the framework of this theory it becomes possible to account for the concentration dependence of T_c in alloys with impurities in the singlet ground state. Some examples are alloys such as $La_{1-x}Pr_xPb_3$, $La_{1-x}Pr_xSn_3$, $(La_{1-x}Pr_x)_3In$, etc. Figure 3.15b, which has been borrowed from [3.104], demonstrates results for $La_{1-x}Pr_xPb_3$ [3.102]. The experimental points fit one of the theoretical curves from Fig.3.15a perfectly for which $\delta/T_{co} = 4.4$. It is worth noting that only one fitting parameter determining the initial slope of the curve at the point arrowed in Fig.3.15b was used for this purpose.

Thus, the splitting of the energy levels of an ion in a crystalline field substantially varies the effect of suppression of superconductivity by the impurity. The deviation of the curve for T_c from the predictions of the Abrikosov-Gor'kov theory is particularly great for impurities having a singlet ground state. In the case of a Kramers doublet or a degenerate magnetic state of an ion these deviations are much less pronounced. For further details we refer the reader to the fundamental review by FULDE [3.104].

3.6 The Kondo Effect in Superconductors

In the ABRIKOSOV and GOR'KOV theory [3.21] for superconductors containing paramagnetic impurities, the scattering of electrons on impurity atoms is considered to a first Born approximation. The amplitude of scattering in this case does not depend on electron energy and temperature, and this circumstance manifests itself in the concentration dependence of T_c determinable from (3.2.22), from which it follows particularly that there exists a critical impurity concentration suppressing completely the superconducting state of the basic metal.

As shown by KONDO [3.105], in the next perturbation-theory approximation with respect to the s-d interaction, the amplitude of scattering of an electron on an impurity diverges logarithmically near the Fermi energy and this leads to logarithmic temperature corrections in the kinetic properties of a dilute alloy, especially in resistivity.

A large number of papers [3.106-111] were devoted to a theoretical study of this phenomenon (the Kondo effect) in normal metals. It was found that the behaviour of the amplitude of scattering on a paramagnetic impurity depends largely on the sign of the s-d exchange integral. In the case when $J < 0$ (antiferromagnetic coupling) the behaviour of the amplitude changes substantially at some temperature

$$T_K \sim E_F \exp\left(-\frac{1}{JN(0)}\right) \quad ,$$

called the Kondo temperature. From a formal standpoint, the quantity T_K characterizes the temperature at which perturbation theory for amplitude calculation already fails (if one moves from the high-temperature side). From a physical point of view, it means that when $T \ll T_K$ the impurity spin tends to be compensated for by the conduction-electron spin-polarization arising in the vicinity of an impurity, the degree of compensation increasing gradually with decreasing T.

Thus, when $T \lesssim T_K$ a bound state with a binding energy of the order kT_K occurs in the electronic system. At the same time the electron binding energy in Cooper pairs is of the order of kT_{co}. This means that the superconducting properties of a dilute alloy with antiferromagnetic s-d coupling must change drastically provided that $T_K \sim T_{co}$. In particular, it has turned out that in this situation levels corresponding to some quasi-bound states of the electronic system appear in the energy gap [3.112].

A great number of papers have been dedicated to a theoretical investigation of the Kondo effect in superconductors (for the bibliography refer to [3.113]). A complete and self-consistent study has been carried out in a series of papers by MÜLLER-HARTMANN and ZITTARTZ [3.108-111] whose main ideas we shall outline in this section. Their approach is based on NAGAOKA's [3.106] decoupling of the equation for the amplitude of scattering in a normal metal. MÜLLER-HARTMANN and ZITTARTZ's papers are a generalization of Nagaoka and Hamann's technique for a superconducting metal. Since, strictly speaking, the solution of the Kondo problem is far from being complete even for a normal metal, it goes without saying that the results of the Müller-Hartmann and Zittartz theory will eventually be reconsidered. It is to be hoped, however, that a refinement of their results will not qualitatively change the main conclusions of the theory presented.

3.6.1 The Equation for the Scattering Amplitude

First, consider a superconductor with a single impurity atom and calculate the amplitude of scattering of an electron on it. The Hamiltonian of the system H consists of two terms $H_s + H_{imp}$, where H_s corresponds to a pure superconductor and H_{imp} describes the interaction of electrons with an impurity. Since we shall be concerned with a common superconductor, let us choose the Hamiltonian H_s corresponding to the weak-coupling approximation and write it in the form [cf. (2.1.3)]

$$H_s = \sum_{\underline{p}\alpha} \varepsilon_{\underline{p}} a^+_{\underline{p}\alpha} a_{\underline{p}\alpha} - \Delta \sum_{\underline{p}} \left(a^+_{\underline{p}\uparrow} a^+_{-\underline{p}\downarrow} + a_{-\underline{p}\downarrow} a_{\underline{p}\uparrow} \right) \quad , \tag{3.6.1}$$

where $a^+_{\underline{p}\alpha}$ and $a_{\underline{p}\alpha}$ are the operators of creation and annihilation of an electron with a momentum $\underline{p}$ and spin $\alpha = \uparrow, \downarrow$.

The order parameter Δ (which coincides with the gap function in this case) does not depend on the energy of the electron, and is found from the usual BCS theory self-consistency equation

$$\Delta = V \sum_{\underline{p}} \left\langle a_{-\underline{p}\downarrow} a_{\underline{p}\uparrow} \right\rangle \tag{3.6.2}$$

(V being the matrix element of the effective electron attraction on the Fermi surface). In the Hamiltonian H_{imp} we restrict our attention to the s-d exchange interaction since the Coulomb interaction of an electron with an impurity atom will not produce any effect in the phenomena of interest. Thus,

$$H_{\mathrm{imp}} = -\frac{J}{N} \sum_{\underline{p}\underline{p}'} \sum_{\alpha\beta} (\underline{S}\, \underline{s}_{\alpha\beta})\, a^{+}_{\underline{p}\alpha} a_{\underline{p}'\beta} \quad . \tag{3.6.3}$$

Note that the positive sign of the s-d exchange integral corresponds to the ferromagnetic coupling of the electron spin $\underline{s}$ with the atomic spin $\underline{S}$, and the negative sign to the antiferromagnetic coupling.

Since in the presence of an impurity atom the system possesses no translational invariance, the electron Green's function should be specified by two momenta $\underline{p}$ and $\underline{p}'$. We shall follow here a series of papers by MÜLLER-HARTMANN and ZITTARTZ [3.108] and also introduce a one-particle Green's function in Gor'kov-Nambu's representation

$$G_{\underline{p}\underline{p}'}(\tau - \tau') = -\left\langle T \Psi_{\underline{p}}(\tau) \overline{\Psi}_{\underline{p}'}(\tau') \right\rangle \quad , \tag{3.6.4}$$

where $\Psi_{\underline{p}}$ and $\overline{\Psi}_{\underline{p}'}$ are two-component Nambu operators

$$\Psi_{\underline{p}} = \begin{pmatrix} a_{\underline{p}\uparrow} \\ a^{+}_{-\underline{p}\downarrow} \end{pmatrix} \quad , \quad \Psi^{+}_{\underline{p}} = \left(a^{+}_{\underline{p}\uparrow} , a_{-\underline{p}\downarrow} \right) \quad .$$

When there is no s-d interaction, the Green's function is diagonal with respect to the momenta, i.e., $G_{\underline{p}\underline{p}'} = G^{o}_{\underline{p}} \delta_{\underline{p}\underline{p}'}$, and for the Fourier component with respect to the variable $\tau - \tau'$ we have the known expression

$$G^{o}_{\underline{p}}(z) = \begin{pmatrix} z - \varepsilon_{\underline{p}} & \Delta \\ \Delta & z + \varepsilon_{\underline{p}} \end{pmatrix}^{-1} \quad , \tag{3.6.5}$$

corresponding to the Hamiltonian (3.6.1) where $z = (2n+1)\pi T i$.

The equation of motion for the Green's function $G_{\underline{pp}'}$ involves mixed Green's functions containing electron and spin operators. In [3.108-110] use is made of Nagaoka's decoupling [3.106] which reduces composite Green's functions to simple ones. In this approximation the Green's function $G_{\underline{pp}'}$ may be expressed in terms of the $\hat{t}$ matrix describing the electron scattering on the impurity spin

$$G_{\underline{pp}'} = G^{o}_{\underline{p}}\delta_{\underline{pp}'} + G^{o}_{\underline{p}}\frac{J}{N}\hat{t}\,G^{o}_{\underline{p}'} \tag{3.6.6}$$

the $\hat{t}$ matrix obeying a singular integral equation of the type obtained by HAMANN [3.107] when investigating the Kondo effect in a normal metal with the help of Nagaoka's decoupling. The only difference for a superconductor is the replacement of all the quantities involved in HAMANN's equation by two-row matrices.

Analysis of the equation obtained shows that the structure of the $\hat{t}$ matrix is as follows

$$\hat{t}(z) = \begin{pmatrix} t_1(z) & t_2(z) \\ t_2(z) & t_1(z) \end{pmatrix} \quad , \tag{3.6.7}$$

where $t_2(z)$ and $t_1(z)$ are the amplitudes of scattering of an electron with and without spin flip, respectively, and

$$t_1(z) = -t_1(-z) \quad , \quad t_2(z) = t_2(-z) \quad . \tag{3.6.8}$$

It is not difficult to show that the matrix equation for $\hat{t}$ reduces to an ordinary equation for the quantity

$$\hat{t}(z) = t_1(z) - t_2(z) \tag{3.6.9}$$

in terms of which initial amplitudes t_1 and t_2 are expressed through the equation

$$t_1(z) = \frac{1}{2}[\hat{t}(z) - \hat{t}(-z)] \quad , \quad t_2(z) = -\frac{1}{2}[\hat{t}(z) + \hat{t}(-z)] \quad . \tag{3.6.10}$$

The equation for the quantity $\hat{t}(z)$ was obtained and solved by MÜLLER-HARTMANN and ZITTARTZ [3.108-111]. The exact solution is representable in the form

$$\gamma\hat{t}(z) = \frac{1}{2\pi i}\sqrt{\frac{z^2 - \Delta^2}{z + \Delta}}\left(1 - \frac{X(z)}{\phi(z)}\right) \quad , \tag{3.6.11}$$

where $X(z)$ and $\phi(z)$ are complicated expressions including integrals of functions of the type $\mathrm{Im}\{G_p^o(z)\}$. These expressions simplify considerably when the region of energies $z \sim \Delta$ is considered. In this case

$$X(z) = \gamma\left(\tau + \frac{a}{y-1}\right) \quad , \quad \phi(z) = |\gamma|\sqrt{\tau^2 + \pi^2 S(S+1)}\,\frac{y_0 - y}{1 - y} \quad , \tag{3.6.12}$$

where the notation $y = z/\Delta$, $\gamma = JN(0)$ and

$$\tau = \frac{1}{\gamma} + \frac{1}{\lambda} \quad \text{or} \quad \tau = \ln\frac{T_K}{T_{co}} \quad (\text{for } \gamma < 0) \tag{3.6.13}$$

is introduced. The quantities a and y_0 involved in (3.6.12) are equal in magnitude

$$a = \tau - \mathrm{sign}\{\gamma\}\sqrt{\tau^2 + \pi^2 S(S+1)} \quad , \tag{3.6.14}$$

$$y_0 = \mathrm{sign}\{\gamma\}\,\frac{\tau}{\sqrt{\tau^2 + \pi^2 S(S+1)}} \quad . \tag{3.6.15}$$

Thus, over the energy interval $z \sim \Delta$ the following approximate expressions hold for the scattering amplitudes on a paramagnetic impurity [3.111]:

$$\gamma t_1(z) = \frac{\sqrt{y^2 - 1}}{2\pi i}\cdot\frac{y(1 - y_0)}{y^2 - y_0^2} \quad , \quad \gamma t_2(z) = -\frac{\sqrt{y^2 - 1}}{2\pi i}\cdot\frac{y_0(1 - y_0)}{y^2 - y_0^2} \quad . \tag{3.6.16}$$

3.6.2 Impurity States Within the Energy Gap

The quantities $\pm y_0$ are the poles of the $\hat{t}$ matrix and, consequently, of a one-particle Green's function. To ascertain the meaning of these poles it is necessary to calculate the density of one-particle excitations $N(\omega)$ of a superconductor containing one paramagnetic impurity atom. By definition

$$N(\omega) = -\frac{2}{\pi}\frac{1}{N}\,\mathrm{Im}\left\{\sum_{\underline{p}} G_{\underline{pp}}^{11}(\omega + i\delta)\right\} \quad . \tag{3.6.17}$$

The superscript in $G_{\underline{pp}}^{11}$ designates the 11 matrix element of the Green's function $G_{\underline{pp}}$. We substitute the expression for $G_{\underline{pp}}^{11}$ into this formula to get

$$N(\omega) = N_0(\omega) + \frac{1}{N}\,n(\omega), \tag{3.6.18}$$

where $N_0(\omega)$ is the density of one-particle excitations in a pure superconductor,

$$N_0(\omega) = 2N(0)\begin{cases}\dfrac{|\omega|}{\sqrt{\omega^2-\Delta}} & , \quad |\omega| \geq \Delta \\ 0 & , \quad |\omega| < \Delta\end{cases} \tag{3.6.19}$$

and $n(\omega)$ is an addition due to the paramagnetic impurity

$$n(\omega) = 2\gamma \mathrm{Re}\left\{\frac{\Delta^2 t_1(z) - \Delta z t_2(z)}{(z^2-\Delta^2)^{3/2}}\right|_{z=\omega+i\delta} \quad . \tag{3.6.20}$$

Calculation of the last expression through the use of approximate amplitudes (3.6.16) for $z \leq \Delta$ leads to the result [3.110]

$$n(\omega) = \frac{a}{2}[\delta(\omega-\omega_0) + \delta(\omega+\omega_0)] - \frac{1}{2}[\delta(\omega-\Delta) + \delta(\omega+\Delta)] \quad , \tag{3.6.21}$$

where $\omega_0 \equiv y_0\Delta$. Since by virtue of (3.6.15) one has $|y_0| < 1$, it follows from (3.6.21) that in the presence of a paramagnetic impurity two levels with energy values $\pm\omega_0$ symmetric with respect to the middle of the gap split off from the upper and lower boundaries of the gap of the superconductor. These levels correspond to bound states of the electron spin density near the paramagnetic impurity. It is worth noting, however, that the charge distribution of electrons in these states does not vary.

In the approximation (3.3.16) for the amplitude, the coefficient $a \approx 1$, and so it turns out that there is one bound state per impurity. At high impurity concentrations the wave functions of bound states overlap and form impurity bands within the energy gap of the superconductor. The structure of these impurity bands was studied by SHIBA [3.112] who considered them only in the limit of classlical spins $(S \to \infty)$.

The dependence of the energy of the bound state ω_0 (or, to be more exact, of the quantity y_0) determined by (3.6.15) on the magnitude of the s-d exchange interaction γ is qualitatively represented in Fig.3.16. In the case of

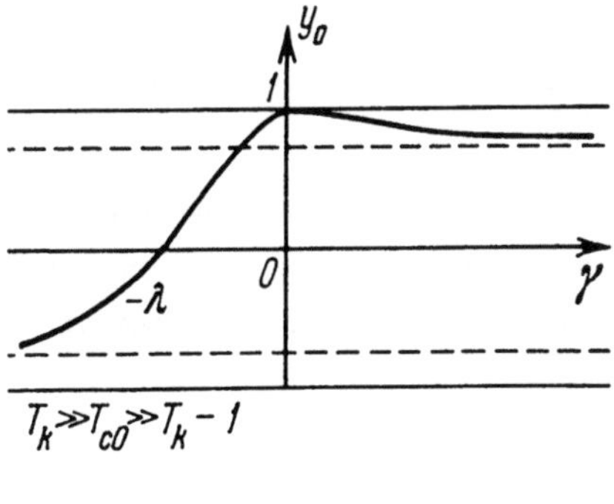

Fig.3.16. The position of the level of the impurity bound state in the gap of the superconductor as a function of the sign and magnitude of the s-d exchange interaction parameter

antiferromagnetic coupling ($\gamma < 0$) the levels of bound states may penetrate into the forbidden gap and arrive at its middle when the Kondo temperature is of the order of the superconducting transition temperature.

3.6.3 The Lowering of the Superconducting Transition Temperature by Magnetic Impurities

Let us now consider the influence of the paramagnetic impurity on the superconducting transition temperature. To this end it is necessary to consider the energy gap equation (3.6.2) which may be rewritten in the form

$$\Delta = \lambda T \sum_n \int d\varepsilon_{\underline{p}} N(\varepsilon_{\underline{p}}) G^{12}_{\underline{pp}}(i\omega_n) \quad . \tag{3.6.22}$$

We now use (3.6.6) for the Green's function through the $\hat{t}$ matrix. Upon substitution of this expression into (3.6.22) and after combining the resulting equation with the equation describing an intrinsic superconductor, we obtain for the superconducting transition temperature

$$\ln\frac{T_c}{T_{co}} = \lim_{\Delta\to 0} \frac{c\gamma}{N(0)\Delta} \int d\varepsilon_{\underline{p}} N(\varepsilon_{\underline{p}}) T_c \sum_n \left(G^o_{\underline{p}}\, \hat{t}\, G^o_{\underline{p}}\right)^{12} \tag{3.6.23}$$

[In (3.6.6) we have replaced $1/N$ by the concentration c, which we consider to be small]. After substituting the expression for $G^o_{\underline{p}}$ here and performing an integration over $\varepsilon_{\underline{p}}$ we come to the following equation for T_c:

$$\ln\frac{T_c}{T_{co}} = -\lim_{\Delta\to 0} \frac{2\pi\gamma c T_c}{N(0)\Delta} \sum_{\omega_n > 0} \frac{1}{\omega_n} \mathrm{Re}\left\{t_2(i\omega_n) + i\frac{\Delta}{\omega_n} t_1(i\omega_n)\right\} \quad . \tag{3.6.24}$$

If one substitutes (3.6.16) here for amplitudes t_1 and t_2 and calculates the sum over frequencies, one can obtain the following approximate expression for the variation of the superconducting transition temperature due to paramagnetic impurities [3.85]

$$\Delta T_c \equiv T_{co} - T_c = \frac{c}{8N(0)} \frac{\pi^2(S+0.5)^2}{(\tau-2.5)^2 + \pi^2(S+0.5)^2} \quad . \tag{3.6.25}$$

The quantity ΔT_c representing the effect of suppression of superconductivity by a paramagnetic impurity as a function of the parameter τ, is roughly described by a Lorentz curve whose width is $\pi(S+0.5)$ and whose maximum at $\tau \approx 2.5$ is equal to

$$(\Delta T_c)_{max} = \frac{c}{8N(0)} \quad . \tag{3.6.26}$$

The Abrikosov-Gor'kov theory, which is restricted to the Born approximation, leads to the following quantity for the suppression of the superconducting transition temperature [3.21]:

$$(\Delta T_c)_{AG} = \frac{c}{8N(0)}\pi^2 S(S+1)\gamma^2 \quad . \tag{3.6.27}$$

The results of (3.6.25,27) are compared in Fig.3.17 where the dependence of $\Delta T_c/(\Delta T_c)_{AG}$ on γ is plotted. It follows from the figure that the Born approximation in the theory of scattering of electrons of a superconductor on a paramagnetic impurity is applicable only in a narrow range of values $|\gamma| \ll \lambda$. Also the strong dependence of the effect on the sign of γ comes to light. In the antiferromagnetic coupling case ($\gamma < 0$) the change of ΔT_c with γ is most drastic when $|\gamma| \approx \lambda$.

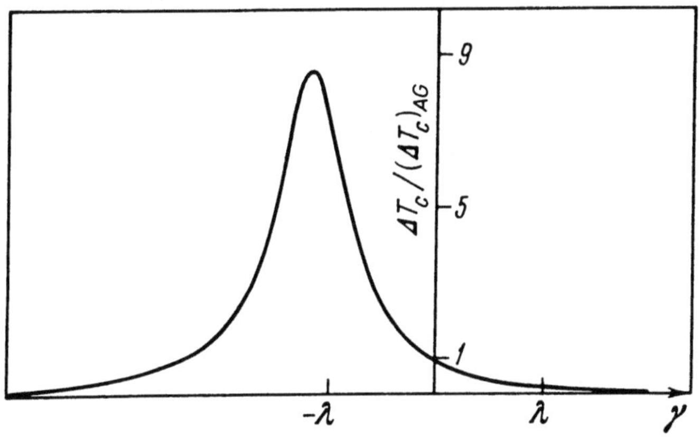

Fig.3.17. Comparison of the effects of suppression of T_c due to the scattering on paramagnetic impurities in theories of MÜLLER-HARTMANN and ZITTARTZ [3.109] and ABRIKOSOV and GOR'KOV [3.21]. The curve has been obtained for $S = 1/2$ and $\lambda = 0.1$

From the very beginning our treatment has been confined to low-impurity concentrations. Since it has been assumed when deriving the basic equation (3.6.25) that $\Delta T_c/T_{co} \ll 1$, we conclude with due reference to (3.6.26) that the concentration c should satisfy the inequality

$$\frac{c}{N(0)T_{co}} \ll 1 \quad . \tag{3.6.28}$$

This makes the limit of admissible concentrations extremely low. In order to be able to extend the theory to a region of larger concentrations and at the same time to remain within the assumption $c \ll 1$, it is necessary from the very outset to consider not an isolated impurity in the crystal but a finite impurity concentration. One should also introduce the Green's function $G_{\underline{p}}(z)$ averaged with respect to impurity concentrations. This should certainly be diagonal with respect to momenta.

We write Dyson's equation for the function $G_{\underline{p}}(z)$,

$$G_{\underline{p}}^{-1}(z) = G_{\underline{p}}^{o^{-1}}(z) - \Sigma(z) \quad , \tag{3.6.29}$$

in which all the quantities are 2×2 matrices. To first order in the impurity concentration the eigenenergy is just $\Sigma(z) = ct(z)$. However, at high concentrations it must be found in a self-consistent manner by substituting the exact functions $G_{\underline{p}}(z)$ for the zero Green's functions in the expression for $t(z)$. Thus, a self-consistent amplitude of scattering $t_{sc}(z)$ can be determinable from [3.111]

$$\Sigma(z) = ct_{sc}(z) \quad . \tag{3.6.30}$$

We introduce the following notation for the matrix elements of the averaged Green's function $G_{\underline{p}}^{-1}(z)$

$$G_{\underline{p}}^{-1}(z) = \begin{pmatrix} \tilde{z}(z) - \varepsilon_{\underline{p}} & \tilde{\Delta}(z) \\ \tilde{\Delta}(z) & \tilde{z}(z) + \varepsilon_{\underline{p}} \end{pmatrix} \quad . \tag{3.6.31}$$

The equations for the determination of $\tilde{z}(z)$ and $\tilde{\Delta}(z)$ are

$$\tilde{z}(z) = z - cJt_1(\tilde{z},\tilde{\Delta}) \quad , \quad \tilde{\Delta}(z) = \Delta - cJt_2(\tilde{z},\tilde{\Delta}) \quad , \tag{3.6.32}$$

where t_2 and t_1 are the amplitudes of electron scattering on an isolated impurity with and without spin flip, respectively.

For further calculation we employ the approximate expressions (3.6.16) for t_1 and t_2 which are valid in the energy range $z \sim \Delta$. Substituting these amplitudes into (3.6.32) and linearizing them with respect to $\tilde{\Delta}$, we obtain the expression for T_c in final form

$$\ln\frac{T_c}{T_{co}} = \psi\left(\frac{1}{2}\right) - \psi\left(\frac{1}{2} + \rho\right) \quad , \tag{3.6.33}$$

where

$$\rho = \frac{c}{(2\pi)^2 N(0) T_c}(1 - y_0^2) \quad . \tag{3.6.34}$$

The factor $1 - y_0^2$, which is a measure of the "unpairing" effect due to electron scattering on magnetic impuritis, equals ($\gamma < 0$):

$$1 - y_0^2 = \frac{\pi^2 S(S+1)}{\ln^2(T_K/T_c) + \pi^2 S(S+1)} \quad . \tag{3.6.35}$$

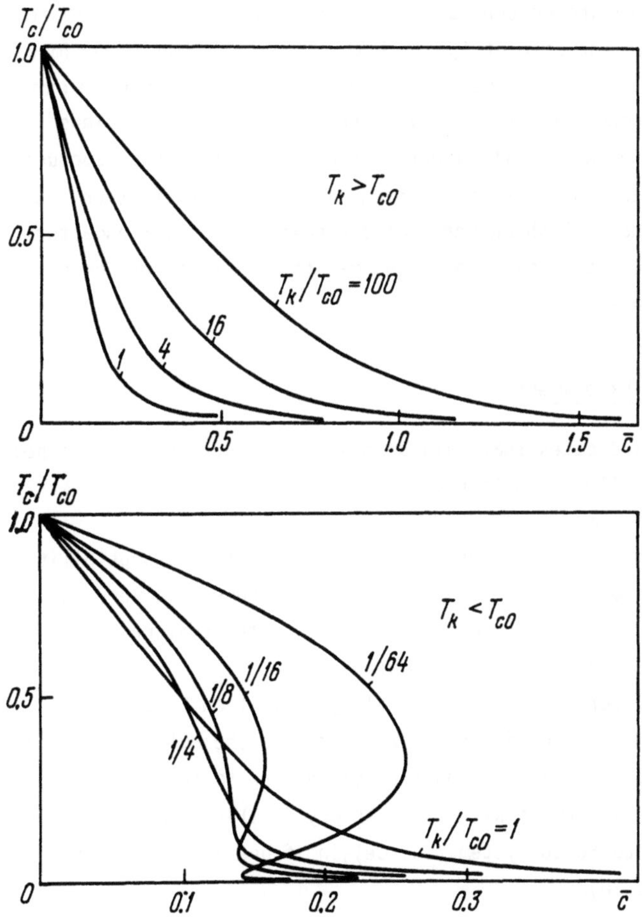

Fig.3.18. The superconducting transition temperature T_c as a function of paramagnetic impurity concentration according to MÜLLER-HARTMANN and ZITTARTZ [3.111]. $\bar{c} = c/[(2\pi)^2 N(0) T_{c0}]$, $S = 1/2$

This expression is obtained when one uses a self-consistent expression, in which $\tau = \ln(T_K/T_{co}) + \ln(T_K/T_c)$, for y_0 instead of (3.6.16). The T_c-dependence of the factor $1 - y_0^2$ leads to a peculiar concentration dependence of the superconducting transition temperature if compared with the Abrikosov-Gor'kov theory. Primarily, it will be observed that at $T_K \ll T_{co}$ $1 - y_0^2 \approx \pi^2 S(S+1)\gamma^2$. As a result, ρ tends to the expression $c \cdot S(S+1)J^2N(0)/4T_c$ appearing in the Abrikosov and Gor'kov theory. At an arbitrary value of T_K the factor $\ln(T_K/T_c)$ is of great importance. Numerical solution of (3.6.33) together with (3.6.34,35) leads to the results shown in Fig.3.18. The results depend largely on which is greater: T_K or T_{co}. When $T_K > T_{co}$ the superconducting transition temperature decreases monotonically with increasing concentration (Fig.3.18) withouth going to zero at any concentration. In the case of $T_K < T_{co}$, starting from some approximate value of the ratio $T_K/T_{co} \approx 0.1$, T_c becomes an ambiguous concentration function.

Thus, in a certain concentration interval there exist three solutions for T_c. This implies that for a definite impurity concentration, a superconducting state arises with decreasing temperature which, as T decreases further, is replaced by a normal phase. Such an unexpected behaviour of superconductors with magnetic impurities is a most pronounced manifestation of the Kondo effect in superconductors.

3.6.4 Comparison with Experiment

Experimental evidence indicates that similar phenomena occur in cerium-doped lanthanum alloys [3.114,115]. Simultaneously, these alloys are both superconductors and Kondo systems.

Low-temperature anomalies in the resistivity and magnetic susceptibility of LaCe and (LaCe)Al_2 show that the Kondo temperature therein is very low ($T_K \lesssim 0.5$ K), much less than superconducting transition temperatures (T_{co} = 6.0 K and 3.24 K for La and $LaAl_2$, respectively).

The dependence of T_c for the (La, Ce)Al_2 alloy [3.114] on Ce content is illustrated in Fig.3.19. The solid curve corresponds to the Müller-Hartmann and Zittartz theory presented above, and the dotted line to the Abrikosov and Gor'kov theory. The agreement with their theory [3.113] is also good for the other lanthanum-based Kondo system (La, Ce)$_3$In if one admits that T_K/T_{co} = 0.125 and $S = 1/2$ (Fig.3.20).

Figure 3.20 also presents data for the Th-U system which shows the behaviour characteristic of the case $T_K/T_{co} \gg 1$ (Fig.3.18). The experimental points lie on the theoretical curve when the parameters $T_K/T_{co} = 32$ and $S = 1/2$ are chosen.

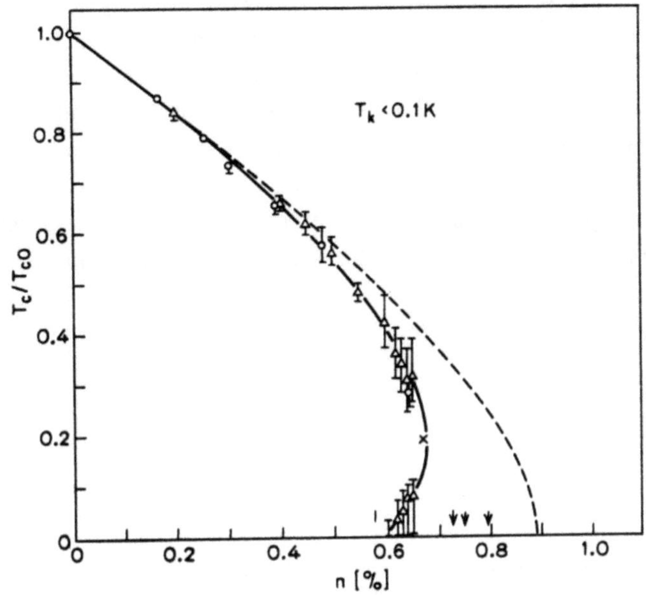

Fig.3.19. The superconducting transition temperature T_c in (La, Ce)Al_2 alloys as a function of the Ce content [3.114]. The solid line corresponds to the theory of MÜLLER-HARTMANN and ZITTARTZ, the dotted one to the theory of ABRIKOSOV and GOR'KOV

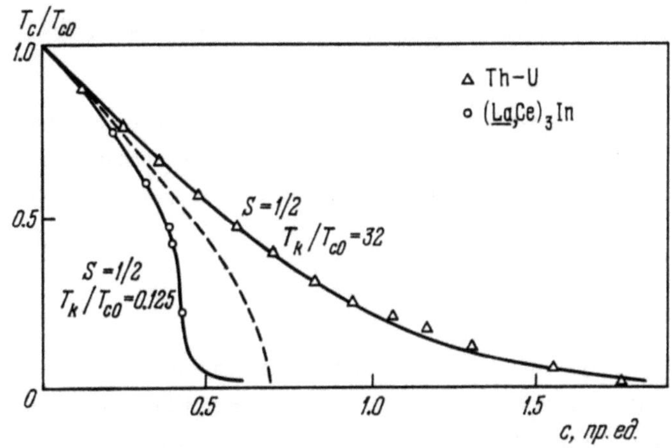

Fig.3.20. Comparison of the theory of MÜLLER-HARTMANN and ZITTARTZ [3.111] with experiments for Kondo systems (La, Ce)$_3$In and ThU. The dotted curve corresponds to the Abrikosov-Gor'kov theory, the triangles denote the alloy ThU, the circles mark the alloy (LaCe)In

However, for the Th-U system such behaviour of T_c versus impurity concentration U may also be explained on the basis of localized spin fluctuations [3.117]. For a more detailed discussion of experimental results on superconducting Kondo systems we refer the reader to the review by MAPLE [3.118].

Apart from Ce the Kondo effect is also exhibited by Sm and Eu ions added to a number of compounds based on rare-earth elements. Two types of dilute systems are known, viz., $(\underline{La}Sm)Sn_3$ and $Eu_xMo_6X_8$, (X = S, Se) in which characteristic Kondo anomalies in the physical properties were found to occur in the superconducting and normal phases [3.119].

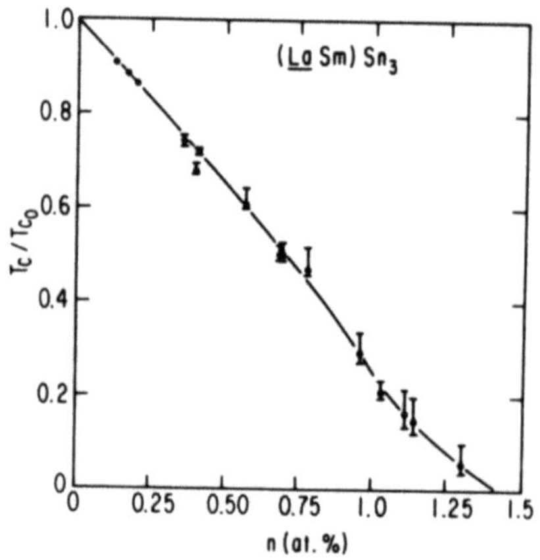

Fig.3.21. The suppression of T_c in the compound $(\underline{La}Sm)Sn_3$ [3.119]

The first evidence that $(\underline{La}Sm)Sn_3$ may be a Kondo system has been obtained from measurements of the suppression of T_c as a function of Sm concentration (Fig.3.21). These measurements have shown that the initial suppression is very large,

$$-\left(\frac{dT_c}{dC}\right)_{C=0} = 4.32\,\frac{K}{at\%\,Sm} \quad ,$$

which is typical of Kondo systems. Measurements of other physical quantities such as the jump in heat capacity at the point T_c, the temperature dependence, resistivity and magnetic susceptibility in the normal phase have confirmed the existence of the Kondo effect in this system, the estimate being $T_K < 100$ K. Similar anomalies have been obtained for the above compounds from Chevrel phases with a europium content of Eu x = 1.2. A peculiar feature of these compounds is the fact that Eu ions are regularly arranged in the lattice.

However, the atomic concentration of Eu ions is not too high in these compounds and amounts to 7%. Concentrated alloys of rare-earth elements also display Kondo-like anomalies in the physical properties, but these anomalies arise not because of the effects of exchange scattering on localized moments but because of the effects of hybridization of localized 4f-electron states and conduction band states.

The above theory, as is seen from (3.6.33,34) is equivalent to the Abrikosov and Gor'kov theory with replacement of the Born amplitude of electron scattering on a paramagnetic impurity by the full amplitude of scattering which describes the Kondo effect to SUHL's approximation [3.120]. Hence the existence of three temperatures of transition to a superconducting state when $T_K \ll T_{c0}$ becomes clear. The first T_c corresponds to the usual behaviour predicted by the Abrikosov-Gor'kov theory, the second arises from a violent increase in the amplitude of scattering with a spin flip as the temperature lowers [see (3.6.35)]. The occurrence of the third point T_c is associated with the decrease of this amplitude due to the compensation of the impurity spin when T drops below the Kondo temperature.

However, it is known in the Kondo effect theory that when calculating the scattering amplitude of an electron with a spin flip it is important to take into account all the electron-hole excitations that are produced near the Fermi surface in the event of a spin flip (infrared catastrophe) [3.121]. Such excitations are disregarded in the Abrikosov-Gor'kov theory. Recently, SCHLOTTMEN [3.122] has developed a new approach to the theory of the Kondo effect in superconductors, which enabled him to take into account the above-infrared problem in scattering and to obtain a more physical result in the description of the concentration dependence of T_c. It has turned out that the curve T_c as a function of concentration yields, at large c, only two transition points. This is in perfect agreement with the experiment represented in Fig.3.19. In the same paper, the dependence of the critical concentration, corresponding to $T_c = 0$, on the magnitude of s-d exchange coupling has been calculated and shown to have a minimum. For lack of space we refer the reader to SCHLOTTMEN's fundamental work for details [3.122] and also to the references given therein. In spite of the appearance of a large number of papers dealing with the theory of the Kondo effect in superconductors, the problem should not be regarded as finally solved since even for a normal metal the approximations are being continuously reconsidered and the greatest achievement in this field is apparently the work by WILSON [3.123]. The qualitative features of this phenomenon in superconductors outlined in this section, will eventually be reaffirmed in investigations to come.

3.7 Nonmagnetic Localized States in Superconducting Alloys

3.7.1 Anderson's Model

It was observed experimentally (see, for example, [3.124]) that the superconductivity in simple, i.e., non-transition metals is suppressed appreciably even when impurities of 3d transition elements (which, however, do not conserve the localized magnetic moment in a diluent metal) are added to them. For a normal metal the phenomenon of the suppression of a localized moment of a transition element impurity was usually studied using Anderson's model [3.125]. In this model, the impurity is treated as an isolated electron level placed into the zone of itinerant electron states of a diluent metal. When the hybridization of the electronic states of the impurity and of the matrix is taken into account, the localized level of the impurity acquires some finite width Γ and corresponds to a resonance state of the entire electronic system. The Coulomb interaction U of electrons with opposite spins (assume for simplicity that the localized level is nondegenerate) splits the resonance state with respect to the spin and, if the magnitude of this splitting is such that one of the levels passes the Fermi level, a localized magnetic moment may be expected to appear on the impurity. This line of reasoning is true, however, only when the width of the resonance level is small compared to its separation from the Fermi level. When $U \gg \Gamma$ a localized moment must exist, while in the opposite limit $U \ll \Gamma$, a localized moment is not expected [3.125].

Thus, nonmagnetic impurities of transition elements suggest the existence of a wide resonance in the electronic system. It will be shown below that such nonmagnetic resonances of d and f states lower the superconducting transition temperature in a ratio which is intermediate between its enormous suppression by magnetic impurities and its weak lowering by nonmagnetic impurities of non-transition elements.

We explore this problem with the help of Anderson's Hamiltonian to which the operator of effective conduction electron interaction in the BCS theory approximation [3.126] is added

$$H = \sum_{\underline{p}\alpha} \varepsilon_{\underline{p}} a^{+}_{\underline{p}\alpha} a_{\underline{p}\alpha} + \sum_{j\alpha} \varepsilon_d a^{+}_{j\alpha} a_{j\alpha} + \sum_{j\underline{p}\alpha} (v_{d\underline{p}} a^{+}_{j\alpha} a_{\underline{p}\alpha} + v_{\underline{p}d} a^{+}_{\underline{p}\alpha} a_{j\alpha})$$

$$+ U \sum_{j} a^{+}_{j\uparrow} a_{j\uparrow} a^{+}_{j\downarrow} a_{j\downarrow} - \sum_{\underline{p}\underline{p}'} v_{\underline{p}\underline{p}'} a^{+}_{\underline{p}\uparrow} a^{+}_{-\underline{p}\downarrow} a_{-\underline{p}'\downarrow} a_{\underline{p}'\uparrow} \quad . \qquad (3.7.1)$$

Here $\varepsilon_{\underline{p}}$ is the energy of a conduction electron with a momentum $\underline{p}$, ε_d stands for the energy of the isolated state of an impurity atom. The third term in the Hamiltonian describes the hybridization of conduction electrons with d-states of impurities (the s-d mixing according to [3.125]); the fourth describes the Coulomb interaction of d-electrons with opposite spins on an impurity atom, and the last term is the effective attraction of Cooper-pair electrons. Summation over j is taken over all lattice sites occupied by impurities. v_{dp} and v_{pd} are s-d mixing matrix elements and $v_{\underline{p}\underline{p}'}$ is the matrix element of electron attraction via phonons.

Except for the last term, the Hamiltonian (3.7.1) describes ANDERSON's model for the localized state of a normal metal. In order to take into account the superconducting state of conduction electrons, we replace the two-particle BCS Hamiltonian by a self-consistent expression involving the sources of Cooper pairs

$$\sum_{\underline{p}\underline{p}'} v_{\underline{p}\underline{p}'} a^+_{\underline{p}\uparrow} a^+_{-\underline{p}\downarrow} a_{-\underline{p}'\downarrow} a_{\underline{p}'\uparrow} \rightarrow -\Delta \sum_{\underline{p}} (a^+_{\underline{p}\uparrow} a^+_{-\underline{p}\downarrow} + a_{-\underline{p}\downarrow} a_{\underline{p}\uparrow}) \tag{3.7.2}$$

where

$$\Delta = v \sum_{\underline{p}}{}' \langle a_{-\underline{p}\downarrow} a_{\underline{p}\uparrow} \rangle \quad . \tag{3.7.3}$$

v is the mean value of the matrix element $v_{\underline{p}\underline{p}'}$ on the Fermi surface, the prime on the summation over momenta denotes summation in a narrow layer, being ω_D thick, near the Fermi surface. As is known, the use of expression (3.7.2) instead of the BCS effective interaction Hamiltonian is equivalent to the weak-coupling approximation in superconductivity theory and also to the Green's functions decoupling in Gor'kov's technique.

We perform a similar decoupling in the Hamiltonian describing the Coulomb interaction of d-electrons on an impurity, but in doing so we retain the conventional self-consistent field Hartree-Fock corrections

$$\begin{aligned} a^+_{j\uparrow} a_{j\uparrow} a^+_{j\downarrow} a_{j\downarrow} \rightarrow & \langle a^+_{j\downarrow} a_{j\downarrow} \rangle a^+_{j\uparrow} a_{j\uparrow} + \langle a^+_{j\uparrow} a_{j\uparrow} \rangle a^+_{j\downarrow} a_{j\downarrow} \\ & + \langle a^+_{j\uparrow} a^+_{j\downarrow} \rangle a_{j\downarrow} a_{j\uparrow} + \langle a_{j\downarrow} a_{j\uparrow} \rangle a^+_{j\uparrow} a^+_{j\downarrow} \quad . \end{aligned} \tag{3.7.4}$$

The idea of taking into account the Cooper pairing in a system of electrons on impurities has been proposed by RATTO and BLANDIN [3.116], proceeding from the considerations given below. Because of the s-d mixing, the eigenfunctions of the one-particle Hamiltonian are a superposition of Bloch functions of conduction electrons and localized d-states. The pairing of conduction electrons as a result of the effective attraction via phonons should lead to the nonzero of $\langle a^+_{j\uparrow} a^+_{j\downarrow} \rangle$ type anomalous averages, which will be verified by self-consistent computations.

Thus, as a result of simple approximations (3.7.2,4) for two-particle interactions it is convenient to rewrite the Hamiltonian of the model under consideration (3.7.1) in the following form

$$H = H_s + H_d + H_{sd} \quad , \tag{3.7.5}$$

where

$$H_s = \sum_{\underline{p}\alpha} \varepsilon_{\underline{p}} a^+_{\underline{p}\alpha} a_{\underline{p}\alpha} - \sum_{\underline{p}} \Delta (a^+_{\underline{p}\uparrow} a^+_{-\underline{p}\downarrow} + a_{-\underline{p}\downarrow} a_{\underline{p}\uparrow}) \quad , \tag{3.7.6}$$

$$H_d = \sum_{j\alpha} E_d a^+_{j\alpha} a_{j\alpha} - \sum_{j} \Delta_d (a^+_{j\uparrow} a^+_{j\downarrow} + a_{j\downarrow} a_{j\uparrow}) \quad , \tag{3.7.7}$$

$$H_{sd} = \sum_{\underline{p}j\alpha} (V_{d\underline{p}} a^+_{j\alpha} a_{\underline{p}\alpha} + V_{\underline{p}d} a^+_{\underline{p}\alpha} a_{j\alpha}) \quad . \tag{3.7.8}$$

Here E_d is the energy of d-states which is renormalized due to the Coulomb interaction

$$E_d = \varepsilon_d + U\langle n \rangle \quad ,$$

$\langle n \rangle = \langle a^+_{j\uparrow} a_{j\uparrow} \rangle = \langle a^+_{j\downarrow} a_{j\downarrow} \rangle$ is the mean value of the density of d-electrons on an impurity site, on the condition that no localized magnetic moment arises on the site. The quantities Δ and Δ_d are the order parameters in the system of s and d-electrons, from which Δ is defined by (3.7.3), and

$$\Delta_d = -U \langle a_{j\downarrow} a_{j\uparrow} \rangle \quad . \tag{3.7.9}$$

The Hamiltonian (3.7.5-8) was used by KIWI and ZUCKERMANN [3.127], and also by KAISER [3.128], to study the problem involved.

3.7.2 The Energy Gap and the Order Parameter

Let us introduce the Green's functions of s and d-electrons:

$$G_{\underline{p}\underline{p}'}(\tau - \tau') = - \langle T\Psi_{\underline{p}}(\tau) \overline{\Psi}_{\underline{p}'}(\tau') \rangle \quad , \tag{3.7.10}$$

$$G_{jj'}(\tau - \tau') = - \langle T\Psi_{j}(\tau) \overline{\Psi}_{j'}(\tau') \rangle \quad , \tag{3.7.11}$$

where $\Psi_{\underline{p}}$ and Ψ_j are two-component field operators

$$\Psi_{\underline{p}} = \begin{pmatrix} a_{\underline{p}\uparrow} \\ a^+_{-\underline{p}\downarrow} \end{pmatrix} \quad , \quad \Psi_j = \begin{pmatrix} a_{j\uparrow} \\ a^+_{j\downarrow} \end{pmatrix} \quad . \tag{3.7.12}$$

Expressions (3.7.10,11) define matrix Green's functions in the Gor'kov-Nambu representation. The Green's function for electrons depends on two momenta in view of the heterogeneity of a metal containing impurities. However, upon averaging over impurity configurations, $G_{\underline{p}\underline{p}'}$ as well as $G_{d\underline{p}\underline{p}'}$ (the Fourier transform of the function $G_{jj'}$) should be diagonal with respect to momenta

$$\left\langle G_{\underline{p}\underline{p}'}(z) \right\rangle_{imp} = G_{\underline{p}}(z)\delta_{\underline{p}\underline{p}'} \quad , \quad \left\langle G_{d\underline{p}\underline{p}'}(z) \right\rangle_{imp} = G_d(z)\delta_{\underline{p}\underline{p}'} \quad . \tag{3.7.13}$$

To compute the functions $G_{\underline{p}}(z)$and $G_d(z)$ we resort to perturbation theory. We shall consider the energy of the s-d mixing (3.7.8) as the perturbation, then according to (3.7.6,7) the zero-approximation Green's function will be of the form

$$G^0_{\underline{p}}(z) = \begin{pmatrix} z - \varepsilon_{\underline{p}} & -\Delta \\ -\Delta & z + \varepsilon_{\underline{p}} \end{pmatrix}^{-1} \quad , \quad G^0_d(z) = \begin{pmatrix} z - E_d & -\Delta_d \\ -\Delta_d & z + E_d \end{pmatrix}^{-1} \quad . \tag{3.7.14}$$

Expanding (3.7.10,11) in a power series of the operator H_{sd} and averaging the expression obtained with respect to impurity configurations, we find graphical series for the quantities $G_{\underline{p}}(z)$ and $G_d(z)$ from which the irreducible self-energy parts $\Sigma_{\underline{p}}(z)$ and $\Sigma_d(z)$ may be singled out. They are related to the corresponding Green's functions by the equations

$$\left.\begin{aligned} G^{-1}_{\underline{p}}(z) &= z\tau_0 - \varepsilon_{\underline{p}}\tau_3 - \Delta\tau_1 - \Sigma_{\underline{p}}(z) \\ G^{-1}_d(z) &= z\tau_0 - E_d\tau_3 - \Delta_d\tau_1 - \Sigma_d(z) \end{aligned}\right\} \quad . \tag{3.7.15}$$

On the other hand, we use the standard matrix representation for the quantities G^{-1} in terms of the functions Z and ϕ:

$$\left.\begin{aligned} G^{-1}_{\underline{p}}(z) &= zZ(z)\tau_0 - \varepsilon_{\underline{p}}\tau_3 - \phi(z)\tau_1 \\ G^{-1}_d(z) &= zZ_d(z)\tau_0 - E_d\tau_3 - \phi_d(z)\tau_1 \end{aligned}\right\} \quad , \tag{3.7.16}$$

from which we find the expressions for the Green's functions

$$G_{\mathbf{p}}(z) = \frac{zZ(z)\tau_0 + \varepsilon_{\mathbf{p}}\tau_3 + \phi(z)\tau_1}{[zZ(z)]^2 - \phi^2(z) - \varepsilon_{\mathbf{p}}^2} \qquad (3.7.17)$$

and a similar expression for G_d with the replacement of Z and ϕ by Z_d and ϕ_d, respectively.

Now consider a graphical representation for the quantities $\Sigma_{\mathbf{p}}(z)$ and $\Sigma_d(z)$. Since H_{sd} is a one-particle perturbation operator, the diagram technique, which is described in the book by ABRIKOSOV et al. [2.4] corresponds to the inhomogeneous external field case. In particular, the frequency on the only electron line of the graph for Σ is conserved everywhere and therefore no integration over frequencies occurs in the expressions. In the lowest order of expansion in H_{sd} the self-energy parts Σ_p and Σ_d are given by the graphs

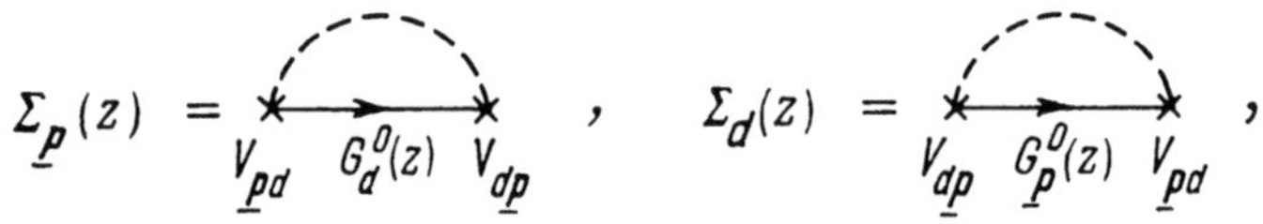

(3.7.18)

in which the arrowed lines show the Green's functions G_d^0 and $G_{\mathbf{p}}^0$ and the crosses mark the expression $V_{\mathbf{p}d}\tau_3$ or $V_{d\mathbf{p}}\tau_3$. The dotted line connecting the vertices indicates that impurity scattering takes place at the same site. Averaging over different site positions is done according to ABRIKOSOV and GOR'KOV [3.30,21].

As in the derivation of the basic equations in the phonon model of the superconductor, we will thicken the electronic lines in (3.7.18). Then we obtain a couple of equations for $\Sigma_{\mathbf{p}}(z)$ and $\Sigma_d(z)$:

$$\Sigma_d(z) = -c\int \frac{d\mathbf{p}}{(2\pi)^3} V_{d\mathbf{p}}\tau_3 G_{\mathbf{p}}(z)\tau_3 V_{\mathbf{p}d} \quad , \qquad (3.7.19)$$

$$\Sigma_{\mathbf{p}}(z) = -V_{\mathbf{p}d}\tau_3 G_d(z)\tau_3 V_{d\mathbf{p}} \quad , \qquad (3.7.20)$$

where c stands for the concentration of impurities. We substitute (3.7.17) for $G_{\mathbf{p}}(z)$ into (3.7.19) and perform the integration over momenta by reducing it according to the relation (2.2.17) to an integration over the energy ξ_p. Computing the standard integral in (3.7.19) and using identities (3.7.15,16) we obtain the following combined algebraic equations for the quantities Z, ϕ, Z_d and ϕ_d [3.116]

$$Z(z) = 1 + \frac{c|V_{pd}|^2 Z_d(z)}{D(z)} \quad , \quad \phi(z) = \Delta + \frac{c|V_{pd}|^2\phi_d(z)}{D(z)} \quad , \qquad (3.7.21)$$

$$Z_d(z) = 1 + \frac{|V_{pd}|^2 \pi N(0) Z(z)}{\sqrt{\phi^2(z) - z^2 Z^2(z)}} \quad , \quad \phi_d(z) = \Delta_d + \frac{|V_{pd}|^2 \pi N(0) \phi(z)}{\sqrt{\phi^2(z) - z^2 Z^2(z)}}$$

where (3.7.21)

$$\mathcal{D}(z) = \phi_d^2(z) + E_d^2 - z^2 Z_d^2(z) \quad . \tag{3.7.22}$$

When writing the combined equations in (3.7.21) we have neglected the dependence on the momentum in the matrix element $|V_{pd}|^2$ of the s-d mixing by taking its mean value on the Fermi surface.

Now write out the order parameters Δ and Δ_d using their definitions according to (3.7.3,9). Substituting into these the expression $G_p(z)$ from (3.7.17) and the analogous expression for $G_d(z)$ we obtain two basic equations:

$$\Delta = \lambda \pi T \sum_n{}' \frac{\phi(i\omega_n)}{\sqrt{\phi^2(i\omega_n) + \omega_n^2 Z^2(i\omega_n)}} \tag{3.7.23}$$

and

$$\Delta_d = -UT \left[\sum_n \frac{\Delta_d}{\mathcal{D}(i\omega_n)} + \sum_n{}' \frac{|V_{pd}|^2 \pi N(0) \phi(i\omega_n)}{\mathcal{D}(i\omega_n) \sqrt{\phi^2(i\omega_n) + \omega_n^2 Z^2(i\omega_n)}} \right] \tag{3.7.24}$$

where $\lambda = N(0)V$ is the usual parameter of the BCS theory. The prime on the summation over frequencies denotes that summation is restricted to Debye frequencies, i.e., $|\omega_n| < \omega_{\mathcal{D}}$. This condition arises due to the restriction of summation over momenta in (3.7.3) within the Debye layer.

In order to investigate the nature of the superconducting state in the model under analysis (particularly, to check whether there is gapless superconductivity) it is necessary to calculate the energy gap. That information is contained in the function $\tilde{\Delta}(\omega) = \phi(\omega)/Z(\omega)$ for real values of ω. Indeed, the density of one-particle states in the superconductor, $n(\omega)$, is expressed as follows

$$n(\omega) = \frac{1}{\pi} \mathrm{Im} \left\{ \sum_p G_p^{11}(\omega) \right\} = N(0)\, \mathrm{Re} \left\{ \frac{\omega}{\sqrt{\omega^2 - \tilde{\Delta}^2(\omega)}} \right\} \quad . \tag{3.7.25}$$

Here $G_p^{11}(\omega)$ is the 11th matrix element of the Green's function (3.7.17).

With the aid of (3.7.21) one can obtain the following expression for the gap function $\tilde{\Delta}(z)$ (within an accuracy of $\sim |V_{dp}|^2$):

$$\tilde{\Delta}(z) = \frac{\Delta[1 - H(z)d]}{1 + H(z)} \quad , \tag{3.7.26}$$

where

$$d = -\frac{\Delta_d}{\Delta} \quad , \quad H(z) = c|V_{pd}|^2/\mathcal{D}(z) \quad . \tag{3.7.27}$$

It is now necessary to perform summation in expressions (3.7.23,24). First consider the quantity $\mathcal{D}(i\omega_n)$ which with the help of (3.7.21) may be transformed to

$$\mathcal{D}(i\omega_n) = E_d^2 + \Delta_d^2 + \Gamma^2 + \omega_n^2 + 2\Gamma\frac{\Delta_d\phi(i\omega_n) + \omega_n^2 Z(i\omega_n)}{\sqrt{\phi^2(i\omega_n) + \omega_n^2 Z^2(i\omega_n)}} \quad . \tag{3.7.28}$$

Here

$$\Gamma = \pi|V_{dp}|^2 N(0) \tag{3.7.29}$$

is the width of the d state due to the s-d mixing processes. The typical value of Γ for transition element alloys in a nontransition matrix will amount to about 1 eV so that $\Gamma \gg \omega_\mathcal{D}$. For $|\omega_n| < \omega_\mathcal{D}$

$$\mathcal{D}(i\omega_n) \approx E_d^2 + \Gamma^2 \quad . \tag{3.7.30}$$

To this approximation the second sum in (3.7.24) is the same as that in (3.7.23), therefore (3.7.24) may be written as

$$\Delta_d\left[1 + UT\sum_n \frac{1}{\mathcal{D}(i\omega_n)}\right] = -\frac{\Gamma U\Delta}{\left(\Gamma^2 + E_d^2\right)\pi\lambda} \quad . \tag{3.7.31}$$

Since $\Gamma \gg \omega_\mathcal{D} \gg \Delta_d$, the frequencies $|\omega_n| \gg \Delta_d$ are essential in the sum over frequencies in (3.7.31). For such large frequencies, (3.7.28) is approximated by

$$\mathcal{D}(i\omega_n) = E_d^2 + (\Gamma + |\omega_n|)^2 \quad , \tag{3.7.32}$$

and we thus obtain

$$T\sum_n \frac{1}{\mathcal{D}(i\omega_n)} = \frac{1}{\pi E_d}\arctan\frac{E_d}{\Gamma} \quad .$$

Finally, we find the quantity of interest, d, with (3.7.27) and (3.7.31)

$$d = \frac{N_d(0)U_{eff}}{\lambda} \quad . \tag{3.7.33}$$

Here

$$N_d(0) = \frac{1}{\pi}\frac{\Gamma}{E_d^2+\Gamma^2} \qquad (3.7.34)$$

represents the density of d-states on the Fermi surface, and

$$U_{eff} = \frac{U}{1+\frac{U}{\pi E_d}\arctan\frac{E_d}{\Gamma}} \qquad (3.7.35)$$

is the effective Coulomb interaction of d-electrons. It is useful to note that the same magnitude of effective interaction arises when account is taken of correlations in Anderson's model with the low density of electrons on the impurity [3.129].

We now write the final expression (3.7.26) for the gap $\tilde{\Delta}(i\omega_n)$. Because of (3.7.30) for frequencies $|\omega_n| < \omega_D$, the quantity $H(i\omega_n)$ does not depend on ω_n at all so $\tilde{\Delta}(i\omega_n) \equiv \tilde{\Delta}$ is also constant. Equation (3.7.26) may then be written in the form

$$\tilde{\Delta} = f\Delta \quad , \quad f = \frac{1-ad}{1+a} \quad , \qquad (3.7.36)$$

where the notation

$$a = cN_d(0)/N(0) \qquad (3.7.37)$$

is introduced.

The quantity f measures the deviation of the energy gap $\tilde{\Delta}$ from the order parameter. It is determined by two dimensionless quantities d (3.7.33) and a (3.7.37). The constant f is temperature-independent, which indicates that the order parameter Δ and the energy gap $\tilde{\Delta}$ have the same temperature dependence. Thus, when the gap vanishes the order parameter (superconductivity) disappears. This is indicative of the fact that in the case of nonmagnetic impurities (though with resonance states in them) no gapless superconductivity occurs. This conclusion has been drawn by KAISER [3.128], in contrast to KIWI and ZUCKERMANN [3.127] who have erroneously assumed that gapless superconductivity exists in such alloys.

3.7.3 Suppression of the Superconducting Transition

Let us compute the variation of T_c in an alloy with a concentration of nonmagnetic impurities. To this end we modify (3.7.23) for the order parameter

Δ to obtain

$$\Delta = \lambda\pi T \sum_n{}' \frac{\tilde{\Delta}}{\sqrt{\tilde{\Delta}^2 + \omega_n^2}} \quad . \tag{3.7.38}$$

By virtue of the relation $\tilde{\Delta} = f\Delta$ this equation may be rewritten as

$$\tilde{\Delta} = f\lambda\pi T \sum_n{}' \frac{\tilde{\Delta}}{\sqrt{\tilde{\Delta}^2 + \omega_n^2}} \quad . \tag{3.7.39}$$

This is the usual BCS formula for an energy gap where $f\lambda$ acts as the pairing interaction parameter. Therefore, we may immediately write an expression for the superconducting transition temperature of the alloy involved:

$$T_c = 1.14\omega_D \exp\left(-\frac{1}{f\lambda}\right) \quad . \tag{3.7.40}$$

With increasing concentration the quantity f decreases from unity and this leads to an exponential decrease in T_c. The variation of T_c obeys the BCS formula with gradually decreasing coupling interaction. In this respect the behaviour of the alloys with nonmagnetic impurities differs crucially from the case of an alloy with magnetic impurities (Abrikosov-Gor'kov theory) where the variation of T_c with concentration is associated with the destruction of Cooper pairs.

If we denote the superconducting transition temperature in a pure metal by T_{co}, (3.7.40) may be rewritten as

$$\ln\frac{T_c}{T_{co}} = \frac{f-1}{f\lambda} \quad . \tag{3.7.41}$$

In the limit of low impurity concentrations this equation yields

$$\frac{\Delta T_c}{T_{co}} = \frac{a(1+d)}{\lambda} = c\frac{N_d(0)}{N(0)\lambda}\left[1 + \frac{N_d(0)U_{eff}}{\lambda}\right] \tag{3.7.42}$$

which determines the effect of the lowering of T_c. This effect is described by two multiplicative factors.

The first factor $N_d(0)/N(0) \sim 1/\Gamma$ (when the resonance is broad, $\Gamma \gg |E_d|$) is due to the resonance level electron scattering which leads to a variation of the local density of states on the impurity (the dilution effect, following the terminology of MARKOWITZ and KADANOV [3.130]). This effect was first obtained in ZUCKERMANN's pioneer paper [3.126].

The second factor on the right-hand side of (3.7.42) takes into account the influence of the Coulomb repulsion of d-electrons on an impurity site. It was first obtained by RATTO and BLANDIN [3.116], and proves to be important when calculating the lowering of T_c. Thereby the magnitude of U_{eff} and also the occupation number $\langle n\rangle$ of the d-electrons play a role. The latter dependence arises through the quantity $N_d(0)$ which, in turn, depends in a resonant kind of way on E_d, i.e., on the position of the d-level relative to the Fermi energy [see (3.7.34)]. This is determined by $E_d = \varepsilon_d + U\langle n\rangle$. In the so-called symmetric limit, which is frequently used, it is assumed that $\varepsilon_d = -U/2$, then

$$E_d = U(\langle n\rangle - 1/2) \tag{3.7.43}$$

depends solely on the parameter U.

Thus, it is evident that $\Delta T_c/T_{co}$ depends resonantly on the quantity $\langle n\rangle$ with a peak width determined by Γ. The maximum is reached when the state is half-filled.

It was assumed in the model under analysis that the localized state on the impurity atom is nondegenerate. If the orbital degeneracy is to be taken into account, then expression (3.7.37) for a should be multiplied by a factor of $(2l+1)$. This results in the factor of $(2l+1)$ appearing in (3.7.42) for $\Delta T_c/T_{co}$. For a half-filled d-shell ($l = 2$) it is $\langle n\rangle = 5$. Thus, depending on the impurity element, $\Delta T_c/T_c$ must have a maximum for manganese.

This, indeed, is the case in experiments with aluminium as dilutent [3.131] (Fig.3.22). This figure also portrays the theoretical dependence of $\Delta T_c/T_c$ on the number of d-electrons in an isolated impurity atom which has been calculated according to the formulae of Ratto and Blandin. One can see that the agreement between theory and experiment is good.

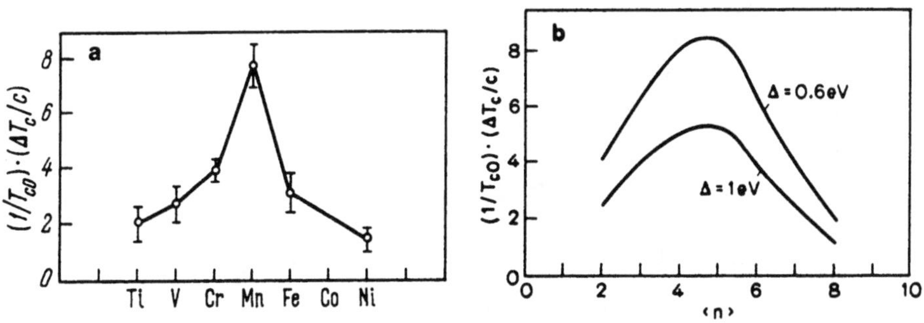

Fig.3.22. The dependence of the effect of suppression of T_c in Al on the impurity of the 3d-transition element: a) experiment, b) RATTO and BLANDIN's [3.116] theory. The adopted value is $U = 10$ eV. The curves have been calculated for two values of Δ

We have treated the case of low impurity concentrations. At high concentrations one should make use of Kaiser's formula (3.7.41), and substitute into it the expression (3.7.36) for f. A comparison of the concentration dependence of T_c/T_{c0} for the system $\underline{Th}$ Ce [3.132] with theoretical values obtained on the basis of Kaiser's formula is given in Fig.3.23. The experiment was carried out at different pressures. The dependence of the effect on pressure is due to the fact that the parameters governing the concentration dependence of T_c, primarily the quantity E_d (the position of the d-level relative to the Fermi energy), vary with pressure. The curves presented in Fig.3.23 have been obtained on the basis of Kaiser's formula (3.7.41), with parameters selected by the least squares method. In this manner, the following values of the parameters were estimated: $N_d(0) = 4.86$ states/(eV·atom) [in a reasonable combination with the value of 7.6 states/(eV·atom) derived from low-temperature measurements of the specific heat capacity in the normal phase] and $U_{eff} = 7.10^{-2}$ eV. This leads to the estimate of the parameters of Anderson's model:

$$\Gamma \gtrsim 1.3 \cdot 10^{-2}\ \mathrm{eV}, \quad E_d \gtrsim 7.6 \cdot 10^{-2}\ \mathrm{eV}, \quad U \gtrsim 1.2 \cdot 10^{-1}\ \mathrm{eV}.$$

As stated in [3.124], the value of Γ is in good agreement with previous estimates [3.133], but U seems to be rather small.

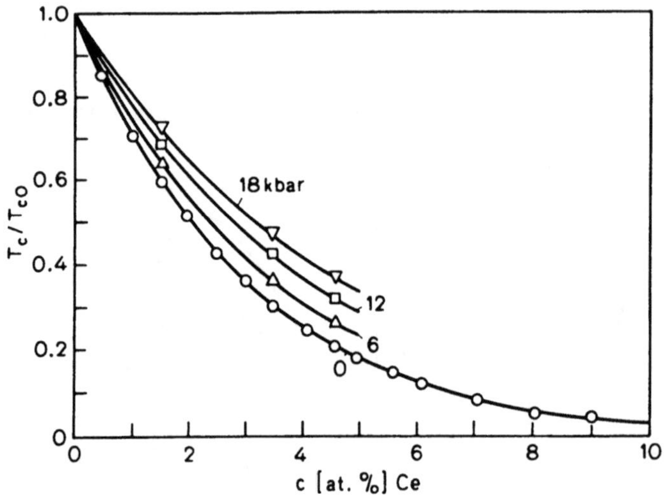

Fig.3.23. The lowering of T_c in the $\underline{Th}$ Ce alloy as a function of the Ce content. The curves have been obtained on the basis of Kaiser's formula (3.7.41) with the parameters chosen according to the least squares method

3.8 Localized Spin Fluctuations in Superconducting Alloys

3.8.1 The Anomalous Behaviour of a Number of Alloys with Nonmagnetic Impurities

One more question arises in connection with the influence of nonmagnetic impurities. The point is that dilute solutions of transition elements in simple metal superconductors exist in which the impurity atoms do not conserve the localized magnetic moment. However, the behaviour of such alloys both in the superconducting and normal phases differs from the case considered above, i.e., from an alloy with nonmagnetic localized states. Examples of such alloys are $\underline{Al}Mn$ and $\underline{Th}U$. Despite the absence of localized magnetic moments in an alloy such as $\underline{Al}Mn$, some features of the alloys which have clearly marked magnetic moments on impurities manifest themselves in its properties and may even be likened to Kondo systems. Thus, CAPLIN and RIZZUTO [3.134] have found that the temperature curve of resistivity $\rho(T)$ of dilute alloys possesses a minimum $\rho(T)$ described by the formula

$$\rho(T) = \rho_0\left[1 - \left(\frac{T}{T_0}\right)^2\right] + AT^3 \quad , \tag{3.8.1}$$

where $T_0 = 500$ K, and the term $\sim T^3$ originates from the electron-phonon interaction. The magnetic susceptibility of $\underline{Al}Mn$ appreciably exceeds the Pauli susceptibility; a similar enhancement occurs in the magnitude of the electronic heat capacity and thermal electromotive force. The decrease of the superconducting transition temperature appears to be particularly significant in this alloy and it turns out to be greater than expected from RATTO and BLANDIN's theory [3.116] (outlined above in Sect.3.7) for an alloy with nonmagnetic resonances on impurities.

The established concept based on these anomalies is that such alloys exhibit localized spin fluctuations (LSF) at which a localized magnetic moment may emerge on impurity atoms for a time of the order of τ_s. This localized magnetic moment fluctuates both in magnitude and direction on this time scale.

One expects that with respect to their physical properties, systems with LSF are intermediate to alloys with nonmagnetic impurities and those with magnetic ones. The limit $\tau_s \to 0$ corresponds to the first of these cases, and the limit $\tau_s \to \infty$ to the second. Thus, alloys with LSF should be specified by the energy parameter

$$\Gamma_s = \frac{\hbar}{\tau_s} \quad \text{or} \quad T_0 = \frac{\hbar}{\tau_s k} \quad , \tag{3.8.2}$$

where the temperature T_0 plays the same role as the Kondo temperature does in the description of Kondo systems (Sect.3.6). The temperature T_0 which is involved in the empirical law (3.8.1) for resistivity should be identified exactly with the T_0 defined in (3.8.2).

The influence of the LSF on the superconducting transition temperature is particularly interesting in the region $T_0 \sim T_{co}$. For $T_0 \ll T_{co}$, the system is well described by the Abrikosov-Gor'kov theory of localized magnetic moments. The case $T_0 \gg T_{co}$ corresponds to nonmagnetic impurities described by the Zuckermann-Ratto-Blandin theory. Thus, if the parameters of the system were varied so as to allow the quantity T_0 to vary continuously it would be possible to make a continuous transition from the case of a magnetic impurity to a completely nonmagnetic one by passing the region of the occurrence of LSF. Such a transition may be performed through the use of pressure. Figure 3.24 shows the dependence of the suppression of the superconducting transition temperature, i.e., the quantity $\Delta T_c/c$ (c stands for the impurity concentration) for the alloy La Ce (1.3% Ce) on applied pressure [3.135]. Two types of behaviour of ΔT_c are observed which merge at a pressure of about 30 kbar. The two solid lines describe the effect according to Abrikosov and Gor'kov's theory (curve A) and Ratto and Blandin's theory (curve B). Both theories become inconsistent in the transition region which corresponds to the appearance of localized spin fluctuations.

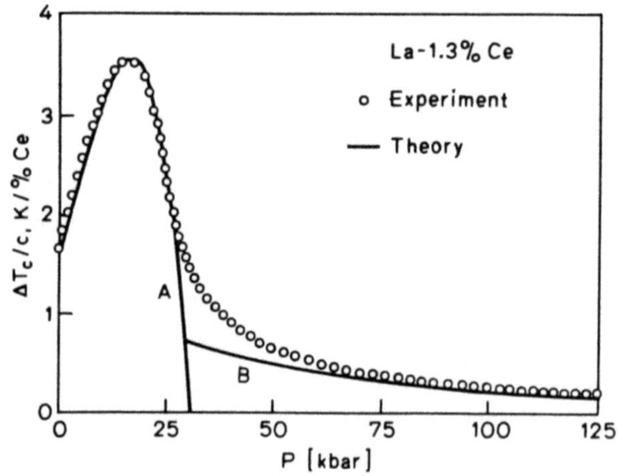

Fig.3.24. The variation of T_c/c in the La Ce alloy with pressure. Curve A corresponds to ABRIKOSOV and GOR'KOV's [3.21] theory, Curve B to RATTO and BLANDIN's [3.116] theory

We now attempt to make clear why the modes A and B correspond to the above theories. According to Abrikosov and Gor'kov's theory, the effect of suppression of T_c by magnetic impurities is proportional to the square of the s-d exchange integral J: $\Delta T_c/c \sim J^2$ [3.21]. On the other hand, there is the expression for J represented as a sum of two terms $J = J_0 + J_1$, where J_0 (>0) is a Heisenberg integral and J_1 (<0) occurs due to the mixing localized electron states on an impurity and conduction electrons of the matrix. Accoring to the transformation of SCHRIEFFER and WOLFF [3.136], Anderson's Hamiltonian transforms to a s-d exchange Hamiltonian and we get the following expression for J_1:

$$J_1 \backsim \frac{|V_{pd}|^2 U}{E_d(E_d + U)} \tag{3.8.3}$$

in the case when $|E_d| \gg \Gamma$. By the action of pressure the d-level of a Ce atom comes closer to the Fermi surface, which leads to an increase in the contribution of J_1 to the s-d exchange integral with increasing pressure. This corresponds to the initial spreading portion of curve A in Fig.3.24. When $|E_d|$ decreases we should use a different expression for the effective s-d exchange integral [3.137]:

$$J_1 \backsim -\frac{2|V_{pd}|^2 |E_d}{E_d^2 + \Gamma^2} \tag{3.8.4}$$

It has a maximum when $E_d = -\Gamma$ and goes to zero when $E_d = 0$. Thus assuming $|E_d|$ to decrease linearly with pressure, we see that J_1 must pass through a maximum, and this leads to curve A in Fig.3.24.

At pressures exceeding the transition region where LSF appear, ΔT_c can be described by the Ratto-Blandin formula (3.7.42) which shows a weak decrease with a further increase in pressure (as $|E_d|$ decreases).

Generally speaking, another point of view [3.124] exists concerning the observed transition in the La Ce system under pressure. At zero pressure the alloy La Ce is a Kondo system. As the ratio T_k/T_{co} increases, as shown in Sect.3.6, the quantity ΔT_c passes through a maximum which may be identified with the maximum given in Fig.3.24.

The substantial difference between these models of the pressure dependence of ΔT_c in the alloy La Ce is described in the following. In the first case the abrupt decrease of ΔT_c after going through a maximum is due to the demagnetization of the localized state when the energy of the 4f-level of Ce tends to the Fermi energy. Thus, the suppression of T_c at low pressures, which is associated with the destruction of Cooper pairs by the magnetic state of the impurity, is replaced by the process of weakening of the pairing energy when the 4f-level passes through the Fermi level. In the second case, the 4f-level of Ce never crosses the Fermi level and the impurity atom never loses its magnetic moment. The nonmonotonic variation of T_c in this case comes from the Cooper pair destruction. The first mechanism seems to be more probable since the second one, which is due to the Kondo effect, requires unrealistically large parameters [3.112] for La Ce.

In conclusion, we note that there are a number of other superconducting alloys in which a similar transition from the magnetic to the nonmagnetic impurity state apparently occurs: Y Ce under pressure [3.138] and (Zn, Al)Mn and (Y, Th) Ce when the composition of the binary matrix varies [3.124,139].

3.8.2 Theoretical Study of the Effect of LSF on T_c

A general theory of transition from a magnetic to a nonmagnetic impurity state in a superconductor should describe the continuous transition from the mechanism of suppression of superconductivity due to the destruction of Cooper pairs by impurities to the mechanism of weakening of the pair coupling energy. Several attempts have been made to construct a general theory of the influence of transition metal impurities on T_c which would transform in the limiting cases into the Abrikosov and Gor'kov's or the Zuckermann, Ratto, and Blandin's theory and, in the intermediate case, would describe localized spin fluctuations. We give a brief description of these attempts below.

THEUMANN [3.140] has employed the functional integral method in the problem of impurities in a superconductor in the Anderson model. Her method is a direct generalization of the approach of WANG et al. [3.141] developed for Anderson's model in a normal metal. It allows the calculation of the partition function of the system in terms of an integral over fields which depend on "thermodynamic time". In the static approximation, in which only the Fourier component of a zero-frequency field is taken into account, the functional integral reduces to a one-dimensional integral. Then the partition function as well as other quantities, e.g., ΔT_c may be calculated for arbitrary ratios of Γ/U.

In the limit $\Gamma/U \ll 1$ it follows that

$$\frac{\Delta T_c}{T_{co}} = -c\,\frac{1}{2N(0)T_{co}}\left(\frac{\Gamma}{U}\right)^2 = -c\,\frac{3\pi^2}{16.8}\,\frac{N(0)}{T_{co}}\,J^2 \quad . \tag{3.8.5}$$

Here J is the effective exchange integral determinable by (3.8.3). In the symmetric limit ($E_d = -U/2$) (3.8.3) is equal to $-4[\pi N(0)]^{-1}\Gamma/U$, from which one obtains a second version of the representation of (3.8.5) (up to a numerical factor of the order of unity) corresponding to the Abrikosov-Gor'kov theory.

In the opposite limiting case $\Gamma/U \gg 1$ we obtain the expression

$$\frac{\Delta T_c}{T_{co}} = -c\frac{1}{\pi N(0)\Gamma}\left(\ln\frac{2\gamma\Gamma}{\pi T_{co}} - 1\right) \tag{3.8.6}$$

which agrees with ZUCKERMANN's result [3.139].

In the intermediate case $\Gamma/U \sim 1$, when the fluctuations of the localized spin are strong in magnitude and direction, the resulting expression for ΔT_c may be calculated numerically. The results presented in Fig.3.25 agree qualitatively with the experimental curve for the alloy La Ce (Fig.3.24) and demonstrate the behaviour of ΔT_c in the transition mode as a function of the applied pressure. Of course, the theory of LSF based on a static approximation is only an interpolation between two known limits, but it is interesting in the respect that it demonstrates the possibility of passing from one limit to the other by varying a limited number of parameters (one parameter in this case: Γ/U).

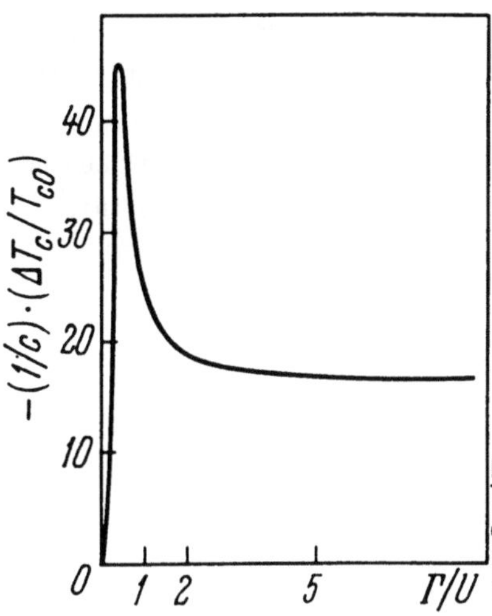

Fig.3.25. The decrease of the critical temperature of a dilute alloy as a function of the parameter Γ/U according to THEUMANN's [3.140] theory based on the continuous integration method

Another attempt to describe the continuous transition from the magnetic case to the nonmagnetic one has been undertaken by RIVIER [3.142] and is based on the use of a theory of relaxation of fluctuations in a magnetic system. RIVIER has introduced the spin fluctuation lifetime τ_s. It is defined as the time during which the localized spin on the impurity exists and after which the correlation of the spin with its initial value is lost. He obtained an equation for the superconducting transition temperature

$$\ln\left(\frac{T_c}{T_{co}}\right) = \alpha \int_0^{1/\eta} du\, e^{-\alpha u} \ln \tanh\left(\frac{uT_c}{2T_{co}}\right) \quad , \tag{3.8.7}$$

in which two parameters are involved:

$$\alpha = \frac{1}{2\pi T_{co}\tau_{ex}} \quad , \quad \eta = \frac{1}{2\pi T_{co}\tau_s} \quad . \tag{3.8.8}$$

Here $\tau_{ex}^{-1} = cN(0)J^2S(S+1)$ is the inverse electron spin exchange relaxation time appearing in the Abrikosov-Gor'kov theory, and τ_s is the phenomenological spin fluctuation lifetime.

It is easy to see that when $\tau_s \to \infty$, (3.8.7) goes into the Abrikosov and Gor'kov equation and when $\tau_s \to 0$, $T_c \to T_{co}$, which corresponds to Anderson's theorem of independence of a T_c on nonmagnetic impurities. The intermediate range of the parameter η corresponds to the existence of localized spin fluctuations. The results of a numerical solution of (3.8.7) are given in Fig.3.26. It is evident from the diagram that at large values of η corresponding to the nonmagnetic regime, T_c decreases monotonically with concentration. At values $\eta \ll 1$ corresponding to the magnetic regime, the curves for T_c tend towards the dotted curve of the Abrikosov and Gor'kov theory, but no critical concentration exists. Of course, in experiments it is impossible to notice that

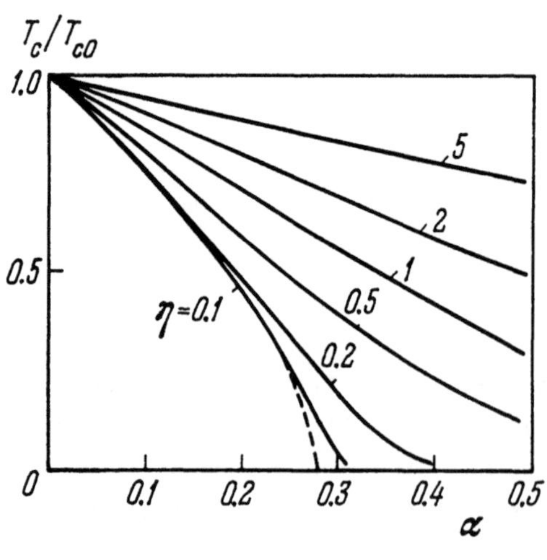

Fig.3.26. The dependence of T_c/T_{co} in RIVIER's [3.143] theory on the parameter α proportional to impurity concentration at different values of the parameter $\eta = 1/\tau_s$. The dotted curve gives the result of ABRIKOSOV and GOR'KOV

there is no critical concentration in this situation. Its existence is connected with the interaction of impurities which we have completely neglected here.

Thus, RIVIER's phenomenological theory also describes the continuous transition from the magnetic regime to the nonmagnetic one in the behaviour of impurities, but the main parameter τ_S remains undecoded in this theory. A series of papers [3.143-146] has been devoted to the microscopic theory of localized spin fluctuations in superconductors.

KRISHNA-MURTHY et al. [3.147] have recently made a numerical analysis of Anderson's model for the spin $S = 1/2$. They found the temperature dependences of the impurity magnetic susceptibility in a broad interval of parameters Γ and U, including the regime of critical fluctuations ($\Gamma/U \sim 1$). This method is based on his earlier method for investigating the Kondo effect by means of the renormalization group technique. Since, as shown by SCHRIEFFER and WOLFF [3.136], Anderson's Hamiltonian may be converted to a s-d model Hamiltonian by a canonical transformation, the Kondo problem and the problem of the localized state in Anderson's model have much in common. It appears that for Anderson's model a critical temperature exists; an analog of the Kondo temperature (Sect.3.6) at which the temperature dependence of the magnetic susceptibility changes

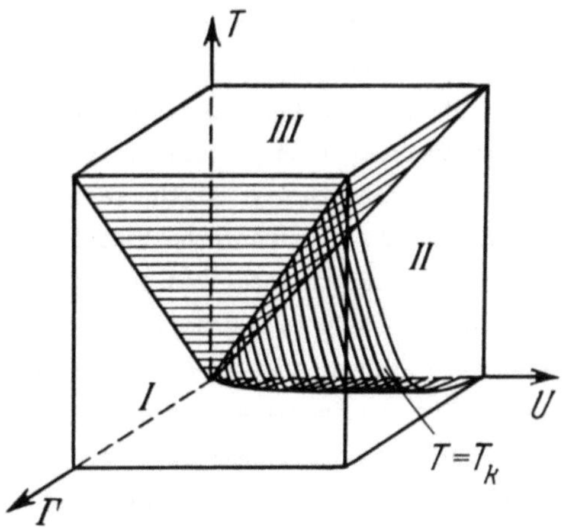

Fig.3.27. A schematic representation of different regimes in the behaviour of the magnetic susceptibility in ANDERSON's [3.147] model
I) the region of the full compensation of the paramagnetic spin with the spin density of conduction electrons, II) the region of the occurrence of a localized moment at the impurity, III) the region of the existence of an orbital moment

sharply. Namely, the increase of χ with decreasing temperature according to the Curie-Weiss law $[\chi \sim (g\mu_0)^2/4kT]$ at some temperature T_K is replaced by another regime which corresponds to a gradual disappearance of the localized moment due to the compensation for the impurity spin by the spin density of the electron gas where $\chi \approx 0.1(g\mu_0)^2/kT$ at $T \ll T_K$.

Conversely, at very high temperatures the Curie-Weiss regime corresponding to the existence of an atomic localized moment is replaced by a free orbital moment regime which leads to the magnetic susceptibility $\chi = (g\mu_0)^2/8kT$. The various regimes of the behaviour of χ in the space of the parameters are shown in Fig.3.27. The surfaces separating them denote only the conventional boundary of the regimes since the transition from one regime to another proceeds gradually and there are no clear-cut boundaries between them. From a knowledge of the parameters of the alloy, Γ and U, and of the superconducting transition temperature T_c of the matrix, one can qualitatively predict the nature of the influence of impurities on T_c.

3.8.3 Alloys Based on Transition Metals

In a number of dilute alloys with the matrix consisting of transition elements, effects also occur which indicate localized spin fluctuations. RIBLET [3.148] has discovered that the T_c of iridium is suppressed by impurities such as Fe, Co and Ni. It has been established that in iridium the above impurities do not conserve the localized magnetic moment and hence the suppression of T_c may not be explained on the basis of the mechanism of paramagnetic impurities. What is more, the decrease of T_c with increasing concentration proceeds more slowly than ought to be expected from the Abrikosov-Gor'kov theory. At the same time, these systems display a concentration growth of the electronic heat capacity. RIBLET has assumed that the observed effects are due to localized spin fluctuations and has tried to describe their influence on T_c by a modification of McMillan's formula (2.3.27). If one takes into account that electron scattering on LSF leads to the destruction of Cooper pairs and increases the eigenenergy of an electron at the same time, thus leading to an additional (with respect to the electron-phonon interaction) enhancement factor, one can obtain the following generalizations of the formulae of strong-coupling theory:

$$\gamma = \gamma_0(1 + \lambda + \lambda_m) \quad , \tag{3.8.9}$$

$$T_c = \frac{\omega_D}{1.45} \exp\left[-\frac{1.04(1 + \lambda + \lambda_m)}{\lambda - \lambda_m - \mu^* (1 + 0.62\lambda)}\right] \quad , \tag{3.8.10}$$

where λ_m is an electron-LSF coupling constant. These formulae may apply if the characteristic frequencies of LSF $\sim 1/\tau_s$ are of the same order as the Debye frequency ω_D.

Assuming that λ_m varies somewhat with concentration, we obtain the exponential behaviour of T_c as a function of c from (3.8.10), just as in KAISER's theory [3.128]. RIBLET has found from (3.8.10) that the values of $d\lambda_m/dc$ for Fe, Co and Ni diluted in Ir are in reasonable agreement with the values derived from electronic heat capacity and magnetic susceptibility measurements.

It is also worth noting that similar effects of the manifestation of LSF were observed in alloys of 4d-transition metals: Mo Ir, Mo Rh, Ru Ir, Ru Pt, Ru Rh, Ru Co, Ru Ni studied by RIBLET and JENSEN [3.149].

3.9 Spin Fluctuations in Pure Transition Metals

3.9.1 Itinerant Spin Fluctuations of a Nonferromagnetic Metal (Paramagnons)

Localized spin fluctuations arise, as we have seen, on transition-metal impurities diluted in the matrix of a simple metal, the effect becoming particularly great when the electronic system is in the vicinity of the condition for the emergence of a localized magnetic moment. In a pure transition metal, in which the atomic magnetic moments are absent, itinerant spin fluctuations are liable to occur if the state of the electronic system of the metal is in the vicinity of the criterion for the occurrence of ferromagnetism. In such a metal the situation for the occurrence of superconductivity is less favourable. In fact, the transition metals at the end of the period, such as Pd and Pt, are not superconductors although the atomic magnetic moments are absent. At the same time it has been long known that in a number of magnetic properties palladium, for example, exhibits an anomalous behaviour which consists, in particular, of an abrupt enhancement of the magnetic susceptibility, It is thought that in such a metal the electronic system is close to the transition to a ferromagnetic state, therefore magnetic moment fluctuations are strong as is always the case in the vicinity of a phase transition. This idea was first put forward by DONIACH [3.150], and the theory of the phenomenon was developed by BERK and SCHRIEFFER [3.151].

To begin with, consider the problem of spin fluctuations in the normal phase of the metal. It is convenient to make use of, e.g., the Hubbard model allowing for the Coulomb interaction of electrons on one site only. In this way one can take into account the strong correlations in a transition metal with narrow d-bands. Since the s-electrons screen the Coulomb interaction of d-electrons the interaction is local in character. Neglecting the degeneracy of the d-states, we write the electron interaction Hamiltonian in the following form

$$H_C = U \sum_j n_{j\uparrow} n_{j\downarrow} \quad , \tag{3.9.1}$$

where $n_{j\sigma} = a^+_{j\sigma} a_{j\sigma}$ is the operator of the number of electrons in a site j with a spin σ.

We calculate the magnetic susceptibility of the system, $\chi(\underline{q},\omega)$, which is related to a two-particle temperature Green's function of the following type

$$K(\underline{R}_j - \underline{R}_{j'},\tau) = \left\langle T\tilde{a}^+_{j\uparrow}(\tau)\tilde{a}_{j\downarrow}(\tau)\tilde{a}^+_{j'\downarrow}(0)\tilde{a}_{j'\uparrow}(0)\right\rangle \quad , \tag{3.9.2}$$

where $\tilde{a}_{j\sigma}(\tau)$ is Heisenberg's representation of the operator $a_{j\sigma}$ with the total Hamiltonian. In the interaction representation the Fourier component $K(\underline{q},\tau)$ in this function may be written as

$$K(\underline{q},\tau) = \sum_{\underline{p}\underline{p}'} \left\langle T a^+_{\underline{p}+\underline{q}\uparrow}(\tau) a_{\underline{p}\downarrow}(\tau) a^+_{\underline{p}'-\underline{q}\downarrow}(0) a_{\underline{p}'\uparrow}(0)\sigma(\beta)\right\rangle \quad . \tag{3.9.3}$$

The magnetic susceptibility $\chi(\underline{q},\omega)$ is obtained by analytical continuation to the real axis of the Fourier components $K(\underline{q},\omega_n)$ of (3.9.3). Thus, the calculation of $\chi(\underline{q},\omega)$ reduces to the calculation of $K(\underline{q},\omega_n)$ in perturbation theory. In the diagram technique corresponding to the interaction Hamiltonian, Coulomb interaction lines link electron propagators with opposite spins.

To a zero approximation we have

$K_0(\underline{q},\omega_n) =$ 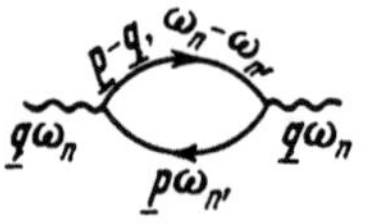

$$= -T\sum_{n'}\sum_{\underline{p}} G_0(\underline{p},\omega_{n'})\, G_0(\underline{p}+\underline{q},\omega_n+\omega_{n'}) \quad , \tag{3.9.4}$$

where $G_0(\underline{p},\omega_n)$ is a Fourier component of the Green's function of a normal met al. Now write the expression for $K(\underline{q},\omega_n)$ in the random phase approximation (RPA). As is known [3.152], this approximation consists of summing the series which takes into account the multiple electron-hole scattering:

$K_{RPA}(\underline{q},\omega_n) =$ 

$$= \frac{K_0(\underline{q},\omega_n)}{1 - UK_0(\underline{q},\omega_n)} \quad . \tag{3.9.5}$$

After applying the analytical continuation in (3.9.5) we obtain the following expression for the susceptibility in RPA:

$$\chi(\underline{q},\omega) = \frac{\chi_0 U(\underline{q},\omega)}{1 - UN(0)U(\underline{q},\omega)} \quad , \tag{3.9.6}$$

where $\chi_0 = 2\mu_0^2 N(0)$ is the Pauli suceptibility of the electron gas and $U(q,\omega)$ is Lindhard's function [3.153].

The system becomes ferromagnetic if at zero values of the arguments $\underline{q}$ and ω, $\chi(0,0) \to \infty$. Since the Lindhard function possesses the property $U(\underline{0},0) = 1$, the condition of the magnetic instability of the system, which follows from (3.9.6), transforms to Stoner's criterion for the occurrence of ferromagnetism

$$UN(0) = 1 \quad . \tag{3.9.7}$$

The behaviour of the magnetic susceptibility $\chi(g,\omega)$ is easy to obtain when q and ω are small. If $q/2p_F \ll 1$ and $\omega/qv_F \ll 1$, the Lindhard function is of the form

$$U(\underline{q},\omega) = 1 - \frac{1}{3}\left(\frac{q}{2p_F}\right)^2 + i\frac{\pi}{2}\frac{\omega}{qv_F} + \ \dots \quad , \tag{3.9.8}$$

so that the static susceptibility is approximated by an Ornstein-Zernicke type expression

$$\chi(\underline{q},0) = \frac{\chi_0}{1 - UN(0) + \frac{1}{3}(q/2p_F)^2} \quad . \tag{3.9.9}$$

This indicates that the radius r of the correlation of fluctuations increases according to the law $r \sim 1/\sqrt{1 - UN(0)}$ when approaching the point of ferromagnetic instability. As shown by SCHRIEFFER [3.152] [who analyzed the frequency dependence of (3.9.6)], the imaginary part of the magnetic susceptibility (3.9.6) as a function of q at low frequencies, has more or less sharp peaks whose position determines the relationship between ω and q. This relationship appears to be linear

$$\omega_{\underline{q}} \sim [1 - UN(0)]v_F q \quad . \tag{3.9.10}$$

It should be regarded as a dispersion relation for intinerant excitations of the spin system which are called paramagnons.

In the vicinity of the Stoner criterion (3.9.7) of ferromagnetic instability the velocity of paramagnons turns out to be much lower than the velocity of electrons, v_F, on the Fermi surface. Thus, paramagnons are excitations similar to highly decaying phonons. From the law of conservation of frequencies at the vertex and from the form of the graphs in the series (3.9.5), they are Bose-like excitations of the electronic system.

Evidently, near the point of ferromagnetic instability the interaction of electrons with paramagnons should lead to a high renormalization of the electronic mass. To this end, calculate the corresponding correction to the electron eigenenergy portrayed in Fig.3.28a. In accordance with RPA, these graphs allow for the multiple scattering of electrons and holes. If these graphs are redrawn as in Fig.3.28b the above process, represented by an infinite series of graphs, may impart a sense of virtual emission and absorption of a Bose-like quasi-particle whose propagator is represented by a series shown in Fig.3.29. To this series corresponds the analytical expression

$$t(\underline{q},\omega_n) = -\frac{K_0(\underline{q},\omega_n)U^2}{1-UK_0(\underline{q},\omega_n)} = -U^2K_{RPA}(\underline{q},\omega_n) \quad , \qquad (3.9.11)$$

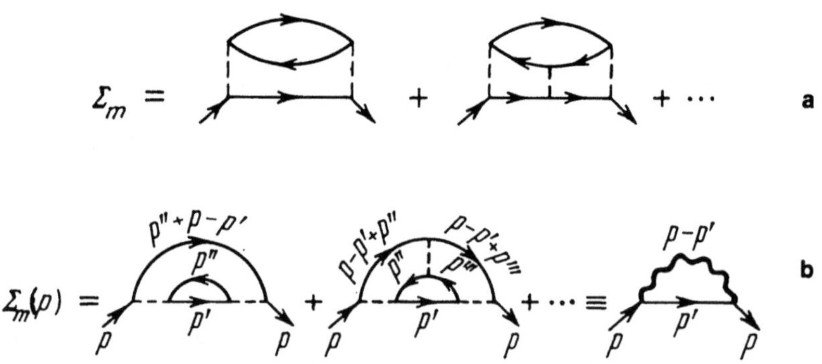

Fig.3.28. The correction from the Coulomb interaction to the self-energy part of an electron in the RPA method: (a) the interpretation of the virtual emission and absorption of a paramagnon, (b) $p = \{\underline{p},\omega_n\}$ 4-momentum

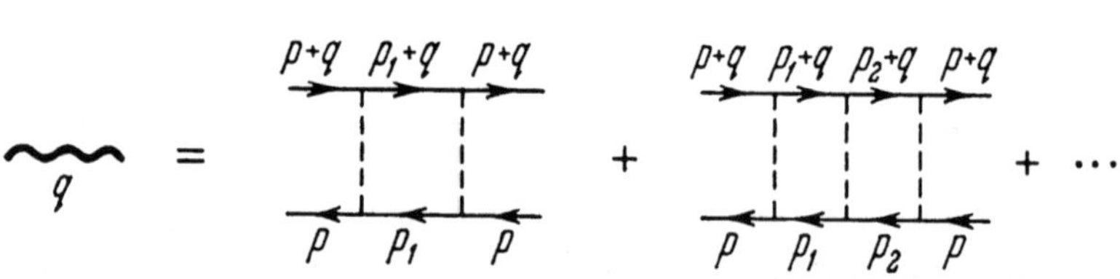

Fig.3.29. The series of graphs describing the propagation of a paramagnon for the matrix of the scattering of an electron on a hole

from which it follows that the poles of the scattering matrix $t(\underline{q},\omega)$ coincide with those of the dynamic susceptibility $\chi(\underline{q},\omega)$ calculated in the RPA.

The eigenenergy correction for an electron (Fig.3.28b) is determined by the expression

$$\Sigma_m(\underline{p},\omega_n) = -T\sum_{n'}\int\frac{d\underline{p}'}{(2\pi)^3}G_0(\underline{p}',\omega_{n'})t(\underline{p}-\underline{p}',\ \omega_n-\omega_{n'})\quad . \tag{3.9.12}$$

It is not difficult to see that this expression leads to a renormalization of the effective electron mass, but this problem will be treated later after we have discussed the influence of the electron-paramagnon interaction on the superconducting transition temperature.

3.9.2 Virtual-Paramagnon Creation and Superconductivity

We derive equations for the Green's function of a superconductor accounting for the interaction of electrons with paramagnons. To this end, in addition to the usual graph describing the electron-phonon interaction, it is necessary to introduce the graph shown in Fig.3.28b into the expression for Σ. Its contribution to the self-energy part Σ may be shown to be equal to

$$\Sigma_m(\underline{p},\omega_n) = -T\sum_{n'}\int\frac{d\underline{p}'}{(2\pi)^3}t(\underline{p}-\underline{p}',\ \omega_n-\omega_{n'})\tau_1 g(\underline{p}',\omega_{n'})\tau_1\quad . \tag{3.9.13}$$

This expression is obtained if one uses, firstly, Nambu's representation for the Coulomb Hamiltionian (3.9.1):

$$H_{int} = U\sum_{\underline{p}\underline{p}'\underline{q}}\left(\Psi^+_{\underline{p}+\underline{q}}\tau^+\Psi_{\underline{p}}\right)\left(\Psi^+_{\underline{p}'-\underline{q}}\tau^-\Psi_{\underline{p}'}\right)\quad , \tag{3.9.14}$$

where $\tau^+ = \begin{pmatrix}1&0\\0&0\end{pmatrix}$, $\tau^- = \begin{pmatrix}0&0\\0&-1\end{pmatrix}$. Secondly, in the series of the graphs determining the quantity Σ_m (Fig.3.28b) the gap function of a superconductor in the electron Green's functions forming the paramagnon propagator (Fig.3.29) should be allowed to tend to zero. The reason for this is the same consideration as in the graphs arising from the electron-phonon interaction in the derivation of Eliashberg's equations, viz., the phonon Green's function in a superconductor is taken to be equal to that for a normal metal. This comes from the smallness of the parameter ω_D/E_F. Therefore, the above approximation for the electron-paramagnon contribution to Σ suggests that the limiting frequency of the paramagnon spectrum, ω_m, is small with respect to the Fermi energy. Following the above approximation, the series of graphs given in Fig.3.28b lead to (3.9.13). Its difference from the similar expression (2.2.2) originating from the electron-phonon interaction lies in the appearance of the matrices τ_1 instead of τ_3. This is so because when a paramagnon is emitted or absorbed by an electron, the spin of the latter varies. This gives rise to a destructive effect on superconductivity, as will be seen later.

Proceeding in the same way as in Sect.2.2 when deriving Eliashberg's equations, we shall obtain the Berk and Schrieffer equations which we shall write in the following form[2]

$$[1 - Z(\omega)]\omega = - \int_{-\infty}^{\infty} dz' [K_{ph}(z',\omega) + K_m(z',\omega)] \operatorname{Re}\left\{\frac{z'}{\sqrt{z'^2 - \Delta^2(z')}}\right\} , \quad (3.9.15)$$

$$Z(\omega)\Delta(\omega) = \int_{-\infty}^{\infty} dz' [K_{ph}(z',\omega) - K_m(z',\omega)] \operatorname{Re}\left\{\frac{\Delta(z')}{\sqrt{z'^2 - \Delta^2(z')}}\right\} . \quad (3.9.16)$$

Here the electron-paramagnon interaction kernel $K_m(z',\omega)$ is determined by an expression similar to (2.2.26) for $K_{ph}(z',\omega)$:

$$K_m(z',\omega) = \int_0^{\infty} dz g_m(z) \frac{1}{2}\left[\frac{\tanh\frac{z'}{2T} + \operatorname{ctanh}\frac{z}{2T}}{z' + z - \omega - i\delta} - \frac{\tanh\frac{z'}{2T} - \operatorname{ctanh}\frac{z}{2T}}{z' - z - \omega - i\delta}\right] . \quad (3.9.17)$$

where $g_m(z)$ is the spectral density of paramagnons averaged with respect to the Fermi surface

$$g_m(z) = \int \frac{d^2\underline{p}}{v_{\underline{p}}} \int \frac{d^2\underline{p}'}{v_{\underline{p}'}} \left[-\frac{1}{\pi} \operatorname{Im}\{t(\underline{p} - \underline{p}',z)\}\right] \bigg/ \int \frac{d^2\underline{p}}{v_{\underline{p}}} . \quad (3.9.18)$$

The linearized Berk-Schrieffer equations may be written as one equation

---

[2] The Coulomb interaction of electrons enters this equation through the t matrix which is represented as a series beginning with a second-order term (Fig.3.29).

The first-order graph should be incorporated into the usual scheme of allowing for the Coulomb interaction in Eliashberg's equations and leads to a pseudopotential taking into account the electron-electron scattering of strongly excited electronic states. This term is not written out in (3.9.15, 16) since its influence on the superconducting temperature is well known. We emphasize once again that, when deriving the Berk-Schrieffer equations, only the vertex corrections from a Coulomb interaction of a special type which describe the multiple scattering of electrons and holes are kept.

$$\Delta(\omega)\left[1+\frac{1}{\omega}A_{ph}(\omega)+\frac{1}{\omega}A_{m}(\omega)\right]=\int_{-\infty}^{\infty}dz'\left[K_{ph}(z',\omega)-K_{m}(z',\omega)\right]\mathrm{Re}\left\{\frac{\Delta(z')}{z'}\right\},\quad(3.9.19)$$

where $A_{ph}(\omega)$ is the earlier expression (2.3.31) and $A_m(\omega)$ is an expression of an analogous structure which is related to the paramagnon contribution

$$A_m(\omega)=\int_{-\infty}^{\infty}dz\,K_m(z,\omega)\quad.\qquad(3.9.20)$$

The quantity $A_{ph}(\omega)$, and hence $A_m(\omega)$ divide into real and imaginary parts in view of (2.3.32) from which it follows that for low frequencies

$$A_{ph}(\omega)=-\frac{\omega}{\pi T}\int_{0}^{\infty}dz\alpha^2(z)F(z)\,\mathrm{Im}\left\{\psi'\left(\frac{1}{2}+i\frac{z}{2\pi T}\right)\right\}\approx\omega\lambda\quad,\qquad(3.9.21)$$

with λ being the electron-phonon coupling constant (2.3.4). Similarly, $A_m(\omega)\approx\omega\lambda_m$, where

$$\lambda_m=2\int_{0}^{\infty}dz\frac{g_m(z)}{z}\quad.\qquad(3.9.22)$$

These estimates are valid for frequencies $\omega\lesssim\omega_D$ and $\omega\lesssim\omega_m$, respectively.

We calculate T_c from (3.9.19) to an approximation in which the kernels of electron-phonon and electron-paramagnon interaction are approximated by constants. Namely, for $K_{ph}(z',\omega)$ use is made of expression (2.3.5), and for $K_m(z',\omega)$ the situation is similar:

$$K_m(z',\omega)=\frac{\lambda_m}{2}\tanh\left(\frac{z'}{2T}\right)\theta(\omega_m-|z'|)\theta(\omega_m-|\omega|)\quad.\qquad(3.9.23)$$

Thus we obtain for the superconducting transition temperature

$$T_c=\frac{2\gamma}{\pi}\bar{\omega}\exp\left(-\frac{1+\lambda+\lambda_m}{\lambda-\lambda_m}\right)\quad,\qquad(3.9.24)$$

where

$$\bar{\omega}=\omega_m\left(\frac{\omega_D}{\omega_m}\right)^{\nu}\quad,\qquad\nu=\frac{\lambda^2}{(\lambda-\lambda_m)\left(\lambda-\lambda_m+\frac{\lambda\lambda_m}{1+\lambda+\lambda_m}\cdot\ln\frac{\omega_D}{\omega_m}\right)}\quad.$$

One can see from (3.9.24) that the interaction with paramagnons leads to a weakening of the pairing interaction in superconductors so that at $\lambda_m\gg\lambda$ the superconductivity appears to be completely suppressed. Strictly speaking, (3.9.24) ought to be corrected by the Coulomb pseudopotential μ^* which enters the expression for T_c in the usual manner and weakens the pairing interaction. The preexponent in (3.9.24) is determined by some effective frequency $\bar{\omega}$ dependent on both limiting frequencies of the phonon and paramagnon spectra, with $\omega_D \sim \omega_m$, $\bar{\omega} \sim \omega$ or ω_D.

We now estimate the electron-paramagnon coupling constant λ_m. For a quadratic electron dispersion relation Eq.(3.9.18) for $g_m(z)$ may be represented as

$$g_m(z) = N(0) \int_0^{2p_F} dq \frac{q}{2p_F^2} \left[-\frac{1}{\pi} \mathrm{Im}\{t(\underline{q},z)\} \right] \quad , \tag{3.9.25}$$

where, in view of (3.9.11),(3.9.6) and (3.9.8), the spectral density of paramagnons is equal to

$$-\frac{1}{\pi} \mathrm{Im}\, t\,(\underline{q},z) = \frac{U}{\pi} \cdot \frac{\frac{\pi}{2} UN(0)\, z/qv_F}{\left[1 - UN(0) - UN(0) q^2/12p_F^2\right]^2 + \left[\frac{\pi}{2} UN(0)\, z/qv_F\right]^2} \quad . \tag{3.9.26}$$

Substitute (3.9.25) into (3.9.22). The integral over q may be calculated immediately and integration over z leads to an approximate formula of the form [3.154]:

$$\lambda_m = a\, UN(0) \ln \frac{1}{1 - UN(0)} \quad , \tag{3.9.27}$$

where $a \approx 2$ if $UN(0) \approx 1$ and $a \approx 1$ if $UN(0) \ll 1$.

Thus, near the ferromagnetic instability of metals, the electron-paramagnon coupling constant increases enormously and the superconductivity will be suppressed if $\lambda_m > \lambda$. It should, of course, be borne in mind that this result has been obtained in RPA which, as is known, cannot claim to be a quantitative description of the phenomenon of instability of the system in the vicinity of the phase transition. Just as in the case of localized spin fluctuations described in the previous section, the Berk-Schrieffer theory stated here may give only a qualitative description of the influence of virtual emission and absorption of paramagnons on superconductivity.

As an experimentally verified consequence of this theory, we may point out to the enormous increase of the electron mass in the vicinity of the region of ferromagnetic instability, due to the interaction with paramagnons. Indeed, we have seen that owing to the electron-phonon interaction the increase of the electron mass m^* was described by the equation $m^* = m(1 + \lambda)$. It is clear from the Berk-Schrieffer equation (3.9.15) that the effect of electron-phonon and electron-paramagnon scatterings should be added, so that in the case under consideration

$$m^* = m(1 + \lambda + \lambda_m) \quad . \tag{3.9.28}$$

3.9.3 Comparison with Experiment

A general confirmation of the ideas about the existence of paramagnons and their influence on the superconducting transition temperature, heat capacity and magnetic susceptibility follows from the analysis of experimental data on alloys of transition elements placed at the end of the periodic table.

BUCHER [3.155] has found that χ and γ increase enormously for the Ni-Rh system with increasing Ni concentration (Fig.3.30). Alloys whose Ni content amounts to more than about 63% become ferromagnetic. Near the point of ferromagnetic instability the increase of γ (proportional to the effective mass m^*) obeys a logarithmic law, as follows from (3.9.25,27). However, the value of γ in the region of the peak turns out to be approximately 3 to 4 times less than that suggested by theory [3.151,152]. In spite of this discrepancy it may be assumed that the experimental results confirm the main idea of a paramagnonic increase of the mass.

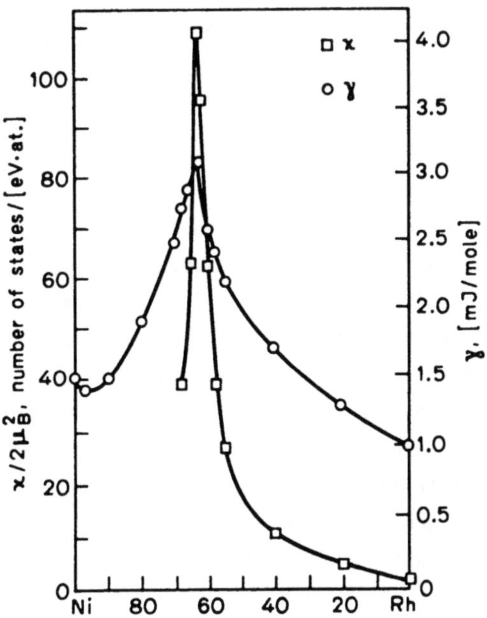

Fig.3.30. The experimental dependences of the magnetic susceptibility χ and of the electronic heat capacity coefficient γ for Ni - Rh alloys

In the system of iridium-based alloys the decrease in superconducting transition temperature is correlated with the increase of χ and γ with varying composition (Fig.3.31). From the abrupt increase of χ it may be predicted for Ir-Pt alloys that T_c should actually fall to zero if only 10% of Pt is added to Ir. However, for Ir-Pt alloys the quantity T_c decreases as quickly for Ir-Os alloys which possess considerably smaller values of χ.

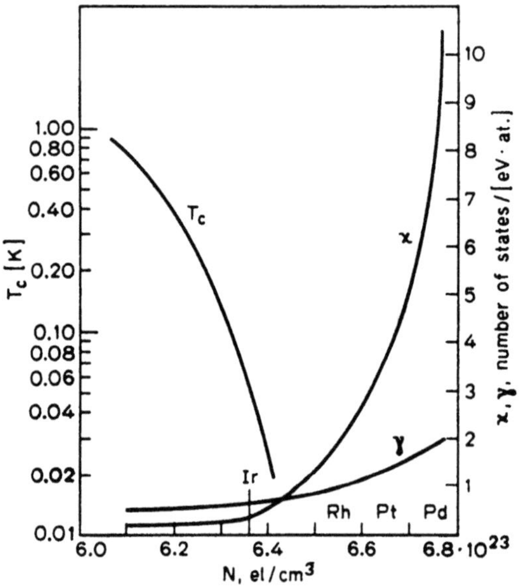

Fig.3.31. The qualitative dependence of T_c, χ and γ on the density of valence electrons, N, for fcc transition-metal alloys

At present we may not, therefore, regard the problem of superconductivity suppression in metals situated near Pd as finally settled, i.e., it is not clear whether paramagnons or just the Coulomb repulsion of electrons play the main role. All the effects of the decrease of T_c and of the increase of χ and γ due to the interaction of electrons with paramagnons are undoubtedly present in the behaviour of the above alloys. However, there is no doubt that the RPA overestimates their value. As in the case of localized spin fluctuations, it is necessary to go beyond the framework of RPA and make use of modern phase transition theory; this is an urgent problem for the near future.

One of the ways to ascertain the role which paramagnons play in the suppression of superconductivity in Pd may be the study of alloys of Pd with Ag

[3.156]. Addition of silver leads to a decrease in the density of states on the Fermi surface [3.156] and accordingly, to a weakening of the Stoner factor $S = [1 - UN(0)]^{-1}$. At the same time, as shown by calculations of the electron-phonon coupling constant λ from first principles [3.156] following the method described in Sect.4.5 λ in the alloy Pd_cAg_{1-c} depends weakly on the palladium concentration and it may be considered to be equal to ≈ 0.4 (as obtained for pure Pd [3.157]). In pure Pd $(\lambda - \lambda_m) < 0$, but at some Ag concentration λ_m becomes less than λ, and this opens up possibilities for the occurrence of superconductivity in the alloy. According to calculations [3.156], superconductivity should occur in the alloy when the Pd concentration is $c \lesssim 0.8$. Such alloys should have a measurable superconducting transition temperature of $0.01 < T_c < 0.5$ K. Confirmation of this prediction would permit the determination of the value of the electron-paramagnon coupling λ_m in pure Pd.

An attempt has been made [3.158] to take into account the influence of paramagnons on T_c in Nb and V within the framework of the Berk-Schrieffer theory described here. Until now it was customary to assume that the role which the paramagnon effect plays in superconducting transition metals is negligible. However, calculations of T_c from first principles on the basis of solutions of Eliashberg equations give an overrated value for T_c and for the electron-phonon coupling constant λ. Thus, for Nb, the use of the function $\alpha^2F(\omega)$ obtained in [3.159] leads to $\lambda = 1.19$ [3.158], and for V to $\lambda = 1.04$ [3.158,160]. In order that the observed values of T_c may be obtained at these values of λ it is necessary to have anomalously large values of the Coulomb pseudopotential: $\mu^* = 0.38$ for Nb and $\mu^* = 0.55$ for V [3.158]. Such values are absolutely unrealistic, and therefore it has been assumed in [3.158] that the virtual paramagnon exchange processes make an essential contribution. This hypothesis may be supported by the value of the Stoner factor S obtained from measurements of the spin contribution to the magnetic susceptibility of Nb and V, and also by straightforward evaluation of S from band calculations [3.161]. For both metals, it turns out that $S \approx 2$. This is much less than the value $S \approx 6$ for Pd, but nevertheless the paramagnon effect should make a certain contribution to T_c.

The solutions of Eliashberg's equations for Nb and V are given in [3.158], supplemented with the corresponding terms of the Berk-Schrieffer theory. This reference also gives values of T_c as a function of the factor S. The values of T_c decrease sharply with increasing S. The observed values of T_c for Nb and V correspond to the values $S \approx 2$ which agree with their measured and calculated values. These values of the factor S lead to values of the electron-paramagnon coupling constant λ_m which are equal to 0.21 for Nb and to 0.34

for V. These values suffice to reduce the calculated quantities to the observed values. It may be assumed that the paramagnon effect will make an appreciable contribution to T_c for other transition metals as well.

4. Superconductivity in Transition Metals

4.1 The Principal Parameters of Superconducting Metals

In this chapter we consider empirical correlations between T_c and physical properties of transition metals (electronic heat capacity, magnetic susceptibility, Debye temperature, melting temperature, pressure). We analyze results of band calculations and experimental data on electronic structure of bcc and and hcp transition metals. Dispersion curves of the phonon spectrum and the phonon density of states are presented, obtained from neutron scattering and tunnel spectroscopy. The contribution of electronic properties and lattice factors to the electron interaction parameter is discussed and empirical estimates are given for different kinetic characteristics of metals (resistivity, heat capacity, tunnel and optical properties). A calculation of the electron-phonon interaction constant from first principles concludes this chapter.

4.1.1 The Superconducting Transition Temperature

There are 24 known elements with incomplete d-shells which form three series of transition d-metals:

the first series (3d-elements) from Sc to Ni,

the second series (4d-elements) from Y to Pd,

the third series (5d-elements) from La(Lu) to Pt.

More than half of these metals (namely, 14) are superconducting at low temperatures, in bulk and at normal pressures [4.1] (Table 4.1).

These superconductors are transition metals which are nonuniformly distributed within the series. In the first series there are only two superconducting elements (Ti and V); in the second series there are five elements and in the third transition series seven elements out of nine are superconductors. The highest transition temperatures have been found in 4d-metals: $T_c = 9.25$ K for Nb and $T_c = 7.8$ K for Tc. Thus, in the series of pure transition metals there are no high-temperature superconductors whose T_c exceeds 10 K.

Table 4.1 Superconducting properties of transition metals

Element	Crystal structure	T_c [K]	H_o [Oe]	$\theta_{\mathcal{D}}$ [K]	γ $\left[\frac{mj}{mole\ K^2}\right]$	$\chi \cdot 10^6$ $\left[\frac{cm^3}{mole}\right]$	α	μ^*	$\frac{\partial T_c}{\partial P}$	P [kbar]
Ti	hcp	0.40± 0.04	56	415	3.3	155			0.06	0-14
V	bcc	5.40± 0.05	1408	383	9.82	300			0.63	0-25
Cr^+	bcc	1.5								
Y^{++}	hcp	2.7				183			3.7	110-160
Zr	hcp	0.61± 0.15	47	290	2.77	129	0± 0.05	0.17	1.5	0-20
Nb	bcc	9.25± 0.02	2060± 50	276	7.80	212.1			-0.20	0-25
Mo	bcc	0.915± 0.005	96± 3	460	1.83	89.2	0.37± 0.04	0.09	-0.14	0-25
Tc	hcp	7.80± 0.1	1410	411	6.28	270			-1.25	0-15
Ru	hcp	0.49± 0.015	69± 2	580	2.8	39.2	0± 0.15	0.15	-0.23	0-18
Pd^{+++}	fcc	3.2								
La	hcp	4.88							14.1	0-23
Hf	hcp	0.128	12.7	252	2.21	70			-0.26	0-10
Ta	bcc	4.47	829± 6	258	6.15	162				
W	bcc	0.0154± 0.0005	1.15± 0.03	383	0.90	53.3				
Re	hcp	1.697± 0.006	200± 5	415	2.35	68.7	0.38	0.10	-0.23	0-18
Os	hcp	0.66± 0.03	70	500	2.35	13.1	0.21	0.12		
Ir	fcc	0.1125± 0.01	16	425	3.19	24.1				
Ce^{++}	fcc	1.7								

\+ superconductivity detected in thin films

++ superconductivity detected only under high pressure (for Y $T_c = 2.7$ K at 160 kbar, for Ce $T_c = 1.7$ K at 55 kbar)

+++ superconductivity detected in thin films when irradiated by α-particles

Most of the superconducting d-metals (eight) have a hexagonal close-packed structure, five metals crystallize in the body-centred cubic structure, and only iridium is a face-centred cubic structure (Fig.4.1). If one analyzes the properties of transition metals from the point of view of their crystalline structure, the familiar conclusion that cubic symmetry favours high T_c may not be drawn here. Hcp metals may also have comparatively high values of T_c for this class of substances (for example, technetium has $T_c = 7.8$ K), and among cubic transition metals low-temperature superconductors also occur (for example, for W $T_c = 0.0154$ or for Ir $T_c = 0.11$ K).

3d	Ti	V					
4d	Zr	Nb	Mo	Tc	Ru		
5d	Hf	Ta	W	Re	Os	Ir	
N_e	*4*	*5*	*6*	*7*	*8*	*9*	
	hcp	bcc		hcp		fcc	

Fig.4.1. The superconducting transition elements

In Table 4.1 the value of the isotope effect for the transition elements in which relevant measurements have been made is also given [4.2-11]. The magnitude of the effect is characterized by a parameter α relating the variation of T_c to the variation of the ion mass [see (1.2.10)]. According to the BCS theory, $\alpha = 0.5$, and the experimentally observed deviations from 0.5 are due to the dependence of the Coulomb pseudopotential μ^* on the limiting frequency of the phonon spectrum ω_D. It can be seen from the table that the parameter α in transition metals differs appreciably from the value $\alpha = 0.5$ predicted by the BCS theory. For all transition metals $\alpha < 0.5$, and in some cases T_c does not depend on the ion mass at all and $\alpha = 0$(Ru, Zr). Unfortunately the experimental data available do not suffice so far as to analyze the variations of the parameter α within the transition series. It may only be stated for the present that for the transition metals Ru and Zr in which $\alpha = 0$ the T_c does not exceed 0.6 K. For Ru and Zr the values of μ^* calculated on the basis of (2.3.27) (Table 4.1) are the largest. Thus, considerable Coulomb interaction in these metals is an obstacle to attaining high values of T_c.

4.1.2 The Energy Gap

Along with T_c, an important characteristic of the superconductor is the energy gap. In the BCS theory the value of the gap for zero temperature $\Delta(0)$ is related to T_c by the following equation: $2\Delta/T_c = 3.52$. Energy gap measurements for the transition metals V, Nb, Ta, Mo, Re were carried out with the help of tunnel measurements and ultrasound decay [4.12-14] (Table 4.2). It follows from these data that for the transition metals which have been investigated the values of the energy gap are close to those predicted by the BCS theory. An interesting result of all these measurements is the fact that the energy gap in Re is anisotropic (Table 4.2). For V, Nb and Mo the variation of $2\Delta/T_c$ in different crystallographic directions is not large and lies practically within the experimental error. It is not yet clear how the detected anisotropy of the energy gap in a transition metal may be explained. ZAVARITSKII [4.25] put forward a hypothesis according to which the anisotropy of the energy gap in superconductors arises from the fact that contributions to the tunnel current are from electrons of several energy bands of different gap widths. The relative contribution of the electrons of different bands in the direction chosen is proportional to the radius of curvature of the Fermi surface in each of the bands. Thus, the energy gap anisotropy in tunnel measurements is attributed to the peculiarities of the Fermi surface of a given superconductor (Sect.4.2). However, the attempt to account for the anisotropy of the gap in Nb proceeding from the peculiarities of its electron energy spectrum was a failure [4.26].

Another point of view concerning the nature of the above effect proceeds from the assumption that the phonon spectrum of the superconductor is the main source for the anisotropy of the gap. Within the framework of this approximation, BENNET [4.27] has calculated the anisotropic portion of the gap and obtained good agreement with experiments. No similar calculations have yet been performed for transition metals.

4.1.3 Empirical Correlations Between T_c and Physical Properties of Transition Metals

It is of interest to consider the dependence between the number of d-electrons (n_d) and the superconducting properties of transition metals.

However, when constructing this dependence one runs into difficulties associated with the determination of n_d in the crystalline state.

Table 4.2 Energy gap in superconducting transition metals

Metal	Crystallographic direction	$2\Delta/T_c$	Method	References
V	[110]	3.4±0.2	Ultrasound decay	[4.17]
	[111]	3.2±0.2	Ultrasound decay	
	[100]	3.1±0.2	Ultrasound decay	
	[110]	3.5±0.2	Ultrasound decay	[4.18]
	[110]	3.6±0.2	Ultrasound decay	[4.19]
	Polycrystal	∿3.4	Tunneling	[4.12]
Nb	[111]	3.6±0.1	Ultrasound decay	[4.20]
	[110]	3.55±0.1	Ultrasound decay	
	[100]	3.35±0.1	Ultrasound decay	
	[111]	3.74±0.04	Ultrasound decay	[4.24]
	[110]	3.68±0.04	Ultrasound decay	
	[100]	3.77±0.03	Ultrasound decay	
	[110]	3.5±0.3	Ultrasound decay	[4.19]
	Polycrystal	∿3.6	Tunneling	[4.12]
	Polycrystal	3.84±0.06	Tunneling	[4.13]
	Polycrystal	∿3.79	Tunneling	[4.14]
Ta	[110]	3.5±0.2	Ultrasound decay	[4.19,21]
	Polycrystal	∿3.5	Tunneling	[4.12]
	Polycrystal	3.6±0.1	Tunneling	[4.13,15]
Mo	[100]	3.5±0.2	Ultrasound decay	[4.22]
	[110]	3.5±0.2	Ultrasound decay	[4.23]
	[100]	3.3±0.2	Ultrasound decay	
	[111]	3.1±0.2	Ultrasound decay	
Re	[0001]	2.9±0.1	Ultrasound decay	[4.23]
	[1010]	3.0±0.1	Ultrasound decay	
	[1120]	3.5±0.1	Ultrasound decay	
	[0001]	3.35±0.13	Tunneling	[4.16]
	[1120]	3.91±0.15	Tunneling	

It is possible to construct a purely phenomenological dependence of T_c on the total number of valence electrons (sd) per transition-element atom N_e, a dependence which would not require any model assumptions as to the redistribution of electrons between the states of different symmetry [4.28]. This dependence was first proposed by MATTHIAS [4.29] who showed that high values of T_c in transition metals group around two values: $N_e = 5$ e/a for V, Nb, Ta and $N_e = 7$ e/a for Tc (Fig.4.2).

To analyze the superconducting properties of transition metals it is important to consider the characteristics of the electron and phonon spectra. Of the electron-spectrum characteristics, we shall be concerned firstly with the density of electron states on the Fermi level N(0). Information on this parameter may be obtained from heat capacity $C_V = \gamma T$ and paramagnetic susceptibility χ measurements. In Table 4.2 the values of γ and χ in transition metals are given [4.1,30]. It can be seen from comparison of Figs.4.3,4 that for the overwhelming majority of transition metals (except for Ir) the variation of T_c within the 3d, 4d or 5d transition series correlates rather closely with analogous variations of γ and χ.

The correlation of one of the most important phonon-spectrum characteristics, viz., the Debye temperature θ_D and T_c, is also of interest. For this purpose, construct the dependence of $1/\theta_D$ versus N_e for the transition metals under consideration (Fig.4.5). From comparison of this figure with Fig.4.2

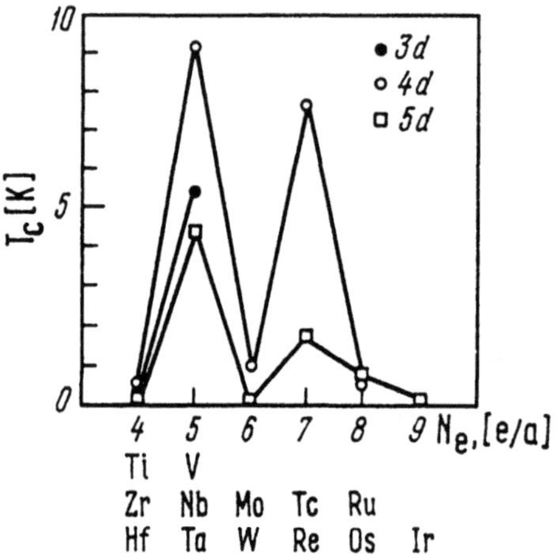

Fig.4.2. The dependence of T_c on the number of valence electron, N_e, in transition metals

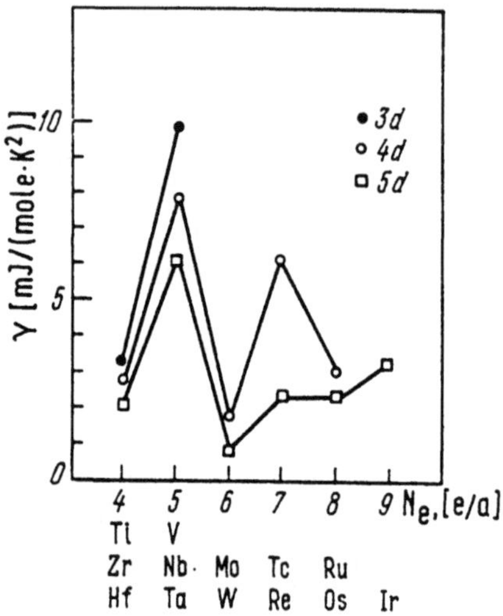

Fig.4.3. The dependence of γ on the number of valence electrons, N_e, in transition metals

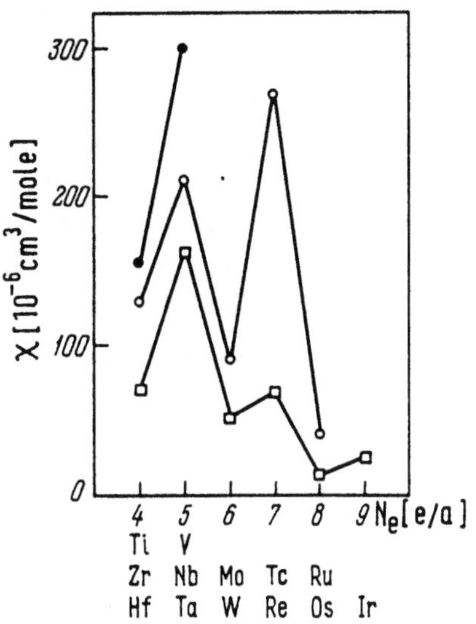

Fig.4.4. The dependence of χ on the number of valence electrons, N_e, in 3d, 4d and 5d-metals

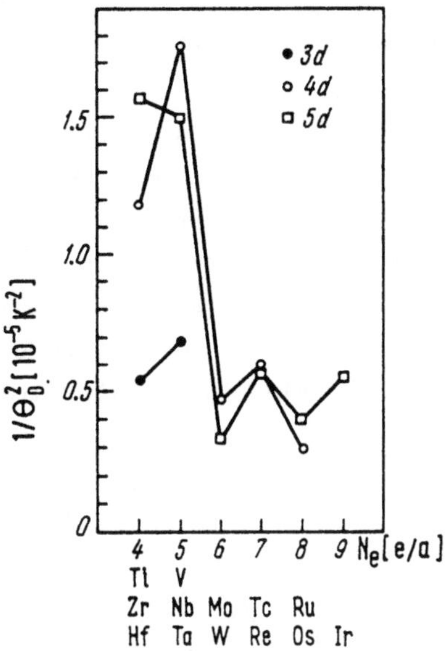

Fig.4.5. The dependence of $1/\Theta_D^2$ on the number of valence electrons, N_e, in transition elements

one notices a certain similarity in the behaviour of T_c and $1/\theta_D^2$. The largest values of the parameter $1/\theta_D^2$ in Nb and Ta are accompanied by high T_c. On the other hand, for Mo, W, Os and Ru the value of $1/\theta_D$ is appreciably less, which also agrees with the low T_c in these metals. The transition metals at the beginning and end of the periods (Hf, Zr, Ir) "fall out" of the correlation noted above. The physical sense of this correlation will be considered in Sect.4.4.

The existence of an empirical correlation between T_c and the sign of the Hall coefficient in superconducting elements including transition metals, has been reported in [4.31a,b].

According to these data, superconductivity occurs in transition metals with a positive Hall coefficient at room temperature and with a mixed nature of conductivity with a predominance of hole conductivity. Despite the fact that the established correlation holds for a large number of elements, their alloys and compounds, its physical sense and connection with the theory of superconductivity in transition metals still need to be clarified.

MATTHIAS [4.31c] turned his attention to the fact that the superconducting transition temperature is related, in a definite fashion, to the melting temperature. That is, as a rule low melting points correspond to high T_c. Let us see how true this correlation is in the case of transition metals. In Fig.4.6 the number of valence electrons are enumerated for the melting temperatures of 3d, 4d and 5d transition metals versus N_e. It can be seen that the transition metals situated in the centre of each series, viz., Cr, Mo, W have the highest melting temperatures. When we move to the left or to the right of these metals along a series, the melting temperatures drop. It is significant that the most refractory metals (Mo and W) really have low T_c (Table 4.1 and Fig.4.2). On the contrary, the lowering of the melting points of V compared to Cr or of Nb and Tc compared to Mo, is accompanied by a substantial increase in the superconducting transition temperature (Fig.4.2).

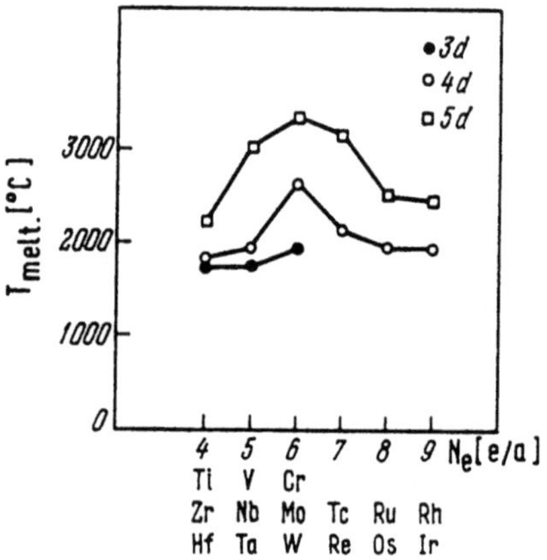

Fig.4.6. The dependence of the melting temperature on the number of valence electrons, N_e, in 3d, 4d, and 5d-metals

This point of view was subsequently developed by STERN [4.32] who suggested that the so-called "broken covalency" concept should be used as a criterion promoting the attainment of high T_c in transition metals. This concept is based on the assumption that the d-d bonds in the metals situated in the middle of the transition series are localized and are of a covalent nature. High melting points, brittleness and hardness of these metals speak in favour of the existence of such bonds. Some indication of the strong localization of d-electrons in covalent bonds are the low values of the density of electron states on the Fermi level for Mo and W; this follows immediately from electron heat capacity measurements (Fig.4.3). If the d-electrons in the above transition metals are localized in covalent bonds, then it is most likely that they form valence pairs (electron pairs with opposite spins, "condensed" in coordinate space) instead of Cooper pairs (electron pairs with opposite momenta and spins, "condensed" in momentum space) [4.33]. Thus, the formation of covalent bonds may, to a certain extent, compete with superconductivity, which was pointed out earlier in [4.33].

4.1.4 The Pressure Dependence of T_c

Except for hafnium and tungsten, the pressure dependence of T_c was studied for all transition superconducting metals [4.34-39]. This variation of $\partial T_c/\partial P$ turned out to be different in different metals. In the metals placed at the beginning of the transition series (Ti, Zr, V and others) the sign of $\partial T_c/\partial P$ was, as a rule, positive, whereas in a number of superconducting transition metals negative values of $\partial T_c/\partial P$ were observed (these data are tabulated in Table 4.1).

One of the first attempts to account for this behaviour in transition metals was undertaken by BRANDT [4.35]. He paid attention to the fact that the increase of T_c in Ti and Zr due to pressure is accompanied by a variation in the inclination of the temperature dependence of the critical field. Measurement of the temperature dependence of the critical field H_c under pressure enables one to estimate the electronic portion of heat capacity and consequently, the density of states on the Fermi level. The values of γ for titanium and zirconium determined in this manner showed that when these metals are compressed an increase in the density of states on the Fermi level occurs which, in its turn, may be connected with the experimentally observed growth of T_c under pressure. BRANDT assumed that the pressure dependence of N(0) in transition metals would be determined by the dependence of N(0) on the number of valence electrons N_e. Then the sign of the derivative $\partial T_c/\partial P$ should be determined by the position of the transition metal involved on this curve. In a superconductor with $\partial N(0)/\partial N_e > 0$ the T_c should rise with compression, whereas in superconductors with $\partial N(0)/\partial N_e < 0$ it should drop. The dependence of N(0) on N_e in transition metals may be judged by the dependence of γ on Ne (Fig.4.3). According to these date, $\partial T_c/\partial P$ in Ti and Zr should be positive, and in Ta and Nb negative, and this agrees exactly with experiments. However, subsequent experiments on V and Mo showed dependences of $\partial T_c/\partial P$ which contradicted the above correlation.

Recent experiments by BRANDT and collaborators [4.38,39] have shown that higher pressures may give rise to a more complicated dependence of $\partial T_c/\partial P$ than that observed at pressures of up to 30 - 40 kbar. Of particular interest in this respect was the pressure dependence of T_c over the interval between 0 and 250 kbar in Nb. According to previous investigations, under compression

T_c dropped over the pressure range from 0 to 25 kbar (Fig.4.7). As the pressure rises further T_c passes through a minimum at 40 kbar and then increases almost linearly with $\partial T_c/\partial P = +(2.6 \pm 0.2) \cdot 10^{-6}$ K/Bar. In Ta, T_c also decreased $[\partial T_c/\partial P = -(2.6 \pm 0.1) \cdot 10^{-6}$ K/Bar] to zero in the range of 0 to 100 kbar and then did not vary up to 250 kbar (Fig.4.7). For V the extension of the high pressure range to 250 kbar was not accompanied by any change in the behaviour of the dependence $\partial T_c/\partial P$.

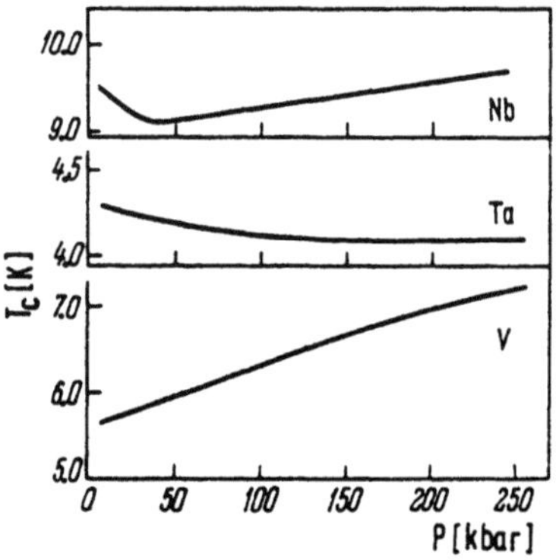

Fig.4.7. The variation of T_c with pressure in V, Nb, and Ta

4.1.5 Superconductivity in Thin Films

Until now we have been considering the superconducting and some other properties of bulk transition metals. Interesting experimental data concerning the superconducting properties of transition metals in thin films have recently been published [4.40.41].

The first group of these results belongs to the so-called "amorphous" films of transition elements which were produced by evaporation of the metal on a cooling substrate [4.40]. The heating of the metal, at which its evaporation proceeded, was effected by an electron beam and its structural state was controlled using an electron microscope. The films of pure transition metals thus obtained showed a diffraction pattern typical of liquid or amorphous materials. The superconducting properties of amorphous metallic materials turned out to be different from those in bulk crystalline samples. This can readily be seen by comparing Figs.4.8 and 4.2 where T_c is presented as a function of the number of electrons N_e per atom. High T_c occur not at the two values $N_e = 5$ and $N_e = 7$ e/a which we got in the case of crystalline samples, but at one value, viz., $N_e = 6$ e/a for 4d metals and $N_e = 7$ e/a for 5d metals (Fig.4.8). Note that the transition to an amorphous state affected the superconducting prop-

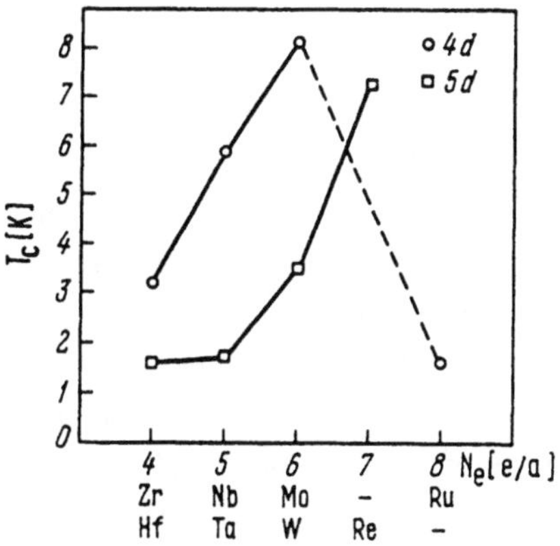

Fig.4.8. The superconducting transition temperature in amorphous films of transition metals as a function of the number of valence electrons

erties of the basic transition metals differently. Thus, in Zr, Mo and W an increase in the values of T_c occurred, whereas for Ta and Nb, T_c decreased (as compared to the crystalline state). The increase of T_c in W and Mo was particularly appreciable, viz., from 0.0154 K to 3.5 K and from 0.915 K to 8.15 K, respectively.

The second group of results applies to films produced by precipitation of transition metals on various substrates using the ionic sputtering technique (noble gases such as Ne, Ar, Kr and Xe were used as ion source). The superconducting transition temperature in films produced using this method also differed from that of a bulk material. For example, in thin W films it has been possible to obtain $T_c = 5.5$ K which is a factor of 350 higher than that in a bulk sample ($T_c = 0.015$ K); in Mo it has been possible to increase T_c from 0.915 K to 7.2 K and in Ti from 0.42 K to 2.52 K, etc. The detection of superconductivity in thin chromium films ($T_c \sim 1.5$ K), i.e., in a metal with atomic magnetic order (an antiferromagnet) was quite unexpected.

Spectral analysis showed that when transition metals are sputtered by an ion beam the films thus produced contain a definite amount of ions of noble gases whose presence makes the crystal lattice of the basic material increase in size. It also turned out that the T_c of the films produced depends on the atomic radius of the dissolved ion of the noble gas and rises as this radius increases. It is of interest that high T_c (in comparison with a bulk material) were found in metals having a low density of electron states on the Fermi level (Ti, Cr, Zr, Mo, W). In transition metals with a high value of this parameter, the transition from bulk samples to thin films, on the other hand, was accompanied by a decrease in the values of T_c. It is probable that the failure in experiments with Y and Pd (no superconductivity was detected in the samples of thin films of these metals synthesized with the help of the technique described above) also arises from the high values of the density of electron states on the Fermi level in these metals. The mechanism of the increase or decrease of T_c in thin films of transition metals has not been ascertained yet. Nevertheless, the effects revealed appear to be highly interesting and further experiments aimed at investigating in detail the electronic and lattice properties in the above films are required.

An increase in the superconducting transition temperature from ∿ 9 to 12.9 K and from 0.61 to 4.65 K was registered when implanting thin films of metallic niobium (∿ 500 Å) by nitrogen ions with a dose of $3 \cdot 10^{17}$ N_2/cM^2 [4.43] and zirconium by H(D) [4.42], respectively. Structural measurements showed an fcc structure in implanted layers with a lattice parameter of 4.5 Å, which is evidence of the formation of a δ-NbN compound (whose T_c varies from 12 to 17 K depending on synthesis conditions [4.44]) as a result of the implantation of Nb by nitrogen ions.

In summarizing the discussion of the superconducting properties of d transition metals it may be noted that the variation of T_c with atomic number or the number of valence electrons within the 3d, 4d, and 5d-series is almost the same. It is also worth noting that the electron and phonon spectrum parameters γ, χ and $1/\theta_D^2$ have similar Z (or N_e) dependences. Thus, it is possible, in principle, to express Z (or N_e) dependence of T_c either in terms of the density of states on the Fermi level or in terms of phonons, or through the electron-phonon interaction. It is of interest to analyze the contributions of the electron and phonon spectrum to the interaction mentioned above. For this purpose we should familiarize ourselves with the electron structure and the lattice properties of transition metals, which will be considered in the following two sections in more detail.

4.2 The Electron Structure of Transition Metals

4.2.1 Theoretical Band Structure Calculations

Success in solid state theory and the development of computing equipment have led to considerable progress in our understanding of the electronic structure of transition metals. By now electron-energy spectrum calculations have been made for almost all d metals. In Fig.4.9 curves of the density of electron states for bcc metals (Mo, Ta, W) are presented [4.45-47]. In the occupied part of the spectrum one notices three individual density of state peaks A,

B, C which are separated from the maximum A' in the free part of the conduction band by a deep-lying minimum. A similar electron-energy spectrum structure occurs for other bcc metals, V and Nb [4.48-52]. These metals have been investigated most elaborately, and most of the available electron structure calculations for them are cited in Figs.4.10 and 11. It may be noted that the results of all the calculations for V and Nb (carried out using different techniques) correlate on the whole quite satisfactorily with each other. Just as in Ta, Mo and W, four major subbands A', A, B and C may be singled out in the spectrum of V and Nb, the character of their relative arrangement and the relations between the peak heights almost coinciding. For V and Nb which possess five valence electrons per atom (as in their analogue in the group Ta), the Fermi level passes in the vicinity of peak A. These metals have comparatively high values of the density of electron states on the Fermi level. For Mo and W which have one more electron, the Fermi level has a higher energy and passes in the region of the minimum of the density of states. Analysis of the data given in Figs.4.9-4.11 shows that for bcc metals with close numbers of valence electrons available for each atom the rigid-band approximation works rather well. According to this approximation, the energy band structure of the neighbouring metals is assumed to be invariable, and the position of the Fermi level is determined by the electron concentration (the number of valence electrons).

If one considers bcc metals having the same concentration of valence electrons but belonging to different series (V, Nb and Ta or Mo, W), one can see some increase in the width of the d band with increasing atomic number of the transition metal.

Thus, electron spectrum calculations for bcc transition metals have revealed the following regularities:

i) There is a rather narrow region of energies, usually about 5-10 eV, in which a multitude of very narrow bands (chiefly of the d type), are concentrated.

ii) The density of states has a clearly marked dip somewhere in the middle part of energies corresponding to the interval of energetic localization of d-bands.

iii) The N(E) curves for different metals are very much alike even in the details of the structure. This gives good reason to consider the rigid-band model as being sufficiently acceptable.

All these conclusions following from band calculations are corroborated by experiments and will be discussed in detail below.

Does modern theory of metals permit at least a qualitative explanation of these regularities?

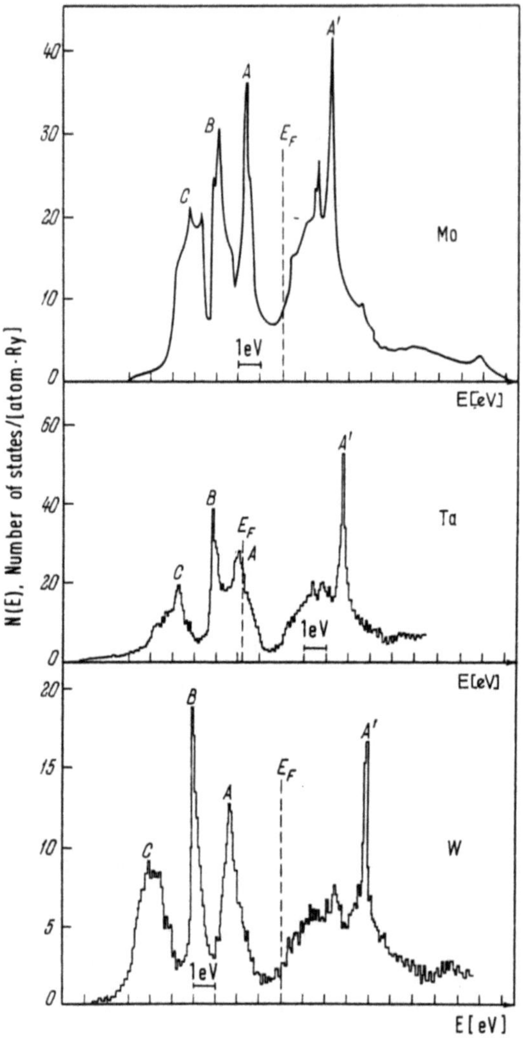

Fig.4.9. The density of states for bcc metals

An answer concerning the first of the points mentioned above follows immediately from MOTT's [4.59] well-known concept. If we brought the atoms of a given element closer together on the condition that initially the distance between them is infinitely large, then upon transition to a metallic state we would discover that the wave functions of the outer s-electrons become itinerant forming an s-band. If we continued bringing the atoms closer together, then the d-electrons which are the remotest from the nucleus would also form a band. This d-band would be narrower than the s-band since the wave functions

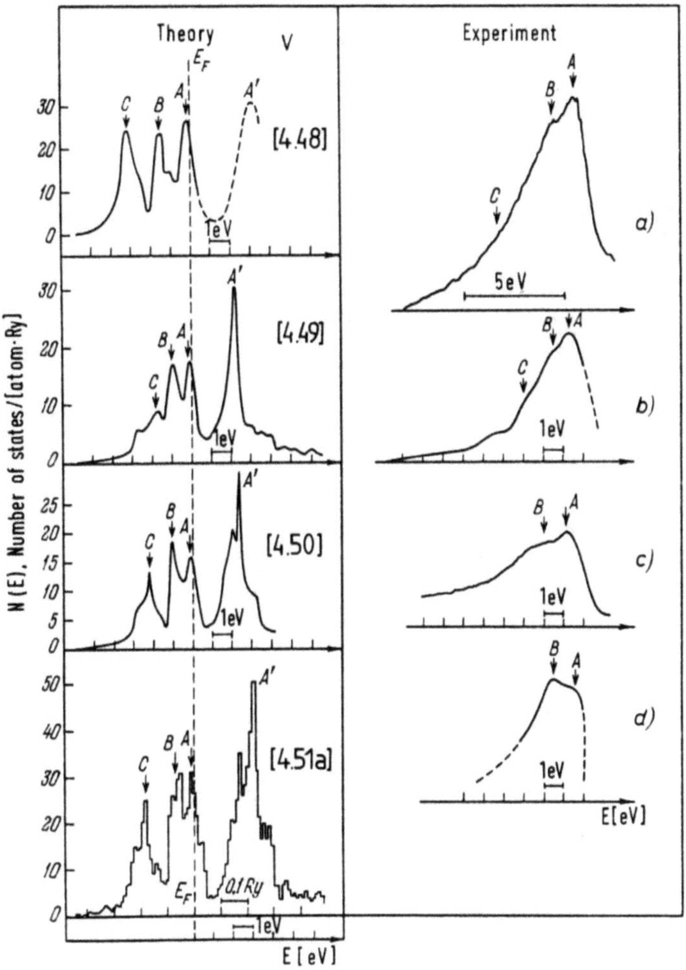

Fig.4.10. Results of calculations and experimental investigations of the electron energy spectrum of metallic vanadium. Experiments: a) $VM_{II,III}$: the X-ray emission band [4.53]; b) $VL_{II,III}$: the X-ray emission band [4.54]; c) the valence band of the X-ray electron spectrum [4.55]; d) the curve of the optical density of states, obtained from photoemission [4.56]

of d-electrons overlap to a lesser degree. Hence we have the following qualitative picture of the density of states in transition metals: there is a narrow peak in the density of states for d-electrons against the background of the broad and narrow densities of states of s-electrons.

The structure of the peak corresponding to d-electrons may be accounted for by resorting to crystallochemical considerations and to the idea of the hybridization of states. Indeed, in a bcc metal the potential acting on the

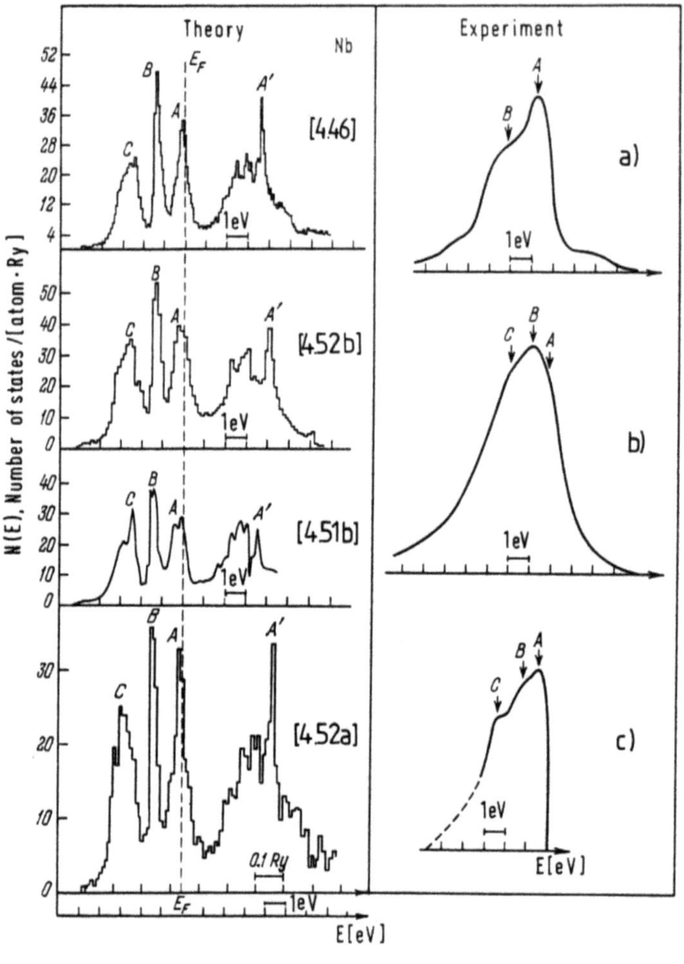

Fig.4.11. Results of calculations and experimental investigations of the electron energy spectrum of metallic niobium. Experiments: a) $NbM_{IV,V}$: the X-ray emission band [4.57]; b) NbL_{III}: the X-ray emission band [4.58]; c) the curve of the optical density of states, obtained from photoemission data [4.56]

electron located on the ion site and created both by this ion and by the neigbouring ions possesses cubic symmetry. We will proceed from the atomic picture of the genesis of electron states in the metal, bearing in mind the tight-binding method. It is known that in a cubic symmetry field the five-fold degenerate d type term splits up into two terms corresponding to the symmetry Γ_{12} and Γ'_{25} (irreducible representations of point group O_h in the nomenclature adopted in [4.60,61]). In the crystallochemical literature these states are often labeled by e_g and t_{2g}, respectively. The doubly degenerate e_g type states possess the symmetry of x^2-y^2 and y^2-z^2 and have maxima extended towards the cube edges, whilst the triply degenerate t_{2g} type states have the symmetry of the functions xy, yz and xz directed along the diagonals of the edges. Since in the bcc structure (Fig.4.12) the neighbouring atoms lying on the cube edges are second-order neighbours, the overlap of the functions should be less than that of the e_g functions of two neighbouring atoms disposed in the vertex and the centre of the cube. This means that the level of states e_g (i.e. Γ_{12}) lies higher than that of states t_{2g}(i.e. Γ'_{25}).

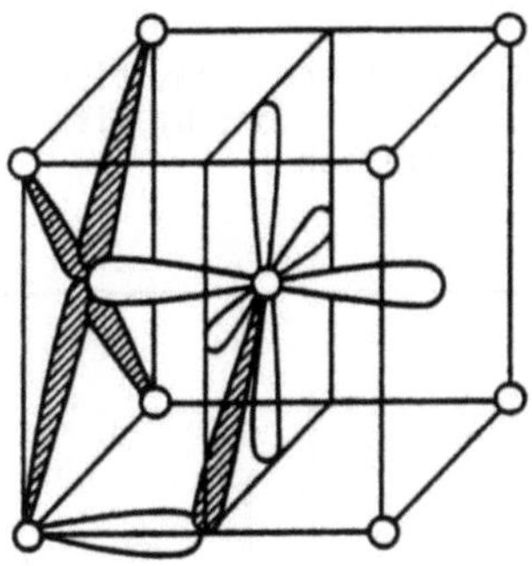

Fig.4.12. The mutual arrangement of the angular lobes of the wave d functions of the atoms in the bcc lattice. The hatched lobes pertain to t_{2g}-states, the clear ones to e_g-states

Each of these levels in the crystal merges into a band. If we neglect, for a moment, the hybridization between the states arising from the d-states (e_g and t_{2g}) and from the s states in the tight-binding method, we can obtain a spectrum structure like the one shown in Fig.4.13a. Due to the overlap of the band arising from d-states (e_g and t_{2g}), the off-diagonal (with respect to band indices) matrix elements from the crystal potential are not small at the points of intersection of the dispersion curves. As a result the picture shown by the dotted lines in Fig.4.13b is obtained. Finally, if we take into account the hybridization of the d-states with a wide band of s-states thus produced, we get the spectrum structure shown in the same figure by the solid lines. According to the estimates made in [4.62] the value of the matrix element of sd mixing V_{sd} in transition metals should amount to 1 - 1.5 eV.

As is easy to see from Fig.13b, the density of states of a transition bcc metal should have a number of sharp peaks with a deep dip at the centre of the d band, and this is the case in band calculations. Thus we arrive at the conclusion that crystal field effects are essentially responsible for the dip in N(E) in bcc metals mentioned above.

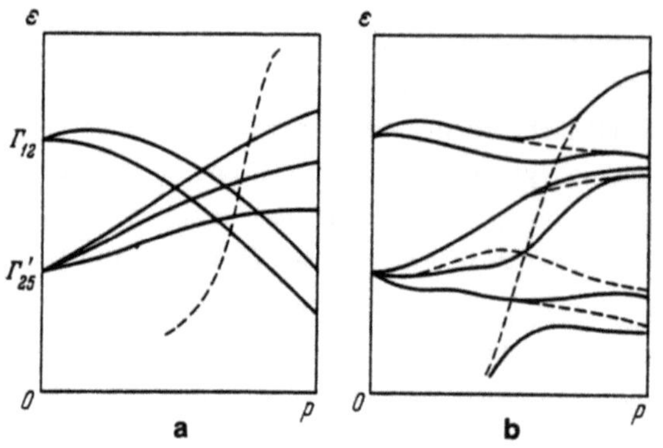

Fig.4.13. The dispersion curves of the electron spectrum in bcc transition metals (a qualitative picture): a) disregarding the hybridization; b) with hybridization of the d-states (---) and with s-d hybridization (—)

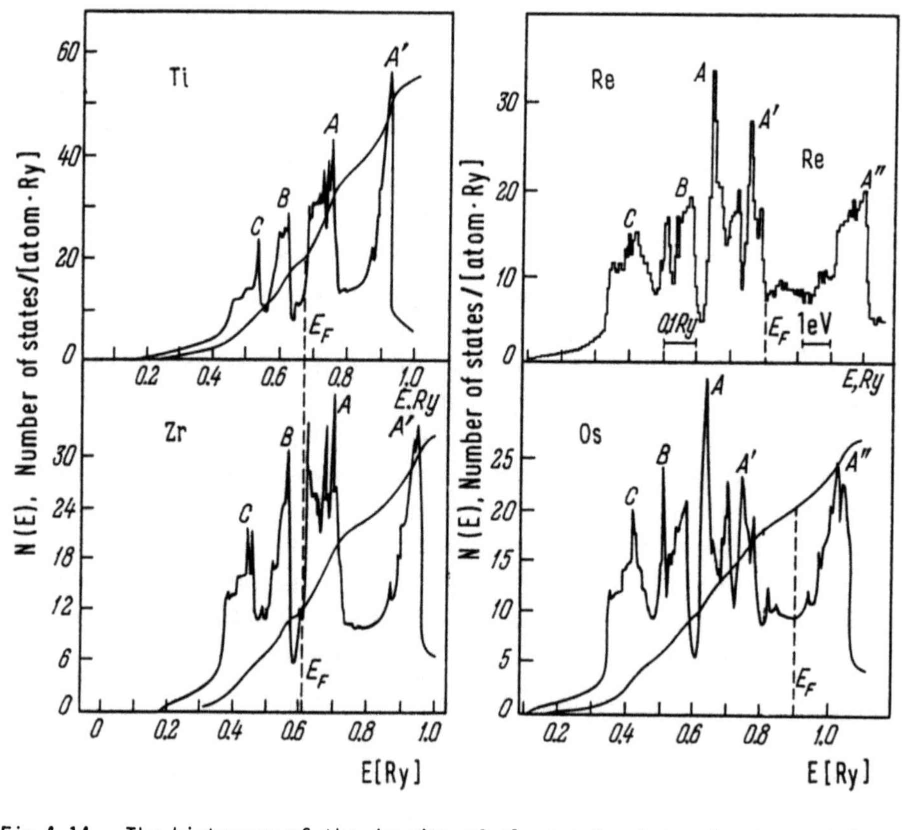

Fig.4.14. The histogram of the density of electronic states for hcp metals

The structure of the energy spectrum of hcp metals (Ti, Zr, Hf, Tc, Ru, Os, Re) is of a more complicated nature [4.63-70]. Compared with bcc metals, here one has to deal with a larger number of spectrum subbands which are separated by a smaller energy interval and have a fine structure rich in details. Analysis of the density of electron state calculation results for the metals enumerated above (Fig.4.14) leads essentially to the same conclusions which we have drawn for bcc metals. For Re which has seven valence electrons, the Fermi level passes near the peak of the density of states, and for Os with eight valence electrons the degree of filling of the energy bands increases and the Fermi level shifts towards higher energies to the region of minimum density of states. Thus, the rigid band approximation can also be used to consider the electron structure of neighbouring hcp metals (Re, Os).

4.2.2 The Fermi Surface

Just as in simple metals, the topology of the Fermi surface in transition metals is determined by the degree of filling of the energy bands. Since these bands are enormously overlapped in d-metals, the Fermi level always crosses several bands. Therefore the Fermi surfaces in transition metals are multisheeted. It is worth noting that one and the same band may generate several sheets of different shapes centred in different points of the Brillouin zone. In addition, each given sheet recurs at all equivalent points of $\underline{k}$-space if not centred at the Brillouin-zone centre.

One may expect the following general topological features of the Fermi surfaces of transition metals. If the surface forms near the bottom of some band, then it has a closed shape. If the number of occupied states is small, this, as a rule, is an ellipsoid. However, in the general case the form of these closed surfaces differs appreciably from a sphere. For example, in cubic lattices of d-metals these surfaces are close to octahedral, highly distorted surfaces of cubes, etc. If the Fermi level goes through the middle part of the zone, open electron or open hole Fermi surfaces arise which cross the boundaries of the Brillouin zone. Finally, the Fermi surfaces which are close to the corners of the Brillouin zone boundary have the form of closed hole surfaces (which contain unoccupied states) which are also ellipsoidal in form when the number of holes is small.

As noted above, Fermi surfaces of isoelectronic elements are similar when their crystal structure is the same. However, when there are critical points of the electron spectrum near the Fermi surface, these may give rise to changes in the details of the latter. A critical point is a point of high symmetry of extrema in dispersion curves. The relative locations of these critical points with respect to the Fermi level are determined by the metal. The changes caused in the form of the Fermi surface can lead to the appearance or disappearance of small pockets of carriers even within one isoelectronic group of transition metals.

What is the relationship between the properties of Fermi surfaces and the possibility of the occurrence of a superconducting state in the crystal of a transition metal? From analysis of the facts taken as a whole, superconductivity is favoured in the cases when the Fermi surface is formed from states of different types, viz., s, p and d. At any rate, the majority of transition metals are superconductors (apart from metals with magnetic order). At the same time, metals with one-sheet Fermi surfaces whose genesis is completely determined by s-electrons (alkali and noble metals) exhibit no superconductivity down to the lowest temperatures now reached. In the case of d-metals possessing no magnetic order, unfavourable factors for superconductivity are the cases when the Fermi level passes either close to the bottom of a d-like energy band or near its top. Thus, for example, for the elements of the third group of Mendeleev's table (Sc, Y, La and Lu), superconductivity has been detected in lanthanum and lutetium only. It occurs in yttrium only at a high external pressure of 140 kbar.

If we now turn to transition d-metals possessing superconductivity, they may not be said to have any disposition to a definite crystal structure or to a definite topology of the Fermi surface. Experiments show that superconductivity occurs in d-metals with the prevalence of both hole (the elements of group 5 - vanadium and other) [4.46,51] and electron current carriers (for example, in iridium). The other superconducting transition d-metals have an equal number of holes and electrons. In exactly the same manner, in superconducting d-metals all the sheets of the Fermi surface may be closed, for example, in molybdenum (Fig.4.15) and tungsten in the sixth group [4.45,71,72] or in iridium in the eighth group [4.73]. But there may also be open surfaces, for example, the elements of the fifth group - vanadium (Fig.4.16) and others, and all the hcp lattice metals - technetium, rhenium, etc. [4.74-81].

It is essential to the problem of superconductivity to take into account the electron-phonon interaction. In this connection we turn the reader's attention to the estimate of the possible renormalization of the cyclotron effective masses which is due to this interaction and which is given in [4.46, 82]. For a number of cyclotron orbits on the hole ellipsoids N3 and on the open hole surface H3 the renormalization factor in niobium (i.e., the ratio $m_{exp}/m_{calc} = 1+\lambda$) varies between 1.61 and 1.94 [4.82]. The theoretical esti-

Fig.4.15. The Fermi surface of the bcc metals of group VI

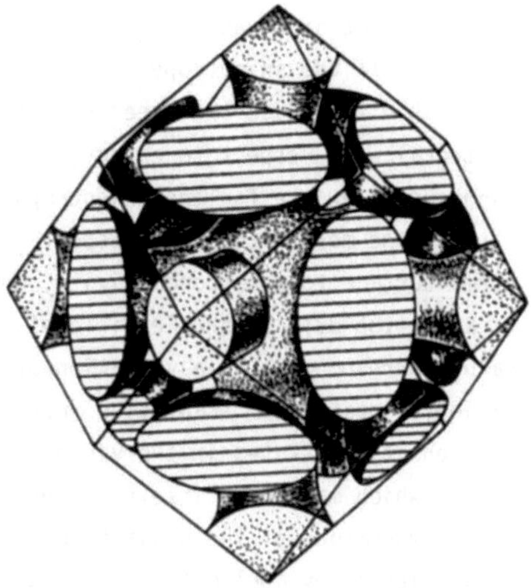

Fig.4.16. The Fermi surface of the bcc metals of group V

mate given by McMILLAN is equal to 1.82 [4.83]. In the case of tantalum the renormalization factor for such orbits varies within 1.61 to 1.94 [4.82], and according to McMILLAN's estimate for Ta we have 1.65 [4.83]. For vanadium, however, experiments yielded values of 2.05 and 2.46 for two orbits on the ellipsoids N3 [4.82], whereas McMILLAN's estimate is equal to 1.6 [4.83].

In the work by KOELLING et al. [4.45] a detailed table of values of electron-phonon renormalization factors λ for many orbits on different cross-sections of the Fermi surface in Mo is given. The quantities $(1+\lambda)$ lie in the interval of 1.15 to 1.47. From comparison of the measured and calculated density of states functions for Mo this factor equals 1.22. At the same time, according to [4.83], we have an estimate of this quantity which is 1.41. It is therefore permature to speak of a good agreement. A more detailed discussion may be found in [4.45]. Similar calculations have been performed for V, Nb and Ta in [4.84,89]. Just as in the case of Mo, an anisotropy of λ has been detected for different orbits of the Fermi surface. The sensitivity of the calculated values of λ to various exchange approximations and the self-consistence in the APW method have also been established.

We see that for many transition metals the Fermi surface is multisheeted and highly anisotropic. At the same time, in the theory of strong-coupling superconductors (Chapt.2) an Eliashberg equation is derived, in which only its energy dependence is taken into account and the momentum is taken on the Fermi surface. For weakly noninsotropic Fermi surfaces it is possible to write

Green's function equations averaged with respect to the angles of the Fermi momentum. If the anisotropy of the Fermi surface is high, it is in general impossible to reduce the Green's function equations to a one-dimensional (integral with respect to energy) Eliashberg equation and the problem of rigorous allowance for the anisotropy remains to be solved [4.85].

Note one more interesting aspect of the relation of the topology of the Fermi surface to superconductivity, which has been developed by BAR'YAKHTAR et al. [4.86-88]. If the Fermi level lies in the vicinity of a critical energy value corresponding to a variation of the topology of the Fermi surfaces (formation of pockets or transition from a closed surface to an open one), then the density of states in the vicinity of the Fermi energy should have a singularity which manifests itself in the variation of the superconducting transition temperature if the position of the Fermi level is varied by doping with impurities or by application of pressure.

4.2.3 X-Ray and Optical Spectra

Consider now the agreement of the results of theoretical electron-structure calculations for transition metals with experiments. The most straightforward experimental techniques of investigating the electron structure of solids are: optical and X-ray photoelectron spectroscopy and X-ray spectroscopy for studying N(E), the de Haas-van Alphen effect, the galvanomagnetic effects, cyclotron resonance and magnetoacoustic effects for determining the topology of the Fermi surface.

By varying the symmetry of the scanning level of the irradiating atom (K, L, M, N, O) and using the selection rules, it becomes possible with the help of X-ray emission spectra to obtain information concerning the energy distribution of states with sd or p-symmetry in the occupied portion of the conduction band in the d-metals being considered [4.90-92]. In contrast to X-ray spectroscopy, ultraviolet and X-ray photoelectron spectroscopy do not select states with respect to symmetry and make it possible to obtain a distribution approximating that of the full density of states [4.93]. However, in the case of X-ray spectroscopy the derived information about the partial density may be distorted by the nonmonotonic energy dependence of the matrix element of the transition probability, and in the case of photoelectron spectroscopy the distribution of the total density of states in energy is fixed to within the differences in the cross sections of the photoionization of electrons of different symmetry. The few X-ray and photoelectron spectrum intensity calculations show that despite the distorting influence of the above factors it is possible from experimental spectra to obtain information concerning the location of the main peaks on the curve of the density of states or of the special Van Hove points.

On the right-hand side of Figs.4.10,11 the distribution of the intensity of experimental X-ray and photoelectron spectra for V and Nb is shown [4.53-58]. Here one can observe a general qualitative agreement with calculations. A comparison of the energy distribution of the spectral subbands for all the transition metals studied is given in Table 4.3. The insufficient energy resolution in spectral experiments does not resolve the closely spaced peaks in the density of states. This makes a general comparison between theory and experiment difficult. Therefore the identification of more or less pronounced maxima on experimental spectra with peaks on the density of states curves, made in Figs.4.10,11 and in Table 4.3, may not be considered strict enough. In each particular comparison it is necessary to smear out the N(E) curve for the net distortion before a comparison with experiment can be drawn. This was done, for example, in [4.58]. It follows from the data given in Table 4.3 that the energy distributions of peaks on theoretical density of states curves and on experimental spectra usually agree to an accuracy of ±0.5 eV. In the light of the above considerations the agreement obtained should be considered quite satisfactory.

Table 4.3 also supplies the density of electron states on the Fermi level computed in [4.45-52,63-69,80,81,95,104]. These results show that high values of this parameter are observed just in those metals (V, Nb, Ta) which have high T_c. Thus, the empirical correlation between the behaviour of γ and T_c or χ and T_c (Figs.4.2-4.4) in transition metals has received a direct confirmation. It is of interest to compare the values of electron heat capacity obtained by calculations (from the density of electron states on the Fermi level) according to the expression $\gamma = 2/3\pi^2k^2N(0)$ with experimentally measured quantities [4.96,97,99-101,103,105-107]. These data, cited in Table 4.3, show that in the case of Ti, Mo, W a fairly good agreement between theory and experiment has been obtained. At the same time the experimental values of γ for V and Nb are considerably higher than the theoretical ones. It turns out that agreement between theory and experiment exists for the transition metals with low T_c.

The values of N(0) obtained from band calculations should be handled with some caution. As a rule, these values of N(0) are calculated from density of states histograms constructed with a rather rough step in the energy scale. For those metals in which the Fermi level passes near the density of states peak, a more exact construction of the N(0) curve involving the use of inter-

Table 4.3 Comparison of calculated data on electron energy spectrum of transition metals with experiment

Metal	Theory					Experiment		
	Method	N(0), number $\left[\frac{\text{of states}}{\text{atoms}\cdot\text{Ry}}\right]$	γ $\left[\frac{10^4\text{cal}}{\text{mole}\cdot\text{K}^2}\right]$	separation of spectrum subbands A - B [eV]	B - C [eV]	γ $\left[\frac{10^4\text{cal}}{\text{mole}\cdot\text{K}^2}\right]$	separation of spectrum subbands A - B [eV]	B - C [eV]
Ti	APW [4.52b]	13.6	5.6		1.0	7.9 [4.96]		
	APW [4.80]	28.5	11.81		0.75	8.15 [4.97]		
	APW [4.81]	22.0	9.18		0.6			0.5 [4.56]
	APW [4.95]	19.0	7.86					0.9 [4.55]
	Green's functions [4.67]	17.5	7.26		0.8			
	MTO [4.63]	12.4	5.1		1.0			
V	Strong coupling [4.49]	32.3	13.4	0.8	0.8	22.1 [4.99]	1.1 [4.55]	
	Green's functions [4.50]	30.1	12.5	0.9	1.1	21.2 [4.100]	0.8 [4.54]	1.5 [4.54]
	APW [4.51a]	26.6	11.0	0.8	1.5	23.7 [4.101]	1.1 [4.56]	
	APW-Xα [4.48]	20.2	8.36	1.3	1.6		1.1 [4.53]	2.8 [4.53]
	APW [4.52b]	35.2	14.5				0.9 [4.51a]	2.1 [4.57]
	Green's functions [4.98]	25.4	10.5	0.9	1.0			
Zr	Cells [4.66]	10.3	4.3		2.2	6.95 [4.97]		1.7 [4.94]
	APW [4.68]	9.1	3.8		1.5	6.62 [4.96]		1.2 [4.102]
	APW [4.52b]	10.9	4.5		1.5			0.7 [4.56]
	MTO [4.64]	13.1	5.4		1.4			

Nb	APW [4.46]	22.0	9.10	1.2	1.2	18.6 [4.96]	1.3 [4.57]	
	APW [4.52a]	28.2	11.67	1.3	1.6	18.45 [4.103]	0.8 [4.58]	1.0 [4.58]
	APW [4.52b]	36.0	14.9	1.2	1.3		0.7 [4.56]	1.3 [4.56]
	Green's functions [4.51b]	24.5	10.1	1.2	1.1			
Mo	APW [4.45]	8.68	3.59	1.2	1.1	4.40 [4.105]	1.1 [4.56]	2.0 [4.56]
	APW [4.104]	8.55	3.54	1.4	1.2	4.37 [4.96]	1.6 [4.57]	2.0 [4.57]
	Green's functions [4.52b]					5.21 [4.106]		
	APW [4.52b]					5.26 [4.107]		
Ru	MTO [4.64]	11.8	4.8			7.1 [4.96]		
Hf	MTO [4.64]	10.9	4.5		1.4	5.1 [4.96]		
Ta	APW [4.46]	17.64	7.30	1.1	1.6	14.3 [4.96]		
	APW [4.104]	29.2	12.1	1.4	1.2	13.6 [4.106]		
						13.5 [4.107]		
W	APW I [4.47]	7.6	3.15	1.4	1.2	3.10 [4.106]	2.0 [4.65]	2.7 [4.65]
	APW II [4.47]	4.35	1.8	0.9	1.1	2.5 [4.107]		
	APW [4.104]	7.48	3.1	1.4	1.2	2.27 [4.96]		
Re	APW [4.69]	10.06	4.16			5.5 [4.101]		
						5.49 [4.96]		
Os	MTO [4.64]	9.2	3.8			5.4 [4.96]		

polation procedures may be appropriate to refine the values of N(0) given in Table 4.3.

4.2.4 Contributions of the Partial Density of States to N(0)

In order to understand the superconducting properties of transition metals we have to estimate the contributions of states of different symmetry to the density of states on the Fermi level. These estimates were obtained in [4.108] for V and Nb (Table 4.4). It turned out that for both metals the contribution of d type states to N(0) substantially exceeds those of s, p and f-states. This shows that mainly d states are responsible for the superconductivity in transition metals.

Analysis of theoretical and experimental investigations into the electron structure of d-metals leads to the conclusion that the density of states on the Fermi level is an important parameter in the formation of their superconducting properties. However, as shown convincingly by calculations, the variation of the density of states alone cannot account for the variation of T_c under pressure in V and Nb. On the other hand, the growth of the density of states on the Fermi level may also yield negative results for superconductivity and under certain conditions leads to ferromagnetism [4.48].

Table 4.4 Contributions of partial densities to the total density of electron states on Fermi level for V and Nb

Metal	N(0)	$N_s(0)$	$N_p(0)$	$N_d(0)$	$N_f(0)$
	[number of states/(atom · Ry · spin)]				
V	15.125	0.172	2.096	12.649	0.208
Nb	12.010	0.321	1.715	9.660	0.291

4.3 Lattice Dynamics of Transition Metals

As discussed previously, the value of the electron-phonon interaction parameter λ in metals is determined by two factors: $\alpha^2(\omega)$, the effective matrix element of electron-phonon interaction, and $F(\omega)$, the phonon density of states [see (2.4.15)].

The function of phonon distribution in frequencies $F(\omega)$ and the dispersion relations $\omega(\underline{q})$ may be determined experimentally with the aid of inelastic coherent and, accordingly, incoherent neutron scattering.

In most of the experiments on the phonon spectra of transition metals the dispersion curves $\omega(\underline{q})$ were determined. Straightforward measurements of $F(\omega)$ were made for a limited number of transition elements: V [4.109-115], Ni [4.115-117]. For the other metals whose cross-section of incoherent neutron scattering is small, the distribution of $F(\omega)$ was determined from dispersion curves $\omega(\underline{q})$. The most complete investigation has been carried out for the phonon spectra of 4d superconducting metals (Y, Zr, Nb, Mo, Tc); for 5d metals, sufficiently detailed data are only available for Ta, W and Hf.

In Figs.4.17 and 4.18 dispersion curves $\omega(\underline{q})$ for transition bcc metals of the fifth (V, Nb) and sixth (Cr, Mo, W) groups [4.118-122] of the periodic system are shown. From these figures one notices a general correspondence in the behaviour of the dependences $\omega(\underline{q})$ for the metals within groups V and VI and some differences between them. The dotted line marks the dispersion curves for these metals calculated on the basis of the Born-Karman approximation.

Taking a sufficiently large number of interacting atoms (and, consequently, force constants), one can get quite a satisfactory description of experimental spectra. The significance of the Born-Karman analysis is that by fitting the calculated and the experimentally measured values of $\omega(\underline{q})$ in definite directions it is possible, using the force constants thus obtained, to describe the spectrum of lattice oscillations in any given direction and to calculate,

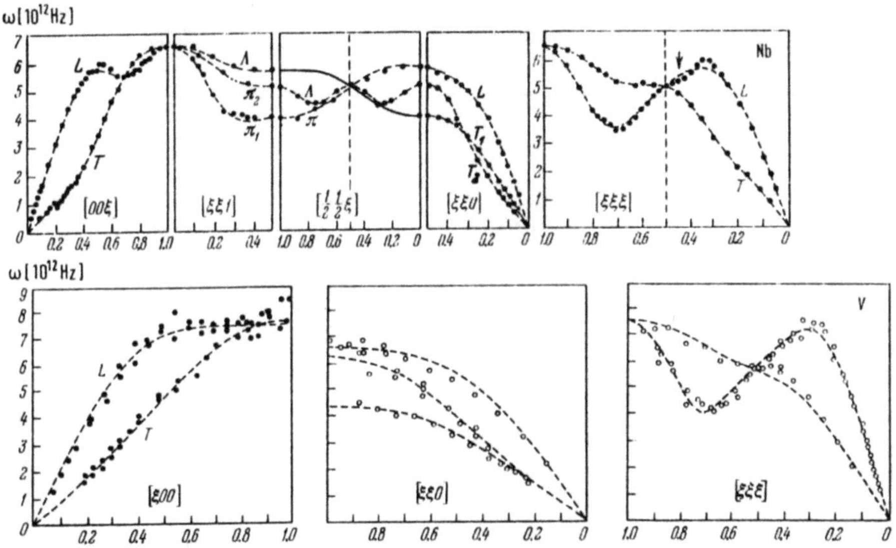

Fig.4.17. The dispersion curves of the phonon spectrum of Nb [4.119] and V [4.118]. L stands for longitudinal modes, T for transverse modes; $\xi = aq/2\pi$

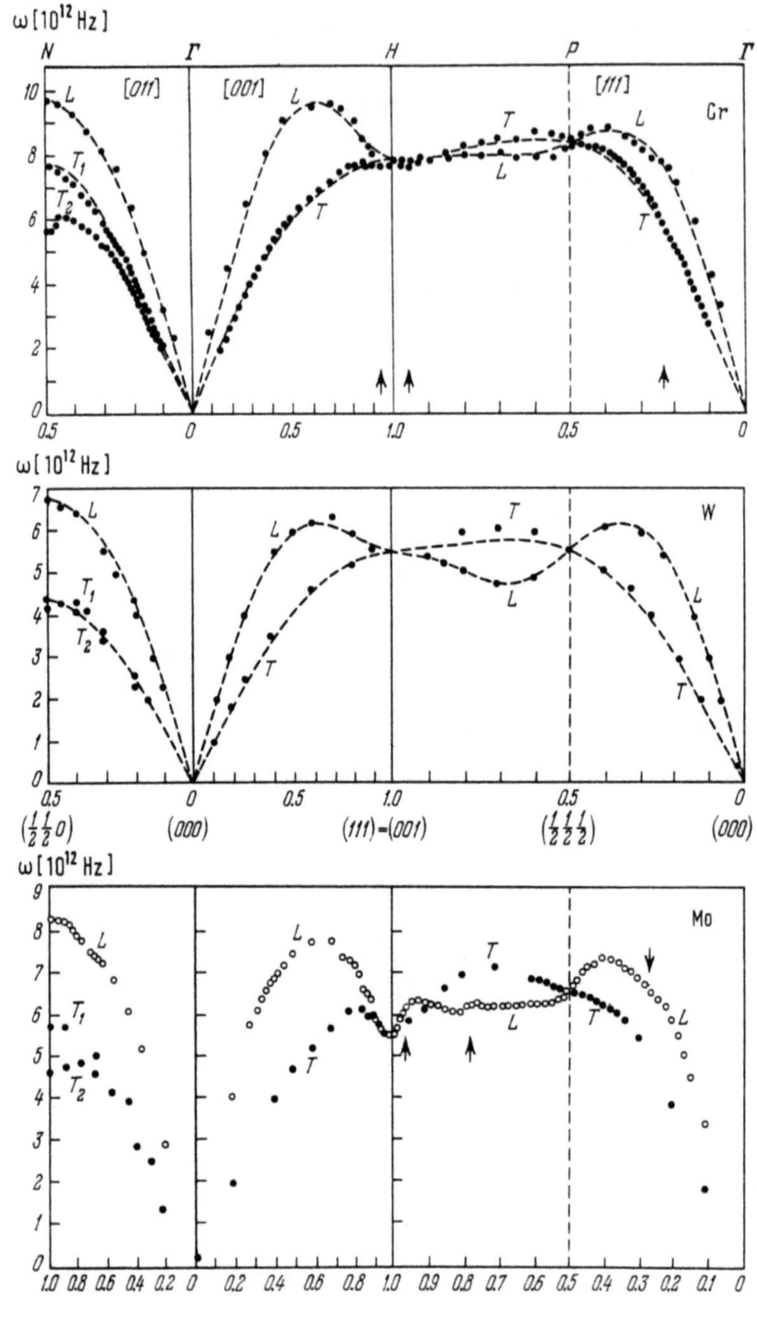

Fig.4.18. The dispersion curves of the phonon spectrum of Cr [4.121], W [4.120] and Mo [4.122]. L stands for longitudinal modes, T for transverse modes

in the final analysis, the phonon density of states. The density thus calculated for V [4.118] is compared in Fig.4.19a with that measured experimentally from inelastic neutron scattering (dotted curve) [4.115]. The agreement between these curves is satisfactory. This gives us some confidence that we can rely on the theoretical curves $F(\omega)$. This is particularly important in those cases when straightforward measurements of $F(\omega)$ are impossible because of the small cross section of incoherent neutron scattering and the $F(\omega)$ obtained by analysis of dispersion curves according to Born and Karman are the only source of information concerning the phonon density of states. Figure 4.19b shows a $F(\omega)$ curve obtained in this manner for metallic Nb [4.119]. One notices similarity in the distribution $F(\omega)$ for various bcc metals and their difference from that for fcc metals. Figure 4.20 shows $F(\omega)$ for the hcp transition metals Ti and Zr which is obtained in [4.123,124] (see also [4.125,136]).

The Born-Karman analysis described above may not be considered exact. It does not take into account the long-range forces arising from the inter-ion interaction modulated by conduction electrons. Charge density fluctuations near a single atom inevitably lead to fluctuations of forces acting on the atoms located far away from a given atom. This screening of the electrostatic charge of the ions by conduction electrons may be expressed in terms of the dielectric permeability $\varepsilon(\underline{q})$ [4.126]. It is worth noting that the dielectric permeability $\varepsilon(\underline{q})$ has a logarithmic singularity at $q = 2k_F$, i.e., when the wave vector is equal to the diameter of the Fermi surface. The function itself remains continuous, but its first derivative in this point diverges logarithmically. Though this is a very weakly pronounced feature, it appears to be important to many properties, primarily to the phonon spectrum. The corresponding anomalies in the oscillation spectrum have the name "Kohn singularities" [4.127]. Physically, they arise from the interaction of conduction electrons with lattice oscillations. Due to the occurrence of a singularity in the screening when one of the Fourier components of the screened pseudopotential goes through the value of the wave vector $q = 2k_F$ the screening efficiency decreases drastically.

The above-mentioned Kohn singularities were found on dispersion curves for Cr, Mo, Nb. In Figs.4.17 and 4.18 they are indicated by arrows.

Analysis of the dispersion curves given in Figs.4.17 and 4.18 allows one to note one more feature peculiar to the phonon spectrum of transition metals: the mean level of the phonon frequencies in the transition metals of the fifth group (V, Nb) is on the whole lower than that in the transition metals of the sixth group (Cr, Mo, W).

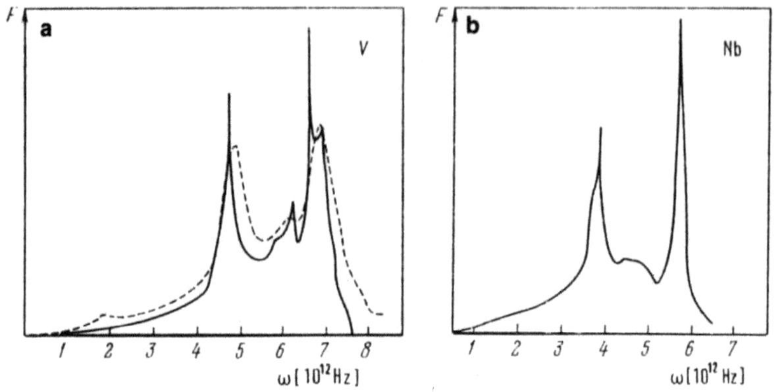

Fig.4.19

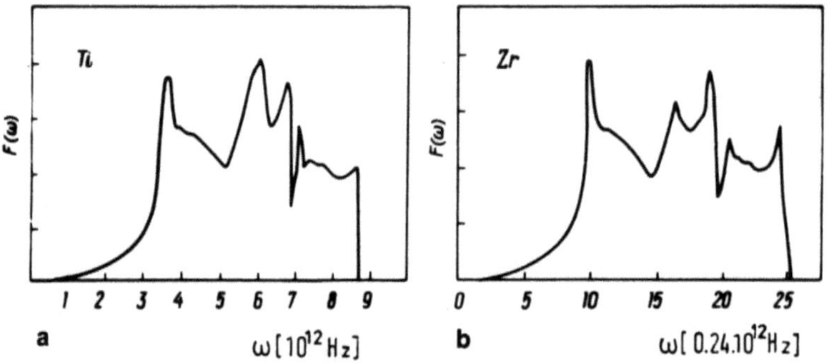

Fig.4.20

Fig.4.19. The phonon density of states in Nb [4.119] and V [4.118], constructed from dispersion curves. The dotted line shows $F(\omega)$ obtained by direct measurements [4.115]

Fig.4.20. The phonon density of states in Ti and Zr

BOHM and STAVER [4.128] have shown that in a system with screening of the electrostatic charge of ions by conduction electrons, the frequency of longitudinal oscillations in the long-wave region is inversely proportional to the density of electron states on the Fermi level

$$\omega \sim [N(0)]^{-1/2} \quad .$$

Thus, proceeding from the phonon spectra given in Figs.4.17 and 4.18, one may expect that the density of states $N(0)$ in the metals of the fifth group exceeds those in the metals of the sixth group. The data presented in Fig.4.3 and Table 4.3 show that this really is so.

More experimental information on the phonon spectrum of a metal may be derived from tunnel measurements. The uniqueness of these measurements is that it becomes immediately possible to obtain the product of both factors involved in the expression for λ, namely (2.4.15), using the experimental-data processing procedure proposed by McMILLAN and ROWELL [4.129].

Unfortunately up to now such measurements have been very laborious and so far, have been carried out elaborately enough only for Ta [4.130a,b,132b] and Nb [4.131a,b,132a,b]. In Fig.4.21 results of tunnel measurements and calculations of $\alpha^2(\omega)\cdot F(\omega)$ are given on the basis for Ta [4.130a]. Comparison of $\alpha^2(\omega)\cdot F(\omega)$ with the phonon density of states which has been obtained from neutron diffraction measurements (Fig.4.21b) shows that the effective matrix element of electron phonon interaction $\alpha^2(\omega)$ depends on frequency only weakly. So, at least for bcc metals, the data obtained from tunnel measurements practically yield the density of phonon states.

SHEN [4.130a] ascribes the jump at $\omega = 1.78 \cdot 10^{12}$ Hz in the curve $\alpha^2(\omega)\cdot F(\omega)$ for Ta (marked with an arrow in Fig.4.21) to Kohn's anomaly. But, as is seen from the data on neutron diffraction measuremtents, the above peculiarity of the phonon spectrum is absent in the curve $F(\omega)$. This problem seems to require further clarification.

Similar tunnel measurements have been carried out for Nb [4.131a,132a]. The positions of the two dips in the second derivative of the tunnel current for Nb - O - Au contacts are also in good agreement with the peaks in the phonon density of states determined by the neutron diffraction technique [4.119].

In his review WOOD (1972) [4.133] noted that the present state of the theory of lattice dynamics does not enable one to establish a direct relationship

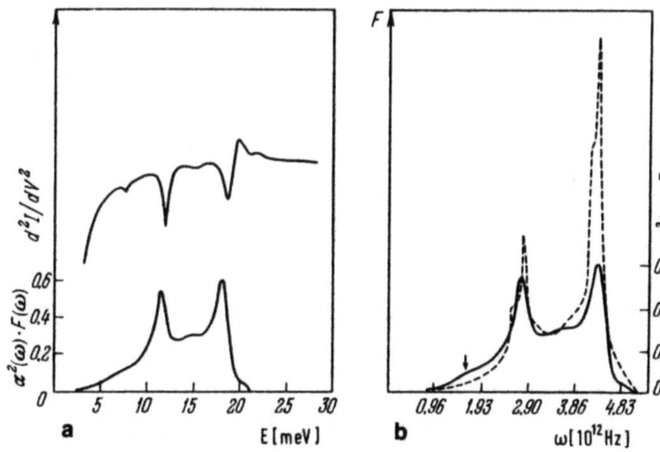

Fig.4.21. The effective phonon spectrum $\alpha^2(\omega)\cdot F(\omega)$ in Ta and the second derivative of the tunnel current with respect to voltage. In b) $F(\omega)$ (dotted line) is constructed from the dispersion curves, $\alpha^2(\omega)\cdot F(\omega)$, according to tunnel measurements data [4.130]

between the measured dispersion relations $\omega(\underline{q})$ and the superconducting properties of transition metals (see also [4.163]). Nevertheless, attempts have been made in literature to compare the dispersion curves in transition metals having highly different values of T_c and to qualitatively consider the possible correlations [4.134,135]. Thus, for example, Fig.4.22 shows dispersion curves of the phonon spectrum of hcp metals for phonons propagating in the c direction [4.134,135]. The authors particularly emphasize the fact that the frequency of the LO (longitudinal optical) mode at zero wave vector is very low in Tc as compared to that in Zr and Y, and this, in their opinion, indicates "softening" of the phonon spectrum which may be responsible for the high T_c in this metal.

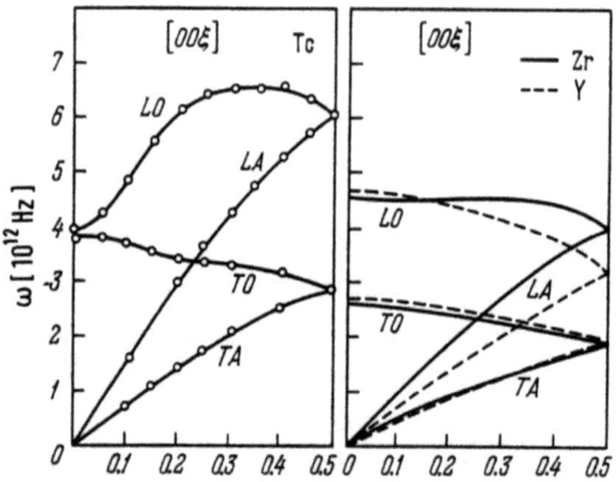

Fig.4.22. The dispersion curves of the phonon spectrum of hcp metals [4.134]. LA stands for longitudinal acoustic modes, LO for longitudinal optical modes, TA for transverse acoustic modes, TO for transverse optical modes, $\xi = aq/2\pi$

4.4 The Electron-Phonon Interaction and T_c in Transition Metals

4.4.1 Analysis of Empirical Values of λ

It has been noted repeatedly that a most important parameter characterizing the magnitude of electron-phonon interaction in metals is the coupling constant or the mass enhancement factor λ.

As described above, the transition temperature T_c for strong-coupling superconductors is expressed in terms of λ according to McMillan's formula (2.3.27). Substituting into this formula the experimental values of T_c, θ_D and μ^* (some mean value $\mu^* = 0.13$ was taken for transition metals), McMILLAN found values of λ for the entire series of superconducting transition elements

[4.83]. These results are given in Table 4.5 and in Fig.4.23. In particular, it is seen from the figure that λ correlates with T_c (Fig.4.2). It takes on the largest value 0.82 for Nb having $T_c = 9.2$ K which is the highest among those in all the transition metals. Accordingly, the minimum of λ and T_c occurs for W.

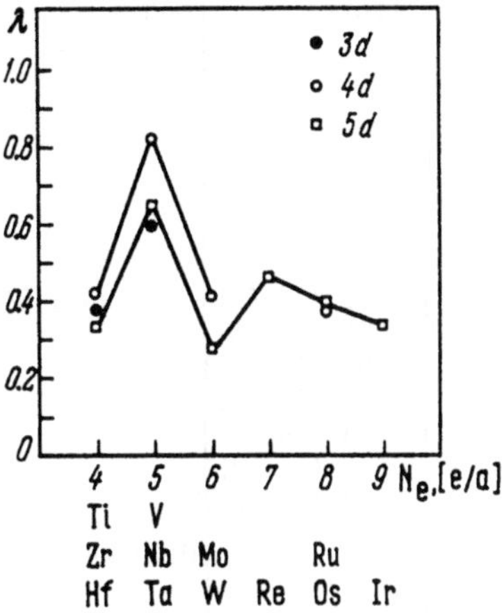

Fig.4.23. The dependence of λ on the number of valence electrons, N_e in transition metals

The importance of [4.83] is that an analysis was given of the role of different factors entering the expression for λ, Eq.(2.4.30). It turned out that the factor $N(0)\langle J^2\rangle$ is almost independent of the electron structure and is ~ 7 eV/Å^2 for most of the transition metals. For nontransition metals (for example, Al, In, Pb), its value is close to 2 eV/Å^2. Thus, an important conclusion was drawn that in the series of similar elements $N(0)\langle J^2\rangle$ is, roughly speaking, constant and therefore T_c is determined primarily by the factor $M\langle\omega^2\rangle$ depending on the ion mass and the phonon spectrum and characterizing the so-called rigidity of the lattice.

This conclusion was somewhat unexpected since, after the creation of the BCS theory, the opinion prevailed that the T_c in metals is primarily determined by the value of the density of electron states on the Fermi level, and the electron attraction matrix element V involved in the expression for T_c, (1.2.5), is almost constant within the series of like metals. A theoretical substantiation for McMILLAN's empirical rule of the constancy of the quantity $N(0)\langle J^2\rangle$ in the series of transition metals was given by HOPFIELD [4.153] on the basis of an analysis of the expansion of the electron-ion interaction matrix element in angular-momentum eigenfunctions. According to [4.153] (Sect.2.4), the magnitude of $N(0)\langle J^2\rangle$ in the series of transition metals

Table 4.5. Electron-phonon interaction parameter in transition metals according to experimental data

Metal	T_c, K	λ according to McMillan [4.83]	λ_{exp} optical measurements	heat capacity measurements	tunnel measurements	resistivity measurements
Ti	0.40	0.38	0.33 [4.137]	0.54^{+}		
V	5.40	0.60	0.62 [4.138] 0.50 [4.140]	1.0 [4.139] 0.97^{+}		
Zr	0.61	0.41	0.37 [4.137]	0.22^{+}		
Nb	9.25	0.82	0.74 [4.141] 0.68 [4.144] 0.93 [4.140, 146] 0.70 [4.137]	0.9 [4.139] 0.87^{+}	0.98 [4.142] 0.81 [4.145] 0.58 [4.131a] 0.93 [4.131b] 0.85-0.9 [4.132b] 0.62 [4.132a]	1.15 [4.143]
Mo	0.915	0.41	0.35 [4.140] 0.38 [4.148] 0.53 [4.149]	0.23^{+}		0.47 [4.143]
Ru	0.49	0.38	0.65 [4.150]	0.47^{+}		
Hf	0.13	0.34		0.14^{+}		
Ta	4.47	0.65	0.48 [4.140]	0.9 [4.139] 0.43^{+}	0.69 [4.130a] 0.68 [4.132b] 0.73 [4.130b]	1.08 [4.143]
W	0.0154	0.28	0.21 [4.140]			0.25 [4.143]
Re	1.69	0.46	0.45 [4.151]	0.37^{+}		
Os	0.66	0.39		0.44^{+}		
Ir	0.11	0.34	0.21 [4.152]			

Note: The asterisk marks the values of λ obtained from λ_{exp} and λ_{theor} given in Table 4.2.

with similar crystal and electron structure is constant, and McMILLAN's formula (2.4.30) for λ may be rewritten in the form

$$\lambda = \frac{\eta}{M\langle\omega^2\rangle} , \qquad \eta = N(0)\langle J^2\rangle \quad . \tag{4.4.1}$$

It can be seen from a comparison of the results of the calculation of η according to (4.4.1) using the experimentally determined values of λ and $\langle\omega^2\rangle$, that η varies considerably from element to element. But within one and the same group of transition elements with similar chemical properties, this quantity appears to be constant.

The constancy of the factor $N(0)<J^2>$ for transition metals is also treated in [4.154] by a tight-binding approximation. Such an analysis may apply to narrow d band transition metals only. In this model it is possible to relate the factor $N(0)<J^2>$ to the quantity E_c which gives the direct contribution of tightly bound electrons to the cohesive energy of the crystal,

$$N(0)<J^2> = q_0^2 E_c \quad ; \tag{4.4.2}$$

q_0 here is the Slater coefficient defining the spatial variation of the radial part of the atomic wave function $\sim\exp(-q_0 r)$. The experimental values of the total cohesive energy for transition metals vary weakly within each series and amount on the average to 4.5, 6.5, 7.5 eV/Å for 3d, 4d and 5d-metals, respectively [4.154]. If one assumes that the main contribution to the expression for E_c of the metal is made by d-band electrons, E_c varies weakly in the series of transition metals, and q_0 is a purely atomic parameter, on the average $q = 0.9$ Å^{-1}. Both circumstances ensure rough constancy of the factor $N(0)<J^2>$. More exact estimates for E_c and q_0 yield variations of the quantities $N(0)<J^2>$, and it turns out that the elements from one and the same group of the periodic system have closer values of $N(0)<J^2>$ than elements from different groups. If follows from (2.4.30) and (4.4.2) that the greater E_c is, the greater λ is and hence T_c. This correlation between T_c and the binding energy is stronger than that between T_c and the melting point of the transition metal which has beer propounded by MATTHIAS [4.31].

In [4.84,155,156], the analysis of McMILLAN and HOPFIELD and of BARISIC, LABBE and FRIEDEL and their assessment of the determining role of lattice rigidity in the formation of superconducting properties of transition metals are subjected to criticism. VARMA and DYNES [4.157] note that (4.4.2) is not rigorous since E_c is also proportional to $M<\omega^2>$. On the other hand, it is shown in [4.155-158] that when deriving (2.4.46), HOPFIELD disregarded the contribution to η by d-f scattering which is essential, as straightforward calculations have shown [4.84,158]. It has also been found that HOPFIELD's assumption that $N_p(0)$ varies weakly within transition series contradicts the calculations made for 3d and 4d metals [4.84]. In [4.157,159,160], attention is again drawn to the existence of a correlation between λ and $N(0)$ and to the important role which $N(0)$ plays in the formation of the superconductivity in transition metals.

So far we have considered the values of λ obtained by McMILLAN's formula (2.3.27) from measured quantities T_c. It is of interest to determine λ from independent experiments and to theoretically calculate this parameter from first principles.

It is shown in [4.161,162] that such information may be obtained from the various kinetic characteristics of the metal which are related to the electron-phonon interaction. One of the experimental techniques of determining λ is based on optical measurements in the infrared region of the spectrum. HOPFIELD [4.162] was the first to pay attention to this possibility and to show that, at $T > \theta_D$, the following relationship between λ_{opt} and the effective electron-phonon collision frequency ω_{e-ph} holds

$$\lambda_{opt} = \frac{\hbar}{2\pi} \frac{\omega_{e-ph}}{kT} \quad . \tag{4.4.3}$$

The quantity ω_{e-ph} is estimated from the conduction electron relaxation frequency when measuring $\varepsilon(\omega)$, i.e., the complex dielectric permeability in the infrared region of the spectrum.

The values of λ_{opt} thus calculated (by considering the data of optical measurements [4.138,140,141,144,148,149]) are cited in Table 4.5. It may be seen that general agreement with the values of λ obtained by McMILLAN is very satisfactory. The maximum discrepancy is for Ta and Nb. This is likely to be due to the fact that the Fermi level in transition metals of the fifth group (V, Nb, Ta) passes near the maximum of the density of states, and the contribution of the interband absorption to $\varepsilon(\omega)$ in these metals is appreciable. This makes it difficult to single out the intraband absorption from $\varepsilon(\omega)$ and, consequently, to correctly estimate the effective conduction electron relaxation frequency and λ.

The electron-phonon interaction may also manifest itself in the resistivity. At high temperatures ($T \gtrsim \omega_D$) the thermal phonons lead to an electron decay

$$\Gamma = 2\pi\lambda T \quad . \tag{4.4.4}$$

Its substitution into the known expression for the resistivity R yields

$$R = \frac{4\pi}{\omega_{pl}^2}\Gamma = \frac{8\pi^2}{\omega_{pl}^2}\lambda T \tag{4.4.5}$$

where ω_{pl} is the plasmon frequency. Thus, the slope of the curve R(T) enables one to obtain λ. This has been done for transition metals in [4.143] (Table 4.5).

Another possibility of experimental determination of λ is offered by tunnel-effect measurements. It is possible from tunnel measurements to directly determine $\alpha^2(\omega)\cdot F(\omega)$ and consequently, λ, (2.4.15). Unfortunately, for transition metals such measurements were performed only for Ta and Nb [4.130,131]. The value $\lambda = 0.69$ obtained for Ta is in good agreement with McMILLAN's estimate ($\lambda = 0.65$).

For Nb, on the contrary, the value $\lambda = 0.58 - 0.62$ determined from tunnel measurements [4.131,132] turned out to be substantially smaller than that obtained by McMILLAN ($\lambda = 0.82$). It is believed that a possible reason for this discrepancy was inadequate removal of impurities from the Nb surface before the experiments [4.131]. In [4.139] a method of estimating the value of λ from experimental heat capacity data was proposed. The authors proceeded from the fact that the electron-phonon interaction affects the effective mass only of electrons whose energy lies within $k\theta_D$ from the Fermi level. Consequently, for those electrons whose energy lies within $k\theta_D$ from the Fermi level. Consequently, for temperatures higher than θ_D the electron heat capacity reflects the density of electron states which is not enhanced by the electron-phonon interaction. At low temperatures, the electron heat capacity coefficient enhanced by the electron-phonon interaction contains a factor $(1+\lambda)$. Thus, comparison of high and low-temperature capacity permits determination of λ and N(0).

The values of λ thus obtained for V, Nb and Ta are given in Table 4.5. It is obvious that the electron-phonon enhancement factor estimated from heat capacity data turned out to be the same for these metals and equal to $\lambda \approx 0.9$. This disagrees with the values of λ obtained from optical and tunnel measurements and with McMILLAN's estimates, and is apparently due to the low accuracy in electron heat capacity measurements.

4.4.2 The Contribution of Electron and Lattice Factors to the Quantity λ

The above comparison of the values of the electron-phonon interaction constants determined from McMILLAN's formula (2.3.27) and obtained independently from experiments shows, on the whole, good agreement with the superconducting transition metals under consideration. This gives good reason to assume that (2.3.27) works well and the approximations made in its derivation are admissible. If this is so, the McMILLAN and HOPFIELD analysis of the role of the electron and phonon factors in the formation of T_c in transition metals should also be regarded as correct. It should be borne in mind that the electron factor $N(0)\langle J^2\rangle$ remains practically invariable for all bcc metals, and the variation of λ and consequently, T_c, arises from the variation of the phonon factor $M\langle\omega^2\rangle$.

This agrees with the empirical correlation between T_c and $1/\theta_D^2$ which we have described in Sect.4.1 if it is assumed that $\sqrt{\langle\omega^2\rangle}\sim\theta_D$; on the other hand, the presence of empirical correlations between T_c and γ and T_c and χ remains obscure. Is it fortuitous that the softening of the phonon spectrum is the greatest in those metals which have the highest values of electron heat capacity and magnetic susceptibility? Or has this relationship a more profound physical sense? The presence of a correlation between T_c and γ should be regarded as a reflection of the existence of a more general correlation between T_c or λ on the one hand and $N(0)$ on the other, since γ is a function of $N(0)$.

Note that this conclusion may not be strictly considered since γ depends not only on $N(0)$ but also on λ, (1.2.15). However, the existence of a correlation between T_c and χ and band calculations show that there really is a correlation between T_c and $N(0)$, namely: maximum values of $N(0)$ correspond to high T_c. Incidentally, as we have seen above, the factor $N(0)\langle J^2\rangle$ determining λ is actually independent of $N(0)$. Then, bearing in mind (2.4.30) for λ, we come to the conclusion that the quantity λ is sensitive to the value of the density of electron states on the Fermi surface through the mean square of frequencies $\langle\omega^2\rangle$ in the phonon spectrum. The $\langle\omega^2\rangle$ dependence of $N(0)$ is caused by the renormalization of frequencies due to the electron-phonon interaction. GOMMERSAL and GYORFFY [4.164] have obtained an approximate formula for $\langle\omega^2\rangle$ by expressing it in terms of $\langle\Omega^2\rangle_0$, i.e., the mean-square frequency of lattice phonons without taking into account the electron-phonon interaction. According to their results

$$\langle\omega^2\rangle = \langle\Omega^2\rangle_0 - \frac{4}{5M}E_F N^2(0)\langle J^2\rangle \quad . \tag{4.4.6}$$

The quantity $N(0)\langle J^2\rangle$ is constant for transition metals, it therefore follows from (4.4.4) that the contribution to $\langle\omega^2\rangle$, which is due to the electron-phonon interaction, is proportional to $N(0)$.

It follows from (2.4.30) and (4.4.5) that the λ dependence of $N(0)$ may be represented as

$$\lambda = \frac{\lambda_0}{1 - 4/5E_F N(0)\lambda_0} \tag{4.4.7}$$

where $\lambda_0 = N(0)\langle J^2\rangle/(M\langle\Omega^2\rangle_0)$ is a coupling constant without phonon frequency renormalization effects. If one plots the empirically determined values of $1/\lambda$ as a function of the parameter $E_F N(0)$, then for 5d-metals: W, Hf, Re, Os, Ta the points fall on a straight line. The slope of that line is approximately 4/5 and the interception with the axis of abscissae gives $\lambda_0^{-1} = 4.3$.

Further, the quantity $E_F N(0)$ may be related to the number of electrons per atom if one makes use of band calculation results for W and accepts the rigid-band hypothesis for 5d-metals and their alloys. For W the density of states was calculated by MATTHEISS [4.47]. The results are presented in Fig.4.24 (solid curve); the dotted curve stands for the dependence of the Fermi energy on the number N_e of electrons per atom.

Using the value $\lambda_0^{-1} = 4.3$ and the data given in Fig.4.24 one can obtain the dependence of λ and, consequently, T_c on the electron concentration N_e. In Fig.4.25 the dependence of T_c thus calculated as a function of N_e is presented in the middle portion of the graph in the interval $4.7 \leq N_e \leq 6.8$. At the boundary of this interval the parameter λ_0 becomes negative. This indicates the instability of the bcc lattice at these boundary values of electron concentration; the lattice should be readjusted so that the electron-phonon coupling decreases. In fact, $N_e = 4$ and 7 e/a corresponds to hexagonal close packing.

The behaviour of T_c in the interval $4.7 \leq N_e \leq 6.8$ gives a surprisingly good fit to experiments. The minimum in N(E) corresponds to W, i.e., an almost

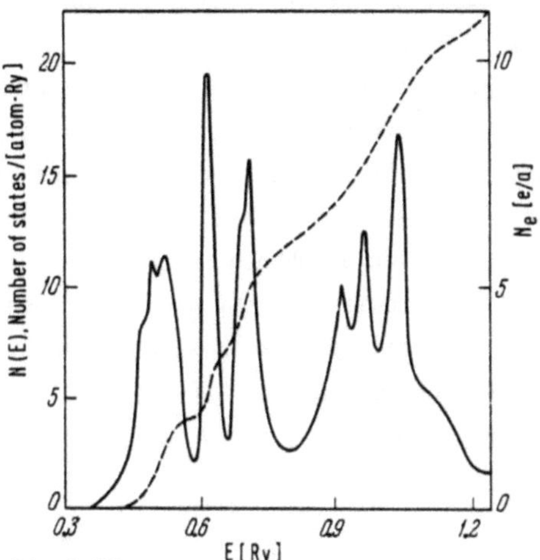

Fig.4.24

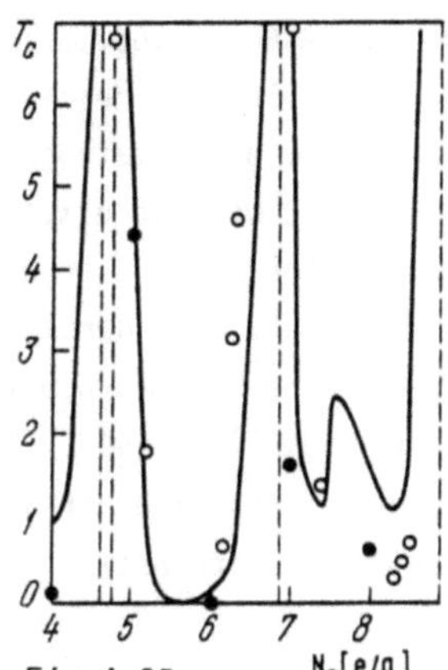

Fig.4.25

Fig.4.24. The density of electron states for W (solid line) and the integral density of states as a function of electron concentration N_e (dotted line)

Fig.4.25. The dependence of T_c on the number of valence electrons, N_e, for transition metals and their alloys. The black points correspond to the experimental T_c for pure metals, the clear ones to alloys. The solid curves result from the GOMMERSAL and JYORFFY [4.164] theory half-filled band. The Fermi level for this metal falls into the valley between two curve peaks which correspond to binding (with lesser energy) and antibinding (with greater energy) d-states. As soon as the Fermi level moves towards one of the peaks, T_c increases drastically due to an increasing λ. Beyond the interval $4.7 \leq N_e \leq 6.8$, when calculating T_c the density of states of Re having a hcp structure was taken. In the point $N_e = 8.8$ the quantity λ diverges again. This corresponds to a transition to a fcc lattice $N_e = 9$ corresponding to Ir.

Thus, (2.3.27) and (4.4.6) give a quantitative behaviour of T_c as a function of the number N_e of electrons in fcc and bcc transition metals using one empirical parameter λ_0. In addition to this, theory correctly predicts the changes in the crystal structure with varying N_e. At some values of N_e which correspond to high densities of states on the Fermi level, the lattice changes its structure to weaken the magnitude of electron-phonon coupling. We note that the values of T_c become larger near such instabilities.

In spite of some progress made in this theory with regard to the coincidence of predictions with a number of properties of transition metals we note a weakness of its starting points in allowing for the electron-lattice interaction in the formation of the phonon spectrum of a metal. The fundamental equation (4.4.5) of the theory is obtained to an approximation in which a correction (as a polarization loop) is added to the initiating phonon frequencies Ω. This was adopted in the scheme by MIGDAL [4.165]. However, as shown by BROWMAN and KAGAN [4.166,167], this approach has an internal defect associated with double consideration of the electron-lattice interaction. If one uses adiabatic perturbation theory it can be seen that major corrections arising on account of double allowance for the electron-lattice interaction are compensated for and the true renormalization of phonon frequencies in the metals becomes weak. Consideration of this circumstance may greatly change the estimate of the bcc and hcp lattice instability boundaries obtained by GOMMERSAL and GYORFFY [4.164]. Thus, the problem requires a fundamental reconsideration, but the idea of a correlation between λ and N_e which underlies their work remains interesting.

4.5 Calculating the Electron-Phonon Interaction Constant from First Principle

4.5.1 Gaspari and Gyorffy's Representation for $\langle J^2 \rangle$

When calculating the parameters specifying the electron-phonon interaction in transition metals one should start from the fact that the electron-ion interaction potential $U(\underline{r})$ leads to d-resonance near the Fermi level. In the case of resonance scattering the electrons whose energy is close to resonance energy stay near the scattering ion for a long time. This confers a local character on the quantities related to the potential $U(\underline{r})$. Below we shall expound the method of calculating the electron-phonon interaction constant λ proposed by GASPARI and GYORFFY [4.168], in which the resonance character of electron scattering by transition metal ions is taken into account from the very outset.

McMILLAN's formula (2.4.30) expresses λ as the product of a number of factors which may be calculated either from first principles [N(0) and $\langle J^2\rangle$] or taken from experiments ($\langle\omega^2\rangle$). Phonon spectrum calculation from first principles is at present impossible for transition metals, therefore $\langle\omega^2\rangle$ should be taken from experiments, for example, from inelastic neutron scattering data.

The quantities N(0) and $\langle J^2\rangle$, which are of a purely electronic nature, can be calculated from first principles on the basis of conventional band calculations and from the representation for $\langle J^2\rangle$ given by [4.168]. It has already been stated above that, according to [4.153], λ in transition metals is determined solely by the local characteristics of the metal on account of the fact that on the Fermi surface states of the same type, namely, with an orbital moment $l = 2$, prevail. This means that $\langle J^2\rangle$, the averaged square of the electron-ion interaction matrix element with respect to the Fermi surface, should be calculated by expanding the electron wave functions in atomic angular-momentum eigenfunctions. Then it is necessary to take exact account of the partial contributions to the scattering of an electron on a separate atom. As has recently been shown by GASPARI and GYORFFY, $\langle J^2\rangle$ may be exactly expressed in terms of partial scattering phases if the lattice potential is taken to be in the muffin tin form. Below we will reproduce the main stages of this theory [4.168,169].

It is convenient to proceed from the following representation for the quantity $\langle J^2\rangle$, see (2.4.32),

$$\langle J^2\rangle = \frac{\int \frac{d\underline{p}}{(2\pi)^3}\int \frac{d\underline{p}'}{(2\pi)^3}\,\delta(\varepsilon_{\underline{p}}-E_F)\,\delta(\varepsilon_{\underline{p}'}-E_F)\sum_j \left|\langle \underline{p}|\underline{e}_j\nabla U|\underline{p}'\rangle\right|^2}{\int \frac{d\underline{p}}{(2\pi)^3}\int \frac{d\underline{p}'}{(2\pi)^3}\,\delta(\varepsilon_{\underline{p}}-E_F)\,\delta(\varepsilon_{\underline{p}'}-E_F)} \;. \qquad (4.5.1)$$

Note that the quantity in the denominator is the square of the density of states on the Fermi surface, i.e., $N^2(0)$. Substitute the explicit expression for the matrix elements with respect to the electron states to transform this formula. Allowing for the normalization of the polarization vectors $\underline{e}_j$ write

$$\langle J^2\rangle = \frac{1}{N^2(0)}\int \frac{d\underline{p}}{(2\pi)^3}\int \frac{d\underline{p}'}{(2\pi)^3}\int d\underline{r}\int d\underline{r}'\,\delta(\varepsilon_{\underline{p}}-E_F)\,\delta(\varepsilon_{\underline{p}'}-E_F)$$

$$\times\ \nabla U(\underline{r})\nabla U(\underline{r}')\psi^*_{\underline{p}}(\underline{r})\psi_{\underline{p}'}(\underline{r})\psi^*_{\underline{p}'}(\underline{r}')\psi_{\underline{p}}(\underline{r}') \quad . \qquad (4.5.2)$$

Here $\psi_p(\underline{r})$ is the Bloch wave function of an electron with a momentum $\underline{p}$.

In terms of the electron Green's function

$$G(\underline{r},\underline{r}';E) = \int \frac{d\underline{p}}{(2\pi)^3}\,\frac{\psi^*_{\underline{p}}(\underline{r})\psi_{\underline{p}}(\underline{r}')}{E-\varepsilon_{\underline{p}}+i\delta} \quad . \qquad (4.5.3)$$

We get

$$\langle J^2\rangle = \frac{1}{\pi^2N^2(0)}\int d\underline{r}\int d\underline{r}'\ \mathrm{Im}\{G(\underline{r},\underline{r}';E_F)\}\nabla U(\underline{r})\nabla U(\underline{r}')\ \mathrm{Im}\{G(\underline{r}',\underline{r};E_F)\} \quad , \qquad (4.5.4)$$

where

$$\mathrm{Im}\{G(\underline{r},\underline{r}';E_F)\} = -\pi\int \frac{d\underline{p}}{(2\pi)^3}\psi^*_{\underline{p}}(\underline{r})\psi_{\underline{p}}(\underline{r}')\delta(\varepsilon_{\underline{p}}-E_F) \quad . \qquad (4.5.5)$$

Equation (4.5.4) is a starting point for the development of the $\langle J^2\rangle$ calculation method.

Expression (4.5.3) is the Green's function of a Bloch electron moving in a periodic lattice. To calculate it we employ the multiple scattering theory [4.170] which yields G in terms of the scattering matrix T. This theory has a general expression of the Green's function

$$\mathrm{Im}\{G(\underline{r},\underline{r}';E)\} = \sum_{LL'}\Delta_L(\underline{r})\Delta_{L'}(\underline{r}')\ \mathrm{Im}\{T^{ii}_{LL'}(E)\} \quad , \qquad (4.5.6)$$

in which $\underline{r}$ and $\underline{r}'$ belong to one and the same muffin-tin sphere on the site i. Summation over L here denotes

$$\sum_L = \sum_{l=0}^{\infty}\sum_{m=-l}^{l} \quad ,$$

and

$$\Delta_L(\underline{r}) = -\frac{\sqrt{E_F}R_l(r,E_F)Y_L(\theta,\varphi)}{\sin\delta_L(E_F)} \quad , \qquad (4.5.7)$$

where $Y_l(\theta,\varphi)$ is a spherical function (θ and φ stand for the polar and azimuth angles of the vector $\underline{r}$), $R_l(r,E_F)$ is the radial function of the Schroedinger equation in the presence of a muffin-tin potential $U(\underline{r})$ and is so normalized that when $r > a$ (the radius of the muffin-tin sphere)

$$R_l(r,E) = \cos\delta_l(E)\cdot j_l(\sqrt{E}r) - \sin\delta_l(E)\cdot n_l(\sqrt{E}r) \quad . \tag{4.5.8}$$

Here j_l and n_l are Bessel and Neumann's spherical functions, $\delta_l(E)$ is the scattering phase for the energy E and orbital moment l.

The quantity $T^{ii}_{LL'}(E)$ involved in (4.5.6) for the Green's function is a scattering matrix describing the scattering of partial waves on point centres and obeys the equation

$$T^{ij}_{LL'}(E) = t^i_L(E)\delta_{ij}\delta_{LL'} + \sum_{s(\neq j)}\sum_{L''} t^i_L(E)G^{is}_{LL''}(E)T^{sj}_{L''L'}(E) \quad , \tag{4.5.9}$$

in which $t^i_L(E)$ is an operator of scattering on one centre located in the site i. It is expressed in terms of the scattering phase corresponding to a given energy E:

$$t^i_L(E) = -\frac{1}{\sqrt{E}}\sin\delta^i_l(E)\cdot e^{i\delta^i_l(E)} \quad . \tag{4.5.10}$$

The quantity $G^{is}_{LL''}(E)$ defines the matrix of structural constants which depends only on the lattice and the given energy of the electron and can be tabulated in advance. Thus, (4.5.6) defines the Green's function of a Bloch electron in terms of quantities determining the exact solution of the scattering problem for one muffin-tin centre and for point centres of the crystal lattice.

The expression for $\langle J^2\rangle$ may be simplified considerably if the quantity $\mathrm{Im}\{T^{ii}_{LL'}\}$ is assumed to be diagonal with respect to the momentum indices L and L'. As shown in [4.171], this assumption holds for symmetric bands, i.e., when ε_p depends only on the module of the vector $\underline{p}$. Substituting now the expansion (4.5.6) into (4.5.4), we obtain the following expression:

$$\langle J^2\rangle \approx \frac{1}{\pi^2N^2(0)}\sum_{LL''}\int d\underline{r}\int d\underline{r}'\Delta_L(\underline{r})\Delta_L(\underline{r}')\,\mathrm{Im}\{T_{LL}(E_F)\}$$

$$\times\nabla U(\underline{r})\nabla U(\underline{r}')\Delta_{L'}(\underline{r})\Delta_{L'}(\underline{r}')\,\mathrm{Im}\{T_{L'L'}(E_F)\} \quad . \tag{4.5.11}$$

We have omitted the superscripts of the quantity T^{ii}_{LL}, assuming that the crystal consists of atoms of one and the same sort. Integration over $\underline{r}$ and $\underline{r}'$ in (4.5.11) is performed within one muffin-tin sphere.

$$\langle J^2\rangle = \frac{E_F}{\pi^2 N^2(0)} \sum_l 2(l+1) \cdot \left| \int_0^a dr\, r^2 R_l(r,E_F) \frac{dU}{dr} R_{l+1}(r,E_F) \right|^2$$
$$\times \frac{N_l(0) N_{l+1}(0)}{N_l^{(1)}(0) N_{l+1}^{(1)}(0)}$$

As shown by [4.168], the integral involved in this equation may be expressed by the scattering phases on the muffin-tin potential,

$$\int_0^a dr\, r^2 R_l \frac{dU}{dr} R_{l+1} = \sin(\delta_{l+1} - \delta_l) \quad , \tag{4.5.17}$$

so that the equation may be written in the final form:

$$N(0)\langle J^2\rangle = \frac{2mE_F}{\hbar^2 \pi^2 N(0)} \cdot \sum_l 2(l+1) \sin^2(\delta_{l+1} - \delta_l) \frac{N_l(0) N_{l+1}(0)}{N_l^{(1)}(0) N_{l+1}^{(1)}(0)} \quad . \tag{4.5.18}$$

Up to now we have been using a system of units [see, for example, (4.5.8)] in which $2m/\hbar^2 = 1$. In (4.5.18) this factor $2m/\hbar^2$ is written out explicitly.

We see that the averaged square of the electron-ion interaction matrix element is defined by two types of quantities, viz., on the one hand, by quantities dependent on one centre only (δ_l and $N_l^{(1)}$) and, on the other hand, by quantities characterizing the scattering of electrons on the crystal lattice [$N_l(0)$], both being taken at the Fermi energy.

Let it be noted that in the muffin-tin potential model the only approximation made in the derivation of (4.5.18) (the diagonality of the quantities $\mathrm{Im}\{T_{LL'}\}$) is exact for spherical bands. However, as demonstrated by JOHN [4.171], for cubic crystals it is possible to use the expansion of Bloch's functions in basis functions of cubic-group irreducible representations; the first three term in (4.5.18) (for $l = 0, 1$ and 2) are then unchanged. Thus, at least for cubic crystals, the constraint which is placed on the result (4.5.18) and which is due to the sphericity of the bands is lifted.

Consider (4.5.18) in the so-called local limit, when the scattering of electrons on a single centre only is taken into account so that $N_l = N_l^{(1)}$. In this case

$$N(0)\langle J^2\rangle = \frac{2mE_F}{\pi^2\hbar^2}\frac{\sum\limits_l 2(l+1)\sin^2(\delta_{l+1}-\delta_l)}{\sum\limits_l N_l^{(1)}(0)} \quad . \tag{4.5.19}$$

For transition metals in which there is a strong d-type scattering, this expression may be simplified since among all the phases δ_2 dominates and varies quickly with energy in the vicinity of the Fermi energy. In this situation the "one-centre" density of states $\sum\limits_l N_l^{(1)}(0)$ is given to a good approximation by the expression $5\delta_2'/\pi$ (δ_2' is the derivative of the scattering phase with respect to energy) so that (4.5.19) becomes

$$N(0)\langle J^2\rangle \approx \frac{2mE_F\sin^2\delta_1}{\pi\hbar^2\delta_1'} \quad . \tag{4.5.20}$$

Assume further that d-waves in transition metals have a resonance scattering and the Fermi energy comes within the half-width of this resonance. In this case the following equation may be written for the scattering phase

$$\tan\delta_2(E_F) = \frac{\Gamma}{E_d - E_F} \quad , \tag{4.5.21}$$

where E_d and 2Γ denote the position and the width of the resonance. It follows from this formula that $\sin^2\delta_2/\delta_2' = \Gamma$ and so (4.5.20) goes into

$$N(0)\langle J^2\rangle \approx \frac{2mE_F\Gamma}{\pi\hbar^2} \quad . \tag{4.5.22}$$

The width of the d-resonance apparently varies little if we move from one transition element in the periodic table to another, and this makes the expression $N(0)\langle J^2\rangle$ approximately constant for all d-elements in accordance with McMILLAN's empirical results. It may be seen from the above derivation of (4.5.22) that this result is of a qualitative nature. The actual variation of the quantity $N(0)\langle J^2\rangle$ for the entire series of transition metals may be established from numerical calculations according to (4.5.18).

4.5.2 Calculation of $N(0)\langle J^2\rangle$ for Transition Metals

Exact calculation of $\langle J^2\rangle$ according to (4.5.18) requires band calculations of the quantities $N_l(0)$, which are rather difficult. Instead, EVANS et al.

[4.169] have carried out approximate calculations for transition metals, proceeding from the following assumptions: the s and p-states in transition metals should not differ much from their one-centre expressions and therefore $N_l(0) \approx N_l^{(1)}(0)$ for $l = 0,1$. They restricted the sum over l to $l \leq 2$ and took a simplified expression

$$N(0)\langle J^2\rangle = \frac{2mE_F}{\pi^2\hbar^2 N(0)}\left\{2\sin^2(\delta_1-\delta_0) + 4\left[\sin^2(\delta_2-\delta_1) + 6\sin^2\delta_2\right]\frac{N_2(0)}{N_2^{(1)}(0)}\right\} \tag{4.5.23}$$

instead of the exact equation (4.5.18). The expressions $N_l^{(1)}(0)$ are easily calculated on the basis of (4.5.15) and $N_2(0)$ may be obtained from the total density of states on the Fermi surface $N_2(0) = N(0) - N_0^{(1)}(0) - N_1^{(1)}(0)$. Thus, in order to calculate $\langle J^2\rangle$ according to (4.5.23), it is necessary to find the quantities determining the scattering on one centre, and the overall density of states N(0) which, for most of the transition metals, may be taken from experiments.

In this manner, $\langle J^2\rangle$ and λ were calculated for ten transition metals: Zr, Nb, Mo, Ta, W, Re, Rh, Pd, Ir and Pt. The phase shifts were calculated for several energy values by numerical integration of the radial Schrödinger equation. As an example, Fig.4.26 gives the result for $\delta_2(E)$ for three transition elements: Zr, Nb and Mo. We see that, strictly speaking, there is no d-resonance in these elements, but the existence of a strong scattering of d-states on the atomic potential is undoubted. The resonance behaviour of the quantity δ_2 manifests itself in Rh and particularly in Pd, as well as in Ir and Pt. Comparison of the numerical values of $\delta_2(E)$ for transition metals shows that the d-resonance is sharper at the beginning and at the end of the transition series period and is very wide in the middle.

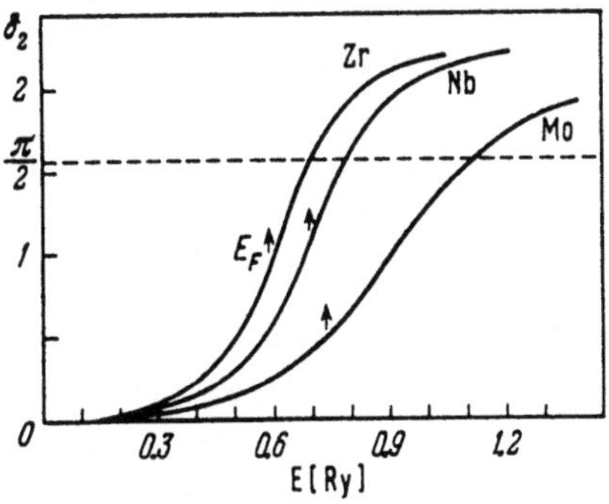

Fig.4.26. The phase shift of the d-states in Zr, Nb, Mo [4.169]

Further, if one passes from 4d to 5d-metals the resonance becomes wider, too. This behaviour agrees with band calculations if one adopts the concept that sharp d-resonance on an isolated atom leads to narrow d-bands in the metal [4.173]. Then the quantities $\delta_2(E)$ should behave so that the d-bands are narrow in metallic Zr and Pd, but wide in metallic Mo. On the other hand, the bands should widen when one passes from 4d to 5d metals. ANDERSON's calculations [4.174] made for the group of metals Rh, Pd, Ir and Pt give a qualitative confirmation of these features of band behaviour.

As far as the scattering phases for s and p-states are concerned, they turn out to be negative for these metals. Their values on the Fermi level are not very small, but their derivatives with respect to energy are negative and small in magnitude. Thus, calculation of phase shifts shows the dominating role of d-states for the transition metals mentioned above. The densities of states on the Fermi surface for the majority of metals were empirical (taken from McMILLAN's work [4.83]). An exception were Rh, Pd, Ir and Pt where these values were taken immediately from ANDERSON's band calculations.

The above calculations have shown that in the general case the third term in (4.5.23) involving $\sin^2\delta_2$ makes the greatest contribution, but the first two terms are comparable in magnitude, so all the phase shifts play an important role.

Similar calculations were carried out for the entire series of 3d and 4d-metals in [4.175] without using artificial simplifying assumptions adopted in the derivation of (4.5.23). These calculations were performed on the basis of the exact formula (4.5.18), and they incorporated band calculations of partial contributions to the density of states on the Fermi surface. The calculations involved have shown that when estimating the values of $\eta = N(0)\langle J^2\rangle$ in transition metals the contribution of d-f scattering may not be neglected, as done so by HOPFIELD [4.153]. Apart from this, it was found that p and f-states are highly hybridized with d-states, so that $N_p(0)$ and $N_f(0)$ substantially change their values within the transition series and the assumption that $N_p(0) = \text{const}$ is not correct [4.153]. A main result of those calculations was that the parameter η varies in the transition metal series and is proportional to N(0). This contradicts the data of McMILLAN and HOPFIELD [4.83,153] and bears witness to the DYNES and VARMA empirical correlation between λ and N(0), and hence between T_c and N(0). Another important conclusion which followed from the calculations was that the calculated λ and η, as a rule, exceed the empirical ones by 40 to 50%. It was shown in [4.155,156] that this might result from neglecting the nonspherical contributions and screening effects in the calculation of d-f scattering matrix elements in the GASPARI-JYORFFY approximation, which makes the d-f contribution approximately twice as high.

Table 4.6 shows T_c calculation results for transition 3d and 4d-metals according to [4.175]. T_c was calculated according to McMILLAN's formula (2.3.27) using the experimental values of the Debye temperature [4.176] and the estimates of the value of the Coulomb pseudopotential to BENNEMANN and GARLAND [4.177]. For μ^* and λ, the values are given taking account of the screening effects in the calculation of d-f scattering matrix elements. In accordance with experiments it was found that V and Nb have comparatively high superconducting transition temperatures, and Ti and Zr possess low temperatures. According to calculations Sc is a superconductor, but this, as is known, is not the case in experiments. In [4.175] this contradiction between theory and experiment is accounted for by the fact that the calculation of T_c was based on a calculation of the electron spectrum of bcc scandium, whilst Sc actually has a hcp structure. The same effects account for the extremely low theoretical transition temperature T_c in hcp metals, as compared to a true hcp metal having $T_c = 7.8$ K.

Table 4.6 Results of calculation of $\eta = N(0)\langle J^2\rangle$, λ and T_c for 3d and 4d-transition metals and comparison with experiment

Element		N(0) $\left[\frac{\text{states}}{\text{atom spin}}\right]$	μ^*	η [eV/Å^2]	λ	λ emp. [4.83]	T_c [K]	T_c exp. [K]
Sc	360	16.02	0.183	2.530	0.489		0.51	0.0
Ti	420	12.38	0.168	3.200	0.426	0.38	0.28	0.40
V	380	12.70	0.169	4.370	0.669	0.60	4.62	5.40
Cr	630	4.74	0.107	2.400	0.131		0.0	0.0
Mn	410	12.02	0.166	2.980	0.364		0.04	0.0
Fe	467	26.11	0.206	2.920	0.270		0.0	0.0
Co	445	14.34	0.176	2.050	0.197		0.0	0.0
Ni	450	27.18	0.208	0.880	0.084		0.0	0.0
Y	280	10.94	0.160	1.700	0.276		0.0	0.0
Zr	291	8.58	0.145	3.520	0.514	0.41	1.53	0.61
Nb	275	9.71	0.153	5.410	0.867	0.82	8.77	9.25
Mo	450	4.42	0.102	3.440	0.199	0.41	0.0	0.915
Tc	411	7.86	0.139	4.590	0.312		0.03	7.80
Ru	600	7.39	0.135	3.520	0.109	0.38	0.0	0.49
Rh	480	8.68	0.146	2.490	0.118		0.0	0.0
Pd	274	14.85	0.179	1.050	0.148		0.0	0.0

4.5.3 Calculation of the Pressure Dependence of T_c

HOPFIELD [4.178] was the first to try to interpret the pressure dependence of T_c on the basis of McMILLAN's formula (2.3.27), assuming that the Coulomb pseudopotential μ^* depends weakly on volume. In this case the volume dependence of T_c is due to the mean frequency of the phonon spectrum and due to the electron-phonon interaction constant λ which we write in terms of the electron and lattice factors

$$\lambda = \frac{\eta}{M\langle\omega^2\rangle} \quad , \quad \eta \equiv N(0)\langle J^2\rangle \quad . \tag{4.5.24}$$

Differentiating (2.3.20) with respect to $\ln V$ (V is the volume of the system) we obtain

$$\frac{d\ln T_c}{dP} = \frac{1}{B}\left\{\gamma_g + \left[\frac{1.04\lambda}{\lambda - \mu^*(1+0.62\lambda)} - \frac{1.04(1+\lambda)(1-0.62\lambda)}{[\lambda - \mu^*(1+0.62\lambda)]^2}\right]2\gamma_g + \frac{d\ln\eta}{d\ln V}\right. \quad , \tag{4.5.25}$$

in which B is the volume module and γ_g the effective Grüneisen constant

$$\gamma_g = -\frac{d\ln\sqrt{\langle\omega^2\rangle}}{d\ln V} \quad . \tag{4.5.26}$$

To estimate the value $d\ln\eta/d\ln V$, HOPFIELD used the experimental data for dT_c/dp and B. Unfortunately, γ_g cannot be directly obtained from experiments, and use was made of tabulated values of γ_g [4.97].

This resulted in values of $d\ln\eta/d\ln V$ for the entire series of superconducting transition metals ranging from -2.6 to -4.4. It is with these data that the theoretical values of $d\ln\eta/d\ln V$ should be compared [4.179-181].

Since for η there is (4.5.18) permitting its calculation from first principles, the problem reduces to calculating η for several volumes. The full computation program demands calculation of the potential and scattering phases as a function of volume, and also band calculations indispensable for calculating the partial contributions of $N_l(0)$ to the density of states. RATTI et al. [4.181] carried out calculations in the limit of one-centre scattering, neglecting the structural effects in the scattering of electrons on the lattice. Moreover, in their calculations they restricted their attention to d type scattering phases only, assuming the s and p type phases to be small. Thus, the working calculation formulae were expressions (4.5.20,22) which should be calculated at various given volumes. It is necessary to calculate how the scattering phase $\delta_2(E)$ and the Fermi energy E_F vary with volume.

As an example, in Fig.4.27 results of a calculation of the d-phase of scattering $\delta_2(E)$ in V for four values of relative volume V/V_0 are given, where V_0 is the volume of a metal under normal conditions. It follows from the results that the d-resonance width increases as the volume is compressed. The same applies to other transition metals that have been investigated: Nb and Fe, Pd and Cu. Their Fermi energy for given compression was calculated from FRIEDEL's summation rule

$$Z_d = \frac{10}{\pi}\,\delta_2(E_F) \quad , \tag{4.5.27}$$

which relates the number of d-electrons per atom to the scattering phase on the Fermi surface [4.182].

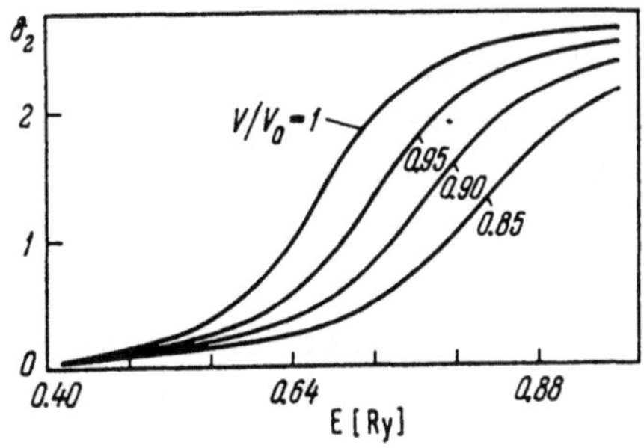

Fig.4.27. The energy dependence of of the d-phase of scattering for V at different compressions

Assume that the dependence of the quantities Γ and E_F determining η is a power type dependence, i.e., $\Gamma \sim V^n$ and $E_F \sim V^{n'}$. Then according to (4.5.22) we have

$$\frac{d\ln\eta}{d\ln V} = n + n' \quad . \tag{4.5.28}$$

For a number of transition elements the calculated values of n and n' [4.181] are presented in Table 4.7.

HOPFIELD's empirical values for the quantity $d\ln\eta/d\ln V$ in superconducting transition metals lie in the same range as the theoretical values given in Table 4.7. Estimates which are close to those obtained by HOPFIELD were also given by GARLAND and BENNEMANN [4.183] on the basis of the latest experimental dT_c/dP data [4.36]. Thus, from this comparison of the results it may be concluded that the d resonance widening during compression is responsible for the

Table 4.7 Exponents determining the volume dependence of the quantities Γ, E_F and η [4.181]

	V	Nb	Fe	Pd	Cu
n	-2.20	-2.30	-2.34	-4.35	-4.05
n'	-1.30	-1.26	-1.36	-2.11	-1.98
n + n'	-3.5	-3.6	-3.7	-6.5	-6.0

large negative values of the quantity $d \ln \eta / d \ln V$ in transition metals.

The problem concerning the variation of the parameter dT_c/dP when one passes from element to element in the periodic table cannot yet be treated unambiguously since the information concerning the Grüneisen constants for transition metals is very vague. However, it may be thought [4.181] that for the transition metals located on the right-hand side of the periodic table γ_g is greater than 1.7 and therefore $2\gamma_g + d \ln \eta / d \ln V$ is positive in view of the expression (4.5.25) $dT_c/dP < 0$. On the other hand, for the elements in the left-hand half of the table (Zr, V, Ti) the γ_g are small and $d \ln \eta / d \ln V$ is a large and negative quantity, therefore $dT_c/dP > 0$. In general, it is this situation that occurs in experiments (see Table 4.1).

The behaviour of T_c versus pressure for V and Nb should be compared separately. According to the calculation data (Table 4.7) the values of $d \ln \eta / d \ln V$ for both metals are close to -3.5, therefore, assuming the Grüneisen constants to have close values, it may be expected that the sign of the quantities dT_c/dP in these elements will be the same. It is seen from Fig.3.27 that the T_c vs. P curve for Nb is nonmonotonic, and $dT_c/dP > 0$ only at high pressures, just like in V. At low pressures $dT_c/dP < 0$ so that the behaviour of Nb and V differs very much. The presence of such a nonmonotonic dependence in one element calls for experiments on other transition metals in the region of higher pressures. The anomaly in the T_c vs. P dependence has not been explained yet.

5. Superconductivity in Transition-Metal Alloys

The alloys of transition metals constitute a wide family of superconductors. The electron structure and the phonon spectrum of an alloy differ from their counterparts in pure metals and therefore their superconducting properties merit individual consideration.

When transition elements are alloyed with each other and with nontransition elements disordered solid solutions and ordered intermetallic compounds form. The latter will be treated in Chapters 6 and 7; here we focus our attention on metallic solid solutions.

5.1 Types of Solid Solutions and Conditions of Their Formation

Of greatest interest among solid solutions are binary and ternary alloys whose components are transition metals[1].

5.1.1 Conditions for the Formation of Solid Solutions

The nature of the formation of a solid solution depends on the structural factor (the similarity or the difference in the crystal lattices of the solvent and of the element being dissolved), the dimension factor (the relation between the atomic radii of the components), and on the chemical factor (the difference in the electronegativity of the elements alloyed). The condition of high mutual solubility of transition metals consists of the closeness of the position in the periodic system and consequently, of minimal

---

[1] Solid solutions of transition metals in nontransition metals were considered partly in Sect.3.7. The superconducting properties of solid solutions of nontransition elements in transition elements are rather incompletely understood. The few experimental data [5.1] indicate that no particularly interesting results have been obtained so far.

differences in their electronegativeness, of the isomorphism of crystal structures, and of inconsiderable differences in the dimensions of atoms, i.e., of the components of the alloys. In all other cases the elements being alloyed tend to form intermetallic compounds [5.1].

5.1.2 Types of Solid Solutions and Phase Diagrams

Let us now give concise information concerning the types of binary solid solutions and the features peculiar to phase diagrams which will be helpful in the analysis of their superconducting properties.

One usually distinguishes solid solutions with complete and partial solubility in liquid and solid states. Examples of the first type are binary Nb - Ta and Ti - Zr systems whose phase diagrams are given in Figs.5.1 and 2. In the case of Nb - Ta alloys both components undergo no polymorphous transformation and the phase diagram of such systems looks very simple (Fig.5.1 [5.1]). Both components of Ti - Zr alloys undergo temperature transformation from bcc to hcp structure and the phase diagram (Fig.5.2 [5.1]) has a peculiarity which shows that in the high-temperature range there is a continuous solubility of components in the bcc structure, and at low temperatures the full solubility is the same but occurs in the hcp structure.

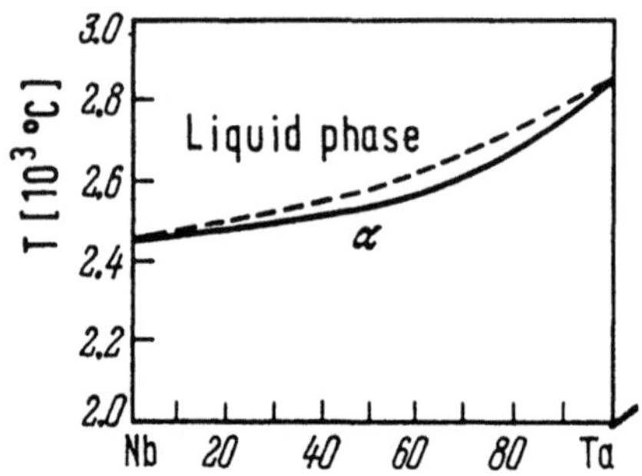

Fig.5.1. The phase-rule diagram for alloys of the Nb - Ta system

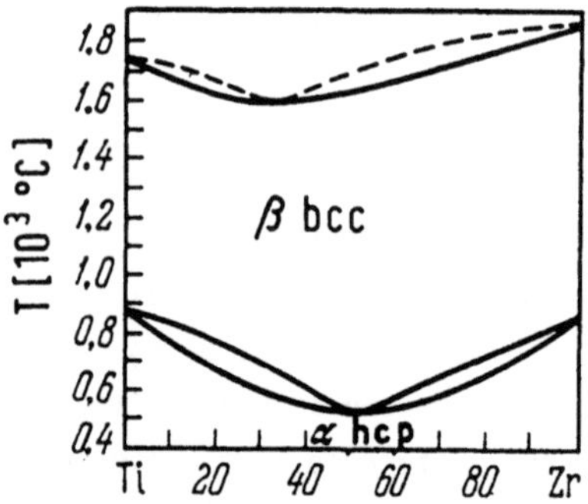

Fig.5.2. The phase-rule diagram for alloys of the Ti - Zr system

If only one of the metals being alloyed possesses polymorphism the phase diagram is of a more complicated form. For example, when hcp metals of group IV are alloyed with bcc metals of groups V and VI, a continuous series of bcc β-solid solutions are produced only at high temperatures (see, for example, the phase diagram for Ti - Mo alloys which is shown in Fig.5.3 [5.1]).

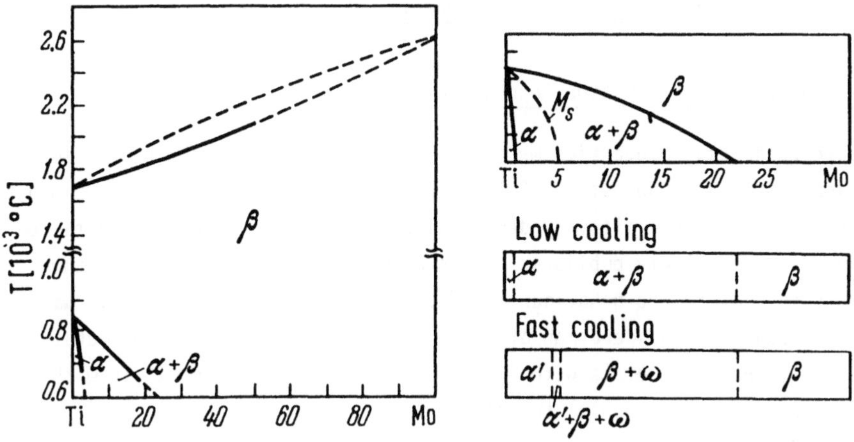

Fig.5.3a,b The phase-rule diagram for alloys of the Ti - Mo system

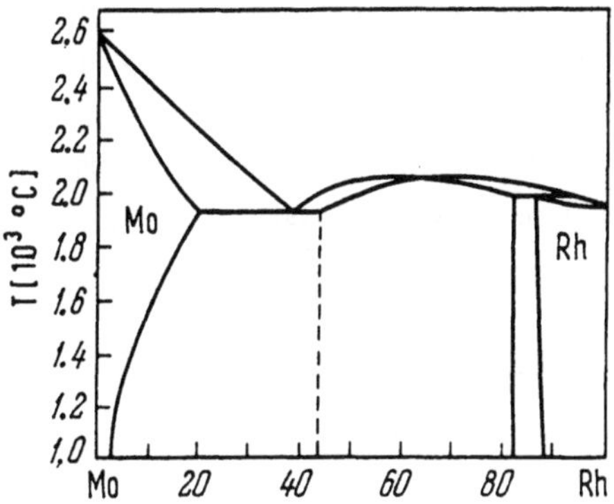

Fig.5.4. The phase-rule diagram for alloys of the Mo - Rh system

When titanium-rich alloys are cooled slowly the β-solid solution breaks down and produces an α-phase of hcp structure and a mixture of (α + β) phases. In the case of fast cooling the β → α transformation proceeds in a diffusionless manner (according to the martensite mechanism) and an α'-martensite phase and a β-phase form. The latter is instable and normally breaks down yielding an ω-phase (Fig.5.3).

An example of the other type of solid solutions is the Mo - Rh system. It is evident from the phase diagram given in Fig.5.4 [5.2] that there are restricted regions of solid solutions based on each of the components.

5.2 The Superconducting Transition Temperature of Alloys

5.2.1 Types of Concentration Dependences of T_c

The superconducting properties were investigated in most of the binary solid solutions of transition metals [5.1,3-31,52,86-88]. The available experimental evidence for T_c in the alloys under consideration is presented in a detailed manner in the regularly published tables by ROBERTS [5.32-34]. One of the attempts to analyze the behaviour of T_c during the formation of alloys was made in [5.35] by consideration of different families of transition metals: "central" transition metals with $4 \le N_e \le 8$, "close" transition metals with $3 \le N_e \le 4$ and "remote" (noble) transition metals with $8 \le N_e \le 10$.

A special feature of our consideration is that from the whole variety of T_c variations during the formation of alloys we have singled out the most prominent and typical concentration dependences. They turned out to be of three types:

1) When an element with a lower value of T_c is added, the superconducting transition temperature of the alloy drops linearily or almost linearily (Fig.5.5 [5.1]).

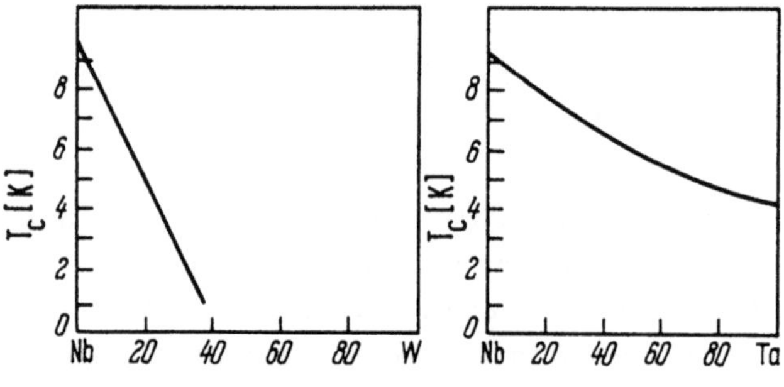

Fig.5.5. The superconducting transition temperature T_c of the alloys Nb - W and Nb - Ta

2) The T_c in solid solutions of transition metals reaches a minimum in the intermediate concentration range, showing an explicit "sagging" from the linear dependence (Fig.5.6 [5.19]).

3) The values of T_c in alloys exceed (in a certain concentration region) those in the basic components (Fig.5.7 [5.1]).

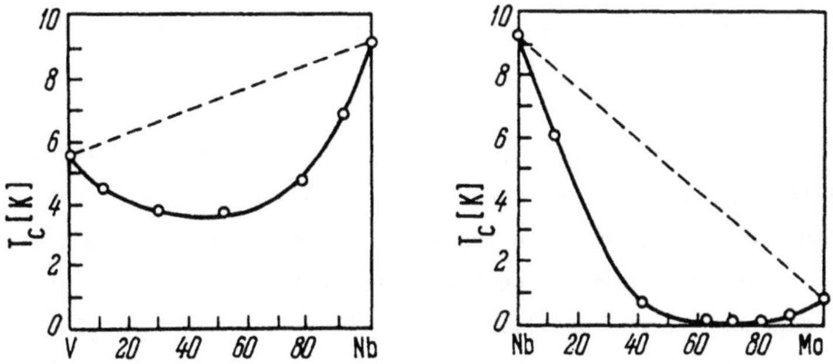

Fig.5.6

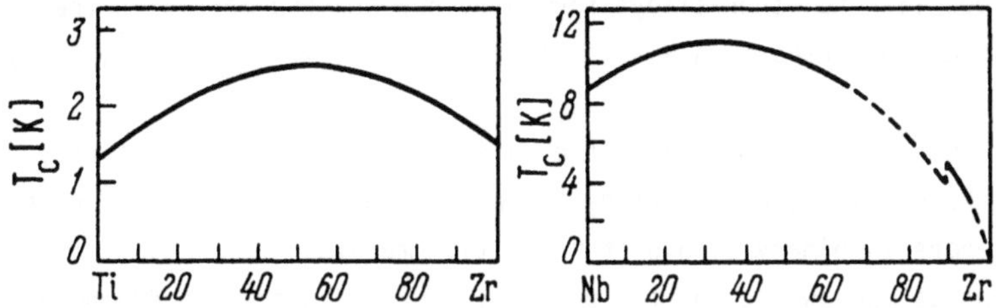

Fig.5.7

Fig.5.6. The superconducting transition temperature T_c of the alloys V - Nb and Nb - Mo

Fig.5.7. The superconducting transition temperature T_c of the alloys Ti - Zr and Nb - Zr

5.2.2 Binary Alloys

Let us now see which groups of binary alloys correspond to the above concentration dependences of T_c. To begin with, consider alloys whose components are transition metals from different groups.

The features of types 1) and 2) are characteristic of solutions in transition metals of the fifth group (V, Nb, Ta) with high T_c, and of elements with a greater number of valence electrons per atom, i.e., those standing to the right in the periodic system - in the sixth, seventh and eighth groups. For example, the dependence of the first type occurs in the alloys Nb - W, Nb - Ta (Fig.5.5), V - Mo, Ta - W and others, whereas that of the second type occurs in Nb - Mo alloys (Fig.5.6). The dependence of the third type occurs, as a rule, in the alloying of metals with low values of T_c, [arranged in the fourth (Ti, Zr, Hf) and sixth (Mo, W) groups], with metals found on the right-hand side in the periodic table of elements, or, conversely, when alloying metals of the fifth (V, Nb, Ta) and seventh (Tc, Re) groups with metals whose number of valence electrons is less. Such a situation takes place in Zr - Nb (Fig.5.7), Ti - Nb, Hf - Nb and other systems.

Thus, the mutual arrangement of the elements alloyed in the periodic table affects the concentration dependence of T_c in solid solutions to a certain extent.

Some peculiarities in the behaviour of T_c during the formation of alloys may be understood from phase diagrams. For example, the disturbance of the smooth concentration dependence of T_c in the Zr - Nb system (Fig.5.7) in the region of alloys rich in Zr is due to the breakdown of the bcc solid solution when cooled. Such dependences are peculiar to all alloys of this type (Ti - Nb, Ti - Mo, Ti - V, etc.) where one of the components possesses polymorphism [5.89, 90], and also peculiar to other alloys formed by hcp and bcc metals in the region of electron concentrations $N_e = 6 \div 7$[e/a] (Mo - Tc, W - Re, Mo - Re, etc.) [5.87,91].

There are a number of exceptions to the rules formulated above, but they are basically these rules that determine the above-mentioned types of concentration dependences in binary alloys of transition metals.

The situation is much worse for the prediction of concentration dependence of T_c when transition metals of one and the same group (Ti - Zr, V - Nb, Nb - Ta, V - Ta, Zr - Hf, Mo - W, etc.) are alloyed. All the three types of the specified dependences may be found here. For example, alloys of the Nb - Ta system have a concentration dependence of T_c which is of the first type (Fig.5.5), the dependence for V - Nb alloys is of the second type (Fig.5.6), and that in Ti - Zr alloys is of the third type (Fig.5.7).

It is worth noting that the phase diagrams are alike for the first two cases. Thus, the type of the concentration dependence of T_c in solid solutions produced by metals of one and the same group may not be predicted by proceeding only from the phase diagram of an alloy and from the mutual arrangement of the components in the periodic system.

From the concentration dependences of T_c in solid solutions of transition metals it can be inferred that in some systems alloying leads to an improvement in the superconducting characteristics of pure metals and the dependence of the third type (Fig.5.7) is valid. Such systems are of greatest practical interest and it is not accidental that there are alloys of this group (Nb + 60% Ti and Nb + 33% Zr with $T_c \sim 10$ K) which have been extensively used in the production of superconducting technical devices.

Detailed experimental data on ternary alloys of transition metals are contained in [5.1], and the papers [5.10,36-45] also discuss the influence which the structural state of ternary solid solutions exerts on T_c. Overall, it may be noted that the initial interest in studying ternary solid solutions of transition metals has declined somewhat in recent years. This is apparently due to the fact that alloying of binary systems has not produced much improvement of the superconducting properties of these alloys.

5.2.3 The Influence of Pressure on the T_c of Alloys

The investigation of the superconductivity in alloys of transition metals under pressure was a continuation of similar work done on pure metals which we have already mentioned in Chapt.4. By investigating alloys one can trace a smooth transition in the properties from one metal to another and thereby verify the hypotheses which account for the dependences of $\partial T_c/\partial P$ for pure elements.

First of all, a test was made of Brandt's idea (Sect.4.1) to the effect that the variation of the density of states on the Fermi surface plays a decisive role in the variation of T_c at uniform compression for transition metals. In doing so it was assumed that the pressure dependence of the density of states is determined by the dependence of the density of states on the number of valence electrons per atom N_e. Within the framework of the above assumptions, the signs of the derivatives $\partial T_c/\partial P$, $\partial N(0)/\partial P$, and $\partial N(0)/\partial N_e$ coincide and, consequently, in alloys with $\partial N(0)/\partial N_e > 0$ the T_c should increase when subjected to compression and decrease when $\partial N(0)/\partial N_e < 0$.

First experiments [5.46,47] showed that this is really the case. In full agreement with the dependence of the electronic heat capacity $\gamma(N_e)$ in alloys

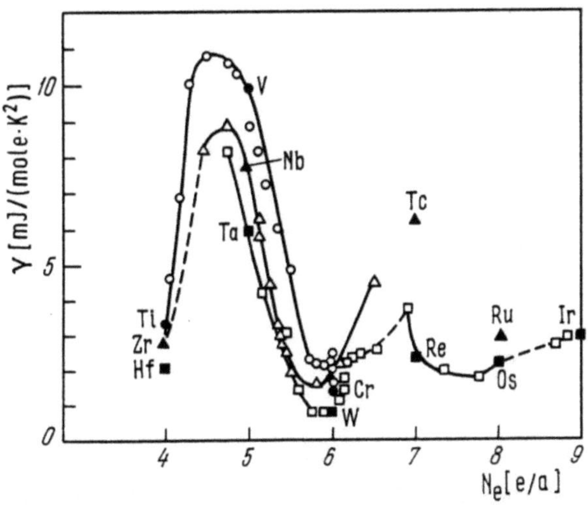

Fig.5.8. The dependence of γ on N_e in solid solutions formed by neighbouring transition 3d(○), 4d(△) and 5d (□) metals. The dark symbols correspond to pure metals

of neighbouring elements (Fig.5.8), T_c increased in alloys $Nb_{0.75}Zr_{0.25}$ and $Mo_{0.90}Re_{0.10}$, and decreased in $Nb_{0.75}Mo_{0.25}$ alloys when high pressure was applied. Similar results were also obtained [5.48,49] for alloys of the Ta - W and Re - Os systems.

However, more complete investigations into the superconductivity in transition-metal alloys under pressure [5.50,51] showed that the simple correlation between $\partial T_c/\partial P$ and $\partial N(0)/\partial N_e$ does not occur in all cases. For example, instead of the expected increase in T_c under pressure for $Ta_{0.90}Hf_{0.10}$, T_c was found to decrease [5.48], etc. Apart from this, it was found that the type and even the sign of the dependence of $\partial T_c/\partial P$ for one and the same alloy may vary at different pressures [5.49,50]. For example, $\partial T_c/\partial P = 0$ for $Nb_{0.81}Mo_{0.19}$ over the pressure range of 0 to 20 kbar, and $\partial T_c/\partial P < 0$ when subsequently compressed, to 45 kbar [5.50].

The concentration dependences of $\partial T_c/\partial P$ for a number of transition-metal alloys [5.48,51,52] are given in Fig.5.9. Alloys of 5d-metals exhibit the same form as analogous alloys of 4d-metals.

In a number of alloys T_c was found to vary after exposure to high pressure [5.53,54]. This residual effect was observed in alloys formed by hcp and bcc metals (such as Ti - Nb and Zr - Nb) in which ω phase forms at atmospheric pressure and under certain heat-treatment conditions. It was found that pressure promotes the formation of the ω phase and, compared with heat treatment

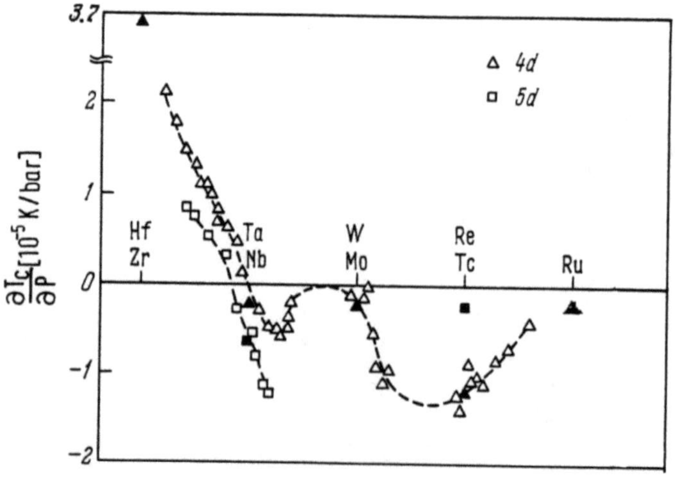

Fig.5.9. The concentration dependence of $\partial T_c/\partial P$ in alloys of 4d and 5d-metals. The intermediate values of $\partial T_c/\partial P$ between pure metals correspond to the variation of the concentration of the components from 0 to 100%. The legend is the same as in Fig.5.8

at normal pressure, widens the interval of compositions where this phase occurs. The formation of the ω phase leads to a lowering of T_c in these alloys.

5.3 The Empirical Correlations Between T_c and the Electron and Lattice Properties of Solid Solutions

It is also important to establish which criteria govern the behaviour of T_c in solid solutions of transition metals. Several points of view may be found in the literature on this matter. Let us consider these in brief.

5.3.1 Matthias' Rule

When analyzing the properties of pure transition metals we have already noted in Chapter 4 that MATTHIAS [5.55] found the dependence of T_c on electron concentration, i.e., on the number of valence electrons per atom N_e, of an oscillatory nature. Two values $N_e = 5$ and $N_e = 7$ e/a turned out to be "favourable" to high values of T_c in transition metals. Investigation into the superconducting properties of solid solutions between neighbouring transition elements permitted a more exact determination of the position of maxima and minima on the dependence curve $T_c(N_e)$ constructed for pure elements (Fig.4.2) since in this case the values of N_e could be varied continuously and smoothly. It turned out that the position of the maxima corresponds not to 5 and 7 elec-

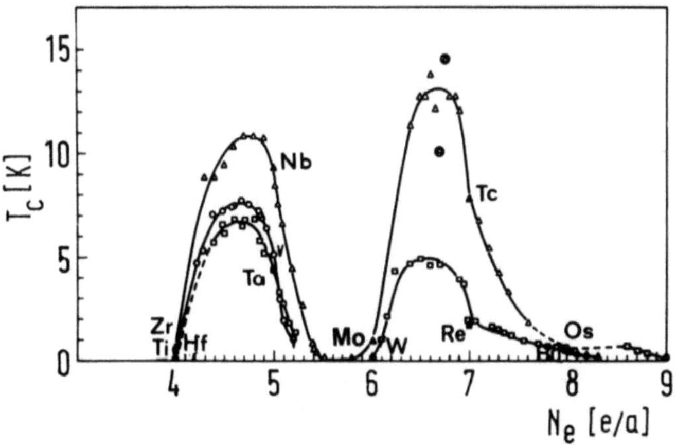

Fig.5.10. The dependence of T_c on the number of valence electrons, N_e, in solid solutions formed by neighbouring transition metals. The legend is the same as in Fig.5.8

trons per atom but to approximately 4.7 and 6.5 e/a (Fig.5.10). It should be noted that the T_c vs. N_e curves in solid solutions formed by transition metals of all three seris turned out to be surprisingly alike.

In MATTHIAS' opinion [5.55], the constructed curves $T_c(N_e)$ enable one to understand the behaviour of T_c in solid solutions of transition metals. In fact, all three types of concentration dependences of T_c shown in Figs.5.5-7 may be derived from the criterion of electron concentration and from the existence of "favourable" and "unfavourable" regions N_e for T_c. For example, the dependence of the second type in an Nb - Mo alloy (Fig.5.6) may be accounted for by the fact that the electron concentration in this alloy shifts in a direction unfavourable for high T_c when molybdenum is substituted for by niobium. Then this concentration passes through a minimum and reaches the beginning of a second region of favourable values of N_e. Here one observes a complete analogy to the dependence of T_c on the Mo content (Fig.5.6) with a section of the T_c vs. N_e dependence (Fig.5.10).

A similar conclusion may be drawn concerning the alloys of the Zr - Nb system which show a concentration dependence of the third type (Figs.5.7,10), etc.

The dependence of T_c on electron concentration in solid solutions formed by metals of different series and groups (Fig.5.11) (including those remotely spaced in the series: Zr - W, Cr - Ir, Mo - Ru, Nb - Os, Nb - Ru, Zr -Rh, and others) turned out to be similar to the electron concentration dependence of T_c constructed for solid solutions between neighbouring elements with maxima and minima at the same values of N_e. For example, addition of 8 at.%

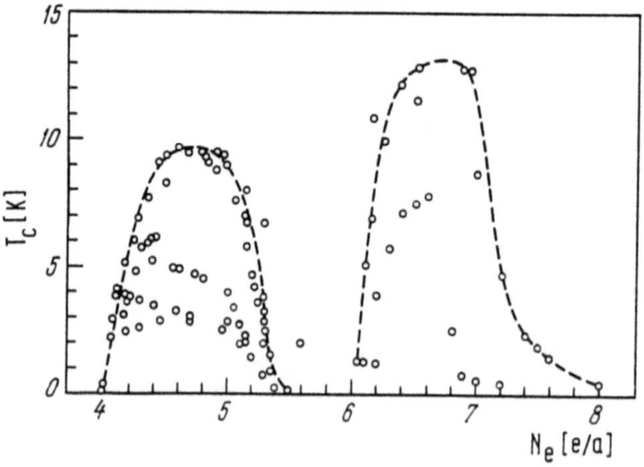

Fig.5.11. The dependence of T_c on the number of valence electrons, N_e, in solid solutions formed by transition metals belonging to different series and groups

Rh to Zr brings about as high an increase in T_c (up to 9 K) as that caused by the addition of 40 at.% Nb. Note that in both cases the average valence electron concentration is equal to 4.4 e/a. When added to niobium, Mo and W or Ru and Os arranged in the same group of the periodic system and having identical electron configurations of atoms influence the lowering of T_c of niobium to just the same degree.

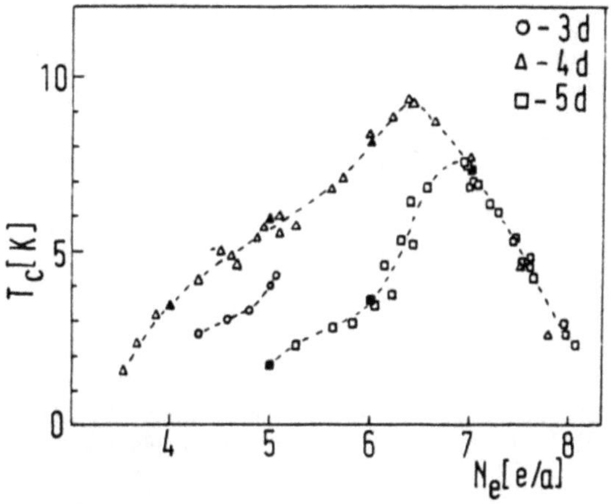

Fig.5.12. The dependence of T_c on the number of valence electrons, N_e, in amorphous alloys of 4d and 5d-metals. The legend is the same as in Fig.5.8

For amorphous alloys of transition metals the N_e dependence of T_c (Fig.5.12 is different from that for normal crystal alloys (Fig.5.10). Instead of two values of electron concentrations favourable for high T_c, a value of 6.4 e/a for 4d-metals and 7.0 e/a for alloys of 5d-metals remains [5.56,92].

5.3.2 The Rigid-Band Approximation

Let us now clarify the physical sense of the above N_e dependence of T_c. To calculate the number of valence electrons available for an atom in an alloy, MATTHIAS made use of the method first proposed in [5.57] where the value of N_e in alloy AB is taken as an average over the components: $N_{e(AB)} = C_A N^0_{e(A)} + C_B N^0_{e(B)}$ ($N^0_{e(A)}$ and $N^0_{e(B)}$ here are the numbers of valence electrons in the free atoms). This presupposes a complete collectivization of the number of valence electrons between the component atoms and completely rules out the possibility of preferential redistribution of electrons from one component to another or, on the contrary, of conservation of electronic configurations of free atoms in a solid solution.

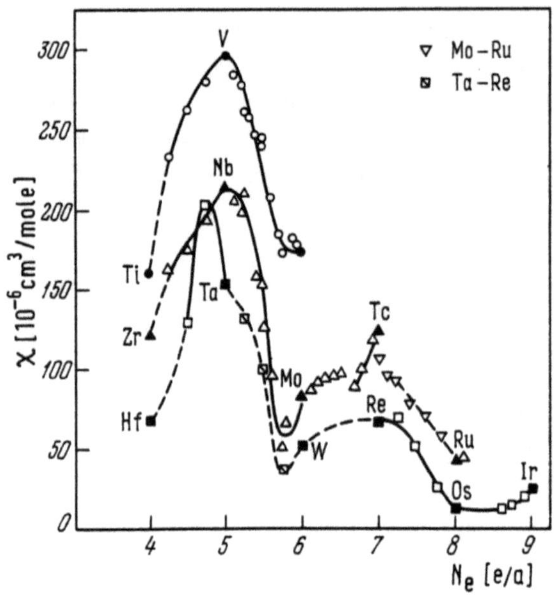

Fig.5.13. The dependence of χ on the number of valence electrons, N_e, for solid solutions of transition metals formed by neighbouring 3d, 4d and 5d-metals. The legend is the same as in Fig.5.8. For lack of data for the systems Tc - Ru and Ta - W we cite data for the alloys Mo - Ru and Ta - Re which have the same N_e

Independently of the representation of T_c as a function of N_e for solid solutions of transition metals, electron concentration dependences of other physical quantities were constructed. For example, when analyzing the variation of magnetic susceptibility in alloys the dependences $\chi(N_e)$ were determined [5.58-63] given in Fig.5.13. Similar dependences were also found for the electronic heat-capacity coefficient γ [5.64] (Fig.5.8). It may be noted that these dependences are very much alike and have maxima and minima at close values of N_e. Since both physical parameters χ and γ characterize the density of electron states on the Fermi level N(0), the similarity of their N_e dependence suggests the existence of that N_e dependence of N(0) which was constructed from $\gamma(N_e)$ in many papers (see, for example, [5.65]) according to the method described in [5.66]. The dependence of N(0) on N_e essentially denotes the presence of some rigid band, universal for this particular group of alloys, whose form remains invariable, and the position of the Fermi level is determined by the degree of filling or by electron concentration. Thus, the alloying reduces to varying the number N_e of valence electrons and these are assumed to be fully collectivized.

The similarity between the T_c, χ and γ vs. N_e curves (Figs.5.8,10,13) permits one to understand the role which the criterion of electron concentration plays in the formation of T_c in solid solutions of transition metals. The universality of the N_e dependence of T_c in an alloy, which is taken as an average of the components, actually denotes the applicability of the rigid-band model. The electron concentration here appears as some universal criterion determining the entire complex of electric and magnetic properties of solid solutions.

Note that the rigid-band approximation and the above method of determining N_e, strictly speaking, are possible only for solid solutions formed by neighbouring transition elements. From calculations of the electron energy spectrum of metals which are neighbours in the periodic table, for example, Nb and Mo (Figs.4.9,11), it could be seen that their electron structure differs only in the occupancy of energy bands. In this case the calculation of the number of valence electrons per atom of the alloy on the assumption that they are completely collectivized and the use of the rigid-band approximation give rise to no objections. What may be objected to is the application of this concept to alloys of any metals, including those with highly different valences. Here the rigid-band approximation is inapplicable in the general case. The different degrees of localization of electron d-states of the components permit no description of the electron states of an alloy on the basis of a full collectivization of valence electrons. A difference in crystal structures of basic metals, as we have seen in Chapt.4, leads to a substantial difference in their electron energy spectrum and prohibits the rigid-band approximation. The difference in electronegativity in the final analysis leads to charge transfer or to an electron redistribution which may differ from the complete collectivization (normally assumed in the calculation of N_e) of electrons of the components.

In spite of some progress made in accounting for the behaviour of T_c in a number of alloys formed by metals of different series and groups, it should be noted that, taken as a whole, the variation of T_c with N_e has no clear-cut functional dependence when metals with greatly different valences are alloyed. From these results which are presented in Fig.5.11, it follows that the experimental values of T_c fill the regions of "favourable" electron concentrations almost uniformly and are practically absent in the region of values of N_e over the interval from 5.4 to 5.9 e/a. Note also that what remains beyond the scope of MATTHIAS' empirical correlation are alloys formed by metals from one and the same group of the periodic system (Ti - Zr, V - Nb, Nb - Ta, Mo - W, Tc - Re, Y - La, etc.) in which N_e remains invariable over the whole concentra-

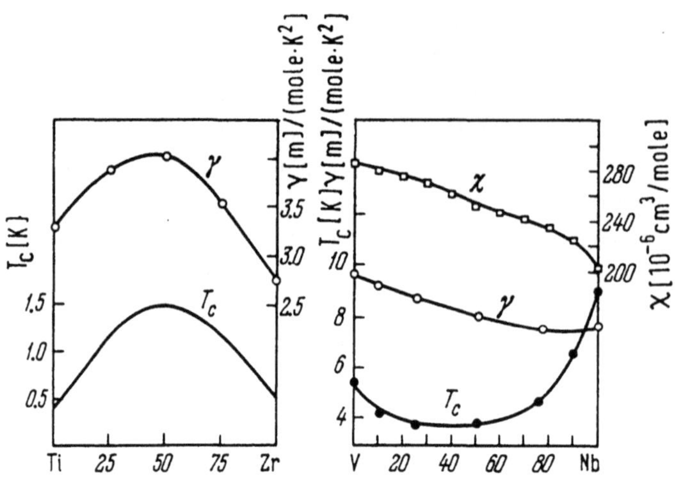

Fig.5.14. T_c, γ and χ in solid solutions of Ti - Zr and V - Nb

tion range. If, however, it is assumed due to the similarity between the N_e dependences of T_c, χ and γ that the density of states on the Fermi level plays a decisive role in the formation of T_c in solid solutions and if one compares the concentration dependences of T_c, χ and γ in these alloys it may be seen that in the general case the correlation between T_c and γ holds good only in some systems, for example, in Ti - Zr (Fig.5.14) [5.67]. In other alloys, V - Nb (Fig.5.14), V - Ta, La - Y, etc. [5.19], the concentration dependences of γ and T_c differ substantially.

5.3.3 The Ralation of T_c to the Lattice Instability

Apart from the above empirical correlation of T_c with the electronic properties of alloys of transition metals, a similar correlation of T_c with lattice properties also exists. COLLINGS et al. [5.30,68] paid attention to the fact that the peaks on the electron-concentration dependence of T_c for alloys of transition metals at $N_e = 4.6 \div 4.7$ [e/a] and $N_e = 6.5$ e/a (Fig.5.15) correspond to the boundaries of instability of the bcc structure. Between these peaks there is a region of maximum instability of the bcc structure to which minimum values of T_c in alloys correspond. When $N_e < 4.6$ e/a and $N_e > 6.5$ e/a the structure of the solid solution varies from bcc to hcp, and T_c decreases again and now reaches a minimum in the region of maximum instability of the hcp structure.

As for the phase diagram for alloys of hcp-transition metals of the fourth group (Ti, Zr) with bcc metals belonging to the fifth and sixth groups (V, Nb, Ta, Mo) it may be noted that alloys rich in titanium and zirconium (at $N_e \leq$ 4.6 e/a) show some structural instability. This instability manifests itself

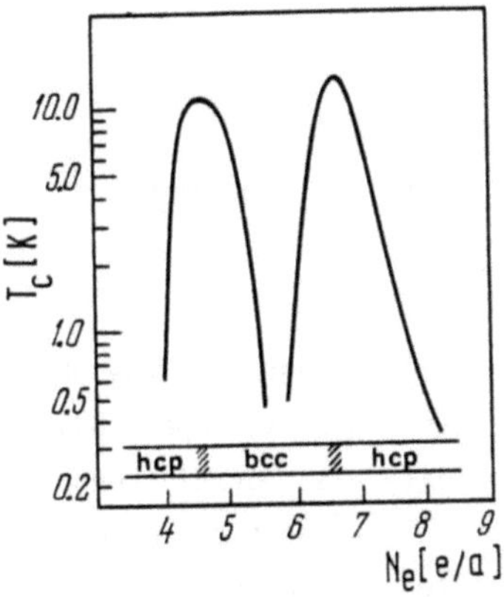

Fig.5.15. The stability regions of crystal structures and of T_c in solid solutions of transition metals

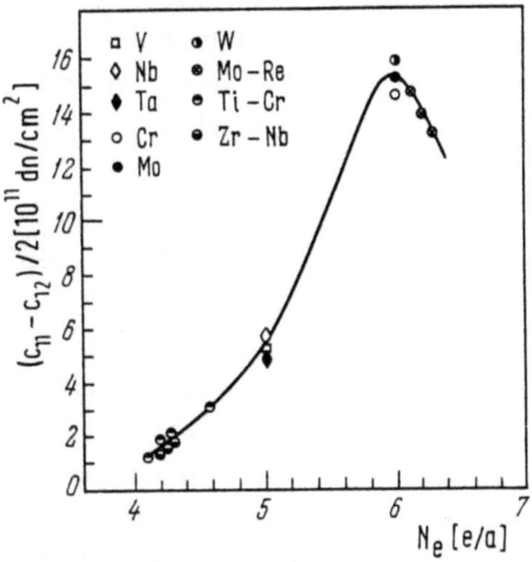

Fig.5.16. The elastic shear modulus for different transition metals and their alloys

in martensitic bcc-hcp transformation and in the breakdown of the bcc-solid solution with the precipitation of the ω phase when the alloy is cooled from high temperatures. A measure of relative instability of the bcc structure is the elastic shear modulus $C' = 1/2\ (C_{11} - C_{12})$ characterizing the ease with which the "cubic-hexagonal" displacement may come about. In Fig.5.16 [5.68] the values of C' for bcc transition metals of the fifth and sixth groups and of some of their binary alloys are given [5.69-74] as a function of electron concentration. It is seen that the highest stability of the bcc structure occurs for metals with $N_e = 6$ e/a (Cr, Mo, W). They have the lowest T_c values among all the transition metals. Transition metals of the fifth group (V, Nb, Ta) have a limited stability of the bcc structure. Alloying these with metals of the fourth group reduces the stability of crystal structures, and this, according to [5.30,68], is the main reason for the increase of T_c in solid solutions of Ti and Zr with transition metals of the fifth and sixth groups. Straightforward measurements of the tunnel spectrum in the alloy Nb + 25% Zr and the comparison with the spectrum of pure niobium have revealed a softening of the phonon spectrum of the alloy in the range from 2 to 10 meV [5.92]. The estimates made in [5.92] show that the above softening of the phonon spectrum leads to an incrase of λ from 0.92 to 1.3 and to a growth of T_c from 9.0 to 10.9 K. Note that here we reveal the correlation between T_c and lattice instability for the first time.

When considering the superconductivity of transition-metal compounds in Chapters 6 and 7 we shall see that this correlation occurs very frequently, there being a great variety of forms in which the lattice instability favouring high T_c may manifest itself.

5.4 Miedema's Empirical Method of Describing Concentration Dependences of the Properties of Metals

An empirical method for calculating the concentration dependence of the physical characteristics of metals such as electronic heat capacity, superconducting transition temperature and magnetic susceptibility has recently been proposed [5.75-77]. To check this method, about 150 alloys composed of diverse transition metals were examined. The agreement between the calculated characteristics and those measured experimentally was found to be surprisingly good. At present this method is rather a recipe than a physically substantiated technique. Nevertheless, its universality which has been established on the basis of experimental verification is a conspicuous feature and deserves serious attention. One is also compelled to look for physical reasons to substantiate it (in this connection the reader is referred to Sect.5.5 where and attempt is made to give such a substantiation).

The method under consideration, proposed by MIEDEMA [5.75-77], in contrast to the rigid-band model used earlier, proceeds from the idea of a quasichemical approach to the alloy which primarily takes into account the atomic characteristics of the system. Below we will dwell upon the working scheme of this method in the same manner as it was proposed in [5.75 - 77].

Assume that the densities of electron states of AB alloys on the Fermi level are constituted by additive contributions of the densities of states of components A and B

$$N(0)_{AB} = C_A N(0)_A + C_B N(0)_B \quad . \tag{5.4.1}$$

A relevant expression may also be written for the electronic heat capacity

$$\gamma_{AB} = C_A \gamma_A + C_B \gamma_B \quad . \tag{5.4.2}$$

The contributions made by the densities of states of the components, $N(0)_A$ and $N(0)_B$, and by the heat capacities, γ_A and γ_B, are not equal to the corresponding values in pure metals but have some other values which are determined in the following fashion. It is assumed that there is a certain dependence of N(0) or γ on the number of valence electrons per atom Z which is universal for a given transition-metal series (3d, 4d and 5d series). The contribution $N_i(0)$ or γ_i from the ith component of the alloy is determined from these dependences for a given number of valence electrons available in the ith component of the alloy. It is believed that Z_i differs from the number of valence electrons in the free atom Z^o by some quantity ΔZ_i which corresponds to the number of electrons transferred between the component atoms during alloying

$$Z_i = Z^o + \Delta Z_i \quad . \tag{5.4.3}$$

Thus, the whole problem reduces to calculating ΔZ_i and constructing N(0) and γ curves as functions of Z. The quantity ΔZ_i depends on the difference between eletronegativities $\Delta\Phi^*$ (it was suggested by MIEDEMA that the experimental values of the work function of pure metals, which are cited in Table 5.1, should be used as such electronegativities) and C_i, the concentration of the ith component of the alloy [5.75]

$$\Delta Z_i = P \cdot \Delta\Phi^*(1 - C_i) \tag{5.4.4}$$

where P is a constant.

Making use of Table 5.1 one can estimate the difference between the electronegativities of any two components of an alloy and determine, according to (5.4.4) the number of electrons transferred, ΔZ_i. This enables the calculation of the number of valence electrons available for the atom of components A and B when they enter an AB alloy.

The dependence $\gamma(Z)$ is constructed as follows. First plot the experimental values corresponding to pure metals in which Z is taken to be the same as in the free atom and then plot the values of γ measured in solid solutions between neighbouring transition metals. However, the derived dependences for $\gamma(Z)$

Table 5.1 Work function in transition metals

Metal	Φ^* [eV]	Metal	Φ^* [eV]	Metal	Φ^* [eV]	Metal	Φ^* [eV]
Sc	3.0	Zr	3.15	Rh	5.4	W	4.8
Ti	3.4	Nb	4.0	Pd	5.75	Re	5.6
V	4.25	Mo	4.65	Lu	2.8	Os	5.55
Cr	4.65	Tc	5.3	Hf	3.3	Ir	5.55
Y	2.95	Ru	5.55	Ta	4.05	Pt	5.55

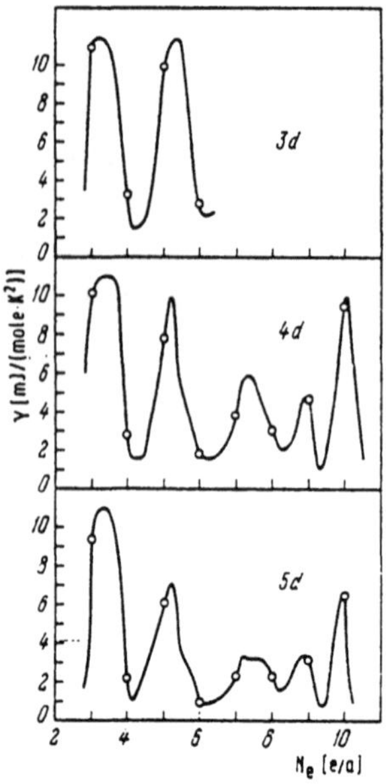

Fig.5.17. The electronic heat capacity γ as a function of Z [e/a] for 3d, 4d and 5d metals

(Fig.5.17) differ from the dependences $\gamma(N_e)$ (Fig.5.8) which we obtained earlier in that there is some distortion along the abscissae. When assessing N_e we simply averaged the number of valence electrons in the components on the assumption of their complete collectivization, while in the calculation of Z allowance was made for the possibility of certain preferential redistribution as a function of the relationship of electronegativities of the component atoms. Therefore the $\gamma(Z)$ and $\gamma(N_e)$ dependences, though similar in form, differ in that the maxima and minima do not correspond to the same values of electron concentration. The two major maxima in $\gamma(N_e)$ (Fig.5.8) which appear at $N_e = 4.6$ and 6.5 e/a, have corresponding minima in the $\gamma(Z)$ curve (Fig.5.17) for the same electron concentration.

Using the above method MIEDEMA [5.75] calculated the values of γ in 116 solid solutions of transition metals and only in a few cases did he observe a discrepancy from experimental data. A comparison between the calculated and experimentally measured values of γ in the solid solutions under analysis is shown in Fig.5.18. It may be seen that the agreement is extremely good.

A similar approach was also proposed for calculation of T_c [5.76,77]. Proceeding from (5.4.1) and from the BCS theory formula (1.2.5) for T_c the following equation was obtained for the electron-phonon interaction parameter λ in a binary AB alloy:

$$\lambda_{AB} = C_A\lambda_A + C_B\lambda_B \qquad (5.4.5)$$

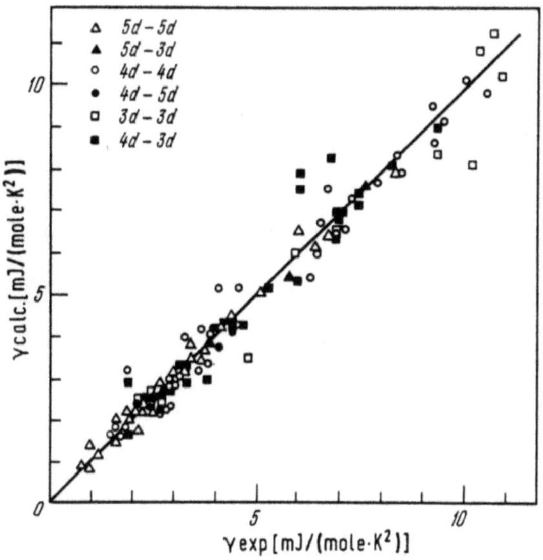

Fig.5.18 Fig.5.19

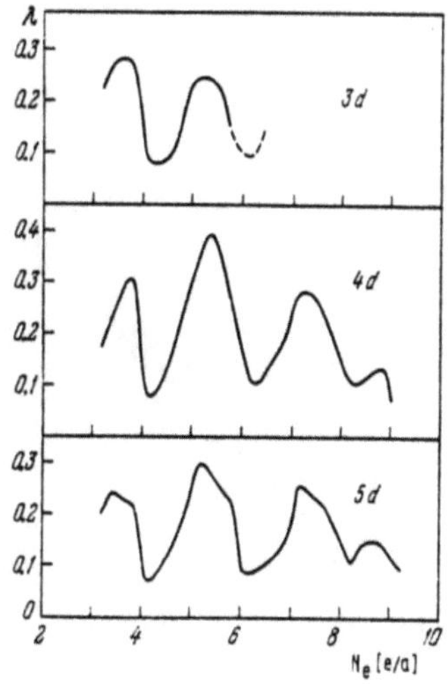

Fig.5.18. Comparison between the experimental and calculated values of γ for binary solid solutions of transition metals

Fig.5.19. The parameter λ as a function of z [e/a] for 3d, 4d and 5d metals

or

$$[\ln(T_c/\theta_D)]^{-1}_{AB} = C_A[\ln(T_c/\theta_D)]^{-1}_{A} + C_B[\ln(T_c/\theta_D)]^{-1}_{B} \quad .$$

Just as in the heat-capacity case, dependences $\lambda(z)$ for the transition metals of the 3d, 4d and 5d series were constructed. These dependences are given in Fig.5.19 [5.77]. On determining the value of Z_i for the ith component of the alloy and the corresponding contributions of λ_A and λ_B with the help of the above method one can find the values of λ and T_c/θ_D in any alloy. Figure 5.20 illustrates a comparison of the values of T_c/θ_D calculated in this manner and of the experimentally measured values of T_c/θ_D for 143 transition metal alloys. It may be noted that a difference between $(T_c/\theta_D)_{calc}$ and $(T_c/\theta_D)_{exp}$ only occurs in some cases (A - Hf + 16%W, B - Zr + 9%W, C - Mo + 60,70,80%Ti, D - V + 70,80, 90%Ru).

Figure 5.21 presents calculated results for γ and T_c in the isoelectron systems V - Nb, Nb - Ta, V - Ta, Ti - Zr. One notices a good general agreement with experiments. This gives conclusive evidence that MIEDEMA's method is also applicable to the decription of the properties of such "difficult" systems in which the behaviour of T_c and γ may be inconsistent and essential deviations from the linear concentration dependence take place.

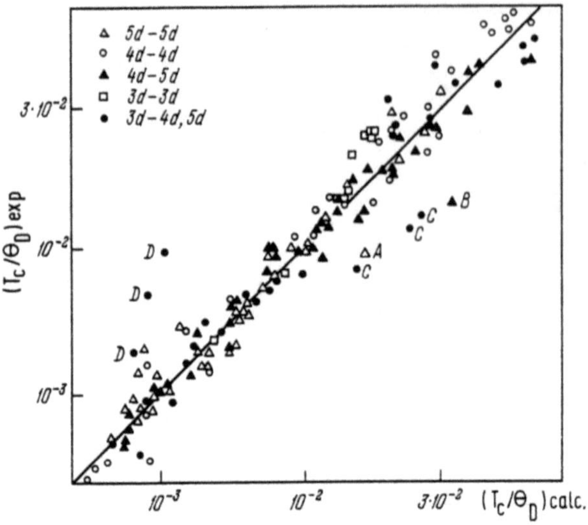

Fig.5.20

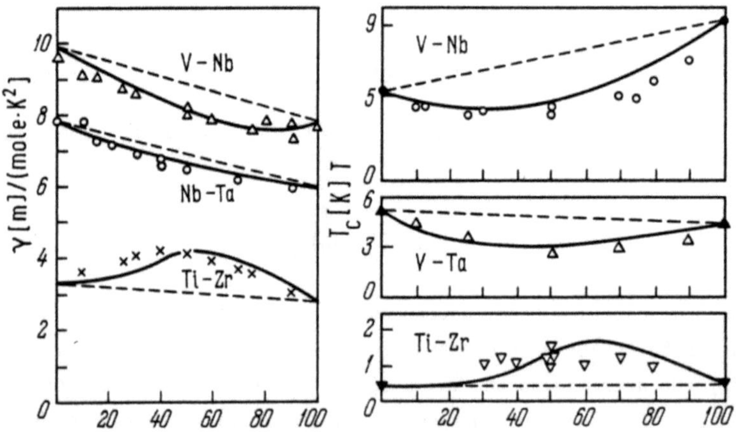

Fig.5.21

Fig.5.20. Comparison between the calculated and experimental values of T_c/θ_D in solid solutions of transition metals. Points A, B, C, D are explained in the text

Fig.5.21. Comparison between the calculated and experimental values of T_c and γ in the alloys: Ti - Zr, V - Ta, Nb - Ta and V - Nb (the dotted line is a linear extrapolation)

Note that MATTHIAS and MIEDEMA's approaches in describing the behaviour of T_c in solid solutions of transition metals have certain features in common which assume the rigidity or universality of the dependences $T_c(N_e)$ or $\lambda(z)$ and $\gamma(z)$. MATTHIAS' approach has a limited application and, as has been noted above, may be employed to analyze T_c in alloys formed by neighbouring transition elements only. For alloys formed by metals remotely spaced in the period table or belonging to different series, this approach is not rigorous in the general case. Besides, alloys formed by metals pertaining to the same group, in which N_e does not vary with concentration, are practically outside the sphere of applicability of the rigid-band model. In this respect MIEDEMA's approach may be considered to be more universal since it covers practically all the solid solutions of transition metals. This approach reflects a realistic situation since it allows for not only the collectivization but also some preferential redistribution of electrons between the component atoms in the formation of a solid solution. Bear in mind that MIEDEMA's approximation for calculation of γ and λ in solid solutions primarily requires a knowledge of the atomic properties of pure metals. This apparently is not accidental. In recent years a new approach (which can also conventionally be called a quasi-chemical one) to describing the various properties, including superconductivity, in alloys and compounds of transition metals has been developed successfully (see particularly the papers by PHILLIPS [5.78]). It is based on the assumption that the properties of such systems are primarily determined by short-range order or atomic interactions.

5.5 Theoretical Analysis of the Concentration Dependence of T_c in Alloys

5.5.1 The Alloy Model

The experimental evidence presented in Sect.5.2 shows that the superconducting transition temperature may vary substantially with the composition of binary alloys formed by transition elements. In these alloys there are no localized moments on the atoms and the components of the alloys do not always differ appreciably in ion mass and electronic properties. Therefore, it is of great interest to ascertain some specific features of the effect of composition on T_c in these alloys.

If we were concerned with dilute alloys then it should be borne in mind that nonmagnetic impurities in a metal do not affect the superconducting transition temperature. Comprehensive formulation of this assertion is contained in the well-known theorem by ANDERSON [5.79] according to which the variation of T_c is possible only at the expense of varying the density of states on the Fermi level of the metal by impurities. In concentrated alloys the variation of the density of states with composition may be considerable. For normal alloys, the only treatment permitting explicit description of the concentration dependence of N(0) rests on the CPA (Coherent Potential Approximation) method

[5.82,83]. The development of this method has made it possible to analyze the dependence of the electronic structure of binary substitution alloys on composition and to understand the concentration dependence of a number of electronic and kinetic characteristics of solid solutions.

The CPA method was generalized to superconducting alloys. WEINKAUF and ZITTARTZ [5.80,81] and others [5.84,85] allowed for the local character of the electron levels in the component atoms of the alloys, and account was taken of the local character of the order parameter in the alloy. It is this feature which puts ANDERSON's theorem outside the range of applicability and yields nontrivial concentration dependences of T_c in alloys.

In itself, the CPA method is associated with the site representation and is thus particularly good for alloys which may be described by a tight-binding approximation. For superconductors, one should primarily bear in mind the alloys of narrow d-band transition metals. Although the theory being developed will henceforth be applied exactly to such alloys, a number of simplifying assumptions (in the first place, the neglect of the orbital degeneracy of electron states) will immediately be made for the model under consideration. Thus, the electronic state of the alloy will be characterized only by the atomic number and electron spin orientation.

Thus, in the site representation of Wannier the Hamiltonian of the alloy in the model involved is written as

$$H = \sum_{i\alpha} V_i a^+_{i\alpha} a_{i\alpha} + \sum_{ij\alpha} t_{ij} a^+_{i\alpha} a_{j\alpha} - \sum_i \lambda_i a^+_{i\uparrow} a_{i\uparrow} a^+_{i\downarrow} a_{i\downarrow} \qquad (5.5.1)$$

where the atomic level V_i of an electron in site i takes on two random values V_A and V_B in a two-component alloy. The quantity t_{ij} stands for the matrix element of transition between sites i and j and can, for simplicity, be regarded as dependent not on the kind of atoms but only on the distance between the sites. The last term takes into account the interaction of electrons on one site, which is of the attraction type ($\lambda_i > 0$). In this manner, in the spirit of narrow-band models, we assume the effective electron attraction to be local. This circumstance, although not very realstic for alloys even of transition metals with high localization of d-states, still enables one to qualitatively obtain undoubtedly true results.

In the BCS-theory approximation it is possible, instead of the effective interaction Hamiltonian, to use the operator taking into account the creation of Cooper pairs

$$\sum_i \lambda_i a^+_{i\uparrow} a_{i\uparrow} a^+_{i\downarrow} a_{i\downarrow} \to \sum_i (\Delta_i a^+_{i\uparrow} a^+_{i\downarrow} + \Delta^*_i a_{i\downarrow} a_{i\uparrow}) \quad , \qquad (5.5.2)$$

where

$$\Delta_i = \lambda_i \overline{\langle a_{i\downarrow} a_{i\uparrow} \rangle} \quad . \tag{5.5.3}$$

Going over from a four-fermion to a two-fermion Hamiltonian involves the replacement of a pair of operators by an average with respect to the ground state of the superconductor. In the case of a disordered system an additional average of the quantities ($\langle a_{i\downarrow} a_{i\uparrow} \rangle$) with respect to impurity configurations must be taken [this is indicated by a bar in (5.5.3)]. However, in order to take into account the local variations of the order parameter Δ_i it is necessary to perform averaging over the configurations of all the sites in (5.5.3), except for the site which is singled out. This procedure is marked by the index of the site singled out. Thereafter, the order parameter Δ_i depends on one site only, and all the effective Hamiltonian parameters become local and thus permit straightforward application of the CPA scheme to it. In the present case the only difference from the CPA scheme for a normal metal is in the occurrence of anomalous terms such as a^+a^+ and aa which can be easily taken into account in Nambu's representation.

In the CPA scheme both averagings carried out for the operators $a_{i\downarrow} a_{i\uparrow}$ (viz., quantum-mechanical and configuration averaging) are equivalent, and the transition (5.5.2) in the effective Hamiltonian denotes the neglect of both quantum-mechanical and alloy composition fluctuations.

5.5.2 The CPA Method

We now introduce the superconductor Green's function using Nambu's representation:

$$G_{ij}(\tau - \tau') = -\langle T\Psi_i(\tau)\overline{\Psi}_j(\tau')\rangle \quad ,$$

$$\Psi_i = \begin{pmatrix} a_{i\uparrow} \\ a^+_{i\downarrow} \end{pmatrix} \quad , \quad \Psi^+_i = (a^+_{i\uparrow}, a_{i\downarrow}) \quad . \tag{5.5.4}$$

We express the Fourier transform of the function with respect to the variable $\tau - \tau'$ in terms of $G_{ij}(z)$ where the function $G_{ij}(z)$ is determined in the points $z_n = i(2n+1)\pi T$ adopted in the method of ABRIKOSOV et al.

The Green's function satisfies the matrix equation

$$\begin{pmatrix} z - H_0 - V & \Delta \\ \Delta^* & z + H_0 + V \end{pmatrix} \begin{pmatrix} G^{11} & G^{12} \\ G^{21} & G^{22} \end{pmatrix} = 1 \quad , \tag{5.5.5}$$

where each of the quantities is a matrix with respect to the site indices

$$[H_o]_{ij} = t_{ij} \quad , \quad [V]_{ij} = V_i\delta_{ij} \quad , \quad [\Delta]_{ij} = \Delta_i\delta_{ij} \quad .$$

It can be readily seen that the following equations of symmetry hold

$$G^{11}(z) = -G^{*22}(-z^*) \quad , \quad G^{12}(z) = G^{*21}(-z^*) \quad . \tag{5.5.6}$$

The Green's function $\overline{G}(z)$ averaged over impurity configurations should possess the same symmetry properties as its self-energy part.

Consider a Green's function corresponding to a perfect crystal made up of atoms of one and the same sort A. Rather than by (5.5.5), it is determined by

$$\begin{pmatrix} z - H_o - V_A & \Delta_A \\ \Delta_A^* & z + H_o + V_A \end{pmatrix} G^o(z) = 1 \quad . \tag{5.5.7}$$

By the symbol $G^o(z)$ we imply a two-dimensional matrix with components $G^{o\mu\nu}(z)$ ($\mu,\nu = 1, 2$). In the case of a dilute alloy the Green's function $G(z)$ determined by the basic equation (5.5.5) could be expanded in power series of the deviation of the parameters Δ and V from their values for a perfect A-type crystal. In the CPA method one, however, should proceed differently.

We introduce a function $\tilde{G}(z)$ corresponding to some hypothetic ideal crystal. The most general form of the equation for it, consistent with the conditions of symmetry (5.5.6), should be as follows

$$\begin{pmatrix} z - H_o - \varepsilon(z) & \Delta(z) \\ \Delta^*(-z^*) & z + H_o + \varepsilon^*(-z\ ) \end{pmatrix} \tilde{G}(z) = 1 \quad , \tag{5.5.8}$$

where $\varepsilon(z)$ and $\Delta(z)$ are scalar matrices with respect to site indices. In particular, the equation obtained for the Green's function averaged with respect to impurity configurations should be of this type. The quantities $\varepsilon(z)$ and $\Delta(z)$ satisfy the symmetry conditions

$$\varepsilon(z) = \varepsilon^*(z^*) \quad , \quad \Delta(z) = \Delta^*(z^*) \quad , \tag{5.5.9}$$

which follow from the general properties of the Green's function. Comparing (5.5.8) and (5.5.7) one can write the relationship between $\tilde{G}(z)$ and $G^o(z)$ as a Dyson equation

$$\tilde{G}(z) = G^o(z) + G^o(z)\Sigma(z)\tilde{G}(z) \quad , \tag{5.5.10}$$

where the quantity

$$\Sigma(z) = \begin{pmatrix} \varepsilon(z) - V_A & \Delta_A - \Delta(z) \\ \Delta_A^* - \Delta^*(-z^*) & -\varepsilon^*(-z^*) + V_A \end{pmatrix} \quad (5.5.11)$$

plays the role of the self-energy part. Again comparing (5.5.8) and (5.5.7) with (5.5.11) we find the following relationship between $\tilde{G}$ and G^o:

$$\tilde{G}(z) = G^o[z - \Sigma(z)] \quad . \quad (5.5.12)$$

Further, we follow the conventional CPA scheme developed for normal alloys. Write the basic equation (5.5.5) for the Green's function G determined for a fixed configuration as

$$G(z) = \tilde{G}(z) + \tilde{G}(z)W(z)G(z) \quad . \quad (5.5.13)$$

In this representation the quantity $\tilde{G}(z)$ plays the part of a zero-approximation Green's function, and $W(z)$ describes the perturbation due to the deviation of the atoms of the impurities in a prescribed configuration from some ideal arrangement:

$$W(z) = \begin{pmatrix} -\varepsilon(z) + V & -\Delta + \Delta(z) \\ -\Delta^* + \Delta^*(-z^*) & \varepsilon^*(-z^*) - V \end{pmatrix} \quad . \quad (5.5.14)$$

Thus, the perturbation $W(z)$ depends on random values of the quantities V and Δ.

It is useful to transform (5.5.13) with the aid of the scattering matrix $T(z)$. Then we have

$$G(z) = \tilde{G}(z) + \tilde{G}(z)T(z)\tilde{G}(z) \quad , \quad (5.5.15)$$

where

$$T(z) = W(z) + T(z)\tilde{G}(z)W(z) \quad . \quad (5.5.16)$$

It is seen from (5.5.15) that the configuration averaging of the Green's function $G(z)$ is related to the averaging of the $T(z)$ matrix. Up till now the quantities $\varepsilon(z)$ and $\Delta(z)$ involved in the expression for the Green's function $\tilde{G}(z)$ have been undefined. We select from the vanishing of the averaged T-matrix,

$$\overline{T}(z) = 0 \quad . \quad (5.5.17)$$

Then from (5.5.15) the basic expression for the averaged Green's function follows, namely

$$\overline{G}(z) = \widetilde{G}(z) \quad . \tag{5.5.18}$$

The T-matrix (5.5.16) can be written in explicit form with respect to site indices

$$T_{ij} = t_i\delta_{ij} + \sum_{l(\neq j)} T_{il}\widetilde{G}_{lj}t_j \quad , \tag{5.5.19}$$

where

$$t_i = W_i\frac{1}{1-\widetilde{G}_{ii}W_i} \quad . \tag{5.5.20}$$

All the quantities involved in the last two equations are 2×2 matrices.

The major coherent-potential approximation is in the replacement of the exact equation (5.5.17) by the approximate equation

$$\overline{t}_i(z) = 0 \quad , \tag{5.5.21}$$

which corresponds to the correlation in the distance on different sites being neglected, i.e., $\overline{t_it_j} = \overline{t}_i\cdot\overline{t}_j$, and to the exact allowance for the scattering on each site.

Thus, (5.5.18) and (5.5.12) together with the equation

$$\overline{W_i(z)[1-\widetilde{G}_{ii}(z)W_i(z)]^{-1}} = 0 \tag{5.5.22}$$

determine the averaged Green's function of the alloy in the CPA method.

5.5.3 The Superconducting Transition Temperature

Equation (5.5.3) for the order parameter in conjunction with the matrix equation (5.5.22) for the averaged Green's function forms a closed system of equations which, upon linearization, enables determination of the superconducting transition temperature. Omitting the lengthy calculus performed in [5.80,81] we give the final expression for T_c, which has the usual BCS form

$$T_c = 1.14\omega_D\, e^{-1/\lambda_{eff}} \quad , \tag{5.5.23}$$

where λ_{eff} is expressed in terms of the imaginary part of two quantities $\overline{F}(\omega)$ and $\overline{F}^{(i)}(\omega)$, taken on the Fermi surface. Here $\overline{F}(\omega)$ is an averaged Green's function of a normal metal and $\overline{F}^{(i)}(\omega)$ is a Green's function averaged with respect to all configurations with the fixed site i occupied by the atom of a given sort, A or B. This function is related to the function $\overline{F}(\omega)$ by the equation

$$\overline{F}^{(i)} = \overline{F} + \overline{F} t_i \overline{F} \quad . \tag{5.5.24}$$

The imaginary parts of the Green's functions $\overline{F}(\omega)$ and $\overline{F}^{(i)}(\omega)$ determine the mean $N(\omega)$ and local $N_i(\omega)$ densities of electron states in the alloy, namely

$$N(\omega) = -\frac{1}{\pi}\,\mathrm{Im}\,\{\overline{F}(\omega)\} \quad , \quad N_i(\omega) = -\frac{1}{\pi}\,\mathrm{Im}\,\{\overline{F}^{(i)}(\omega)\} \quad . \tag{5.5.25}$$

The quantity λ_{eff} is expressed in terms of these densities by

$$N(0)\lambda_{\mathrm{eff}} = C_A\lambda_A N_A^2(0) + C_B\lambda_B N_B^2(0) \quad , \tag{5.5.26}$$

where

$$N(0) = C_A N_A(0) + C_B N_B(0) \tag{5.5.27}$$

is the density of states of the alloy on the Fermi surface, and C_A and C_B stand for the concentration of A and B atoms in the alloy.

This is the final result of the CPA method in the theory of superconducting alloys. In [5.80] a calculation of the quantities $N_i(0)$ for a model density of states described by a Lorentzian function having a distribution width $\mathcal{D}$ (which acts as an electron band width) has been carried out,

$$N_i(0) = \frac{1}{\pi}\,\frac{1}{1 + (V_i/\mathcal{D})^2} \quad . \tag{5.5.28}$$

Substituting this result into (5.5.26) we find

$$\lambda_{\mathrm{eff}} = \left(\frac{C_A}{1 + (V_A/\mathcal{D})^2} + \frac{C_B}{1 + (V_B/\mathcal{D})^2}\right)^{-2} \left(\frac{C_A\lambda_A}{[1 + V_A/\mathcal{D})^2]^2} + \frac{C_B\lambda_B}{[1 + (V_B/\mathcal{D})^2]^2}\right) . \tag{5.5.29}$$

To this expression it is necessary to add the equation expressing the relationship between the total number of electrons and the Fermi level. Let the number of valence electrons for one atom of kind A and B be $1 + Z_A$ and $1 + Z_B$. Counting the energy from the Fermi level we have

$$1 + C_A Z_A + C_B Z_B = 2 \int_{-\infty}^{0} N(\omega)d\omega \quad , \tag{5.5.30}$$

where $N(\omega) = \pi^{-1}\,\mathrm{Im}\{F(\omega + i\delta)\}$. These last two equations define λ_{eff} in terms of the quantities Z_A, Z_B and $\delta \equiv (V_A - V_B)/\mathcal{D}$.

If one assumes that the Debye temperature of the alloy varies weakly with composition, then the concentration dependence of T_c should be determined solely by the quantity λ_{eff}.

As shown by WEINKAUF and ZITTARTZ [5.80,81], for a number of isoelectron alloys of transition metals the experimental data fit the curves obtained from (5.5.23,29,30) for reasonable values of the parameters δ and $Z \equiv Z_A = Z_B$ (Fig.5.22). This indicates that the model under consideration is applicable to such systems.

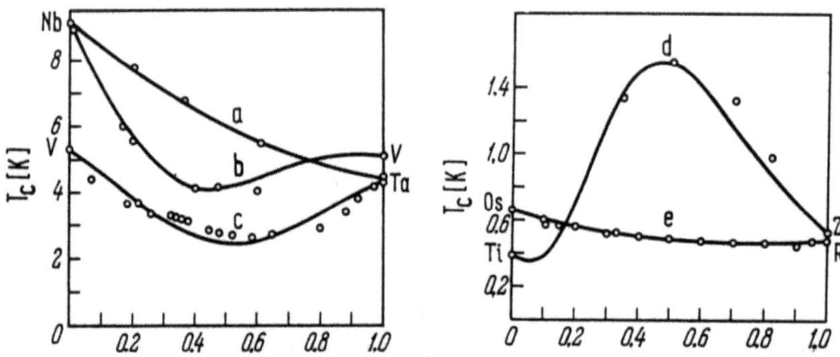

Fig.5.22. The concentration dependence of T_c in transition-metal alloys and a comparison with the theory using the CPA method
a - Nb - Ta: $\delta = 0.33$, $Z = 0$; b - Nb - V: $\delta = 0.86$, $Z = 0.14$; c - V - Ta: $\delta = 0.77$, $Z = 0$; d - Ti - Zr: $\delta = 1.15$, $Z = 0.44$; e - Os - Ru: $\delta = 0.3$, $Z = 0$

In conclusion we note that the fundamental equations (5.5.26,27) of the theory presented here are of the same character as (5.4.1,5) of MIEDEMA's semi-empirical scheme. They reflect the fact that the effective electron-phonon coupling constant and the density of states on the Fermi surface in transition-metal alloys depend on some local characteristics of the system which are related to the atomic components. Such a local characteristic in MIEDEMA's scheme is the effective number of electrons on a given atom in the alloy, whereas in the alloy model treated above it is the energy of the atomic d-state. This confirms the common believe that the superconducting transition temperature is chiefly related to chemical parameters.

6. Compounds with A-15 Structure

The highest superconducting transition temperatures have not actually been found in pure metals or their alloys but in intermetallides and chemical compounds based on transition metals. The superconductivity of chemical and other compounds will be treated in Chapter 7. Here we shall focus our attention on intermetallic compounds of transition elements with A-15 structure.

There are several types of intermetallides in which high-temperature superconductivity (with $T_c > 10$ K) has been found, transition-metal compounds with A-15 structure (β-W), Laves phases, σ phases and α-Mn type compounds. The most prominent of them are undoubtedly A-15 structure compounds comprising the famous V_3Si, Nb_3Sn and V_3Ga, and also Nb_3Ge with a superconducting transition temperature of $T_c = 23.2$ K [6.1,2], the highest known at present. This family of superconductors is widely known and has been studied intensively in recent years. It is with this class of compounds that we shall begin our treatment of intermetallides.

6.1 Physical Properties

6.1.1 The Crystal Structure

We consider the main features peculiar to the crystal structure of A-15 compounds, the chemical formula of which is A_3B. The B atoms in them form a body-centred cubic lattice, and the A atoms are arranged pairwise on the cube edges, parallel to the coordinate axes (Fig.6.1). The unit cell of the compounds under investigation contains eight atoms, and the space group is O_h^3 - Pm3n.

A typical feature of this structure is that A atoms make up families of intersecting linear chains, the distance between the atoms within one chain being the shortest distance between the atoms in the A-15 structure and being 22% less than that between A atoms belonging to different chains.

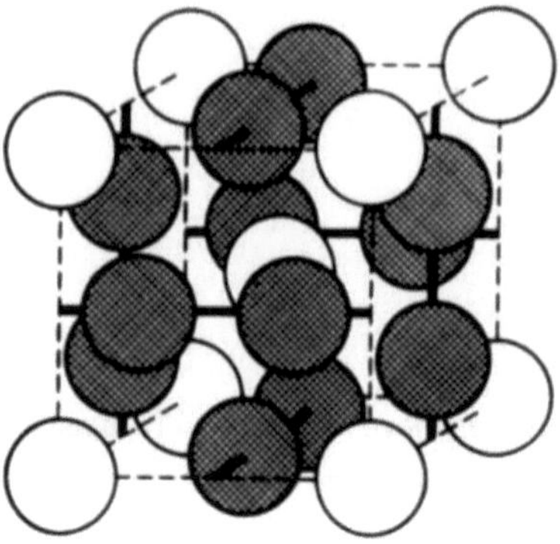

Fig.6.1. The A-15 structure of A_3B-type intermetallic compounds. White balls are B atoms, the grey ones illustrate A atoms

The A atom always constitutes a transition element of groups IV, V or VI (Ti, V, Cr, Zr, Nb, Mo, Ta, W), whereas the B atom may be represented either by a nontransition element (Al, Si, P, Ga, Ge, As, In, Sn, Sb, Pb, Bi) or a transition element (mainly of group VIII: Co, Ni. Ru, Rh, Pd, Re, Os, Ir, Pt, Au).

The majority of these compounds have narrow ranges of homogeneity [6.3]. When there are vanadium compounds with components of almost equal atomic radii the formation of the A-15 structure does not take place according to peritectic reactions, but instead via the ordering of the solid solution in a bcc lattice. These compounds possess sufficiently extended regions of homogeneity [6.4].

At deviations from stoichiometry in both directions in the compounds concerned, atomic substitution occurs. This implies that at the composition $A_{3+x}B_{1-x}$ the linear chains remain intact, and the excessive A atoms occupy the B sites. But in the composition $A_{3-x}B_{1+x}$ a partial breakdown of the A-atom chains is likely to occur.

When A_3B compounds are alloyed, the nontransition B elements, as a rule, replace the B atoms in the bcc positions. On the other hand, transition elements can replace both A atoms in the chains and B atoms in the bcc positions [6.5].

One of the interesting structural features peculiar to A_3B phases with an A-15 type lattice is their proneness to high ordering. However, it appears that the degree of ordering depends on whether the B atom is a transition element or not. A_3B compounds containing B nontransition elements are, as a rule, highly ordered. But for the second group of these compounds the degree of ordering depends on the mutual arrangement of elements A and B in the periodic table [6.6].

We now consider the physical properties of A_3B compounds.

6.1.2 Superconducting Transition Temperature of Pure Compounds and Their Alloys

In Table 6.1 data on superconducting transition temperatures in compounds with A-15 structure are presented as reported in the literature (up to 1981). With the improvement of the technique of synthesizing compounds, the values of T_c have been continuously refined, therefore the table contains results of most recent measurements. For the table use has been made of different sources [6.3,7-31].

A cursory inspection of the results quoted in Table 6.1 shows that the highest superconducting transition temperatures for A_3B compounds of vanadium and niobium with A-15 structure are obtained in those cases when the B atom is represented by a nontransition element: Al, Si, Ga, Ge, Sn. It is among these compounds that the superconductors with the highest superconductin transition temperatures have recently been obtained: Nb_3Ga with $T_c = 20.3$ K [6.21] and Nb_3Ge with $T_c = 23.2$ K [6.1,2]. A number of unusual phenomena may be noted for this group of compounds. Take, for example, compounds of V with Si, Ge and Sn. The three metalloids form one group, the valence electron concentration in all the three compounds is the same, but T_c differs appreciably. It might be thought that the decrease of T_c from V_3Si to V_3Sn is due to the increase in the mass of element B [6.32]. But if we take compounds of niobium with Al and Sn the T_c in them undergoes almost no changes. Thus, the chemical affinity of elements within one group of compounds, in the general case, does not lead to similar T_c, and the difference between them is not caused by the atomic mass.

For the second group of superconducting A_3B compounds with transition B-elements, the superconducting transition temperatures are on the whole lower: T_c reaches 15 K for Mo_3Re ($Mo_{0.65}Re_{0.35}$) which has the highest temperature [6.31]. While for the first group of compounds approximately equal values of T_c occur equally frequently both in vanadium and in niobium, the situation is different for the second group of compounds: niobium compounds have superconducting transition temperatures which, as a rule, are two to three times larger than those for similar vanadium compounds.

Investigations of the superconducting properties of A_3B compounds at deviations from stoichiometry and with the variation of the degree of long-range order S have shown that T_c is very sensitive to these effects in one group of these compounds (V_3Si, V_3Ga, V_3Ge, V_3Au, Nb_3Ga, Nb_3Ge, etc.). They are called "typical" by convention [6.33]. In "typical" A-15 compounds the maximum values of T_c are obtained when approximated to stoichiometric composition and to a high degree of order (see also [6.34,35]. For example, in the compound V_3Si

Table 6.1 T_c in compounds with A-15 structure

B \ A_3	Ti	Zr	Hf	V	Nb	Ta	Cr	Mo	W
Al				11.70	18.80			0.58	
Ga				15.9	20.3		<0.35	0.76	
In				13.9	8.8-9.2				
Tl	<0.35	<0.35		<4.20					
Si				17.1	18.0-19.0		<0.015	1.70	<1.20
Ge				11.2	23.2	8.0	1.20	1.8	
Sn	5.80	0.93		12.3-17.9	18.0	8.35			
Pb		0.76		<4.2	8.0				
P				<1.00					
As				0.20					
Sb	6.5	<1.20		0.80	2.0	0.72			
Bi		3.4		<4.20	3.0				
Tc								13.2^{++}	
Re				8.4^{+}				15.0	11.4
Ru							3.42		
Os				5.7	1.0		4.68	12.65	
Co				<0.015					
Rh				1.0	2.64	10.0	0.07		
Ir	4.30			1.71	3.2	6.6	0.75	8.5	
Ni				0.30					
Pd				0.082					
Pt	0.49			3.7	10.9	0.40	<0.015	4.6	
Cu									
Ag									
Au	<0.015	0.92		3.0	11.5	0.55			
Zn									
Cd				<4.20					
Hg	<0.35	<0.35							
Te					2.5				

+ The composition of the compound V - Re with structure A-15 is $Re_{0.71}V_{0.29}$

++The composition of the compound Mo - Tc with structure A-15 is $Tc_{0.6}Mo_{0.4}$

of stoichiometric composition $T_c = 16.85$ to 17.1 K, whereas at 20.1 at.% Si, $T_c = 9.4$ K [6.36]. Note that the best success in increasing T_c has been achieved by approximating ideal stoichiometry: in Nb_3Ga from 14.9 K to 20.3 K [6.21], in V_3Sn from 3.8 K to 17.9 K [6.25], in V_3Ge from 6.0 K to 11.2 K [6.27] and in Nb_3Ge from 6.9 K to 23.2 K [6.1,37]. As an example illustrating the influence of ordering on T_c we point to the compound V_3Au in which the transition from a disordered state to a highly ordered state was accompanied by an increase of T_c almost by a factor of 300 (from $T_c < 0.0012$ K to $T_c = 2.97$ K [6.38]).

In the other group of A_3B compounds, known as "atypical" [6.33] (Mo_3Ir, Mo_3Pt, Cr_3Os, Cr_3Ir, V_3Ir, etc.), the maximum superconducting transition temperature does not correspond to a stoichiometric composition, and the correlation between T_c and S is less pronounced. For example, at a strictly stoichiometric composition the compound Cr_3Ir has $T_c = 0.16$ K, whereas a higher $T_c = 0.75$ K is reached at a deviation from stoichiometry - at 18 at.% Ir [6.33].

BONGI, FISCHER and JONES [6.39,40] have proposed a model describing the charge transfer between component atoms in typical and atypical A-15 compounds. This model is based on MIEDEMA's quasi-chemical concepts (Sect.5.4) which relates the dependence of the redistribution of electrons between the components of the alloy to the ratio between their electronegativities.

According to this model, in typical A-15 compounds the electrons are redistributed from the B component (nontransition element) to the A transition metal. As for atypical compounds, the reverse is true for this electron redistribution. Straightforward experimental measurements of the electron density distribution in V_3Si and Cr_3Si [6.41-43] have also shown the electron transfer from Si to V or Cr. Some indications of the charge transfer from Si to V have also been obtained in a cluster calculation of the electron structure of the compound V_3Si [6.44].

Consider now the superconducting properties of alloys of A-15 structure compounds. Two cases are possible in the formation of solid solutions when A_3B compounds are alloyed: isomorphous substitution of A' atoms for A atoms and isomorphous substitution of B' atoms for B atoms.

For the first group of solid solutions, T_c has been investigated most elaborately for the systems $V_3Si - Mo_3Si$, $V_3Sn - Nb_3Sn$, $Nb_3Au - V_3Au$, $V_3Sn - Ta_3Sn$, $Nb_3Sn - Ta_3Sn$ [6.3], $V_{3-x}M_xGa$(M - Ti, Cr, Mn, Fe, Co, Ni) [6.3,45] and $V_{3-x}M_xSi$ (M - Ti, Cr, Nb, Mo, Hf, Ta, W) [6.46,47].

In all the enumerated cases and in many others described in [6.3,48-51] T_c varies smoothly and continuously with varying composition of the alloys. However, not a single case is known where the replacement of the A atom by another transition element A' leads to an appreciable increase in T_c whose value exceeds the superconducting transition temperature of the compound alloyed which has the highest temperature. A similar situation takes place when alloying A_3B compounds whose B atom is a transition element and is replaced by B' which is another transition element. A detailed study has been made of the following systems of alloys: $Nb_3Rh - Nb_3X$(X - Co, Ru, Pd, Os, Ir, Pt, Au) [6.52], $Mo_3Ir - Mo_3Ru$, $Mo_3Os - Mo_3Ru$ [6.53] and $Nb_3Ge - Mo_3Ge$ [6.54].

In the $A_3B - A_3B'$ systems where B and B' atoms are nontransition elements, it has been found that solid solutions form $V_3Si - V_3Ga$, $V_3Si - V_3Ge$, $V_3Si - V_3Sn$, $Nb_3Sn - Nb_3Al$, $Nb_3Sn - Nb_3Ge$, $Nb_3Sn - Nb_3Sb$, $Nb_3Al - Nb_3Ge$, $Nb_3Al - Nb_3Sb$ [6.3,46, 55-57], $Nb_3Ge - Nb_3Si$ [6.58,59], $Nb_3Al - Nb_3Si$ [6.60], etc. The situation here is somewhat different from that in the first group of solid solutions. For example, ALEKSEEVSKII et al. [6.61] were the first to find (1966) an increase in T_c in the $Nb_3Al - Nb_3Ge$ system compared to the basic components at the composition $(Nb_3Al)_{0.8}(Nb_3Ge)_{0.2}$. Later this result was repeated by many researchers and their joint efforts made it possible to bring the superconducting transition temperature in this system, initially, to 20.05 K [$Nb_3(Al_{0.75}Ge_{0.25})$] [6.62-64] and, finally, to 20.7 - 21.06 K [$Nb_3(Al_{0.763}Ge_{0.237})$] [6.65,66].

It turned out that the highest values of T_c could be obtained after a continuous annealing at comparatively moderate temperatures, and it was assumed [6.64,65] that this is due to ordering processes. It was also believed that along with the ordering of Nb atoms in the A-15 structure (as indicated by the fact that the highest T_c in the $Nb_{3+x}Al_yGe_{1-y}$ system is reached at $x = 0$) the ordering of Al and Ge atoms would also occur. However, the X-ray structure analysis performed in [6.6] on a polycrystalline $Nb_3(Al_{0.75}Ge_{0.25})$ specimen has established no presence of new superstructural lines in the X-ray diffraction picture.

A study of the superconducting properties in solid solutions of $Nb_3(Al_{1-x}Me_x)$, $Nb_3(Sn_{1-x}Me_x)$ and $Nb_3(Ga_{1-x}Me_x)$ [6.60,67-69] has shown that it has also been possible to obtain some increase in T_c at an isomorphous substitution of the B' atom for a nontransition element B atom not only in the Nb_3(Al - Ge) system but also in other systems, for example, in Nb_3(Al - Ga), Nb_3(Sn - Ga), Nb_3(Al - B), Nb_3(Al - Si), Nb_3(Ge - Si) and also in V_3(Si - C) [6.70].

A number of papers [6.71,72] were devoted to the study of superconductivity in hydrogen-saturated compounds with A-15 structure. The possibility of introducing hydrogen arises from the fact that an A-15 structure has tetrahedral voids in which comparatively small hydrogen atoms may be accommodated. It was registered for hydrogen-saturated compounds of Nb_3Sn, Nb_3Ge and Nb_3Al that the lattice parameter increased and T_c decreased in all the cases.

However, in spite of some optimism which the publication of [6.62,63] has given to the experimentalists, it must be stated that in the past ten years the numerous attempts of alloying with the use of different model criteria have not actually led to any encouraging results (see, for example, [6.73-75]). Moreover, recent developments in the synthesis of the compounds Nb_3Ga (T_c = 20.3 K) and Nb_3Ge (T_c = 23.2 K) give ample grounds for the following contemplation: either carry on experimental attempts at alloying in the more complicated systems or concentrate efforts on developing new methods of synthesis which would make it possible to obtain compounds closer to stoichiometric ones.

The abundant experimental evidence concerning the influence of impurities on the superconducting transition temperature in A_3B compounds with A-15 structure is convenient to systematize using the following principles. Firstly, to compare the effect of impurities on different basic compounds with the transition temperature T_{co} (at a strictly stoichiometric composition); it is convenient to plot the concentration dependence versus the transition temperature $T_c/T_{co} = t$. Secondly, it is important to know what lattice site the impurity occupies, i.e., whether it replaces the A atoms that form chains or replaces the B atoms (nontransition or transition elements) which are located at the edges or at the centre of the cube.

Figures 6.2 and 3 have been drawn up following this principle, the first of them supplying information on ternary systems such as $(A_{0.75-x}A'_x)B_{0.25}$ where $0 < x < 0.75$ [6.3,76-80], and the second applying to systems of the type $A_{0.75}(B_{0.25-x}B'_x)$ where $0 < x < 0.25$ [6.3,52,67,81]. For each separate curve portrayed in the figures, the compounds whose transition temperature is the highest were chosen as basic ones. That is why the quantity T_c/T_{co} is less than 1 everywhere.

Analysis of the results presented in Figs.6.2 and 3 shows that the general feature in all the situations considered is as follows: if two A_3B and A'_3B' compounds are alloyed into a ternary alloy, the superconducting transition temperature T_c becomes lower than the maximum T_c for pure basic compounds. This unfortunate rule has only a few exceptions which have been considered above.

In the region of small concentrations of x of the third component (an impurity situation) we can readily see a considerable lowering of T_c with increasing impurity concentration in most of the alloys pertaining to all the five cases treated above. As a rule, this decrease happens to be particularly drastic in the case when the atoms are substituted in the chains. It is of special interest to study the introduction of magnetic impurities into A-15 type compounds. Much consideration has been given to the systems $Mo_{0.75}Os_{0.25-x}$

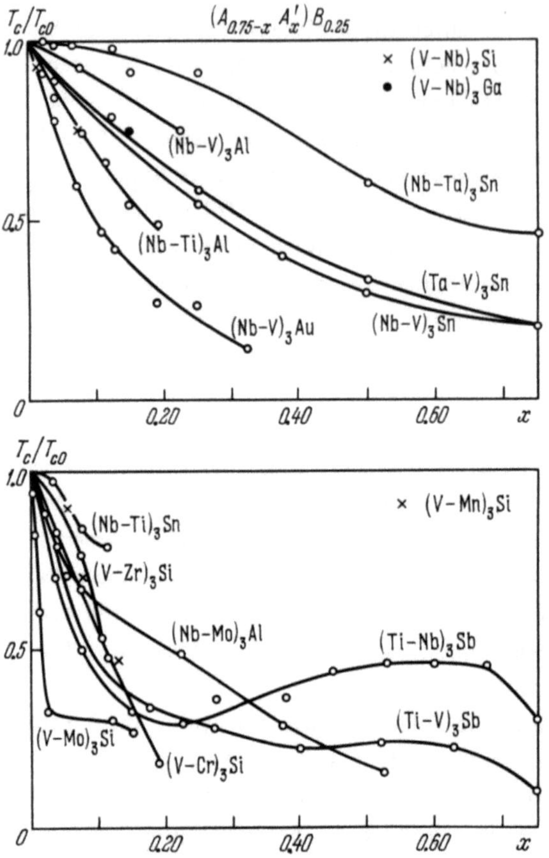

Fig.6.2. The dependence of T_c/T_{co} on the composition of the alloy $(A\text{-}A')_3B$. In the upper figure elements A' belong to the same group of the periodic table as elements A do, in the lower figure A and A' are elements of different groups

Fe_x [6.82], $Mo_{0.77}Ir_{0.23-x}Fe_x$ [6.83], $Nb_{0.76}Pt_{0.24-x}Fe_x$ [6.84] and $Nb_yPt_{1-y-x}Fe_x$ ($y = 0.73 \div 0.78$) [6.85]. It has been established that the lowering of T_c with Fe concentration differs from the predictions of the Abrikosov-Gor'kov theory (Sect.3.2). From the data presented in Fig.6.4 it follows that in the systems $Nb_{0.75}Pt_{0.25-x}Fe_x$, $Nb_{0.75}(Pt-Au)_{0.25-x}Fe_x$, and $Nb_{0.75}Au_{0.25-x}Fe_x$, a positive deviation from the Abrikosov-Gor'kov curve constructed for a system with noninteracting magnetic impurities occurs. On the other hand, for the $Mo_{0.75}Os_{0.25-x}Fe_x$ system, as shown in [6.82], a negative deviation from the Abrikosov-Gor'kov curve takes place.

The difference in the behaviour of T_c has been discussed by FISCHER et al. [6.83] within the framework of a model of interacting pairs of impurities. These authors have shown that depending on whether a given interaction is ferromagnetic or antiferromagnetic, the unpairing parameter λ_m may be less or greater than the parameter λ_m determined for a system of noninteracting

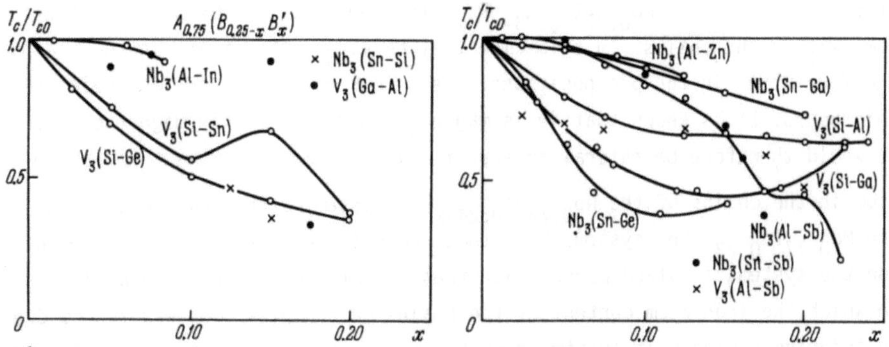

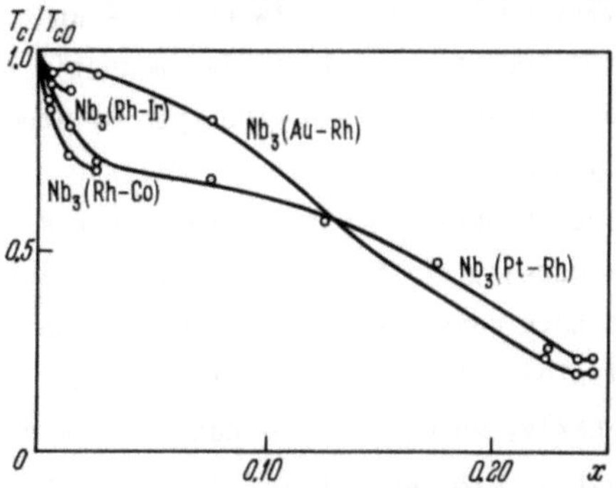

Fig.6.3. The dependence of T_c/T_{c0} on the composition of the alloy $A_3(B-B')$. In the two upper figures B' is a non-transition element, in the first figure B' is taken from the same group as B, in the second figure it is taken from different groups. In the lower figure B and B' are transition elements

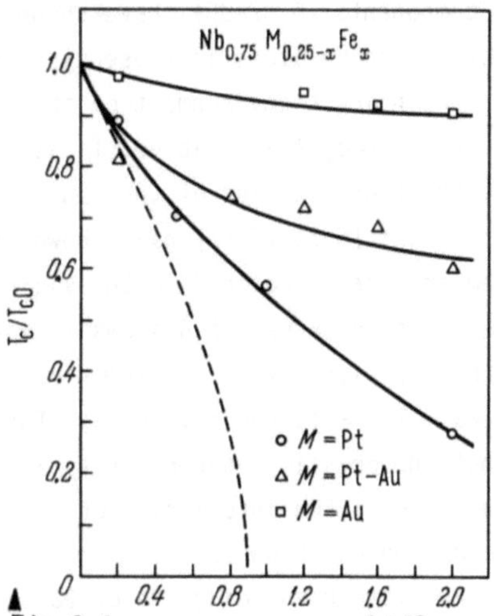

Fig.6.4

Fig.6.4. T_c/T_{c0} as a function of Fe concentration of Nb_3M compounds. The dashed line corresponds to the Abrikosov-Gor'kov theory

impurities. This leads to a positive or negative deviation with respect to the Abrikosov-Gor'kov curve.

Analysis indicates, however, that not all of the Fe atoms in ternary systems such as A-15 make a contribution to the unpairing parameter λ_m. In an A-15 type structure, iron atoms may occupy bcc positions or sites in the chains and, depending on the nature of the closest environment, be magnetic on some sites and nonmagnetic on others. Let us exemplify this. In the systems $Nb_{0.76}Pt_{0.24-x}Fe_x$ and $Mo_{0.77}Ir_{0.23-x}Fe_x$ the Fe atoms in the bcc positions have twelve Nb or Mo atoms as their nearest neighbours. Knowing that Fe atoms are magnetic in pure Mo and nonmagnetic in pure Nb [6.86] it is reasonable to assume that they will be nonmagnetic on bcc sites of the $Nb_{0.76}Pt_{0.24-x}Fe_x$ system and magnetic on the same sites in $Mo_{0.77}Ir_{0.23-x}Fe_x$. The opposite is true when Fe atoms come within the chains. In this case they have two other atoms in the chains and four atoms in the bcc positions, i.e., four Pt or Ir atoms as their nearest neighbours. It is known that Fe is magnetic in Pt and weakly magnetic in Ir. It would therefore be natural to expect a higher magnetic moment for an Fe atom in the chains of the $Nb_{0.76}Pt_{0.24-x}Fe_x$ system rather than in those of the $Mo_{0.77}Ir_{0.23-x}Fe_x$ system. Note that magnetic susceptibility measurements for the systems involved corroborate these assumptions. Special experiments, in which the iron atom content of the chains or bcc positions was varied due to ordering during heat treatment or by varying the ratio between the A and B components [6.82-85], have shown that the formation of localized magnetic moments on Fe atoms in these systems is determined mainly by the type of atoms in the nearest environment of the impurity atom.

As already stated above, T_c is sensitive to deviations from stoichiometric composition. Studies of the region of homogeneity of the compounds V_3Si, Ti_3Sb and Nb_3Pt [6.36,77,87] have shown that in both cases the maximum transition temperature is reached in the vicinity of stoichiometric composition. When there is an excess of B atoms T_c varies weakly, whereas in the case of their shortage (i.e., when the atoms of a transition element replace those of a non-metal) T_c falls sharply [6.88]. The latter circumstance apparently indicates that the drastic lowering of T_c in the alloys concerned is associated with the increase of the interaction of chains via the atoms of the transition element occupying position B.

From the phase diagrams of 28 compounds with A-15 structure [6.89,90] the A_3B compounds may be divided into two groups, according to how their regions of homogeneity are formed. The first group comprises compounds whose region of homogeneity is practically absent or extends only in the direction rich in the A component. Such a situation occurs, as a rule, in compounds with high T_c and is thought to be associated with the conservation of the integrity of the linear A chains in an A-15 structure. The other group contains compounds whose region of homogeneity extends towards the B component or in both directions from stoichiometric composition. In these compounds (having lower T_c) atoms in the chains are likely to be replaced by B atoms and integrity is not conserved.

Thus, the superconducting transition temperature is very sensitive to substitutions of atoms in linear chains, irrespective of whether this takes place during alloying or at deviations from stoichiometric composition.

6.1.3 The Electron Concentration Dependence of Electronic Heat Capacity and Magnetic Susceptibility

Analysis of different empirical criteria of superconductivity shows that a clear correlation occurs between T_c and N_e, i.e., the mean number of valence electrons per atom (electron concentration) [6.3]. In fact, it follows from consideration of Fig.6.5 that the superconductors with lower values of T_c group near two electron-concentration intervals: 4.50 to 4.75 e/a and 6.25 to 6.60 e/a. The first range encompasses compounds whose B element is either

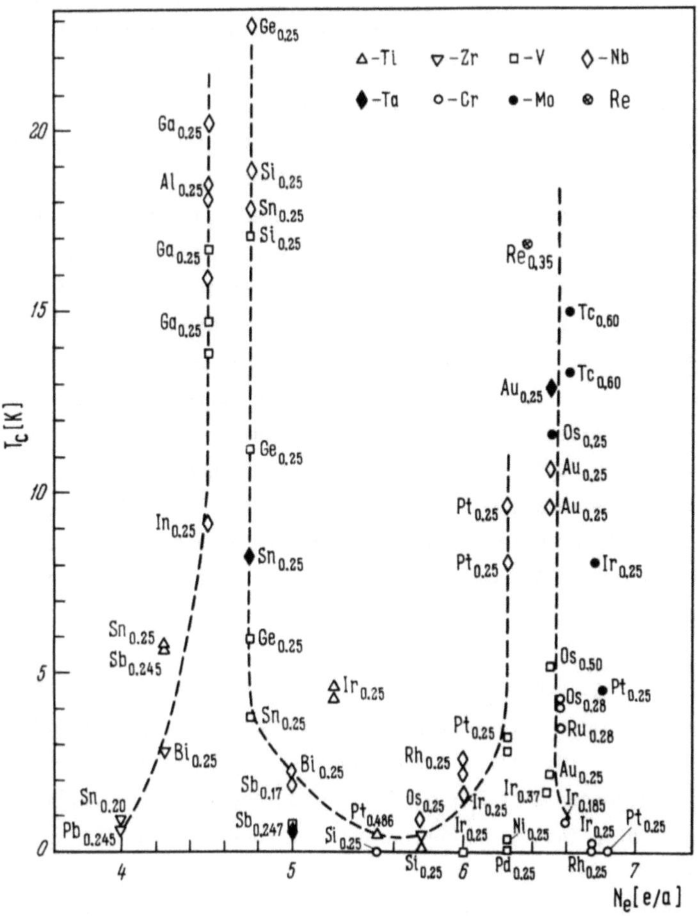

Fig.6.5. The superconducting transition temperature of compounds with A-15 structure as a function of electron concentration N_e. At the experimental points the B components are specified, the symbols given at the top mark the A components

not a metal or a simple metal, and the second region covers compounds with a B atom which is a transition or a noble metal.

Close inspection of Fig.6.5 reveals that the above electron-concentration ranges, favourable to high T_c, are actually regions in which there is a greater number of A_3B compounds than there is in other N_e regions.

For example, in the region of the first "peak" two groups of A_3B compounds with the same electron-concentration stand out: either 4.50 e/a (Nb_3Ga, Nb_3Al, V_3Ga, V_3In, V_3Al, Nb_3In) or 4.75 e/a (Nb_3Ge, V_3Ge, V_3Si, Nb_3Sn, V_3Sn). The same may be said with regard to the second "peak". It is natural that different compounds with the same concentration of valence electrons per atom have different superconducting transition temperatures which are disposed along the vertical line on the T_c vs. N_e curve. The spread in values appears to be the greatest in the ranges of 4.50 to 4.75 e/a and of 6.25 to 6.60 e/a. Pertinence to these regions apparently is a necessary but not a sufficient condition to obtain a high T_c. In this connection we would like to point out that when plotting a T_c vs. N_e curve one often refers to the existence of two maxima, forgetting that in the middle portion of "favourable" electron concentration within which the peak usually falls, there is no A_3B compound of stoichiometric composition (see, for example, the T_c vs. N_e curve given in [6.91]). If the data on ternary compounds, for which a continuous variation of electron concentration can be obtained, were traced on this figure we would see that their T_c take on various values (as a function of composition within one and the same value N_e, see, for example, [6.74,92,93]) that uniformly fill the space between the vertical lines delineating the region of "favourable" electron concentrations. Therefore we are not inclined to give the T_c vs. N_e curve the meaning of an "absolute" parameter, although it is the only guide in the purposeful search for alloys of A_3B compounds with high T_c. However, this parameter proves to be instrumental in treating other physical properties of these compounds.

For A_3B compounds, Figs.6.6 and 7 carry values of electronic heat capacity [6.19] and paramagnetic susceptibility [6.20] as functions of N_e. One can readily notice a certain similarity in the character of the variation of these properties with the dependence of T_c on electron concentration (Fig.6.5). The relative arrangement of points coincides in all the three figures (Figs.6.5-7), except in rare cases.

Since both the electronic heat capacity and the paramagnetic susceptibility in metals is proportional to the density of electronic states on the Fermi level, the presence of the above correlation gives good reason to conclude that high values of T_c in A-15 compounds are accompanied by a higher den-

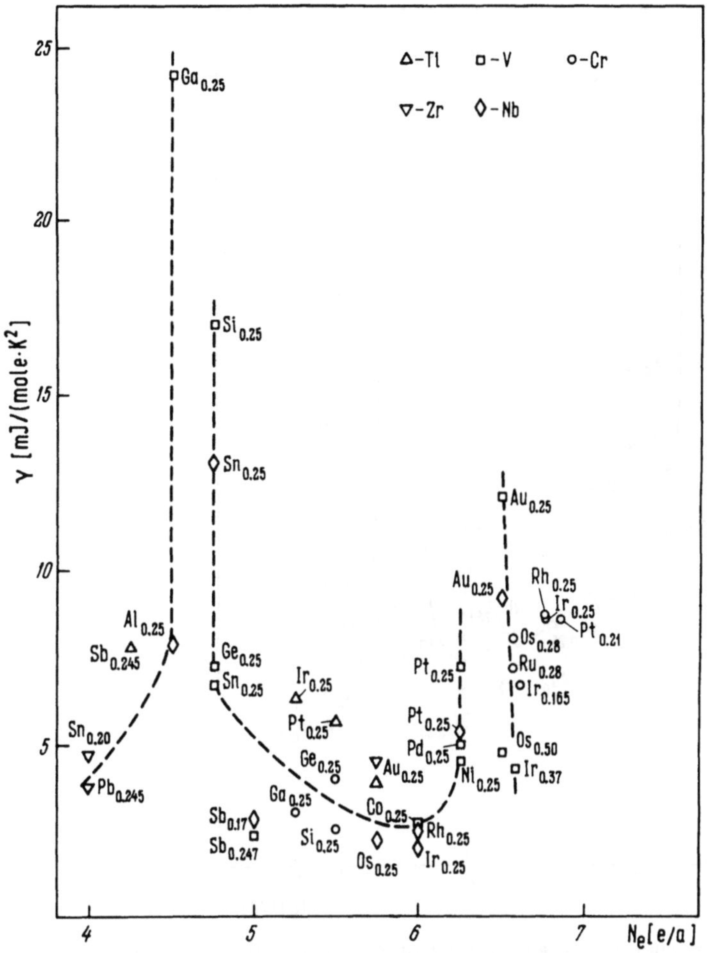

Fig.6.6. The electronic heat capacity of compounds with A-15 structure as a function of N_e. The legend is the same as that in Fig.6.5

sity of electron states on the Fermi level and that the same correlation between T_c, γ and χ occurs also in the region of homogeneity of the compounds concerned [6.20,36].

However, some deviations from the regularity described above take place in stoichiometric compounds of niobium. For example, to the high value of T_c in the compound Nb_3Al (T_c = 18.8 K) there correspond comparatively low values of γ and χ (Figs.6.6,7). In this respect some authors drew the conclusion that (see, for example, [6.10]) the high-temperature properties are not associated with the location of the Fermi level in the vicinity of the peak on the N(E)

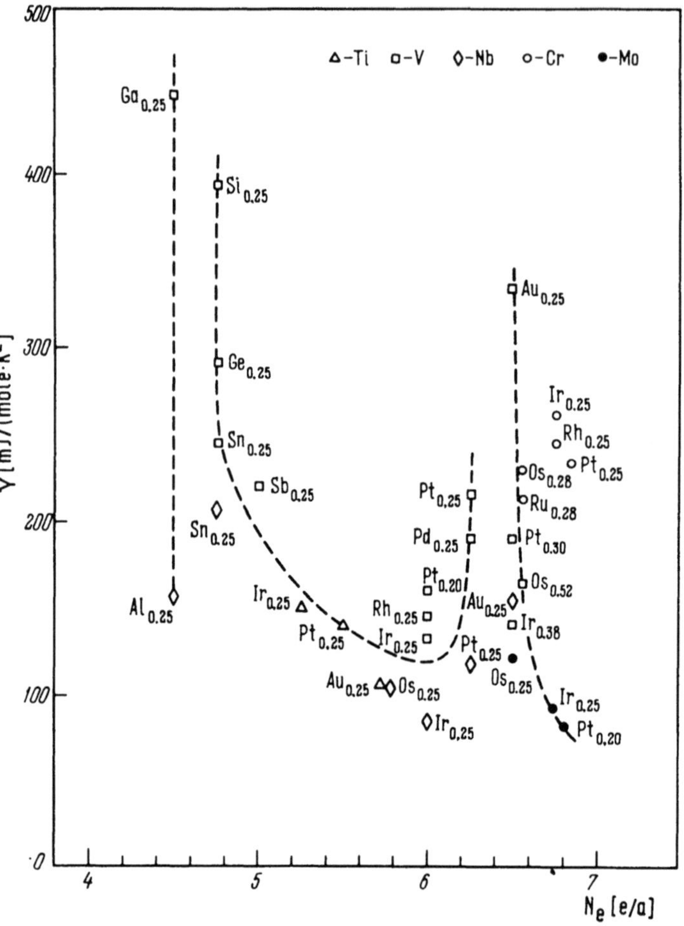

Fig.6.7. The paramagnetic susceptibility of compounds with A-15 structure as a function of N_e. The legend is the same as in Fig.6.5

curve but are possibly determined by other factors such as, for example, the structure of the vibrational spectrum. We wish to draw the reader's attention to the following circumstance. Although the values of γ in A_3B compounds of niobium with high T_c are not as large as might be expected by analogy with vanadium compounds, the correlation between T_c and γ in the series Nb_3Os - Nb_3Ir - Nb_3Pt - Nb_3Au [6.94] (Figs.6.5,6) is nevertheless general, i.e., higher values of γ correspond to higher T_c.

Thus, although the electron concentration is not the key with the help of which one can search for A-15 alloys with higher T_c, it is nevertheless useful to consider a number of physical properties as functions of it. This consideration leads us to the important conclusion that high values of T_c in the compounds are accompanied by a high density of electron states on the Fermi level.

6.1.4 The Temperature Dependence of the Susceptibility, the Knight Shift and X-Ray and Optical Spectra

Many compounds with high T_c exhibit an interesting temperature dependence of such properties as the susceptibility and the Knight shift. It is well known that the usual Pauli susceptibility and the Knight shift (which is proportional to susceptibility) should not depend on temperature for a degenerate electron system. This is the case for most of the metals. The compounds with A-15 structure have been found to exhibit a strong temperature dependence of these characteristics and it is most interesting that this applies exactly to those compounds with high T_c. Does this imply that whereever the temperature dependences of susceptibility and Knight shift are stronger one should necessarily expect high values of T_c? Experimental research [6.95-98] shows that such correlation does occur for V_3X compounds ($X = Si$, Ga, Ge, Sn, As and Sb)(Fig.6.8). For example, the strong temperature dependence of $K_V(T)$ and $\chi(T)$ in V_3Si is accompanied by a high value of T_c (17 K) and, on the contrary, in V_3As ($T_c < 1$ K) and V_3Sb ($T_c = 0.8$ K) the above dependences are almost absent.

However, the temperature dependence of susceptibility in niobium compounds has turned out to be slight, and the Knight shift with temperature is of a more complicated nature (Fig.6.8) than in V_3X compounds [6.10,20,99-104]. It has not been possible to establish any correlation of these characteristics with the superconducting properties of Nb_3X compounds. Thus, it may also be noted for this group of properties that niobium compounds display a behaviour different from that exhibited by vanadium compounds.

Let us return to the treatment of the peculiar dependence of susceptibility and of the Knight shift in V_3X compounds. The strong temperature dependence of $K_V(T)$ and $\chi(T)$ observed in compounds with high T_c indicates that the Fermi level obviously lies near a narrow peak in the density of states, which is so narrow that its width is comparable with the energy kT [6.97]. What type of electron states can produce such a peak? A number of other experiments enable us to obtain such information, primarily X-ray spectra. X-ray photoelectron and X-ray emission spectra of different series of both components of V_3X and Nb_3X compounds make it possible to determine the character of the distribution of states of different types in the occupied portion of the conduction band. Such measurements [6.105-107] have shown that the V3d and Nb4d states emerge on the

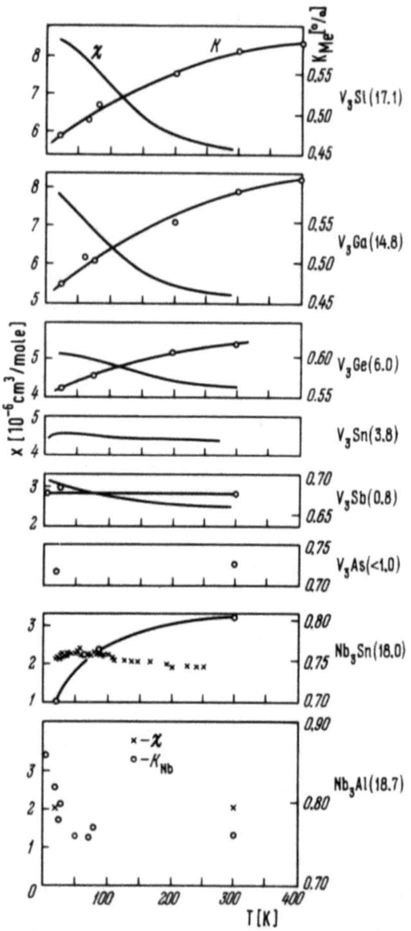

Fig.6.8. The temperature dependence of the susceptibility and the Knight shift in A_3B compounds of vanadium and niobium (B is a nontransition element) according to [6.10,91,94,95]. For Nb compounds χ is shown by crosses, K by circles. The T_c of these compounds is in brackets

Fermi surface in all the V_3X compounds (X - Si, Ga, Ge, Sn, Rh, Pd, Os, Ir, Pt and Au) and Nb_3X compounds (X - Os, Ir, Pt and Au); they form the above-mentioned density-of-states peak. The electron states of the X component normally lie near the bottom of the valence band, and the degree of their localization depends on the arrangement of the X element with respect to vanadium in the periodic table.

Comparison of the X-ray photoelectron spectra of Nb_3Au and V_3Au compounds (Fig.6.9) shows that they are very much alike [6.107]. The separation of the Au5d and Nb4d bands practically coincides with that of the Au5d and V3d bands in V_3Au. In other V_3X and Nb_3X compounds with the same X components (X - Os, Ir and Pt), the partition of energy of spectrum subbands also turns out to be close (see Table 6.2). This shows that the electron structures of Nb_3X and V_3X compounds are of the same character.

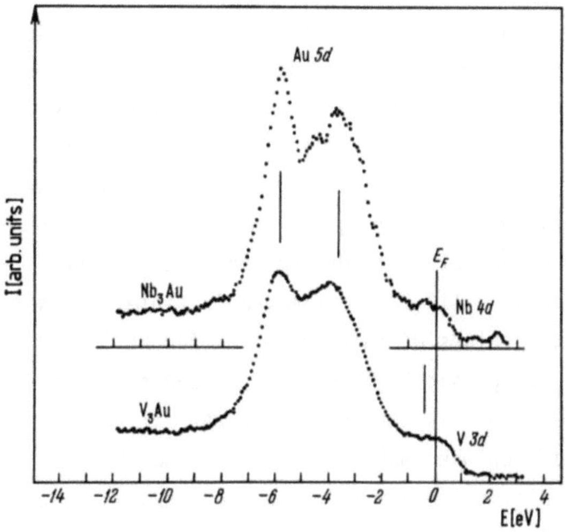

Fig.6.9. The X-ray photoelectron spectra of the valence band in V_3Au and Nb_3Au

Table 6.2 Energy separation of spectrum subbands in Nb_3X and V_3X compounds due to X-ray emission and photoelectron spectra

Compound	$E_{4d} - E_{5d}$ [eV]	Compound	$E_{3d} - E_{5d}$ [eV]
Nb_3Os	2.2 ± 0.4	VOs	1.5 ± 0.3
Nb_3Ir	2.6 ± 0.4	V_3Ir	2.7 ± 0.2
Nb_3Pt	3.3 ± 0.3	V_3Pt	3.8 ± 0.2
Nb_3Au	4.5 ± 0.2	V_3Au	4.5 ± 0.2 4.7 ± 0.2

Obtaining similar information from optical spectra is associated with considerable difficulties arising from ambiguous interpretation of experimental evidence. However, from optical measurements one can obtain other important parameters such as, for example, conduction electron concentration. MOTULEVITCH et al. [6.108,109] have found that for pure Nb, as well as V_3Ga and Nb_3Sn, the concentration of conduction electrons decreases from 4.5×10^{22} to 2.2×10^{22} cm^{-3} and 0.7×10^{22} cm^{-3}, respectively. It is of interest that the T_c of the above-mentioned compounds increases in the same direction, too. Those authors [6.109] expected the subsequent decrease in conduction electron concentration in A-15 compounds below 0.7×10^{22} cm^{-3} to lead to a further growth of T_c, but this decrease should not be very large, so as not to pass over to the region of "semiconductor" concentrations.

Optical measurements are also available because they permit determination of the energy gap in superconductors. Thus, for example, measurement of the optical properties of the compound Nb_3Sn in the infrared spectrum range gave a relative value of the energy gap of $2\Delta/T_c = 3.77$ which is close in magnitude to the value from the BCS theory ($2\Delta/T_c = 3.52$). However, the infrared reflection method is not the only technique of measuring the energy gap, therefore it is of interest to compare the values of $2\Delta/T_c$ determined by different methods. Consideration of these data [6.110-124] collected in Table 6.3 shows that there is a large spread in the values of this parameter for Nb_3Sn and V_3Si. The relative values of the energy gap $2\Delta/T_c$ range from 0.2 to 4.8 for Nb_3Sn and from 1.0 to 3.8 for V_3Si. What is the reason for the different values of the energy gap obtained even when making use of one and the same method of investigation (for example, tunneling)? On the one hand, as shown in [6.113], this may arise from the anisotropy of the energy gap as a function of the orientation of the crystal with respect to the tunnel contact. In the case of polycrystalline specimens a set of different orientations will occur and, as a consequence, several values for the gap [6.116,124]. On the other hand, since tunneling experiments measure the energy gap within the coherence length ξ from the surface, it is probable that in compounds with A-15 structure, which have small ξ ($\xi \sim 100$ Å for Nb_3Sn), the values of the gap are to some extent determined by the state of the surface and by the impurities present there. For example, it has been shown in [6.117] that a disturbance of the surface of the Nb_3Sn specimen leads to a decrease in the value of the gap ($2\Delta/T_c$) from 3.77 to 1.9.

Ingeneous energy gap measurements for superconducting Nb_3Sn have been carried out by AXE and SHIRANE [6.121]. Using inelastic neutron scattering they studied the decay of phonons at low temperatures and found that for some transverse phonons the width of the phonon decay peak decreased sharply at a temperature below T_c. This behaviour of the decay may be attributed to the decay of a phonon into two quasiparticles in the superconductor. This process evidently has a threshold of $2\Delta(T)$, therefore when the energy of the phonon is less than the above-mentioned threshold its decay should decrease drastically. The data for $2\Delta/T_c$ [6.121] given in Table 6.3 correspond to a phonon with a wavevector $\sim$ [110]; that is why the obtained value of the gap corresponds exactly to this direction in k-space. As for other techniques, most of them give some averaging over the anisotropy of the gap characteristic.

Energy gap measurements certainly provide valuable information, but it remains an unfortunate fact that the data obtained, particularly for the com-

Table 6.3 Energy gap in compounds with A-15 structure

Compound	T_c [K]	$2\Delta/T_c$	Measurement temperature [K]	Method of measurements	References
Nb_3Sn	17.5	1.3	1.7	Tunneling	[6.110]
	-	3.6	1.2	Tunneling	[6.114]
	14.0	2.3-3.1	1.5	Tunneling	[6.115]
	17.8-18.3	0.2; 1.0 1.4; 3.0	1.9	Tunneling	[6.116]
	18.3	1.0-2.8	2.0-4.2	Tunneling	[6.113]
	18.3	3.77	5.0; 78	Infrared reflection	[6.117]
	18.3	3.56	2.6-2.8	Heat conduction	[6.118]
	-	3.8	2.0-40	Photoluminescence	[6.119]
	18.0	4.8	1.0-42	Heat capacity	[6.120]
	18.3	4.4	-	Inelastic neutron scattering	[6.121]
V_3Si	17.0	3.8	1.3	Tunneling	[6.122]
		1.8; 3.8	1.2	Tunneling	[6.114]
		1.0; 3.8	4.2	Infrared reflection	[6.123]
	15.7	3.76			[6.111]
	16.9	2.64-3.63	4.2	Tunneling	[6.112]
V_3Ge	-	3.2	1.2	Tunneling	[6.114]

pound Nb_3Sn, are contradictory. If the many-valued character of the results in tunneling experiments is due to the existence of a strong anisotropy of the Fermi surface, it would be desirable to determine on single crystals to which directions of $\underline{k}$ space the obtained data belong.

6.1.5 Structural Transformation

In 1963, SHULL [6.125] found in his neutron diffraction study of V_3Si that the diffraction lines widened below T_c. This is considered to be the first observation of a structural transformation in A-15 compounds. Later, BATTERMAN and BARRETT [6.126] made a thorough low-temperature X-ray analysis of a V_3Si monocrystal and found that at a temperature of $\sim$ 21 K the crystal undergoes a transformation from a cubic to a tetragonal structure. A similar structural transformation was also found in another A-15 compound, Nb_3Sn [6.127,128] (Table 6.4).

Table 6.4 Structural transformation temperature in A-15 type compounds

Compound	Structural transformation temperature [K]	T_c [K]	Tetragonality c/a	References
V_3Si	21	17	1.0024	[6.126]
Nb_3Sn	43	18	0.9939	[6.127]
V_3Ga	>50	14.5	-	[6.129]
Nb_3Al	80	17.9	-	[6.130]
$Nb_3(Al_{0.75}Ge_{0.25})$	105	18.5	0.997	[6.130]
$Nb_{3.1}(Al_{0.7}Ge_{0.3})$	130	17.4		[6.130]

The main characteristic features of this transformation are as follows.
i) Diffusionlessness, i.e., the composition of the initial (cubic) and final (tetragonal) phases is the same. This means that during the transformation only a readjustment of the crystal lattice occurs, without the variation of composition due to the diffusion of the atoms of the components.
ii) No appreciable change in crystal volume has been registered in the process of transformation.
iii) The structural transformation temperature always exceeds the superconducting transition temperature.
iv) The transformation is reversible, i.e., when heated to a temperature exceeding the structural transition point the crystal always returns to its initial orientation (absence of hysteresis effects).
v) The structural transformation temperature may vary from specimen to specimen: for V_3Si between 18 and 30 K, for Nb_3Sn between 35 and 50 K. No structural transformation has been detected in a number of V_3Si and Nb_3Sn specimens.
vi) The single crystallinity of the material is not conserved when transformed from a cubic structure to a tetragonal one. A domain structure twin has been detected in the tetragonal state.

Apart from the enumerated common features of the compounds V_3Si and Nb_3Sn, there is one substantial difference: in V_3Si the ratio of $c/a > 1$, whereas in the tetragonal structure of Nb_3Sn the ratio of $c/a < 1$ (Table 6.4).

The diffusionlessness, the absence of volume variations and some other features resembled the well studied martensitic transformation in the alloy In - Tl [6.131]. In this connection BATTERMAN and BARRETT [6.126] called the structural transformation in V_3Si and Nb_3Sn also a martensitic one.

In addition to V_3Si and Nb_3Sn, a martensitic transformation was registered in the compounds $Nb_3(Al_{0.75}Ge_{0.25})$ [6.130], Nb_3Al [6.130] and V_3Ga [6.129]. In other binary compounds with A-15 structure, no structural transformation was observed at low temperatures. Thus, so far the martensitic transformation has only been observed in those compounds which have high T_c. Has this anything to do with high-temperature superconductivity? This problem will be discussed below.

The first X-ray structure measurements have shown [6.126,127] that at a temperature below the martensitic transformation temperature the ratio of c/a increases rapidly, reaching saturation at a temperature near T_c. (This conveys the impression that the occurrence of a superconducting state inhibits further growth of the tetragonal deformation). No discontinuous jump of c/a was observed in those experiments and it was accordingly assumed that this transformation may be regarded as being a second-order phase transition. However, the more elaborate measurements by VIELAND et al. [6.132] have shown that a region of temperatures where the tetragonal deformation undergoes a jump exists (Fig.6.10), which indicates that this is a first-order phase transition [6.133,134].

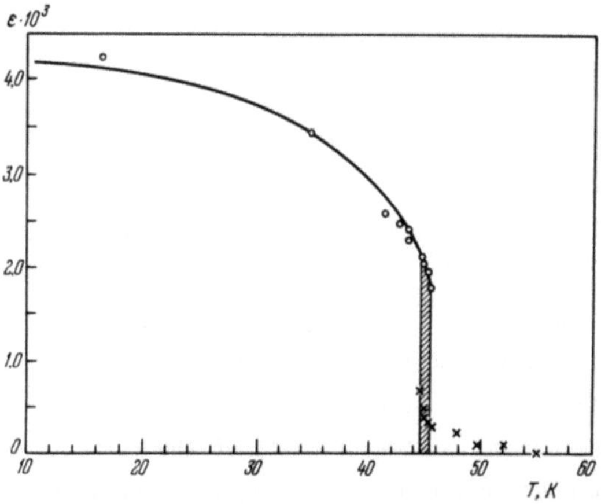

Fig.6.10. The temperature dependence of tetragonal deformation $\varepsilon_0 = c/a - 1$ for the compound Nb_3Sn. The solid line gives the results of the calculation according to the data of [6.132]. On the ordinate the quantities $\varepsilon_0 \cdot 10^3$ are plotted

In [6.135-138] the influence of residual stresses on the transition of a number of binary A-15 compounds and their alloys to a superconducting state has been investigated. It has been found that the superconducting transition is smeared out and that an additional low-temperature step on the transition curves forms, which is attributed to an intensive development of martensitic transformation in specimens by the action of stresses. The low-temperature step on the transition curve is ascribed to the transition of the martensitic phase to a superconducting state. This agrees with the general character of the processes going on at usual martensitic transformations. In the presence of plastic deformations T_m increases. Connected with this is an increase of the martensitic phase at a chosen temperature. However, for a final conclusion concerning the influence of residual stresses or of martensitic transformation in the above compounds and alloys, straightforward X-ray structure investigations are required.

An interesting experimental study of the effect of a magnetic field on the martensitic transformation in V_3Si has recently been made by MAITA and BUCHER [6.139]. Heat capacity measurements have been made in zero field and in a field of 90 kOe applied along the [001] direction of the crystal. In a magnetic field the martensitic transformation point shifts towards lower temperatures: $\Delta T_m = -0.26$ K.

The estimates by DIETRICH and FULDE [6.140] made for V_3Si using the Labbé-Friedel model yielded $\Delta T_m = -0.31$ K in a field of $H = 90$ kOe. A more complete measurement of $T_m(H)$ in fields of up to 156 kOe [6.141] has revealed a quadratic dependence of ΔT_m on H, i.e.,

$$\frac{\Delta T_m}{T_m(0)} = -\alpha\left(\frac{\mu_0 H}{kT_m(0)}\right)^2 \tag{6.1.1}$$

where α has turned out to be equal to 0.10. In an earlier paper by MAITA and BUCHER [6.139] the value obtained was $\alpha = 0.15$. This difference is apparently due to the purity of the V_3Si specimens.

The sign of the effect ($\Delta T_m < 0$) denotes the stabilization of the cubic state in a high magnetic field, which must be accompanied by an increase (at a given temperature) of the shear modulus. Measurements of the velocity of sound corresponding to the soft mode in a nontransforming V_3Si specimen as a function of a magnetic field have shown [6.142] that the velocity does indeed increase with a field according to a quadratic law, in exact agreement with theoretical predictions [6.140].

What is the role which alloying plays in these phenomena? VIELAND [6.143] has investigated the superconducting and structural transitions in the alloy $Nb_3(Sn_{1-x}Sb_x)$ at $0 < x \leq 0.15$. It has turned out that alloying of a Nb_3Sn compound decreases T_c but practically does not change T_m. However, when a small amount of Sb (with $x = 0.15$) is added, c/a becomes >1, whereas in pure Nb_3Sn $c/a < 1$. Thus, in this case the reversal of the sign $(c/a - 1)$ is due to alloying. It is useful to note that theory [6.144] also predicts the reversal of the sign $(c/a - 1)$ in V_3Si when it is placed into a high enough magnetic field: $H = 275$ kOe at $T = 17.4$ and $H = 298$ kOe with $T \to 0$ K.

When Nb_3Sn was alloyed, the nonmetallic atoms occupying the bcc positions in a structure such as A-15 were substituted in $Nb_3(Sn_{1-x}Sb_x)$. What will be the result of the influence of alloying on T_c and T_m if we substitute the atoms in the chains? An answer to this question is given in [6.145] in which the superconducting properties and elastic properties of $(V_{1-x}Ti_x)_3Si$ alloys $(0 < x < 0.10)$ were studied. When the compound V_3Si was alloyed, the lowering of T_c was accompanied by a shift of the minimum in the temperature dependence of the velocity of sound V_{min} towards lower temperatures.

The available information on the influence of alloying on T_m is quite inadequate to permit any conclusions. In contrast to the influence on T_c, which was widely discussed in the literature (see, for example, [6.146]), there are only several isolated papers dealing with the effect of alloying on T_m. It should be noted that the correlation between T_c and T_m in alloyed compounds may prove to be highly indeterminate since, apart from the effect of the variation of the elastic lattice properties influencing T_c, the superconducting transition temperature may also vary due to purely electronic properties, for example, as a result of interband scattering on impurities. However, a study of the variation of the elastic properties of alloyed A-15 compounds itself is of considerable interest, particularly in connection with the theoretical models considered in Sect.6.3.

It is worth noting that at temperatures near the structural transformation temperature anomalies in the behaviour of such physical properties as electronic heat capacity [6.147], magnetic susceptibility [6.139], nuclear magnetic resonance (NMR) [6.148] and the Moessbauer effect [6.149] occur. For example, in studies of NMR spectra of the compound V_3Si, large relative variations of the electric-field gradient were revealed. These variations are two orders of magnitude higher than those in a permanent lattice and differ from the predictions of the point charge model [6.148] almost by a factor of 60. The experimental data concerning the temperature dependence of the probability of resonance absorption of γ quanta in Nb_3Sn indicate that the lattice of this compound is highly anharmonic [6.149].

Measurements of the thermal expansion in V_3Si showed that this parameter behaved anomalously in a region of temperatures exceeding the structural transition temperature by 50 K [6.150,151]. This was ascribed to the effect of a precursor of the martensitic transformation. A similar result was also obtained when measuring the NMR spectrum of ^{51}V in a V_3Si monocrystal [6.152]. X-ray measurements for V_3Si also registered a widening of the (200) and (400) Bragg reflections, starting from 60 K to T_m [6.153]. This effect is assumed to be associated with the fluctuations of the d-interplane distances and points to the forerunner of the martensitic transformation and to structural instability.

6.1.6 Acoustic Properties

Some high-temperature superconducting compounds such as A-15 and, as a matter of fact, those developing a structural transition, exhibit anomalies in elastic properties. The temperature dependences of the velocity of sound for polycrystalline specimens of a number of A-15 structure compounds, measured in the interval of 4.2 to 300 K [6.154], are shown in Fig.6.11. It may be concluded from these data that for compounds with comparatively high T_c (>14 K) the velocities of sound decrease anomalously with decreasing temperature. This indicates the softening of lattice moduli.

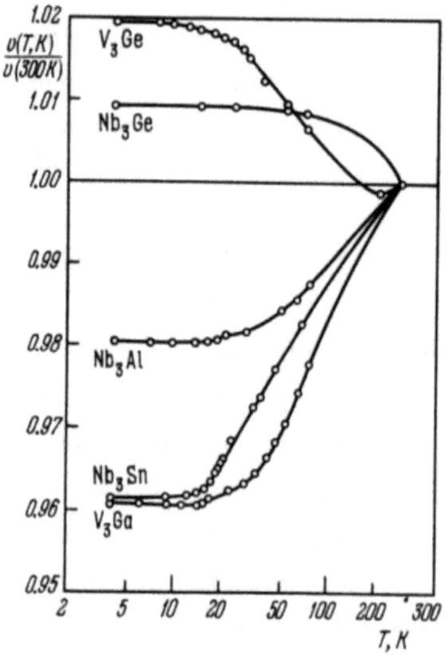

Fig.6.11. Temperature dependence of the velocity of sound for A_3B compounds

The inelastic moduli C_{ij} were calculated from measurements of the velocity of sound in single crystalline compounds V_3Si, Nb_3Sn and V_3Ge for different directions and polarizations. The data[1], borrowed from [6.155-158], are given in Fig.6.12. It is worth to notice the absolutely anomalous temperature dependence of the elastic moduli C_{11} and $(C_{11}-C_{12})/2$ in V_3Si and Nb_3Sn at low temperatures. The temperature dependences of these moduli in V_3Ge is similar to those observed for most of the metals. For all three compounds, the variation of C_{44} turned out to be the slightest. Special mention should be made of the large temperature variation of the shear modulus $(C_{11}-C_{12})/2$ which almost goes to zero in the compound V_3Si and is highly softened in Nb_3Sn. The softening of shear modes was also registered in measurements of the temperature dependence of the derivative of the shear modulus with respect to pressure [6.159], and in TESTARDI's absolutely unique experiment which revealed the generation of second ultrasound harmonics by shear waves in V_3Si [6.160]. It is apparently the lattice softening in V_3Si and Nb_3Sn and the instability of the shear phonon modes which lead to the martensitic phase transition in these compounds.

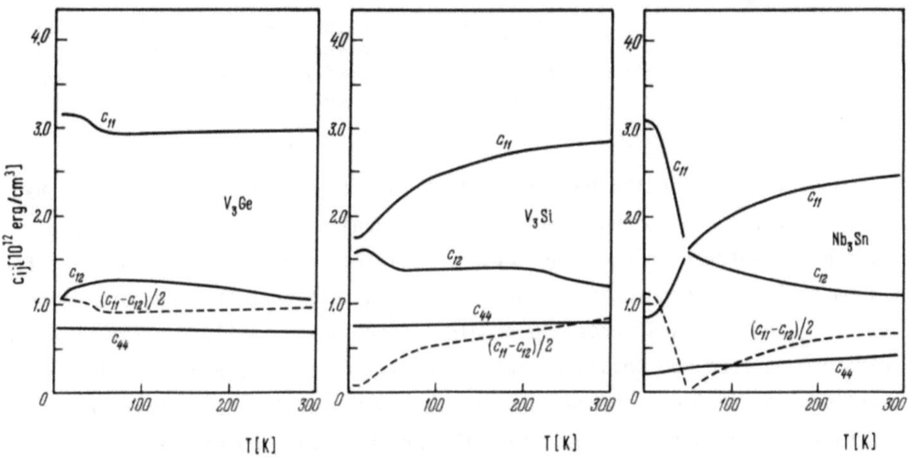

Fig.6.12. Temperature dependence of the elastic moduli for V_3Ge, V_3Si, Nb_3Sn

Measurements of the temperature dependence of the derivative of the shear modulus $C_s = (C_{11}-C_{12})/2$ with respect to pressure, $(\partial C_s/\partial P)$, for V_3Si specimens undergoing and not undergoing a structural transformation [6.161,162] have

[1] In TESTARDI's [6.156] original paper, the temperature of the shear modulus in V_3Si was given in a logarithmic scale (Fig.6.30). It may be inferred that for T appreciably exceeding T_m the logarithmic dependence $C_{ij}(T) = a + b\ln T$ holds

shown that the above dependences coincide only in the range 296-96 K where $\partial C_s/\partial P$ has a positive sign and decreases in magnitude, reaching zero at 80 to 90 K (Fig.6.13a,b). With the temperature increasing further, $\partial C_s/\partial P$ changes sign and decreases continuously to -5 at 13.5 K in a nontransforming V_3Si specimen (Fig.6.13b). For a specimen exhibiting a structural transformation, $\partial C_s/\partial P$ also changes sign at $T < 80$ K but then decreases only to 62 K. With a further decrease in temperature it changes sign again and increases in magnitude already in the range of positive values (Fig.13b).

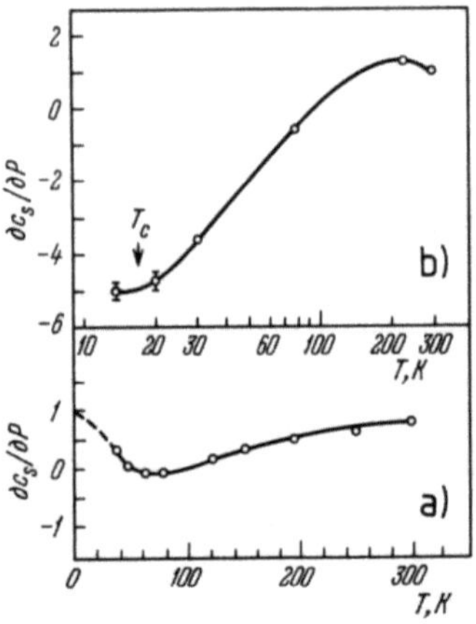

Fig.6.13. Temperature dependence of the first derivative of the shear modulus with respect to pressure in a transforming (a) and a nontransforming (b) V_3Si specimen

This difference in the behaviour of $\partial C_s/\partial P$ at low temperatures means that in one case (in a nontransforming V_3Si specimen) the pressure softens the shear mode, whereas in the other cases (in a transforming specimen) it stiffens the mode. In this connection it was assumed [6.161] that the pressure dependences of T_c in the cases concerned should differ substantially ($\partial T_c/\partial P$ of a nontransforming specimen should be approximately 16 times larger than that of a transforming specimen, according to the estimates of [6.161]). However, direct experiments have not confirmed this. Measurements made on the same specimens which were used to study the temperature dependences $\partial C_s/\partial P$ presented in Fig.6.13 have shown that, instead of an appreciable increase, a decrease in $\partial T_c/\partial P$ for a V_3Si specimen exhibiting no transformation [6.163] takes place.

The presence of negative values of $\partial C_s/\partial P$ at certain temperatures indicates that the shear mode may be "zeroed" and a structural transformation may occur at these temperatures when the hydrostatic pressure is sufficiently high. In this case the behaviour of $\partial T_m/\partial P$ will be determined entirely by the variation of $\partial C_s/\partial P$ in this temperature range. From the condition $C_s(P,T_m) = 0$ it is possible to obtain [6.161]

$$\frac{\partial T_m}{\partial P} = -\left(\frac{\partial C_s}{\partial P}\right)_T \Bigg/ \left(\frac{\partial C_s}{\partial T}\right)_P \quad . \tag{6.1.2}$$

Since for V_3Si, $(\partial C_s/\partial T)_p = 20$ kbar/K [6.161] the sign of $\partial T_m/\partial P$ depends only on the sign of $\partial C_s/\partial P$. For a transforming V_3Si specimen $\partial C_s/\partial P > 0$ for $T < 50$ K, hence $\partial T_m/\partial P < 0$. Experiment [6.164] yielded $\partial T_m/\partial P = -1.5 \cdot 10^{-4}$ K/bar in accordance with this prediction (Fig.6.14). The same figure depicts the variation of T_c with pressure [6.165]. It is seen that in the region of pressures under investigation, $\partial T_c/\partial P > 0$ and $\partial T_m/\partial P < 0$. For the compound Nb_3Sn, the behaviour of T_c and T_m in this pressure range has turned out to be absolutely different. According to [6.164], T_c decreased and T_m increased when pressure was applied (Fig.6.15). It remains an unfortunate fact that for Nb_3Sn no measurements of the temperature dependence of the elastic moduli under pressure have been made. By analogy with V_3Si it may be assumed that $\partial C_s/\partial P$ has a negative sign in this temperature interval.

The variation of the quantity $\partial C_{ij}/\partial T$ for $T = T_c$ at a transition from a normal state to a superconducting state in V_3Si and V_3Ge compounds is also anomalously large. (The normal state in these compounds was maintained by applying a magnetic field of about 23 kOe which destroyed superconductivity at low temperatures). The jump of the derivative of the shear modulus with respect to temperature in V_3Ge and V_3Si has turned out to be, respectively, two and four orders higher than the values typical for most of the superconductors.

The attenuation of sound with temperature for the compounds V_3Si and Nb_3Sn is also anomalous [6.156,166] (Fig.6.16). At temperatures of about several tens of degrees Kelvin the attenuation is inappreciable and increases slowly

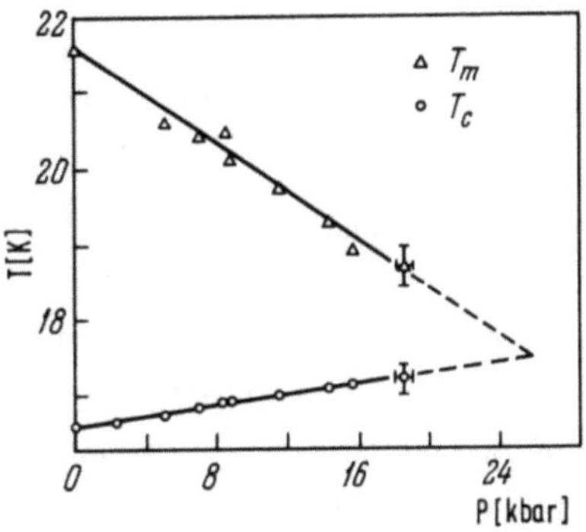

Fig.6.14. The influence of pressure on the superconducting and structural transition temperatures in V_3Si

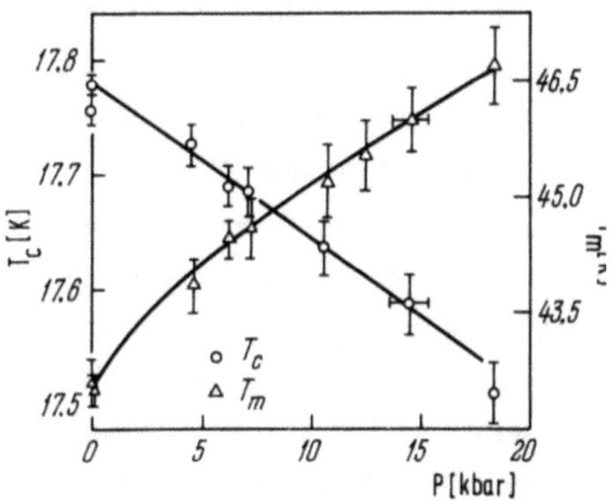

Fig.6.15. The influence of pressure on the temperature of superconducting and structural transitions in Nb_3Sn

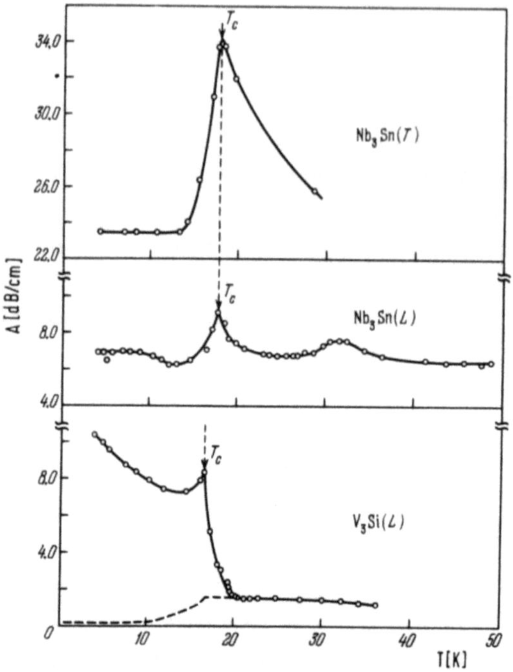

Fig.6.16. Damping of transverse (T) and longitudinal (L) sound waves as a function of temperature in the compounds V3Si and Nb3Sn. The dotted line shows the temperature dependence of the damping of sound for a typical superconductor

with decreasing temperature. At a temperature of near 32 K Nb_3Sn exhibits a sharp peak in the attenuation, which is apparently associated with the structural transformation from the cubic phase to a tetragonal one. In addition, at the superconducting transition temperature (~18 K) another sharp peak of the attenuation occurs. For V_3Si the martensitic transformation temperature T_m and the temperature of transition to the superconducting state T_c are separated by only four degrees Kelvin, so it is possible, that two attenuation peaks may be superimposed on each other.

Thus, it follows from an analysis of the propagation and attenuation of sound that A-15 structure compounds exhibit quite unusual behaviour of the elastic characteristics of the crystal. What is most interesting is that many anomalies in the elastic properties occur in high-temperature superconductors. This means: either a high T_c is associated with the instability of the lattice (in the cubic phase) or both characteristics are a consequence of one and the same cause, for example, anomalies of the electron spectrum.

6.1.7 The Phonon Density of States

The inelastic scattering of neutrons has been measured for five compounds with an A-15 structure (V_3Si, V_3Ge, V_3Ga, Nb_3Sn, Nb_3Al). These measurements, performed at 297 and 77 K, have permitted the determination of the phonon density of states $F(\omega)$ and of the frequency moments of the components $\langle\omega^2\rangle_i$ ($i = A,B$). They made it possible to estimate the effect of phonon mode softening at low temperatures [6.167]. The phonon densities of states of all the compounds enumerated above turned out to be very close. Special measurements of the angular dependence of inelastic scattering for V_3X compounds have made it possible to decipher the spectrum structure and to single out the elements associated with the oscillations of the atoms of the nontransition B and transition A elements. It has turned out that the main peak in $F(\omega)$, which is genetically related to oscillations of the A atoms, in A_3B compounds with $N_e = 4.5$ e/a (V_3Ga, Nb_3Al) is shifted into the high-frequency region of the spectrum, as compared with A_3B compounds with $N_e = 4.75$ e/a (V_3Si, V_3Ge, Nb_3Sn). This indicates that the interatomic forces in A-15 compounds with $N_e = 4.5$ e/a are 20% lower than those in compounds with $N_e = 4.75$ e/a. Another important result of the measurements is that the constituent components of the lattice rigidity factor are approximately equal: $M_A\langle\omega^2\rangle_A \approx M_B\langle\omega^2\rangle_B$, and $M_A\langle\omega^2\rangle_A$ is almost invariable for compounds with the same electron concentration.

If we try to analyze the variation of T_c in A_3B compounds within the framework of McMillan's model on the basis of the data obtained for $\langle\omega^2\rangle_i$, we can arrive at the following conclusions:

i) the compounds V_3Si and V_3Ge should exhibit similar T_c,
ii) compounds with $N_e = 4.5$ e/a should have higher T_c than compounds with $N_e = 4.75$ e/a,
iii) all V_3X compounds should have higher T_c compared to Nb_3X compounds.

All these conclusions contradict the experimental evidence and bring into question McMillan's concepts that $M\langle\omega^2\rangle$ plays a decisive role in the formation of λ and T_c and that the factor $\eta=\langle J^2\rangle N(0)$ is constant for this class of materials.

Temperature measurements have shown that the phonon spectrum softens in A_3B compounds. The softening is most pronounced in high-temperature superconductors. However, the change in $M\langle\omega^2\rangle$ resulting from the softening of the phonon spectrum has turned out to be inadequate in accounting for the difference in T_c of the above compounds in McMillan's approximation.

Dealing with the investigation of the phonon spectra in A-15 compounds, we should also mention the theoretical calculations [6.168,169] of the full density of states $F(\omega)$ and of the partial contributions made by sublattices in V_3Si, V_3Ga, V_3Ge and Nb_3Sn compounds. It is also necessary to recall the experiments concerning the tunnelling and Raman scattering measurements in V_3Si [6.112,170,171]. Taken as a whole, the data on $F(\omega)$ obtained in these papers agree with neutron diffraction measurements [6.167], except for the fact that the peak in $F(\omega)$ in the tunnel and the Raman spectra of the compound V_3Si is not reproduced at 42 MeV.

6.1.8 Effect of Pressure on the Superconducting Transition Temperature

SMITH has made superconducting transition temperature measurements for nearly 30 compounds with A-15 structure using highly uniform hydrostatic pressure [6.172]. His results for $\partial T_c/\partial P$ are presented in Fig.6.17. They show interesting peculiarities in the behaviour of $\partial T_c/\partial P$ in these compounds. In the investigated chrome compounds and in most of the vanadium compounds, $\partial T_c/\partial P$ has a positive sign (except in V_3Pt and V_3Ir). At the same time, T_c has been found to decrease with pressure in all the molybdenum compounds and in most of the niobium compounds.

Worthy of notice is the substantial difference in the behaviour of $\partial T_c/\partial P$ for compounds of vanadium and niobium with the same second components and, hence, with the same electron concentration [V_3Ga and Nb_3Ga, V_3Ge and Nb_3Ge, V_3Sn and Nb_3Sn, V_3Ir and Nb_3Ir, $V_3(Al-Ge)$ and $Nb_3(Al-Ge)$]. One more inter-

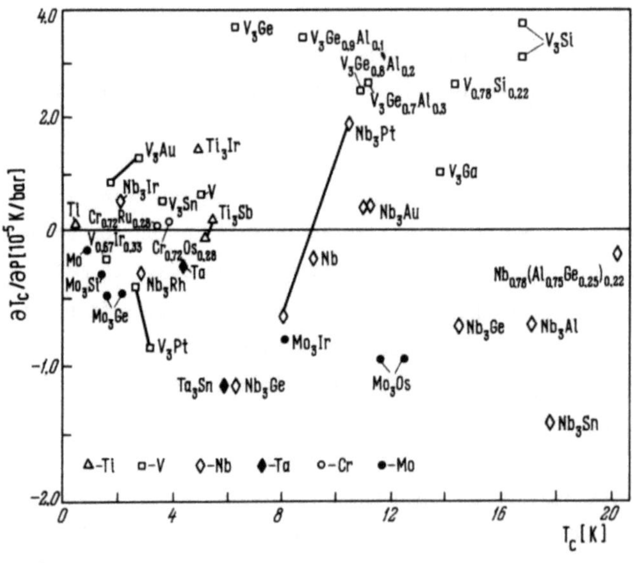

Fig.6.17. The values of $\partial T_c/\partial P$ for some compounds with A-15 structure

esting feature may be noted. The A_3B compounds with A-15 structure, in which 3d-metals (Ti, V, Cr) act as A atoms, show, as a rule, a positive dependence $\partial T_c/\partial P$, whereas for similar compounds of 4d- and 5d-metals (Nb, Mo, Ta) T_c decreases with pressure. The deviation from stoichiometry in the compound V_3Si and the disordering of the compound V_3Au lead to a decrease in the value of T_c with pressure.

Thus, the experimental evidence concerning the influence of pressure on T_c (see also [6.173-176]) which is presented here convincingly demonstrates that there is no universality in the behaviour of $\partial T_c/\partial P$ for various compounds. It is difficult to single out a parameter that might be instrumental in determining a regularity in the variation of $\partial T_c/\partial P$.

It should be noted that the application of hydrostatic pressure also leads to a change in the other characteristics of the crystal, particularly, to a decrease in volume and in the lattice parameter. Let us see whether the same effect may be obtained in a different way, for example, due to alloying. What will the behaviour of the superconducting properties be in this case? Experimental data show that when alloying V_3Si, V_3Ge and V_3Ga compounds it is possible, in contrast to the effect of pressure, to obtain a variation of the lattice parameter in both directions. However, to draw a comparison with experiments involving the use of pressure we shall be concerned only with those data on alloying which refer to the decrease of the lattice parameter. Consider these results. In Fig.6.18 the experimental points correspond to the variation of T_c with the lattice parameter in ternary alloys based on V_3Si, V_3Ga and V_3Ge [6.177]. It is seen that when the lattice parameter decreases the superconducting transition temperature lowers in V_3Ga, but rises in V_3Ge. In other words, from these data one might expect the dependence $\partial T_c/\partial P$ to be positive for V_3Ge and negative for V_3Ga. Measurements of T_c under high pressure show that $\partial T_c/\partial P$ has a positive sign. Thus, the comparison drawn denotes that the influence of alloying should not be reduced to volume variations only.

The dependence of T_c on hydrostatic pressure is a particular case of the problem concerning the deformation dependence $T_c = T_c(\varepsilon)$. In a series of papers [6.155,177,178] TESTARDI has attempted to systematize the ample experimental data on the basis of a phenomenological expansion of T_c in powers of the strain tensor, where ε_j with $j = 1,2,\dots,6$ numbers the independent strain tensor components ε_{xx}, ε_{yy}, ε_{zz}, ε_{xy}, ε_{yz} and ε_{zx}. It was postulated that $T_c(\varepsilon)$ may be represented as a series

$$T_c(\varepsilon) = T_c(0) + \sum_{j=1}^{6} \Gamma_j\varepsilon_j + \sum_{i,j=1}^{6} \Delta_{ij}\varepsilon_i\varepsilon_j + \dots \quad . \tag{6.1.3}$$

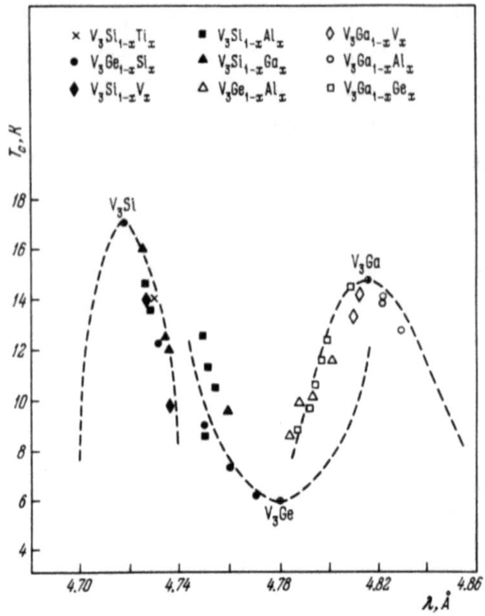

Fig.6.18. The dependence of the transition temperature to the superconducting state for compounds with A-15 structure on the lattice parameter (according to [6.177]

The expansion coefficients Γ_j and Γ_{ij} may be determined independently from a number of experiments. To this end it is necessary to use the expression for the difference of the free energies of the system in the normal and superconducting states near T_c:

$$\Omega^N - \Omega^S = \frac{H_c^2(T,\varepsilon)}{8\pi} = \frac{\alpha^2(\varepsilon)}{8\pi}[T_c(\varepsilon) - T]^2 \quad . \tag{6.1.4}$$

Differentiating this expression with respect to deformations and the temperature we obtain three equations relating the experimentally observed quantities at point T_c to the coefficients Γ_j and Δ_{ij}

$$c_{ij}^S - c_{ij}^N = -\frac{\alpha^2}{4\pi}\Gamma_i\Gamma_j, \quad \frac{dc_{ij}^S}{dT} - \frac{dc_{ij}^N}{dT} = -\frac{\alpha^2}{4\pi}\Delta_{ij}, \quad C_V^N - C_V^S = -\alpha^2 T_c/4\pi \quad . \tag{6.1.5}$$

Here c_{ij}^N and c_{ij}^S are the elastic moduli in the normal and superconducting phases, related to the free energy by

$$c_{ij} = \frac{\partial^2\Omega}{\partial\varepsilon_i\partial\varepsilon_j} \quad , \tag{6.1.6}$$

and C_V^N and C_V^S are the corresponding heat capacities.

After performing complete measurements of the elastic properties of the single crystals V_3Ge and V_3Si [6.155,177], TESTARDI determined the expansion coefficients Γ_j and Δ_{ij},

$$|\Gamma_1| < 50\ \mathrm{K}, \quad \Delta_{11} = -24 \cdot 10^4\ \mathrm{K}, \quad \Delta_{12} = -5 \cdot 10^4\ \mathrm{K}, \quad \Delta_{44} = -1 \cdot 10^4\ \mathrm{K}$$

for V_3Si and

$$|\Gamma_1| < 45\ \mathrm{K}, \quad \Delta_{11} = 9.4 \cdot 10^4\ \mathrm{K}, \quad \Delta_{12} = -1.8 \cdot 10^4\ \mathrm{K}, \quad \Delta_{44} = 0.4 \cdot 10^4\ \mathrm{K}$$

for V_3Ge. Such small (compared to the coefficients Γ_j) values of the coefficients Δ_{ij} obtained for both compounds are indicative of a preferentially quadratic deformation dependence of T_c, which, by convention, is represented by parabolas with the branches down for V_3Si (since $\Delta_{11} < 0$) and with the branches up for $V_3Ge(\Delta_{11} > 0)$. This means that all deformations should increase the superconducting transition temperature in V_3Ge and decrease T_c in V_3Si. Do these predictions agree with the experimental data?

For V_3Si, the various methods of obtaining deformation give a lowering of T_c. Thus, for example, application of uniaxial stress [6.179] as well as alloying by different components specified in Fig.6.18 leads to a decrease in the values of T_c. Further, the spontaneous deformation arising during the structural transition in V_3Si also reduces T_c by 0.1 to 0.4 K [6.177] for specimens undergoing structural transformation, as compared to specimens in which the martensitic transition was not observed. This figure is very close to the value 0.38 K given by TESTARDI's estimate [6.155]. As regards V_3Ge, no direct experiments involving the use of uniaxial pressure have been carried out for this compound, and the decrease in the lattice parameter caused by alloying leads to an increase of T_c (Fig.6.18). It must be noted that the experimental points obtained from the dependence of T_c on the lattice parameter (the data on alloying) give a good fit to the curves predicted by TESTARDI from analysis of the elastic properties of V_3Si and V_3Ge.

In spite of the fact that a number of experimental data have been interpreted successfully it should be noted that TESTARDI's approach is not rigorous or unambiguous. In itself, the possibility of expanding T_c in powers of deformation is problematic since T_c may nonanalytically depend on ε.

V_3Si and V_3Ge, chosen as the object used to experimentally verify the deformation dependence of T_c, are also compounds with extreme values of superconducting transition temperatures. As seen from the above statements, alloying of high-temperature superconductors (including V_3Si and V_3Ga) usually leads to a decrease in T_c. Further, some experimental data, for example, for $(V-Pb)_3Si$ and $(V-Mo)_3Si$ alloys [6.3], which do not fit the parabolic dependences of T_c on the lattice parameter explicitly, are not included in Fig.6.18. Therefore it is possible that the above correlation between the dependence of T_c on the lattice parameter proposed by TESTARDI, and the selected experimental data concerning the variation of T_c in A_3B compounds, is fortuitous.

To draw final conclusions as to the legitimacy of this approach it is worthwhile considering the intermediate case, i.e., the superconductors whose transition temperature amounts to 9 or 10 K and whose alloying could lead both to a decrease and an increase in T_c.

Measurements of T_c and of other physical properties under pressure have made it possible to test the validity of some empirical correlations established for A-15 compounds. For example, it was shown in [6.180] that the defects of the crystal lattice largely affect the properties of these compounds in the normal and superconducting states. It was assumed that V_3Si [6.181] would contain a comparatively high vacancy concentration at room temperature and that hydrostatic pressure would remove the vacancies. The effect of the variation of the vacancy concentration under pressure can be registered by measuring the lifetime of a positron in the material investigated [6.56,182]. Such measurements were carried out for V_3Si and have shown that the lifetime of a positron in V_3Si does not vary at pressure up to 20 kbar [6.183]. This shows that either no vacancies in V_3Si exist or the lifetime of the positrons in a vacancy is equal or approximately equal to the lifetime of the positrons in the volume.

6.2 Band Structure

6.2.1 The Clogston-Jaccarino Model

The experimental evidence outlined above shows that high-temperature superconducting compounds such as V_3Si, V_3Ga, Nb_3Sn, etc. exhibit high values of magnetic susceptibility and electron heat capacity, which indicates a high density of states near the Fermi surface. Apparently, they possess a peak in the density of states and the Fermi level comes within this peak. It is significant that these compounds have appreciable temperature dependences of the magnetic susceptibility and the Knight shift, which indicate that the density of states peak occurring in the vicinity of the Fermi level is extraordinarily narrow, its width being comparable with thermal motion energy kT.

What kind of atomic electron states are capable of forming such a peak? We can investigate this by analyzing all the anomalies of the compounds concerned (as a specific example we will refer to V_3Ga). The anomalies involved are as follows: (i) the Knight shift of vanadium atoms K_V is positive and diminishes with decreasing temperature, (ii) the Knight shift for gallium atoms K_{Ga} is negative and increases in module as T decreases. Recall that the susceptibility increases with decreasing temperature.

To account for these facts, CLOGSTON and JACCARINO [6.97] have proposed the following electron spectrum model for V_3Ga. There is a wide conduction-electron band formed chiefly by 4s- and 4p-states of gallium. This band is overlapped with a narrow band formed by 3d-states of vanadium with the Fermi level passing somewhere within it (Fig.6.19). It is also assumed that 4s-states belonging to Ga atoms form a narrow band lying deep under the Fermi surface. This assumption reflects the fact that the Ga atoms in the lattice are located at considerable distances from each other so that their levels are weakly smeared. Finally, it is assumed that the 4p-states of Ga lie close enough to the Fermi surface so that they mix well with the vanadium bands. To supply an explanation for the strong temperature dependence of susceptibility one has to assume that the width of the density-of-states peak on the Fermi surface amounts to 0.04 eV for V_3Ga. We shall see below how numerical spectrum calculations confirm the correctness of the structure proposed. It will also be observed that the band-structure model proposed for vanadium compounds completely agrees with the X-ray spectroscopy results obtained for these compounds.

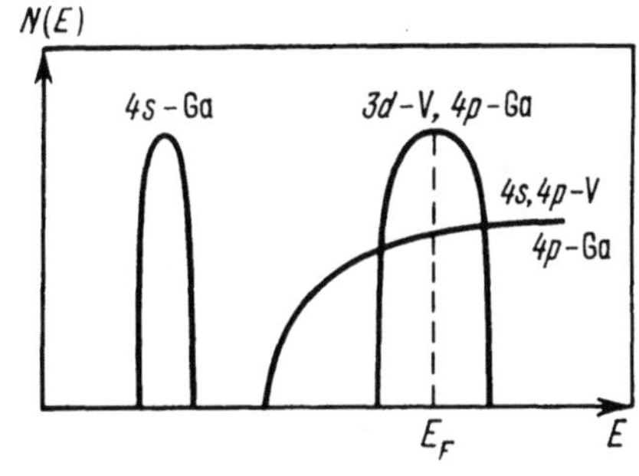

Fig.6.19. Band structure model for V_3Ga according to CLOGSTON and JACCARINO [6.97]

6.2.2 The Noninteracting Chains Model

WEGER [6.184] was the first to attempt to explain the occurrence of a sharp peak on the Fermi surface in A-15 compounds on the basis of the features peculiar to their crystal structure. As stated above, in these compounds the A (transition element) atoms lie on straight lines aligned with [100], the distance between the adjacent atoms on such a straight line being much less than that between the atoms located on different straight lines. In other words, the A atoms form three mutually perpendicular systems of linear chains in the crystal. It is assumed, firstly, that the s-electrons of the transition atoms and the B atoms become collectivized in the crystal, thus forming a wide band (it may comprise valence p-electrons of B atoms), and secondly, that the d-electrons form a narrow band and may be described in the tight-binding approximation. In this manner, a crystal such as V_3Ga is, from the electronic point of view, represented as a set of one-dimensional chains immersed into the s-electrons which stabilize the system [6.185].

In a tight-binding approximation, the d-electron energy for a linear chain may be written as

$$\varepsilon_p = \varepsilon_0 - E_m \cos(pc) \quad , \tag{6.2.1}$$

where ε_0 is the atomic level of the d-electrons, E_m is a doubled matrix element of transition between the next nearest atoms in a chain, which are spaced at a distance c, and p is the quasimomentum.

It can be easily verified that the density of states N(E) of one-dimensional motion has a square root singularity at the band edges. In fact, substituting (6.2.1) into the formula for the density of states accounted for by one atom of the chain

$$N(E) = \frac{c}{2\pi} \int_{-\pi/c}^{\pi/c} \delta(E - \varepsilon_p)\, dp \tag{6.2.2}$$

and integrating, we obtain the E-dependence [6.185]:

$$N(E) = \frac{1}{\pi} \frac{1}{\sqrt{E_m^2 - (E - \varepsilon_0)^2}} \quad . \tag{6.2.3}$$

In the vicinity of the band bottom, (6.2.3) yields a root singularity of the form

$$N(E) = \begin{cases} \dfrac{B}{\sqrt{E - E_m}} \ , & E > E_m \\ 0 \ , & E < E_m \end{cases} \quad , \tag{6.2.4}$$

where the origin of energy is so chosen that $\varepsilon_0 = 0$.

It is now necessary to take into account the angular behaviour of atomic d-states. In the linear chain the five-fold degenerate atomic term breaks down into three degenerate terms uniting the states with magnetic quantum numbers $m = 0, \pm 1, \pm 2$:

$$d_{3z^2-r^2} \quad , \quad \left(d_{xz}, d_{yz}\right) \quad , \quad \left(d_{xy}, d_{x^2-y^2}\right) \tag{6.2.5}$$

(the z axis is aligned with the chain). Each of the orbital levels in the chain is smeared into a one-dimensional band according to the dispersion relation (6.2.1), where the parameter E_m will already depend on the type of orbitals (6.2.5). Assume that the electron spectrum structure for A-15 compounds with high T_c such as V_3Si, Nb_3Sn, etc., has the form of three bands inserted into each other (which correspond to the one-dimensional motion of d-electrons), and of a wide band of s-electrons collectivized throughout the crystal (Fig.6.20).

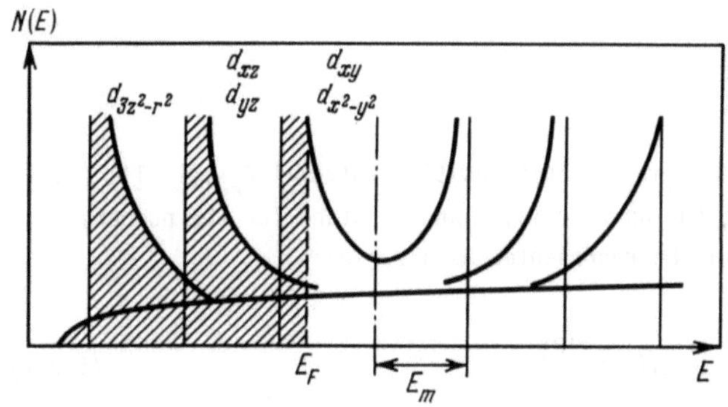

Fig.6.20. Band structure of A-15 compounds according to the Labbé-Friedel [6.185] model

For vanadium compounds, where the number of d-electrons is equal to 3 or 4 per vanadium atom, the d-band is less than half-filled, and the Fermi level may turn out to be near the bottom of the upper-lying d-band, thereby ensuring the highest density of states near the Fermi level. It is precisely this situation which is assumed to take place for V_3Si, Nb_3Sn and a number of other compounds with high superconducting transition temperatures. The contribution to the density of states which is made by the other d-subbands as well as by the s-band in the vicinity of the Fermi level may be neglected. It will thus become possible to approximate the electron spectrum of such compounds by the one-dimensional dispersion relation (6.2.1) and by the density of states in the form of (6.2.4).

6.2.3 Electron Structure in the Presence of Interchain Interaction

Before discussing what relationship the Weger-Labbé-Friedel model bears to reality, we supplement it with the interchain electron transitions which we shall, as before, describe in the tight-binding approximation. Just as before, we assume the d-electrons to be capable of moving only on the sites of the lattice formed by transition metal atoms. However, we shall neglect the real nature of the atomic d-states, viz., their degeneracy, on the assumption that the peculiarities of the A-15 band structure are associated primarily with the space symmetry of the lattice but not with the angular behaviour of atomic wave functions.

In this model the electron spectrum is determined from the conventional secular equation

$$\det\left\{L_{\kappa\kappa'}(\underline{p}) - E\delta_{\kappa\kappa'}\right\} = 0 \quad , \tag{6.2.6}$$

in which the Fourier component of the transition integral $Z_{\kappa\kappa'}(\underline{l}-\underline{l}')$ ($\underline{l}$, $\underline{l}'$ are the radii-vectors of the unit cells, κ and κ' stand for the numbers of the atoms in the cell) may be represented as a series

$$L_{\kappa\kappa'}(\underline{p}) = Z_{\kappa\kappa'}(0) + \sum_{\underline{t}} Z_{\kappa\kappa'}(\underline{t})e^{-i\underline{p}\underline{t}} \quad , \tag{6.2.7}$$

where $\underline{t}$ is the vector of translation to the adjacent cells. Of the matrix elements $Z_{\kappa\kappa'}(\underline{l}-\underline{l}')$, we retain only those which connect either the nearest neighbours ($\underline{1}-\underline{2}$) or the next-nearest ones ($\underline{1}-\underline{3}$). In this fashion, the matrix $L_{\kappa\kappa'}(\underline{p})$ will incorporate only three real parameters, for which we introduce the notation

$$Z_{11}(0) = \varepsilon_0 \quad , \quad Z_{12}(0) = J \quad , \quad Z_{13}(0) = I \quad , \tag{6.2.8}$$

where ε_0 stands for the atomic level, J is the transfer integral along a chain, and I is the transfer integral between the chains.

The structure of the 6×6 matrix $L_{\kappa\kappa'}(\underline{p})$ is as follows

$$L_{\kappa\kappa'} = \begin{pmatrix} A_x & h_y & h_x^+ \\ h_y^+ & A_y & h_z \\ h_x & h_z^+ & A_z \end{pmatrix} , \qquad (6.2.9)$$

where every single element denotes a 2×2 block as, for example,

$$A_x = \begin{pmatrix} \varepsilon_0 & J[1+\exp(ip_x a)] \\ J[1+\exp(-ip_x a)] & \varepsilon_0 \end{pmatrix} ,$$

$$h_x = I[1+\exp(ip_y a)] \begin{pmatrix} 1 & 1 \\ \exp(-ip_z a) & \exp(-ip_z a) \end{pmatrix} . \qquad (6.2.10)$$

The other blocks h_y and h_z are obtained from h_x by cyclic permutation of the coordinates. The symmetry of the matrix (6.2.9) is such that if $E(\underline{p})$ is the root of (6.2.6), then the expression obtained from it by cyclic permutation of the wave vector components $\{p_x, p_y, p_z\}$ will also be the root of the equation.

In the limit $I = 0$, (6.2.6) falls apart into three quadratic equations with the solution[2] (footnote see next page):

$$E_\lambda^\pm(\underline{p}) = \varepsilon_0 \pm 2J \cos\left(\frac{p_\lambda a}{2}\right) \quad (\lambda = x,y,z) \quad . \qquad (6.2.11)$$

Each of the solutions is the energy of an isolated chain oriented in the λ direction. The spectrum determined by (6.2.12) is shown in Fig.6.21 by dotted lines. It may be shown from analysis of the secular equation (6.2.6) with the transition matrix (6.2.9) that allowance for the interchain interaction leads to the splitting of a number of dispersion curves due to the removal of degeneracy and some shifting of the bands (Fig.6.21).

Write out the results of the solution of (6.2.6) for the spectrum branches with an energy close to zero (Fig.6.21).

---

[2] In (6.2.11), a labels the A-15 lattice parameter so that $a/2$ is the distance between the atoms in the chain. This expression yields the electron spectrum of the three-dimensional lattice composed of a set of three mutually perpendicular chains. In contrast to it, (6.2.1) gives the spectrum of an electron in an isolated chain, the closest interatomic distance c coinciding in this case with the period of the one-dimensional lattice.

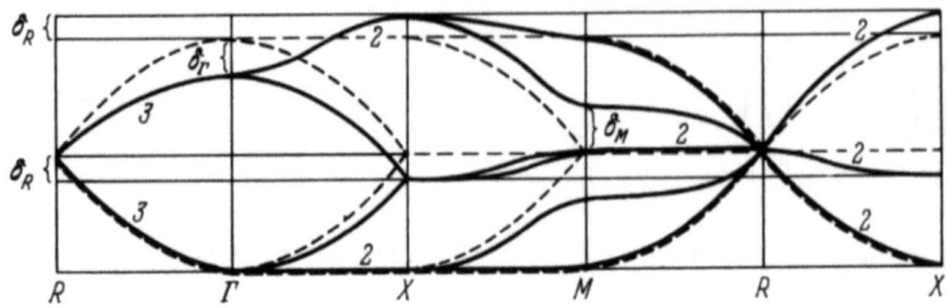

Fig.6.21. Band spectrum in A-15 structure according to the model of weakly interacting chains. The dotted curve represents the spectrum of non-interacting chains. The fortuitous degeneracy in a series of symmetric directions and in symmetric points is removed by the interchain interaction. The energy shifts are $\delta_\Gamma = 8/3\ I$, $\delta_M = 4\ I$ and $\delta_R = 8I^2/J$. The digits near the solid curves indicate the degeneracy multiplicity

For the symmetric directions at the edge of the Brillouin zone we obtain

$$\text{XR:}\quad E_1 = E_2 = -\frac{8I^2}{J}\cos^2\left(\frac{pa}{2}\right) \quad (p = p_x = p_y),$$

$$\text{XM:}\quad E_1 = -\frac{8I^2}{J}\cos^2\left(\frac{p_y a}{2}\right), \quad E_2 = \begin{cases} -8I^2/J \text{ near X} \\ -2J\cos^2\dfrac{p_y a}{2} \text{ near M} \end{cases}, \tag{6.2.12}$$

$$\text{RM:}\quad E_1 = E_2 = 0 \quad .$$

It should, of course, be borne in mind that the present model disregards the angular symmetry of the d-states so that the spectrum related to the matrix in (6.2.9) corresponds, by itself, to the s-states. If account is taken of the five-fold degeneracy of the atomic d-functions, the noninteracting chains spectrum (6.2.12) should contain three parameters ε_{om} and J_m for each value $m = 0, \pm 1, \pm 2$. This will lead to the superposition of the three series of curves portrayed by dotted lines in Fig.6.21. If allowance is made for the matrix elements between the d-states with different m, an additional splitting of states with $m = \pm 1$ and $m = \pm 2$ for the nearest atoms pertaining to different chains occurs. This splitting leads to a complicated picture of the spectrum. The matrix (6.2.6) will then be a 30×30 matrix (six A atoms per cell and five d-states in each atom). In this case the spectrum can only be investigated with the use of computers. However, the use of the tight-binding method is efficient provided that electron spectrum calculations are based on first principles, at least for the symmetric points of the Brillouin zone. We will describe this hybridized technique of investigating the band structure of A-15 compounds below.

6.2.4 Band Structure Calculations from First Principles

In 1965 MATTHEISS [6.186] carried out band structure calculations for a number of V_3X compounds (X = Al, Si, Co, Ga, Ge, As) using the augmented plane waves (APW) method. The spectrum was calculated only at the symmetry points of the Brillouin zone, but for V_3Ga it was also calculated for a number of symmetric directions which permitted construction of the density of states function N(E). The results of his calculations turned out to be in qualitative agreement with the Clogston-Jaccarino model. Indeed, the lowest-lying bands were those preferentially related to the 4s-states of Ga atoms lying deep beneath the Fermi surface. They are separated by an energy gap from the overlapping 3d-bands, formed by the atomic states of V and having a width of about 7 eV, and lie in the middle of the bands which are related to the 4s and 4p-states of V and to the 4p-states of Ga. The 3d-bands are divided into two parts with a deep minimum in N(E) between them. The Fermi level for V_3Ga falls within the peak in N(E), and analysis of the wave functions in the vicinity of E_F showed that they correspond preferentially to the 3d-states of V with a small amount (about 5%) of 4p-states of Ga.

Ten years later MATTHEISS [6.187] performed more elaborate electron structure calculations for four other compounds (V_3Si, V_3Ge, Nb_3Al and Nb_3Sn) on the basis of a non-selfconsistent APW method involving the use of the SLATER-KOSTER [6.188] interpolation procedure. The convergence of these calculations was an order of magnitude better than that of the calculations made in 1965, and this enabled him to obtain more exact values of the electron spectrum characteristics. An important result of that calculation was the establishment of the fact that the value of N(0) in V_3X compounds is appreciably higher than that in Nb_3X compounds with smaller values of T_c. As well as this, he estimated the applicability of the quasi-one-dimensional model of the electron structure in A-15 compounds. One-dimensional effects may show up in the electron structure of these compounds if the matrix elements of transition between the nearest A atoms are much greater than the matrix elements for the next-nearest atoms. MATTHEISS' calculation revealed that, for example, in Nb_3Sn

$$(dd\sigma)_1 = -0.096 \quad ; \quad (dd\pi)_1 = 0.094 \quad ; \quad (dd\delta)_1 = -0.027 \quad ;$$
$$(dd\sigma)_2 = -0.066 \quad ; \quad (dd\pi)_2 = 0.033 \quad ; \quad (dd\delta)_2 = -0.008 \quad , \qquad (6.2.13)$$

where the upper row refers to the nearest neighbours and the lower one pertains to the next-nearest atoms [σ, π and δ are the atomic functions of the d-states with symmetry like $d_{3z^2-r^2}$, (d_{xz}, d_{yz}) and $d_{x^2-y^2}$, respectively]. Thus, the interactions of the d-states of the atoms in the chains and between the chains are of the same order. The same conclusion was drawn earlier by WEGER and GOLDBERG [6.189] who used another (a rougher) interpolation processing of MATTHEISS' APW calculations carried out for V_3Si (see also [6.190]).

Calculation of the wave functions has made it possible to establish the decomposition of the density of states on the Fermi level into partial contributions. It has turned out that δ_1-type states with symmetry as in $d_{x^2-y^2}$ prevail. In the WEGER and GOLDBERG calculations these states are completely absent on the Fermi surface and δ_2-type states of d_{xy} symmetry predominate. In the Labbé-Friedel model ($\delta_1+\delta_2$)-type states are prevalent on the Fermi surface.

In 1978-1980 KLEIN et al. [6.72,191-193] reported on more complete and fundamental data resting on a calculation of the electron structure of ten A-15 compounds (V_3X, Nb_3X; X = Al, Ga, Si, Ge, Sn) and some of their alloys. This calculation was carried out from first principles using the APW method with self-consistence and is an achievement of band theory and computing technique applicable to analysis of complicated crystal structures such as A-15. The data thus obtained refine the results of previous calculations with respect to the position of the low-energy s-bands of the X-components relative to the Fermi level, and also with respect to the shape and width of the bands [6.187,194]. But the chief merit of these data is that they enabled a calculation of η, λ and T_c.

Figures 6.22 and 23 depict dispersion curves as well as net and partial densities of states for V_3Si and Nb_3Sn according to results of KLEIN et al. [6.191]. It is worth noting that the spectra of V_3Si and Nb_3Sn have very flattened sections in the dispersion curves near the Fermi level along the Σ, Λ and Δ directions, which ensures a high density of states and, consequently, high λ and T_c (Table 6.5). On the other hand, the inclination of the dispersion curves in the unoccupied part of the spectrum is very large, and this results in an abrupt decrease of the density of states immediately above the Fermi level. Calculations have confirmed that the temperature dependences of the magnetic susceptibility and the Knight shift in high temperature superconductors with the A-15 structure are associated with a rapidly varying density of states near the Fermi level. This variation of N(E) amounts to about 50% for Nb_3Sn, V_3Si and V_3Ga in the energy range of 150 K around E_F.

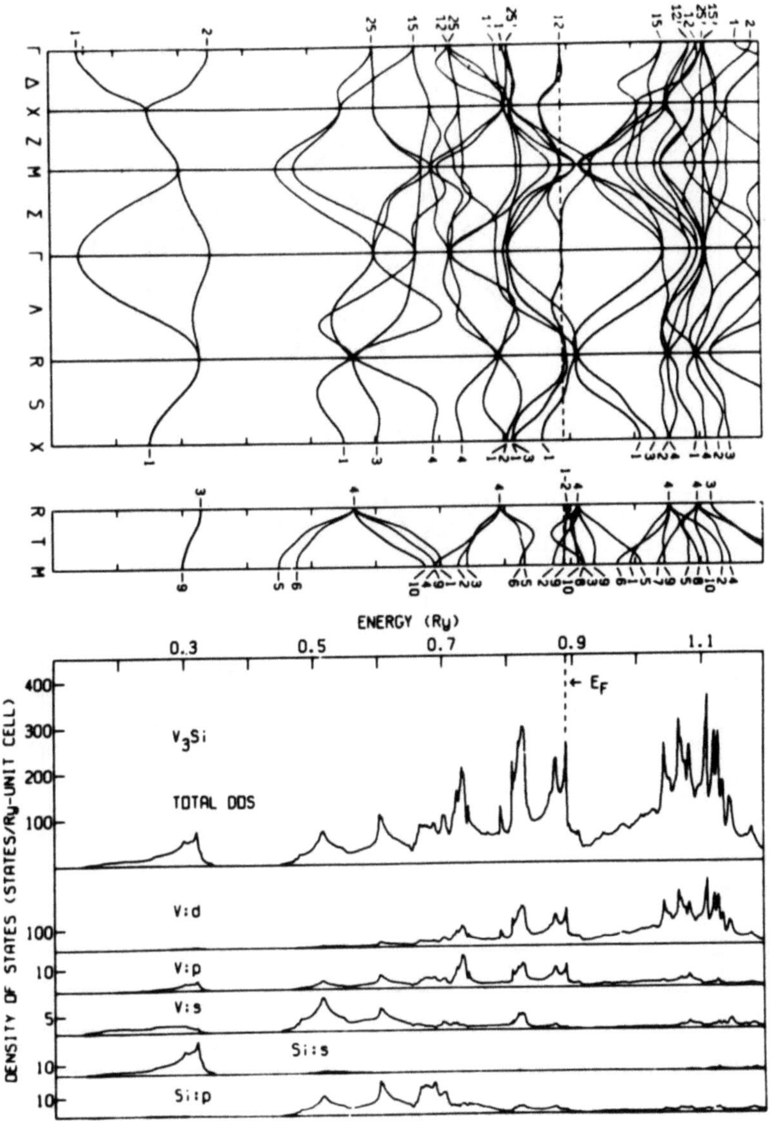

Fig.6.22. Energy bands and density of states for V_3Si according to the self-consistent calculation [6.191]

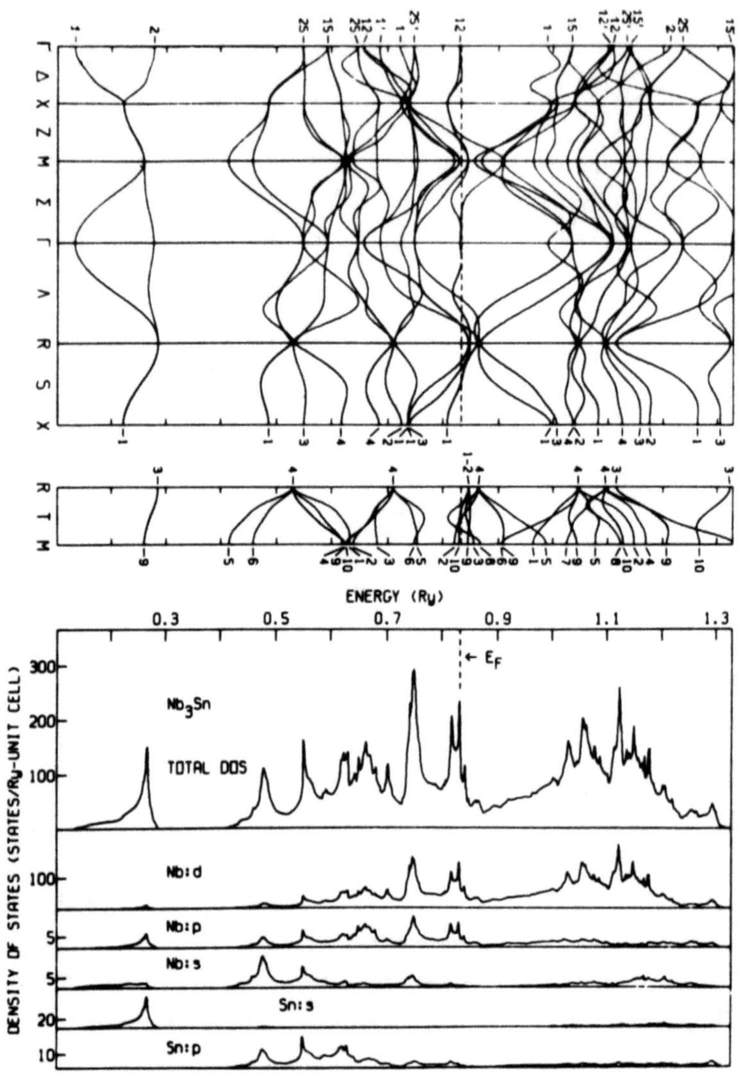

Fig.6.23. Energy bands and density of states for Nb_3Sn according to the self-consistent calculation [6.191]

Table 6.5. Density of states on Fermi level N(0), electron-phonon coupling constant λ and superconducting transition temperature T_c in A-15 compounds according to calculated data [6.192]

Compound	N(0) states/Ry · cell	a) $\lambda_{calc.}$	$\lambda_{exp.}$	b) T_c, K calc.	T_c, K exp.
V_3Al	188.9	1.09	0.87	14.8	9.6
V_3Ga	266.7	1.48	1.17	23.0	16.5
V_3Si	200.2	1.18	1.12	18.4	17.1
V_3Ge	114.7	0.67	0.70	5.3	6.1
V_3Sn	123.0	0.52	0.61	1.9	3.8
Nb_3Al	199.2	2.14	1.60	25.1	18.6
Nb_3Ga	191.8	1.92	1.74	22.5	20.3
Nb_3Si a=5.03 Å	61.42	0.65		3.5	
Nb_3Si a=5.10 Å	69.20	0.64	>1.0	3.4	>10.0
Nb_3Si a=5.20 Å	83.61	0.64		3.4	
Nb_3Ge	106.7	0.89	1.80	8.2	23.2
Nb_3Sn	158.8	1.12	1.44	12.8	18.0

a) The parameter λ was estimated according to (2.4.30) with due account for the data on $\langle\omega^2\rangle$ from neutron diffraction measurements of phonon spectra [6.167].

b) T_c was calculated according to McMILLAN's formula (2.3.27) with due account for the corrections made in [6.195] and the phonon density of states [6.167]

At the same time, the compounds devoid of a sharp structure in N(E) near E_F (V_3Al, V_3Sn, V_3Ge, etc.) display no anomalies in the temperature dependence of the susceptibility, which is in agreement with experiments. The level of agreement between theory and experiment can be seen in Fig.6.24 which gives a comparison of the calculated and experimentally measured temperature dependences of the magnetic susceptibility for Nb_3Sn [6.158,192]. For Nb_3Ge and Nb_3Si, two other high-temperature superconductors, the results were somewhat different. It follows from the calculated results that the Fermi level is located below the flat section of the dispersion curves which corresponds to the peak in N(E). In virtue of this, the values of the density of states on the Fermi level for Nb_3Ge and Nb_3Si are appreciably less than those for

V_3Si and Nb_3Sn, which, as a result, leads to lesser values of λ and T_c (Table 6.5). Special consideration has shown that, within reasonable lattice parameter variations and exchange potential approximations, the band structure N(0), λ and T_c for Nb_3Ge and Nb_3Si, exhibit minimum deviations (Table 6.5). By what then can one account for such a large difference between the calculated and experimental values of λ and T_c for Nb_3Si and Nb_3Ge? It should be recalled here that high T_c for these compounds were obtained in films or specimens manufactured according to special technology, whereas the results of the calculation apply to a bulk material for which γ and T_c have comparatively moderate values [6.196].

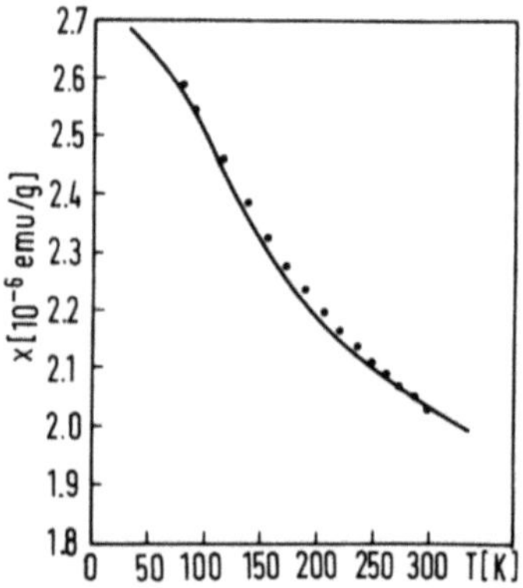

Fig.6.24. Temperature dependence of the magnetic susceptibility in Nb_3Sn obtained from a self-consistent band structure calculation [6.192], and its comparison with experiments [6.158]

Of the recent electron structure calculations made for A-15 type compounds [6.197-204], mention should be made of the data obtained by JARLBORG [6.199-201] for 26 A-15 structure compounds (V_3X, X = Al, Si, Ga, Ge, As, Sn, Sb, Pb, Co, Ni, Au; CrX, X = Ga, Si, Ge; Nb_3X, X = Al, Ga, Ge, In, Sn, Sb, Pb, Bi; Mo_3X, X = Al, Si, Ga, Ge). Although the LMTO calculation method (linearized muffin-tin orbitals method, see [6.205,206]) used by him is inferior to the APW method, the above calculations were carried out in a self-consistent version and are interesting in that they additionally cover a new family of A_3B compounds, viz., Cr_3X and Mo_3X compounds, as well as V_3X compounds in which the X atom is a transition element (X = Co, Ni, Au).

As in [6.191,192], JARLBORG also calculated λ and T_c [6.200,201]. In accordance with experiments, JARLBORG's calculation gave low values of N(0), λ and T_c for Cr_3X and Mo_3X compounds, and also for V_3X compounds with X d-metals (Co, Ni, Au). For V_3Ge, V_3Sn, V_3Au, Nb_3Al and Nb_3Ga, the level of agreement between theory and experiment may be considered as quite satisfactory. JARL-

BORG's calculation has shown that the highest values of T_c (ca. 20-30 K) may be expected for V_3Ga, V_3Al, Nb_3In, Nb_3Al and Nb_3Ga. If these predictions hold true, one should give attention to improvement of the technique of synthesizing these materials, which, when prepared according to the standard procedure exhibit much lower T_c values.

On the basis of the electron structure calculations performed, KLEIN et al. [6.191] and JARLBORG et al. [6.207] calculated the distribution of the intensity of X-ray emission and photoelectron spectra in A-15 compounds. In Fig. 6.25, part of these results are compared with the experimental spectra obtained in [6.208]. The close agreement of the calculated and experimental spectra [6.208-217] over a wide energy range is testimony to the correctness of the above electron structure calculations made for A-15 compounds. It is of great interest to experimentally verify the theoretically calculated topology of the Fermi surface in A-15 compounds. For this class of compounds, measurements were made of the de Haas-van Alphen effect [6.218,219] and of the angular dependence of the electron-positron annihilation spectra [6.220]. These measurements are in agreement with the theoretical results (Fig.6.26).

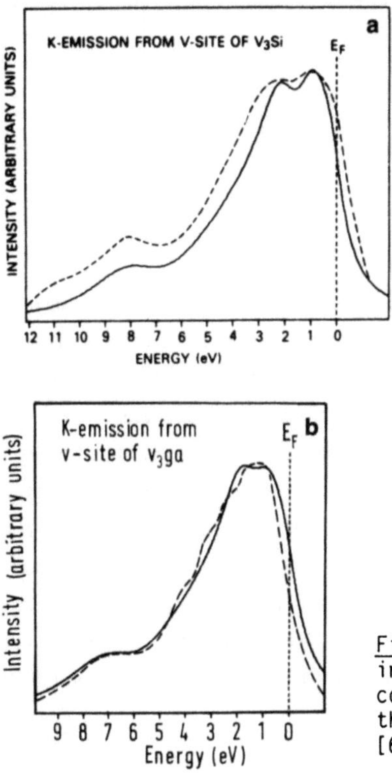

Fig.6.25. The X-ray K emission band of V in V_3Si (a) and V_3Ga (b). The solid curve corresponds to the calculation [6.191], the dotted line pertains to experiment [6.208]

Fig.6.26. Cross sections of the Fermi surface obtained from electron-positron annihilation [6.220] and their comparison with the calculation [6.191]

6.3 Investigation into Theoretical Models

6.3.1 The Weger-Labbé-Friedel Model

The quasi-one-dimensional electron-spectrum model constructed for a number of compounds with an A-15 structure and based on the existence of a system of mutually perpendicular chains of transition-metal atoms in this structure has been developed intensively during the last decade. Its success in accounting for many anomalous properties of such compounds as V_3Si and Nb_3Sn generated new papers in spite of the fact that band structure calculations from first principles, discussed in the previous section, do not confirm the assumption as to the smallness of the interchain interaction for these compounds. To begin with, we consider how it is possible, using this model, to account for the strong temperature dependences of the magnetic susceptibility and the lattice moduli [6.221,222] as well as the high values of the superconducting transition temperature [6.223].

The magnetic susceptibility consists of two major contributions, namely the Pauli term $\chi_d(T)$, which is related to paramagnetism, and the orbital term χ_o:

$$\chi = \chi_o + \chi_d(T) \quad .$$

Since d-electrons make a large contribution to the density of states on the Fermi surface, the contribution of s-electrons to χ may be neglected, $\chi_d(T)$ may be written in the form

$$\chi_d(T) = -2\mu_B \int f'(\varepsilon)N_d(\varepsilon)d\varepsilon \quad , \tag{6.3.1}$$

where $f(\varepsilon)$ is the Fermi distribution function, ε is the energy measured from the Fermi level, and $N_d(\varepsilon)$ is the density of states of the weakly filled d-band which is approximated by (6.2.4). The chemical potential of the system at a given temperature is determined by

$$Q = 2 \int f(\varepsilon)N_d(\varepsilon)d\varepsilon \quad , \tag{6.3.2}$$

where Q stands for the number of electrons in this band. Selected values of the main parameters for V_3Si, viz., the band width $2|E_m|$, the Fermi energy E_F and Q_V (the number of electrons per vanadium atom) are

$$2|E_m| = 9\ \text{eV} \quad , \quad E_F = 1.8 \cdot 10^{-3}\ \text{eV} \quad , \quad Q_V = 0.036 \quad . \tag{6.3.3}$$

Formula (6.3.1) leads to a temperature dependence $\chi_d(T)$ which corresponds to the experimentally observed behaviour of χ over a temperature range from 50 to 400 K. However, the numerical value of $\chi_d(T)$ at room temperatures is a factor of seven lower than the experimental value. The latter amounts to $8 \cdot 10^{-4}$ cm^3/mol whereas $\chi_d(T) \approx 1.1 \cdot 10^{-4}$ cm^3/mol. Thus, the orbital contribution should be around $6 \cdot 10^{-4}$cm^3/mol. In order of magnitude, the orbital contribution is $\chi_o \sim \mu_B^2/\delta$, δ being the mean difference of the energy levels bound by the orbital angular momentum. Since δ is of the order of 1 eV, the orbital contribution should be temperature-independent and amount to $\sim 10^{-4}$ cm^3/mol.

LABBE's [6.221] calculations yielded the value $\chi_o \approx 2.3 \cdot 10^{-4}$ cm^3/mol which is appreciably less than the experimentally observed value. These calculations should be regarded as an estimate of the order of magnitude. Note that at the chosen values of the parameters (6.3.3) for V_3Si the agreement between the estimated value and the experimental value is also satisfactory for the Knight shift [6.222].

To ascertain the qualitative peculiarities of superconducting compounds such as V_3Si using the Labbé-Friedel model we shall first try to employ the usual BCS superconductivity theory, in taking into account only the substantial dependence of the density of states on the energy near the Fermi level [6.223]. The equation for T_c,

$$1 = V \int_{-\hbar\omega_D}^{\hbar\omega_D} N_d(E_F + \varepsilon) \frac{\tanh(\varepsilon/2T_c)}{2\varepsilon} d\varepsilon \qquad (6.3.4)$$

(V is the effective interelectron attraction parameter of the BCS model), should be solved simultaneously with (6.3.2) for the chemical potential of the system. In view of the large density of states in the weakly filled subband, the contribution to (6.3.4) made by other subbands as well as by the s-band is neglected. Thus, in the model being considered [6.223] it is assumed that only the d electrons of the weakly filled d-subband play an important part in the superconductivity of the compounds involved. That is why in the equation under investigation it is necessary to substitute (6.3.4) in place of the total electron density $N(\varepsilon)$.

Allowing for the narrowness of the electron spectrum, i.e., adopting the inequality $E_F - E_m \ll h\omega_D$, the upper limit in the integral in (6.3.4) may be extended to ∞. The results of the numerical solution of (6.3.4) and (6.3.2) are given in Fig.6.27 where the dependence of T_c on the number Q of d-electrons in the weakly filled subband is plotted. It is significant that the curve for T_c has a maximum. Thus, in the limit being considered ($E_F - E_m \ll \hbar\omega_D$) the dependence of T_c on electron concentration Q is weak.

If for V_3Si we take the parameters (6.3.3) used to calculate the susceptibility and the Knight shift and add the value $V = 0.15$ eV to them (rather than the value 0.4 eV normally expected for transition metals), we obtain the transition temperature $T_c = 17$ K. Thus, the adopted quasi-one-dimensional spectrum

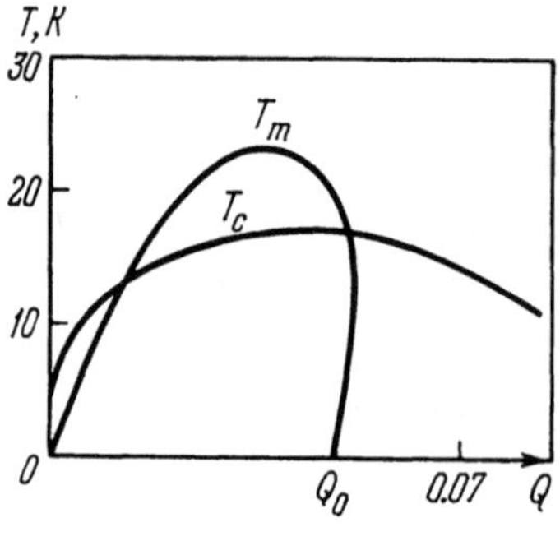

Fig.6.27. Dependence of the temperatures of superconducting (T_c) and structural (T_m) transitions on the number of d-electrons in the band for V_3Si [6.223]

model leads to high values of T_c when the choice of physical parameters is reasonable.

An important remark should be made concerning the applicability of the BCS theory to the noninteracting chains model. It has been rigorously proved that in a one-dimensional system superconductivity is impossible [6.224] due to the destructive effect of electron density fluctuations [6.225] and order-parameter phase fluctuations. Density fluctuations can be suppressed due to the long-range nature of the Coulomb interaction if we consider, for example, a model of a metal consisting of a parallel system of linear chains spaced at a distance which is much greater than the interatomic distance within a chain [6.226]. In such a system the long-wave density fluctuations remain practically the same as they are in a normal three-dimensional metal.

To restrict the influence of phase fluctuations it is necessary to take into account the real transitions from chain to chain which in the quasi-one-dimensional metal model [6.226] are described by a dispersion relation of the form

$$E_p = \frac{p_z^2}{2m} + \alpha[\cos(ap_z) + \cos(ap_y)] \qquad (6.3.5)$$

(the z axis is aligned with the chain), with $\alpha \ll p_F^2/2m$, where p_F is the Fermi momentum. It has turned out that at very small $\alpha \ll T_c$ the formula for the superconducting transition temperature practically coincides with that derived for the case of a purely one-dimensional metal [6.227] (within the framework of a "logarithmic" approximation, disregarding the fluctuations). At large α the superconductivity is described by the usual BCS theory formulae. Since, as we have seen above, the matrix element of transition from chain to chain is not very small even for V_3Si and Nb_3Sn compounds, the BCS theory equations may be used to estimate the quantity T_c. To this end, it is necessary to substitute the expression for $N_d(\varepsilon)$ witn a singularity corresponding to a quasi-one-dimensional spectrum. The qualitative correctness of the results of this approach was substantiated by KATS [6.228]. The latest investigations into this problem have been carried out in [6.229-231].

The chain model permits a fundamental explanation of the lattice phase transition [6.185,221]. As revealed by the experiment described in Sect.6.1, this transition is associated with the softening of the elastic shear modulus $(C_{11} - C_{12})$ whose zeroing should determine the lattice instability temperature identified with the structural transformation temperature T_m. Using a quasi-one-dimensional model it may be shown (detailed calculations of the elastic moduli in a similar model are given below) that the interaction of d-electrons with a distorted lattice leads to a highly temperature dependent term in the free energy, and it is this term which ensures the softening of the modulus $(C_{11} - C_{12})$ when the temperature decreases. The results of calculations of T_m for V_3Si as a function of the number Q of d-electrons are presented in Fig. 6.27 (along with T_c). A similar curve is also obtained for Nb_3Sn [6.232]. It is seen from the figure that the lattice instability only occurs for sufficiently small Q not exceeding Q_o which is as low as 0.01 to 0.1 e/a for typical values of the parameters. The reverse statement does not hold true since large values of T_c are not necessarily associated with lattice softening and the possibility of a phase transition. This is apparently the reason why martensitic transitions are known to occur only in some A_3B-type compounds. As a matter of fact, already from the calculated results given in Fig.6.27 it becomes clear that the martensitic transition temperature strongly depends on stoichiometry. If, for a pure compound, Q is in the vicinity of Q_o, the deviation from stoichiometric composition is liable to bring Q beyond Q_o and thereby lead to the occurrence or disappearance of the phase transition, with T_c varying weakly in this event.

The temperature T_m determined from the condition of instability of the cubic phase is the point of phase transition to a tetragonal structure. At a temperature below T_m a tetragonal-type spontaneous deformation arises. This deformation has different signs for the compounds V_3Si $(c/a > 1)$ and Nb_3Sn $(c/a < 1)$. It is easy to understand how the electron spectrum changes in this case. For V_3Si, the structure of the d-electron spectrum in the cubic phase is described (according to LABBE and FRIEDEL) in Fig.6.20. The upper-lying subband, which is almost empty, turns out to be doubly degenerate with respect to the magnetic quantum number. What is most important now, it is triply degenerate with respect to the direction of the chains. Relate the [100] direction to the tetragonal axis; then the two perpendicular directions [010] and [001] will be equivalent. For V_3Si, the cell turns out to be elongated and therefore the band narrows in the [100] directions and widens in the perpendicular directions. Thus, the spectrum undergoes splitting so that the Fermi level may fall within the span between the lower-lying boundaries of the subbands (Fig.6.28). It is assumed that it is exactly this situation which occurs in V_3Si and Nb_3Sn compounds when $T = 0$ K. It is clear that as soon as the tetragonal distortion arises at a temperature below T_m and the subbands start extending, a fast redistribution of electrons occurs among them, leading to a sharp increase in spontaneous deformation with decreasing temperature. As soon as the Fermi level goes beyond the edge of the uppermost of the split subbands, the electron redistribution terminates and the deformation should

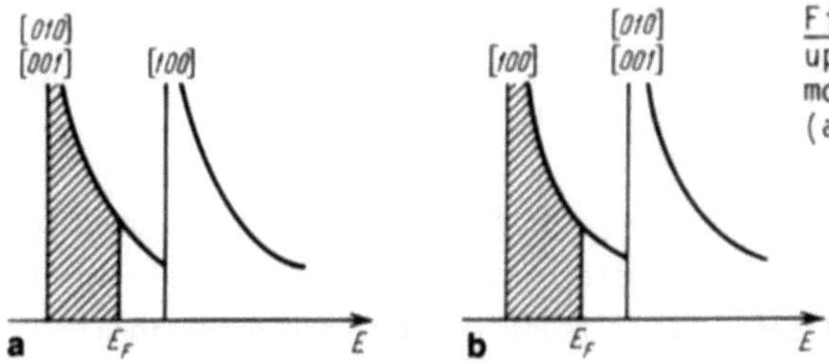

Fig.6.28. The splitting of the upper d-subband for the chain model in the tetragonal phase (a) for $c/a > 1$, (b) for $c/a < 1$

stop increasing. It is exactly this character of the temperature dependence of spontaneous deformation which was observed in experiments.

Comparing the conclusions drawn from the Labbé-Friedel model with experimental evidence relating to compounds such as V_3Si and Nb_3Sn we infer that it seems to qualitatively describe the whole set of their anomalous properties: the temperature behaviour of shear moduli and magnetic susceptibility, the lattice transition, high-temperature superconductivity and their relation to each other.

A number of more particular properties of A-15 compounds can also be explained within the framework of the noninteracting chains model. Here belongs, for example, the dependence of the martensitic transformation temperature T_m on the magnetic field. For V_3Si, estimates made according to the Labbé-Friedel model by DIETRICH and FULDE [6.233] yielded $\Delta T_m = -0.31$ K in a field of $H = 90$ koe. The experimental value of the temperature shift in this field turned out to be fairly close to $\Delta T_m = -0.26$ K [6.139]. It should, however, be borne in mind that the Labbé-Friedel model involves many indeterminate parameters which are used as fitting characteristics in the comparison with experiments, therefore the agreement of theory with experiments may be highly illusory.

6.3.2 The Cohen-Cody-Halloran Model

The physical basis of the anomalies in the properties of A-15 compounds is the sharp variation of the density of states near the Fermi surface. In the Labbé-Friedel model this variation is attributed to the weak filling of the quasi-one-dimensional d-band. If what is most essential here is the fact that the density of states varies sharply in the vicinity of the Fermi level, the model may be simplified. COHEN et al. [6.234] have proposed an electron model for these compounds, postulating the following behaviour of the d-electron density of states as a function of energy

$$N(E) = \begin{cases} N_0 \quad , \quad E > E_0 \\ \\ 0 \quad , \quad E < E_0 \end{cases} \quad , \tag{6.3.6}$$

where N_0 is some constant and E_0 is the edge of the band which is assumed to be three-fold degenerate in the cubic crystal (for the three types of linear chains). Thus, in its physical basis, this model is quasi-one-dimensional, but instead of an exact prescription of the density of states such as (6.2.4) an approximation by a constant is adopted. The Fermi level, in accordance with the chains model, passes near the band edge. It has turned out that such a model enables one to easily obtain analytical results which also agree well with experiments.

We first consider how the temperature behaviour of the transverse lattice moduli is calculated. The crystal is assumed to be below the martensitic transformation temperature and to have a tetragonal structure. To calculate the moduli it is necessary to estimate the free energy Ω of the crystal as a function of the deformation tensor $\varepsilon_{\alpha\beta}$. It is convenient to split the free energy in two parts: $\Omega = \Omega_0 + \Omega_e$, where Ω_0 pertains to the lattice proper, and Ω_e to the system of d-electrons located in the distorted lattice. Ω_0 may be series-expanded in deformations. For a tetragonal crystal, this expansion is usually of the form

$$\Omega_0 = \frac{1}{2}A_{11}\varepsilon_1^2 + \frac{1}{2}A_{22}\left(\varepsilon_2^2 + \varepsilon_3^2\right) + A_{12}\left(\varepsilon_1\varepsilon_2 + \varepsilon_1\varepsilon_3\right) + $$
$$+ A_{23}\varepsilon_2\varepsilon_3 + \frac{1}{2}A_{44}\varepsilon_4^2 + \frac{1}{2}A_{66}\left(\varepsilon_5^2 + \varepsilon_6^2\right) \quad , \tag{6.3.7}$$

where A_{ij} are constants having the sense of shear moduli of a lattice without d-electrons.

The electronic part of the free energy is expressed in terms of the internal energy W and the entropy S: $\Omega_e = W - TS$. For the electron gas we have

$$W = \sum_\lambda \int f_\lambda(\varepsilon)N_\lambda(\varepsilon)d\varepsilon \quad , \tag{6.3.8}$$

$$S = -k\sum_\lambda \int N_\lambda(\varepsilon)\{f_\lambda(\varepsilon)\ln f_\lambda(\varepsilon) - [1 - f_\lambda(\varepsilon)]\ln[1 - f_\lambda(\varepsilon)]\}d\varepsilon \quad , \tag{6.3.9}$$

where $N_\lambda(\varepsilon)$ is the density of states in the λth band ($\lambda = 1,2,3$ corresponds to the three types of orthogonal chains), and $f_\lambda(\varepsilon)$ is the Fermi distribution function for the d-band electrons interacting with the field of the distorted lattice. We shall assume that this interaction is described by a deformation potential, i.e., that the addition of an electron to the energy due to the deformation is $\delta E_\lambda = U_\lambda \varepsilon_\lambda$. The deformation interaction obviously shifts the band edge in a deformed crystal, and therefore in the case when the Fermi level lies close to the edge this interaction leads to a strong temperature dependence of the quantities estimated.

With (6.3.8,9) the free energy of the electron system may then be written as

$$\Omega_e = QE_F - kT \sum_\lambda \int_{-\infty}^{\infty} \ln\left[1 + \exp\left(\frac{E_F - E_\lambda(\varepsilon)}{kT}\right)\right] N_\lambda(\varepsilon) d\varepsilon \quad , \qquad (6.3.10)$$

where $E_\lambda(\varepsilon)$ is the energy of the band related to a chain in the λ direction in the deformed crystal (E_λ^0 stands for the band edge in a nondeformed tetragonal crystal), and $N_\lambda(\varepsilon) = N_0/3$ for $\varepsilon > E_\lambda^0$ and is equal to 0 for $\varepsilon < E_\lambda^0$.

After differentiation of Ω_e with respect to deformations, the integral over energies is easy to calculate since the density of states is a constant. As a result we get

$$C_{ij} = A_{ij} - \frac{N_0}{3} \sum_{\lambda=1}^{3} f(E_\lambda^0) \left[\frac{\partial(E_F - E_\lambda)}{\partial\varepsilon_i} \frac{\partial(E_F - E_\lambda)}{\partial\varepsilon_j}\right]_{\varepsilon=0} \qquad (6.3.11)$$

where $f(E_\lambda^0)$ is a Fermi function for the energy $\varepsilon = E_\lambda^0$, and $E_\lambda = E_\lambda^0 + \delta E_\lambda$.

In the expression thus derived E_λ^0 and E_λ, E_{0F} and E_F stand for the edge of the λth band and for the Fermi energy in the nondeformed and deformed crystals, respectively. The Fermi energy is determined from the condition of conservation of the number of electrons in the λ band which, in this model, becomes

$$\frac{1}{3} N_0 kT \sum_{\lambda=1}^{3} \ln\left[1 + \exp\left(\frac{E_F - E_\lambda}{kT}\right)\right] = Q \quad . \qquad (6.3.12)$$

Differentiating the equation with respect to ε_i we obtain an expression for the derivative of the Fermi energy

$$\frac{\partial E_F}{\partial \varepsilon_i} = \sum_{\lambda=1}^{3} \frac{\partial E_\lambda}{\partial \varepsilon_i} f(E_\lambda) \Big/ \sum_{\lambda=1}^{3} f(E_\lambda) \quad . \qquad (6.3.13)$$

Equation (6.3.11) in combination with (6.3.13) determines the temperature dependence of the shear moduli in the tetragonal phase

$$C_{11} = A_{11} - \frac{2}{3}N_o U_1^2 \frac{f_1 f_2}{f_1 + 2f_2} \quad ,$$

$$C_{22} = C_{33} = A_{22} - \frac{1}{3}N_o U_2^2 \frac{f_2(f_1 + f_2)}{f_1 + 2f_2} \quad ,$$

$$C_{12} = C_{13} = A_{12} + \frac{1}{3}N_o U_1 U_2 \frac{f_1 f_2}{f_1 + 2f_2} \quad ,$$

$$C_{23} = A_{23} + \frac{1}{3}N_o U_2^2 \frac{f_2^2}{f_1 + 2f_2} \quad , \tag{6.3.14}$$

For brevity, the notation $f_\lambda = f(E_\lambda^o)$ is introduced here. The quantities A_{ij} may be considered to be temperature-independent. Thus, the total temperature dependence of the moduli C_{ij} is determined by the factors f_λ.

Since the tetragonality occurring in Nb_3Sn and V_3Si compounds is nonmarked it is possible in (6.3.14) to approximately set [6.235]

$$A_{11} = A_{22} \quad , \quad A_{23} = A_{12} \quad , \quad U_1 = U_2 \equiv U \quad . \tag{6.3.15}$$

The tetragonal crystal itself may be regarded as resulting from the spontaneous tetragonal deformation of the cubic crystal with the volume remaining constant, i.e.,

$$\hat{\varepsilon} = \{-\varepsilon_o, \frac{1}{2}\varepsilon_o, \frac{1}{2}\varepsilon_o\} \quad . \tag{6.3.16}$$

In this case the edges of the energy bands for the chains aligned with the tetragonal axis and for the chains which are perpendicular to it are given by (Fig.6.29)

$$E_1 = E_o - U\varepsilon_o \quad , \quad E_2 = E_3 = E_o + \varepsilon_o/2 \quad . \tag{6.3.17}$$

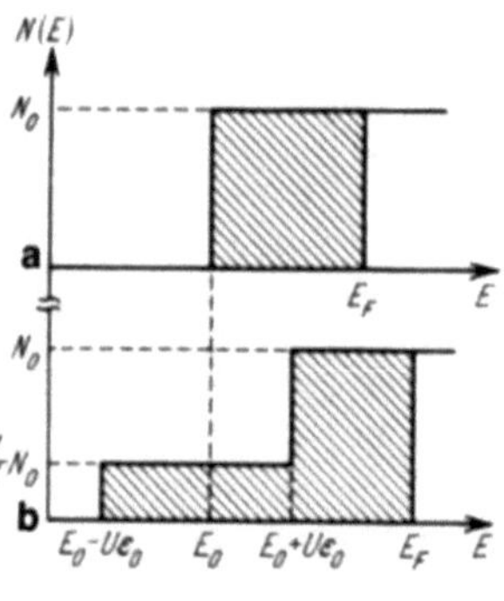

Fig.6.29. Density of states in the d-band of Nb_3Sn according to the COHEN-CODY-HALLORAN [6.234] model. (a) for the cubic and (b) for the tetragonal phase. At $T = 0$, when the spontaneous deformation reaches the highest value, the Fermi level is assumed to lie entirely in the first band

In the cubic phase the position of the edge is the same for all the three bands, so the moduli are still determined by (6.3.14) in which, along with (6.3.15), f_1 and f_2 should be replaced by

$$f_0 = \left[1 + \exp\left(\frac{E_0 - E_F}{kT}\right)\right]^{-1} \quad , \tag{6.3.18}$$

to yield

$$C_{11} = A_{11} - \tfrac{2}{9}N_0U^2f_0 \quad , \quad C_{12} = A_{12} + \tfrac{1}{9}N_0U^2f_0 \tag{6.3.19}$$

whence for the shear modulus we have

$$C_{11} - C_{12} = A_{11} - A_{12} - N_0U^2f_0/3 \quad . \tag{6.3.20}$$

We introduce the temperature T_0 which is a measure of the Fermi energy in the cubic phase: $kT_0 = E_F - E_0$. As follows from (6.3.20) and (6.3.18), the temperature behaviour of the shear modulus is specified by the dimensionless quantity T_0/T since A_{11} and A_{12} are assumed to be temperature-independent. The shear modulus diminishes with decreasing temperature and goes to zero at a point which is determined from the condition

$$\frac{1}{1 + \exp(-T_0/T)} = \frac{3(A_{11} - A_{12})}{N_0U^2} \quad . \tag{6.3.21}$$

Detailed measurements of the temperature dependence of the shear moduli in a single crystal of Nb_3Sn ($T_m = 45$ K) were carried out in [6.158] (Fig.6.12). The experimental data for C_{11} and C_{12} in the cubic phase over the temperature interval $49\ \text{K} < T < 300$ K fit the curves (6.3.19,20) with the values of the theory parameters being as follows:

$$A_{11} = 2.94 \cdot 10^{12}\ \text{erg/cm}^3 \quad , \quad A_{12} = 0.84 \cdot 10^{12}\ \text{erg/cm}^3$$

$$N_0U^2 = (7.47 \div 7.86) \cdot 10^{12}\ \text{erg/cm}^3 \quad , \quad T_0 = 80\ \text{K} \quad . \tag{6.3.22}$$

For the same set of parameters satisfactory agreement is obtained between theory (6.3.14) and experiments for C_{11} and C_{12} in the tetragonal phase. Thereby one has to take from the experiment the temperature dependence of the tetragonal deformation parameter $\varepsilon_0(T)$ [6.132].

To estimate $\varepsilon_0(T)$ we construct an equation by minimizing the free energy with respect to ε_0:

$$\frac{3}{2}(A_{11} - A_{12})\varepsilon_0 - \frac{1}{2}N_0U^2f_0\varepsilon_0 + \frac{1}{3}N_0UkT\ln(f_1/f_2) = 0 \quad . \tag{6.3.23}$$

Together with (6.3.12) it determines the temperature behaviour of spontaneous deformation. We show that (6.3.23) describes a first-order phase transition [6.233,236,237]. First write this equation for the limit $\varepsilon_0 \to 0$:

$$\left[\frac{3}{2}(A_{11} - A_{12}) - \frac{1}{2}N_0U^2f_0\right]\varepsilon_0 = \frac{1}{8}\,\frac{N_0U^3f_0(1 - f_0)}{kT}\varepsilon_0^2 \quad . \tag{6.3.24}$$

There are two solutions: $\varepsilon_0 = 0$ and

$$\varepsilon_0 = \frac{12kT}{N_0U^3}\cdot\frac{C_{11} - C_{12}}{f_0(1 - f_0)} \tag{6.3.25}$$

[we have used (6.3.20) for the shear modulus]. This equation denotes that the temperature T_1, at which the shear modulus goes to zero, is the absolute instability temperature of the cubic phase. On the other hand, the temperature T_2, beyond which the tetragonal phase is absolutely instable, is determined from the condition $(d\varepsilon_0/dT)_{T=T_2} = \infty$. Differentiation of (6.3.23) with respect to T yields

$$\frac{f_1f_2}{f_1 + 2f_2} = \frac{A_{11} - A_{12}}{N_0U^2} \quad , \tag{6.3.26}$$

which, simultaneously with the basic equation (6.3.23), determines T_2 and the magnitude of the spontaneous deformation at this point. Note that if we set $\varepsilon_0 = 0$ in (6.3.26) it goes over into (6.3.21).

The temperature range between T_1 and T_2 corresponds to metastable states; T_1 evidently corresponds to the supercooling and T_2 to the superheating of the cubic and tetragonal phases. The phase transition point T_m which should lie in between is determined by equating the free energies for the tetragonal and cubic phases:

$$(A_{11} - A_{12})\varepsilon_0^2(T_m) = \frac{4}{9}N_0kT_m[2F(-U\varepsilon_0) + F(\tfrac{1}{2}U\varepsilon_0) - 3F(0)] \quad , \tag{6.3.27}$$

where

$$F(\Delta) = \int\limits_{\Delta}^{\infty} \ln\left[1 + \exp\left(\frac{E_F - \varepsilon}{KT_m}\right)\right]d\varepsilon \quad . \tag{6.3.28}$$

The solution to (6.3.27) together with (6.3.12) for the chemical potential determines the tetragonal transition and the jump in spontaneous deformation. Proceeding from the numerical solution of these equations, this model yields the T dependence of ε_0; it is represented in Fig.6.10 for the parameters (6.3.22) together with the experimental points for Nb_3Sn determined from X-ray measurements [6.132]. As we see, the value of the jump of the spontaneous deformation parameter at point T_m is in good agreement with the theoretical predictions. In experiments the region of the coexistence of both phases near T_m (hatched in the figure) is also clearly defined.

Thus the model under consideration offers a reasonable explanation of the shear modulus softening and, as a consequence of this, the occurrence of the lattice phase transition. However, in the calculation of other moduli the agreement with experiment is far from being as good. Thus, for example, it can readily be seen that in this model the modulus $C_{44} = A_{44}$ is not temperature dependent, whereas in actual fact it undergoes appreciable softening. A quantitative disagreement also arises when calculating the magnetic susceptibility, although the characteristic behaviour $\chi(T)$ near T_m (a drastic decrease when $T < T_m$) is obtained from this theory.

6.3.3 Gor'kov's Model I

The anomalous behaviour of a number of A-15 compounds had long been considered to be associated with the Weger-Labbé-Friedel hypothesis about the quasi-one-dimensional character of the electron spectrum and about the closeness of the Fermi level to the bottom of the band. But in reality it is still most unlikely that d-electrons may show this quasi-one-dimensional character since the value of the interchain transition matrix element is high [6.187,189]. At the same time, as shown by GOR'KOV [6.238], an absolutely different explanation may be supplied for the anomalies in A-15 structure. Its space group O_h^3 comprises nontrivial translations, and this leads to a two-fold degeneracy of the electron levels at a point X of the boundary of the Brillouin zone. Simultaneously, the symmetry of the system is such that the dispersion relation turns out to be linear in the vicinity of this point. If, for some of the A-15 compounds, the Fermi level passes through such a point, then in the vicinity of it there are two branches of excitations (electron and hole) with a linear dispersion relation. These small groups of electrons make a highly temperature-dependent contribution to the free energy, and this leads immediately to a logarithmic dependence of the shear modulus on temperature. It is this dependence which was experimentally observed by TESTARDI in V_3Si [6.156]. It has also been shown by GOR'KOV that the electronic properties of the compounds under consideration, for example, the temperature dependence of the magnetic susceptibility, find a satisfactory explanation in his model which reduces to the assumption: for A-15 compounds exhibiting anomalous properties, the Fermi level passes in the vicinity of the point X of the Brillouin zone.

Following GOR'KOV, to find the spectrum in the vicinity of the point X we use the method of invariants [6.239], the gist being as follows: let the energy level belonging to some irreducible representation D of dimension l of group G and its basis functions ψ_i $(i=1,\dots,l)$ be given. The bilinear combination $\sum_i \psi_i \psi_i^*$ is invariant under the transformations of the symmetry group of the Hamiltonian.

It may then be shown that the matrix of the Hamiltonian on the states belonging to irreducible representations D should have the following structure:

$$H = \sum_\nu n^\nu \sum_{i=1}^{l} K_i^{\nu *} T_i^\nu \quad , \tag{6.3.29}$$

where K_i^ν and T_i^ν are some basis functions transforming according to the νth irreducible representation involved in the decomposition of the direct product $D \times D^*$ into irreducible representations of the Hamiltonian symmetry group

$$D \times D^* = \sum_\nu n^\nu D^\nu \quad . \tag{6.3.30}$$

The quantities T_α^ν here are $l \times l$ matrices transforming by the action of group G elements according to the relation

$$\hat{D}(g)T = D(g)TD^{-1}(g) \quad . \tag{6.3.31}$$

The function K_i^ν is made up of quantities characterizing the matrix of the Hamiltonian: momenta, spins, external field vectors, deformation tensors, etc.

Since (6.3.30) involves a direct product of a representation by the conjugate representation, the expansion (6.3.30) should actually be carried out over the irreducible representations of the point group G_p^0 of the wave vector group G_p for any point p of the Brillouin zone. For compounds with A-15 structures, the results of expanding (6.3.30), the direct product of an irreducible representation of the wave vector group of point X over the irreducible representations of wave vector point group G_p^0, are given in Table 6.6. Since the representations in point X are two-dimensional, the T matrices should also be two-dimensional, and it is possible to take the Pauli matrices τ_x, τ_y, τ_z and the unit matrix τ_0 as these matrices. Table 6.6 shows which irreducible representation of group D_{4h} each of these matrices transforms into.

Table 6.6 Representations according to which Pauli matrices transform, and the expansion of $X_i \times X_i^*$ in irreducible representations of the group of the wave vector at point X for A-15 structures

Representation	τ_x	τ_y	τ_z	τ	Expansion
X_2 X_4	B_{1g}	B_{2u}	A_{2u}	A_{1g}	$A_{1g} + A_{2u} + B_{1g} + B_{2u}$
X_1 X_3	B_{1g}	B_{1u}	A_{1u}	A_{1g}	$A_{1g} + A_{1u} + B_{1g} + B_{1u}$

Determine the dispersion relationship in the vicinity of point X for the representations X_2, X_4. Then according to (6.3.29), Table 6.7 specifying the basis functions ψ_i of group D_{4h}, and Table 6.8, the matrix of the Hamiltonian for representations X_2, X_4, will take the form (disregarding the deformation)

$$H = \upsilon p_z \tau_z + C(p_x^2 + p_y^2)\tau_0 + C'(p_x^2 - p_y^2)\tau_x \quad . \tag{6.3.32}$$

Here the momentum components p_x, p_y and p_z are counted from point $X = \{00\pi/a\}$ and are assumed to be small. The matrix (6.3.32) has two eigenvalues

Table 6.7 Basis functions of the irreducible representations of the group D_{4h} composed of momentum, deformation tensor, electron spin and magnetic field components

D_{4h}	Basis functions
A_{1g}	p_z^2; $p_x^2 + p_y^2$; ε_{zz}; $\varepsilon_{xx} + \varepsilon_{yy}$; σH; H_z^2
A_{1u}	$\sigma_z p_z$; $p_y \varepsilon_{xz} - p_x \varepsilon_{yz}$; $H_z p_z$
A_{2g}	σ_z; H_z; $\sigma_x H_y - \sigma_y H_x$
A_{2u}	p_z; $p_x \varepsilon_{xz} + p_y \varepsilon_{yz}$
B_{1g}	$p_x^2 - p_y^2$; $\varepsilon_{xx} - \varepsilon_{yy}$; $\sigma_x H_x - \sigma_y H_y$
B_{1u}	$p_z \varepsilon_{xy}$; $p_x \varepsilon_{yz} + p_y \varepsilon_{xz}$; $p_x H_x - p_y H_y$
B_{2g}	$p_x p_y$; ε_{xy}; $\sigma_x H_y + \sigma_y H_x$
B_{2u}	$p_x \varepsilon_{xz} - p_y \varepsilon_{yz}$; $p_x H_y + p_y H_x$; $p_x \sigma_y + p_y \sigma_x$
E_g	$(\varepsilon_{xz}, \varepsilon_{yz})$; (H_x, H_y)
E_u	(p_x, p_y)

Table 6.8 Characters of the group D_{4h}^{g} of the wave vector at point $X = (0\ 0\ \pi/a)$ for A-15 structures (space group O_h^3)

	$\{\varepsilon\|0\}$	$\{\delta_{2x}\|0\}$	$\{\delta_{2y}\|0\}$	$\{\delta_{2z}\|0\}$	$\{i\|0\}$	$\{\rho_x\|0\}$	$\{\rho_y\|0\}$	$\{\rho_z\|0\}$
X_1	2	2	-2	-2	0	0	0	0
X_2	2	0	0	2	0	2	2	0
X_3	2	-2	2	-2	0	0	0	0
X_4	2	0	0	2	0	-2	-2	0

Note: Group D_{4h}^{g} contains 8 pure mirror rotation elements whose characters are specified in the table and 8 rotation elements accompanied by a nontrivial translation $\tau = (a/2;\ a/2;\ a/2)$

$$\{\delta_{4z}|\tau\}\ ,\quad \{\delta_{4z}^{-1}|\tau\}\ ,\quad \{\delta_{2xy}|\tau\}\ ,\quad \{\delta_{2\bar{x}y}|\tau\}\ ,$$
$$\{\rho_{xy}|\tau\}\ ,\quad \{\rho_{\bar{x}y}|\tau\}\ ,\quad \{\sigma_{4z}|\tau\}\ ,\quad \{\sigma_{4z}^{-1}|\tau\}\ .$$

The characters of these elements are equal to zero. The designation used for the elements is universally accepted: ρ stands for reflection in plane, σ for inversion rotations and δ for rotations

$$E_{1,2}(\underline{p}) = C(p_x^2 + p_y^2) \pm \sqrt{\upsilon^2 p_z^2 + C'^2(p_x^2 - p_y^2)^2} \quad , \tag{6.3.33}$$

representing the electron spectrum in the vicinity of point X which, at the point itself, belongs to the irreducible representation X_2 or X_4. In the immediate vicinity of point X (6.3.33) determines two branches obeying the linear dispersion relation

$$E_i(\underline{p}) = \pm\upsilon p_z \quad . \tag{6.3.34}$$

Since the Fermi level is assumed to pass through point X, (6.3.34) denotes that there is an electron and a hole branch of excitation (υ stands for the velocity of the electrons on the Fermi surface).

Now we investigate the behaviour of the small group of electrons in the vicinity of point X as a function of external deformations. To this end we again make use of the method of invariants. Accounting for small deformations, to the Hamiltonian (6.3.32) there should be added the term

$$H' = D_1(\varepsilon_{xx} - \varepsilon_{yy})\tau_x + [D_2\varepsilon_{zz} + D_3(\varepsilon_{xx} + \varepsilon_{yy})]\tau_0 \quad . \tag{6.3.35}$$

The terms such as $(p_x\varepsilon_{xz} + p_y\varepsilon_{yz})\tau_z$ are not taken into account here since they are quantities of second order of smallness with respect to p and ε. The last term in (6.3.35) leads to a general energy shift of electrons during deformation, therefore we shall omit it. As for the first term, it leads to a transition between two bands degenerate at point X. As a result, H' will become

$$H' = \Delta_z\tau_x \quad , \quad \Delta_z \equiv D_1(\varepsilon_{xx} - \varepsilon_{yy}) \quad . \tag{6.3.36}$$

The eigenvalues of the matrix $H_0 + H'$ give two branches with a gap equal to Δ_z, which determines the splitting of the spectrum in point X [6.240]

$$\varepsilon_{1,2}(\underline{p}) = \pm\sqrt{(\upsilon p_z)^2 + \Delta_z^2} \quad . \tag{6.3.37}$$

Now calculate the free energy of the electron system with the spectrum (6.3.37)

$$\Omega_e = -2kT \sum_{i=1,2} \int \frac{d\underline{p}}{(2\pi)^3} \ln\left[1 + \exp\left(-\frac{\varepsilon_i(\underline{p})}{kT}\right)\right] \quad . \tag{6.3.38}$$

We are interested only in the contribution of electrons in the small vicinity of point X. For the arm lying in the z direction, when the $\underline{p}$'s are small the spectrum depends only on p_z, therefore the contribution to the integral (6.3.38) which corresponds to it, may be written as

$$\delta\Omega_z = -\frac{ckT}{\pi\upsilon}\int_{-\tilde{\omega}}^{\tilde{\omega}}\left\{\ln\left[1 + \exp\left(-\frac{\sqrt{\varepsilon^2 + \Delta_z^2}}{kT}\right)\right] + \ln\left[1 + \exp\left(\frac{\sqrt{\varepsilon^2 + \Delta_z^2}}{kT}\right)\right]\right\} \quad , \tag{6.3.39}$$

where c is the phase volume of the vicinity of point X in the transverse direction

$$c = \int \frac{dp_x dp_y}{(2\pi)^2} \quad ,$$

and $\tilde{\omega}$ is the cutoff energy corresponding to the size of the vicinity in the z direction.

To calculate moduli requires an expansion of (6.3.39) with respect to a series of deformations, i.e., in the quantity Δ_z^2. Taking summation over all star arms, we then write the contribution to the free energy, dependent on deformations, in the form

$$\delta\Omega = -\frac{ck}{\pi\upsilon}\sum_{i=x,y,z}\Delta_i^2\int_0^{\tilde{\omega}}\frac{d\varepsilon}{\varepsilon}\tanh\left(\frac{\varepsilon}{2kT}\right) = -\frac{ckD_1^2}{\pi\upsilon}\ln\left(1.14\frac{\tilde{\omega}}{kT}\right)\sum_{i=x,y,z}\Delta_i^2 . \quad (6.3.40)$$

To determine the shear moduli we shall make use of the expression for the free elastic energy of the cubic crystal

$$\Omega = \frac{1}{2}C_{11}\left(\varepsilon_{xx}^2+\varepsilon_{yy}^2+\varepsilon_{zz}^2\right) + C_{12}\left(\varepsilon_{xx}\varepsilon_{yy}+\varepsilon_{yy}\varepsilon_{zz}+\varepsilon_{xx}\varepsilon_{zz}\right) + \frac{1}{2}C_{44}\left(\varepsilon_{xy}^2+\varepsilon_{yz}^2+\varepsilon_{xz}^2\right) \quad (6.3.41)$$

and compare it with (6.3.40). From this comparison we find the expressions

$$\delta C_{11} = -\frac{2ckD_1^2}{\pi\upsilon}\ln\left(\frac{\tilde{\omega}}{kT}\right) , \quad \delta C_{12} = \frac{ckD_1^2}{\pi\upsilon}\ln\left(\frac{\tilde{\omega}}{kT}\right) , \quad (6.3.42)$$

$$\delta(C_{11}-C_{12}) = -\frac{3ckD_1^2}{\pi\upsilon}\ln\left(\frac{\tilde{\omega}}{kT}\right) , \quad (6.3.43)$$

obtained by GOR'KOV for the temperature-dependent electron contributions to the lattice moduli.

We see a logarithmic increase of negative contributions to the moduli with decreasing temperature. This should lead to a loss in lattice stability at some temperature T when the temperature-dependent electron contribution will compensate for the contribution to the corresponding modulus from the ion-ion interaction. Thus, the assumption that the Fermi surface passes through a point X leads to a logarithmic softening of the shear modulus in the A-15 structure, which agrees with TESTARDI's experimental results presented in Fig.6.30.

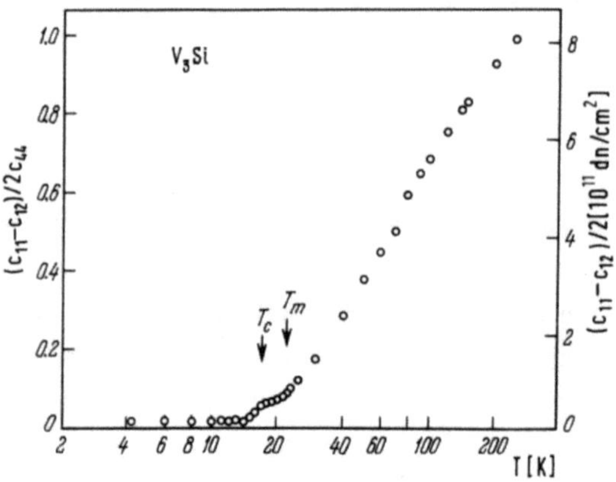

Fig.6.30. Temperature softening of the shear modulus in V_3Si [6.156]

Let us now see what peculiarities in the magnetic susceptibility arise from the assumption that the Fermi surface contacts the Brillouin-zone boundaries at the point X. Using Table 6.7 and the basic formula for the method of invariants (6.3.29) we can write the addition H_m to the Hamiltonian which takes into account the interaction of the selected group of electrons with the magnetic field

$$H_m = B_1(\underline{\sigma}\underline{H})\tau_0 + B_2(\sigma_x H_x - \sigma_y H_y)\tau_x + B_3(p_x H_y + p_y H_x)\tau_y \quad . \tag{6.3.44}$$

The second and third terms in H_m are due to the spin-orbit and the VAN VLECK mechanisms [6.238]; therefore the contributions which they make to the magnetic susceptibility, although growing with decreasing temperature according to the relation $\ln(\tilde{\omega}/T)$, are small.

This concerns the irreducible representations X_2 and X_4. The other representations at point X : X_1 and X_3 lead, as may be shown, to a large value of the logarithmic terms in the temperature dependence of χ. However, the invariant linear one with respect to p_z is $\tilde{\upsilon} p_z \sigma_z \tau_z$ rather than the invariant $\upsilon p_z \tau_z$, as was the case for the representations X_2 and X_4. This invariant does not ensure the required magnitude of the logarithmic term in the shear modulus. Consequently, the hypothesis that the Fermi level falls into the state X_2 or X_4, although explaining the softening of the lattice moduli, experiences difficulties in accounting for the anomalies in the magnetic susceptibility.

In addition to the logarithmic softening of the shear modulus, GOR'KOV's model easily explains the termination of this softening in nontransforming specimens at the transition to a superconducting state [6.241]. The cause of the effect is the opening of the energy gap in the electron spectrum. The presence of the gap drastically changes the temperature dependence of the electron contribution to the free energy associated with deformation. We calculate this contribution accounting for the electron attraction responsible for the occurrence of superconductivity. The Hamiltonian describing the motion of electrons along one chain aligned with the z axis in the presence of external static stresses and of the interelectron interaction should be written as a sum of three terms:

$$H_0 = \sum_{i=1}^{2} \sum_{\underline{p}\alpha} E_i(\underline{p}) a^+_{\underline{p}\alpha i} a_{\underline{p}\alpha i} \quad ,$$

$$H' = \sum_{ij} \sum_{\underline{p}\alpha} \{H'\}_{ij} a^+_{\underline{p}\alpha i} a_{\underline{p}\alpha j} \quad ,$$

$$H_{int} = -V \sum_{i \neq j} \sum_{\underline{p}\underline{p}'} a^+_{\underline{p}\uparrow i} a^+_{-\underline{p}\downarrow j} a_{-\underline{p}'\downarrow j} a_{\underline{p}'\uparrow i} \quad . \tag{6.3.45}$$

Here $E_i(\underline{p})$ is the energy of the electrons of the ith band and $\{H'\}_{ij}$ stands for the matrix elements of the operator (6.3.35). The expression corresponding to the chains aligned with the axes x and y is obtained through cyclic permutation of the indices x, y and z. In accordance with BCS theory, H_{int} allows for the attraction of electrons to each other and of holes to each other at the opposite ends of the Fermi surface. That is why, in accordance with Fig. 6.31, a Cooper pair is formed by the states 1 and 2 taken from different bands. We write the correction Ω_z to the free energy of z-chains in the second-order deformation

$$\Omega_z = -\frac{1}{2}kT \sum_{iji'j'} \{H'\}_{ij}\{H'\}_{i'j'} \times \sum_{\underline{p}\underline{p}'} \sum_{\alpha\alpha'} \int_0^\beta \int_0^\beta d\tau_1 d\tau_2 \left\langle T a^+_{\underline{p}\alpha i}(\tau_1) a_{\underline{p}\alpha j}(\tau_1)\, a^+_{\underline{p}'\alpha' i'}(\tau_2) a_{\underline{p}'\alpha' j'}(\tau_2)\right\rangle \quad . \tag{6.3.46}$$

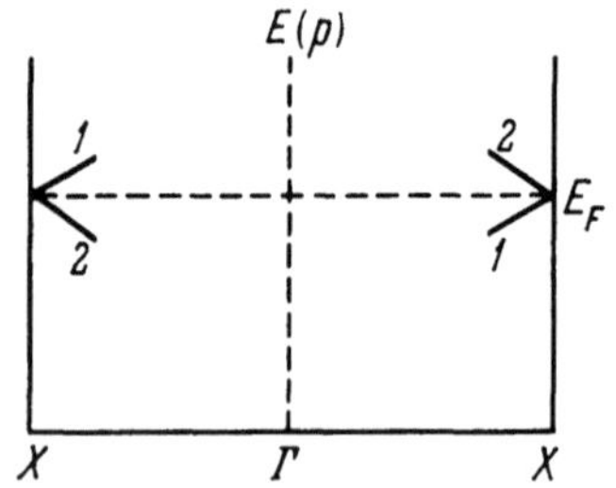

Fig.6.31. Electron spectrum E(p) of A-15 compounds in the vicinity of point X. The spectrum corresponds to irreducible representations of X_2 and X_4

The average of the T product should be calculated according to Wick's theorem with due account to anomalous pairings. Thereupon Ω_z is expressed in terms of the Green's function of the superconductor. It is not difficult to see that the temperature-dependent contribution to Ω_z is

$$\delta\Omega_z = -\{H'\}^2_{12} \frac{ck}{\pi\upsilon} \int_0^{\tilde{\omega}} d\xi \tanh \frac{\sqrt{\xi^2+\Delta^2}}{2kT} \Big/ \sqrt{\xi^2+\Delta^2} \quad . \tag{6.3.47}$$

The value of the energy gap of the superconductor Δ is found from the usual BCS theory equation

$$VN(0) \int_0^{\omega_D} d\xi \tanh \frac{\sqrt{\xi^2+\Delta^2}}{2kT} \Big/ \sqrt{\xi^2+\Delta^2} = 1 \quad . \tag{6.3.48}$$

The expressions (6.3.47,48) jointly determine the temperature dependence of Ω_z. When $T > T_c$, $\Delta = 0$ and (6.3.47) yields a logarithmic behaviour $\Omega_z \sim \ln(\tilde{\omega}/T)$. At $T < T_c$ the integral in (6.3.47) may be replaced, with "logarithmic accuracy", by the integral from the gap equation (6.3.48), since the cutoff energy and the Debye temperature are of the same order. In this situation (6.3.47) turns out to be independent of temperature at all $T < T_c$. Thus, the shear modulus $c_{11} - c_{12}$ should have the following temperature dependence

$$c_{11} - c_{12} = c_{11}^{0} - c_{12}^{0} - 3R(T) \quad ,$$

$$R(T) = \begin{cases} \dfrac{ckD_1^2}{\pi\upsilon} \ln\left(1.14\dfrac{\tilde{\omega}}{kT}\right) \quad , & T > T_c \\ \dfrac{ckD_1^2}{\pi\upsilon} \ln\left(1.14\dfrac{\tilde{\omega}}{kT_c}\right) \quad , & T < T_c \end{cases} \quad . \tag{6.3.49}$$

This result agrees perfectly well with $c_{11} - c_{12}$ measured in a nontransforming V_3Si specimen [6.156] (Fig.6.32). For Nb_3Sn, the agreement for a wide temperature interval is only of a qualitative nature [6.166].

It should be noted that the delay of lattice modulus softening at the superconducting transition may also be explained on the basis of the Weger-Labbé-Friedel model [6.242]. Subject to the assumption that $E_F = 21$ K for V_3Si, $(c_{11} - c_{12})$ turns out to be independent of T in accordance with experiments. However, to account for the observed dependence of the martensitic transition point in V_3Si on the magnetic field on the basis of the same model a value of $E_F = 42$ K is required. At this value the agreement of the moduli calculated according to the Weger-Labbé-Friedel theorem with experiments becomes significantly poorer. The agreement is also poor for Nb_3Sn.

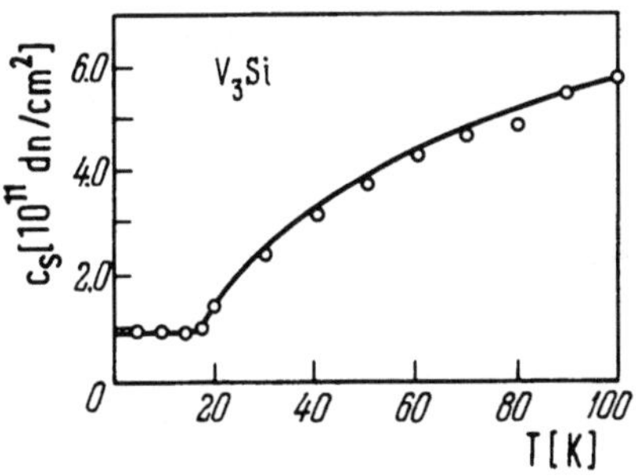

Fig.6.32. Temperature dependence of the shear modulus in a nontransforming V3Si specimen. The experimental points have been borrowed from [6.156] and the curve has been obtained from (6.3.49) with the fitting values of the parameters c_s^0 and $D_1^2\, c/\pi\upsilon$

6.3.4 Gor'kov's Model II

Formula (6.3.33) gives a general expression for the energy spectrum of electrons in the vicinity of the point $X = \{0,0,\pi/a\}$ for A-15 structures. If one calculates the electron spectrum in the vicinity of this point in the model of weakly interacting chains using (6.2.6), one sees that it has the same form (6.3.33) with the parameters C and C' now proportional to I^2/J and hence small compared to υ due to the smallness of I/J. It is now possible to generalize (6.3.33), obtained from the method of invariants, by replacing p_x and p_y with $\sin p_x$ and $\sin p_y$. This is equivalent to extending (6.3.33) from the vicinity of the point X to the entire edge of the Brillouin zone. It is also necessary that the values of energy in the XR direction coincide with the corresponding expressions (6.2.12) obtained in the model of weakly interacting chains. This leads to the condition $C = C'$ so that as a result we come to

$$E_{1,2} = T^*\left\{\sin^2\left(\frac{p_x a}{2}\right) + \sin^2\left(\frac{p_y a}{2}\right) - 2 \pm \sqrt{\left(\frac{\upsilon p_z}{T^*}\right)^2 + \left[\sin^2\left(\frac{p_x a}{2}\right) - \sin^2\left(\frac{p_y a}{2}\right)\right]^2}\right\} , \qquad (6.3.50)$$

where

$$T^* = 4I^2/J \quad , \qquad (6.3.51)$$

which GOR'KOV [6.243,244] used to approximate the spectrum in the interacting chains model. In Fig.6.33 the solid lines represent the spectrum (6.3.50) for a number of symmetric directions. It coincides with the spectrum (6.2.12) obtained from the strong-coupling method in the model of weakly interacting chains in all places except for the lower branch in the XM and MR directions. In these directions, the strong-coupling method yields different branches. However, from the character of the curves in the XM and XR directions we see that GOR'KOV's formula (6.3.50) represents the spectrum of weakly interacting chains in a wide vicinity of point X at the Brillouin edge.

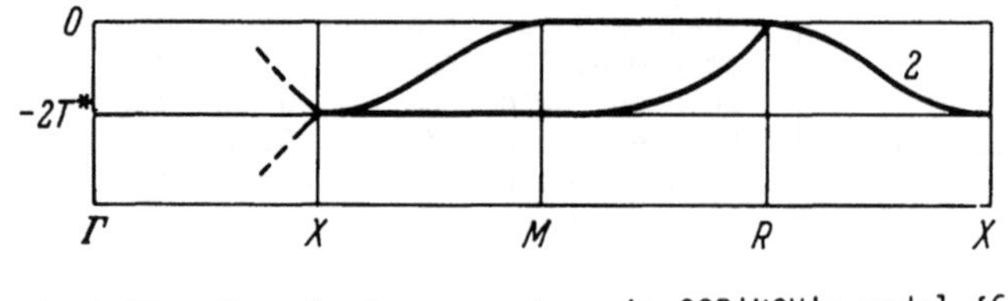

Fig.6.33. The electron spectrum in GOR'KOV's model [6.243,244] of weakly interacting chains, calculated from (6.3.50)

The increasing dispersion curves ensure a high density of states over an interval of about $2T^*$. To be more exact, there is a logarithmic singularity in $N(\varepsilon)$ when $\varepsilon \to 0$ [6.243]:

$$N(\varepsilon) = N(0)\frac{2}{\pi^2}\ln\left(\frac{32T^*}{|\varepsilon|}\right) \quad . \tag{6.3.52}$$

According to GOR'KOV's estimate, T^* in V_3Si and Nb_3Sn compounds amounts to several hundreds of degrees. Here $N(0)$ is the volumetric density of state for a "plane spectrum" of weakly interacting chains. Calculated for one cube edge and one spin we have

$$N(0) = \frac{1}{2\pi \upsilon a^2} \quad . \tag{6.3.53}$$

The same singularity occurs near the energy $\varepsilon = -2T^*$, which is due to the horizontal section on the XM line. This section results from the model character of (6.3.50). In the model of weakly interacting chains this peculiarity must be absent.

GOR'KOV believed that the Fermi level in A-15 compounds exhibiting anomalous properties lies somewhere in between the two peaks previously mentioned. Suppose it falls within the vicinity of the upper-lying peak of the peaks with the energy $\varepsilon = 0$. It may be shown that in the presence of a deformation the electron spectrum is determined by (6.3.50) in which under the root sign we have to add the expression $\sin^2(p_x a/2) - \sin^2(p_y a/2)$ to the quantity $D(\varepsilon_{xx} - \varepsilon_{yy})$. Expression (6.3.39) for the contribution which the chain oriented in the z-direction makes to the free energy may be written in the form of an integral over energy with the density of states (6.3.52)

$$\delta\Omega_z = -\frac{1}{3}N(0)\frac{2}{\pi^2}T\int_{-\infty}^{\infty} d\varepsilon \ln\frac{T^*}{|\varepsilon|}\ln\left[1 + \exp\left(-\frac{\varepsilon + \Delta_z - \mu}{T}\right)\right] \quad , \tag{6.3.54}$$

where Δ_z is the energy shift dependent on deformation and magnetic field. Consider for the moment the deformation so that $\Delta_z = D(\varepsilon_{xx} - \varepsilon_{yy})$. For the two other chains, the corresponding contributions Ω_x and Ω_y are obtained from (6.3.54) by a cyclic permutation of the coordinates. Differentiating the free energy with respect to deformation, we obtain an expression for the shear modulus $c_s = c_{11} - c_{12}$:

$$c_s(T) = -N(0)D^2\frac{2}{\pi^2}\left[\ln\frac{T_m^0}{T} + S\left(\frac{\mu}{T}\right)\right] \quad , \tag{6.3.55}$$

where T_m^o is the temperature at which the shear modulus goes to zero if $\mu = 0$, i.e., T_m^o is the martensitic transformation temperature if the chemical potential falls exactly within the peak in the density of states and

$$S\left(\frac{\mu}{T}\right) = \psi\left(\frac{1}{2}\right) - \mathrm{Re}\left\{\psi\left(\frac{1}{2} + i\frac{\mu}{2\pi T}\right)\right\} \quad . \tag{6.3.55}$$

The magnetic susceptibility can be calculated in a similar way. The temperature-dependent contribution to χ determined by the singularity peak in $N(\varepsilon)$ is equal to [6.244]:

$$\chi(T) = -\mu_B^2 \int_{-\infty}^{\infty} \frac{N(\varepsilon)d\varepsilon}{4T\coth^2\left(\frac{\varepsilon - \mu}{2T}\right)} \quad . \tag{6.3.57}$$

Calculation of the integral yields

$$\chi(T) = N(0)\mu_B^2 \frac{2}{\pi^2}\left[\ln\frac{4\gamma T^*}{\pi T_m^o} + \ln\frac{T_m^o}{T} + S\left(\frac{\mu}{T}\right)\right] \quad . \tag{6.3.58}$$

Comparing (6.3.55) and (6.3.58), we have the result that the temperature dependence of the shear modulus and the magnetic susceptibility occurring for the cubic phase is the same and therefore the following simple equation should hold

$$-\frac{\chi(T) - \chi(0)}{c_s(T)} = \frac{\mu_B^2}{D^2} \quad , \tag{6.3.59}$$

which may be verified by measurements of χ and c_s using the same specimens. The temperature dependence of the quantities c_s and χ is determined by the function

$$\eta\left(\frac{T}{\mu}\right) = \ln\frac{\mu}{T} + S\left(\frac{\mu}{T}\right) \quad , \tag{6.3.60}$$

constructed in Fig.6.34 for a chemical potential μ_0 with $T = 0$. It is this dependence which qualitatively describes the experimentally observed behaviour $\chi(T)$ and $c_s(T)$ in the cubic phase. Formulae (6.3.55) and (6.3.58) are derived on the assumption of μ, $T \ll T^*$. At arbitrary ratios between these parameters as well as when account is taken of the temperature dependence of the chemical potential μ, numerical calculation leads to temperature dependences of χ and c_s which agree with the experimental results if $\mu_0/T \lesssim 0.2$.

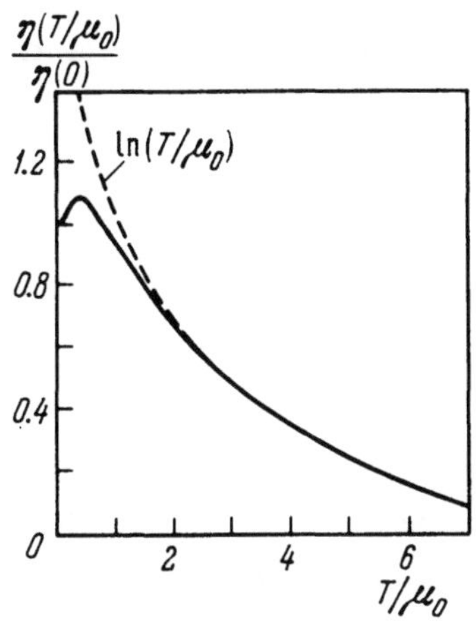

Fig.6.34. The graph of the function $\eta(T/\mu_0)$ determining the temperature dependence of the magnetic susceptibility and of the shear modulus in the cubic phase of A-15 compounds according to Gor'kov's model

The martensitic transformation temperature T_m is determined from the equation $c_s(T_m) = 0$. In the presence of a magnetic field, the chemical potential is displaced by the value of $-\mu_B H$, which should lead to a shift of T_m by an amount of ΔT_m. It is easy to show that this shift is quadratic with respect to the field, just as in the Labbé-Friedel model [see (6.1.1)]. For $\mu = 0$, the coefficient $\alpha \approx 0.21$ is close to the experimental value $\alpha = 0.207$ [6.144].

The equation for determining the structural transition temperature T_m

$$\ln\frac{T_m}{T_m^0} = \psi\left(\frac{1}{2}\right) - \mathrm{Re}\left\{\psi\left(\frac{1}{2} + i\frac{\mu}{2\pi T_m}\right)\right\} \tag{6.3.61}$$

coincides in form with Sarma's equation (3.1.80) in the theory of the paramagnetic effect in superconductors. As already described in Sect.3.1, this equation gives an ambiguous dependence of T_m on the parameter μ (Fig.3.3) which points to the possibility of a first-order phase transition at large values

$$\mu > \mu_c = \frac{\pi}{2\gamma} T_m^0 \quad . \tag{6.3.62}$$

When $T < T_m$ a spontaneous tetragonal deformation in the crystal arises along the z-axis, which is described by the following deformation tensor

$$\varepsilon_{xx} = \varepsilon_{yy} = \frac{\varepsilon_0}{3} \quad , \quad \varepsilon_{zz} = -\frac{2}{3}\varepsilon_0 \quad , \quad \varepsilon_0 = \frac{c}{a} - 1 \quad . \tag{6.3.63}$$

The lattice portion of the free energy (6.3.41) in this case is equal to $\varepsilon_0^2(c_{11}^0 - c_{12}^0)/3$. By adding the electron contribution (6.3.54) for all the three chains subject to the condition (6.3.63) and performing minimization with respect to ε_0, we obtain an equation for the equilibrium value of the spontaneous deformation parameter ($\Lambda = D\varepsilon_0$)

$$\Lambda\left[\ln\frac{T_m^0}{T} + S\left(\frac{\mu}{T}\right)\right] = \frac{1}{4}\int_{-\infty}^{\infty} d\varepsilon\, \ln\frac{T_m^0}{|\varepsilon|}$$

$$\times\left[\tanh\left(\frac{\varepsilon-\mu-\Lambda}{2T}\right) - \tanh\left(\frac{\varepsilon-\mu+\Lambda}{2T}\right) + \frac{\Lambda}{T}\coth^{-2}\left(\frac{\varepsilon-\mu}{2T}\right)\right] \quad . \tag{6.3.64}$$

With $\mu = 0$, this equation yields the typical behaviour of the order parameter during a second-order phase transition. This behaviour agrees qualitatively with experiments for Nb_3Sn. At large values of μ, (6.3.54) should give a jump in the spontaneous deformation ε_0 since, as has been stated above, the transition will be of the first order.

In the tetragonal phase, the shear modulus and the magnetic susceptibility have different temperature dependences:

$$c_s = -N(0)D^2\frac{2}{\pi^2}\left[\ln\frac{T_m^0}{T} + \frac{1}{2}S\left(\frac{\mu+\Lambda}{T}\right) + \frac{1}{2}S\left(\frac{\mu-\Lambda}{T}\right)\right] \quad , \tag{6.3.65}$$

$$\chi = N(0)\mu_B^2\frac{2}{\pi^2}\left\{\ln\frac{4\gamma T^*}{\pi T_m^0} + \ln\frac{T_m^0}{T} + \frac{1}{3}\left[S\left(\frac{\mu}{T}\right) + S\left(\frac{\mu+\Lambda}{T}\right) + S\left(\frac{\mu-\Lambda}{T}\right)\right]\right\} \quad . \tag{6.3.66}$$

Thus, the Gor'kov model under consideration makes it possible to explain, in A-15 compounds, the strong temperature dependence of the major anomalies in the magnetic susceptibility and in the shear lattice moduli. Apart from this, it leads to the observed softening of phonon frequencies in a broad region of the Brillouin zone [6.245]. Again, it needs to be recalled that all these anomalies are due to the logarithmic peak in the density of states, which arises from the splitting of the doubly degenerate levels of the electron spectrum of the isolated chains due to interchain electron transitions.

However, recent experiments by SHIRANE and AXE [6.246] on inelastic neutron scattering in Nb_3Sn have shown that the phonon modes are considerably softened in the [ξξξ] direction when the values of ξ are fairly large and that this softening should be absent in the model [6.243,244]. The neutron diffraction experiment has been performed on a specimen cut out of the same single crystal. It was earlier revealed [6.121] that transverse phonon modes soften in different directions ([ξξ0] and [ξ00]) and this was observed in a wide range of the Brillouin zone including its boundaries (Fig.6.35). Detection of the

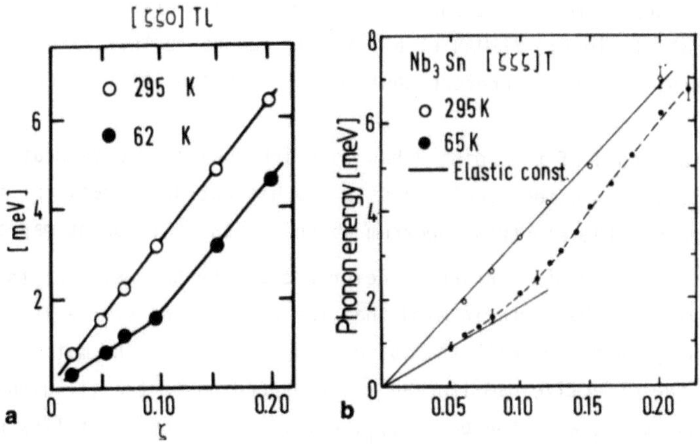

Fig.6.35. The softening of transverse phonon modes in Nb_3Sn according to the neutron diffraction investigations by AXE and SHIRANE [6.246,281]

transverse-mode [ξξξ] softening requires a stricter allowance for the interchain interaction than is made in Gor'kov's model II.

6.3.5 The Birman-Lee-Williamson Model

All the band structure models of A-15 compounds which we have considered above are one-dimensional or quasi-one-dimensional. Although they account for most of the experimental facts showing up as anomalies in the properties of a number of compounds, the idea of a quasi-one-dimensional nature of the electrons in the spectrum of d-states in these compounds is, as stated previously, in poor agreement with numerical band structure calculation results.

BIRMAN, LEE and WILLIAMSON [6.247-249] have proposed a model disregarding the assumption about the quasi-one-dimensionality of the A-15 systems but taking into account their crystal space symmetry described by the group O_h^3. Ideologically, their model is a development of Gor'kov's model for the role of Brillouin-zone symmetry points and their closest vicinities in forming anomalous contributions to the density of states N(E) of the electron spectrum. If the Fermi surface passes near one of these points, the Fermi level falls within an anomalous section N(E), and this gives rise to anomalous behaviour of the properties of a given compound. Gor'kov's model I was based on the assumption that in V_3Si and Nb_3Sn the Fermi surface passes near the point X. BIRMAN et al. [6.247] assumed that in these compounds the Fermi surface passes near the point $R = \{\pi/a, \pi/a, \pi/a\}$ lying in a corner of the Brillouin zone. However, in contrast to GOR'KOV who made the additional assumption that the spectrum is of a quasi-one-dimensional character (this enabled analytic calculations), BIRMAN et al. introduced no further assumption of quasi-one-dimensionality, therefore their model is essentially three-dimensional.

The wave vector group of the point R has four irreducible representations, one of which, R_4, is six-dimensional (this is the highest degeneracy multiplicity in crystals). The electron spectrum which is near this point may be found using the technique of invariants, described at the beginning of this section. It should also be borne in mind that the wave vector star R is a one-arm star and $G_{\underline{k}}$ coincides with the space group O_h^3 itself. The matrix $H_{ik} = (\psi_i^* H \psi_k)$ of the effective Hamiltonian H corresponding to the six-dimensional representation R_4 with the basis $\{\psi_1, \psi_2, \ldots, \psi_6\}$ has the following nonzero elements (as well as their Hermitian conjugations)

$$\begin{aligned}
&H_{11} - Ap^2 - Bp_z^2 = -(H_{55} - Ap^2 - Bp_z^2) = C(p_x^2 - p_y^2)\\
&H_{22} - Ap^2 - Bp_x^2 = -(H_{66} - Ap^2 - Bp_z^2) = C(p_y^2 - p_z^2)\\
&H_{33} - Ap^2 - Bp_y^2 = -(H_{44} - Ap^2 - Bp_z^2) = C(p_z^2 - p_x^2)\\
&H_{12} = H_{45} = Dp_xp_y , \quad H_{13} = H_{46} = Dp_zp_x , \quad H_{23} = H_{56} = Dp_yp_z ,\\
&H_{15} = Fp_z , \quad H_{26} = Fp_x , \quad H_{34} = Fp_y\\
&H_{16} = Gp_y , \quad H_{24} = Gp_z , \quad H_{35} = Gp_x \quad .
\end{aligned} \tag{6.3.67}$$

The dispersion relations for the six energy bands are determined from the secular equation $\det|H - E| = 0$. BIRMAN et al. make some assumptions respecting the constants A, B, C, D, E, F and G defining the energy which, as it turns out [6.249], does not tangibly affect the accuracy of the results obtained. These assumptions are: D and F are equal to zero, or D and G are equal to zero. In both cases one obtains identical expressions for the six energy bands

$$E_z^{\pm}(\underline{p})/T^* = p_x^2 + p_y^2 \pm \sqrt{\upsilon^2 p_z^2 + \lambda^2 (p_x^2 - p_y^2)^2} \quad . \tag{6.3.68}$$

Here T^*, υ and λ are parameters of the proper dimension, which are expressed in terms of the constants A, B, ..., but these expressions will be different for two versions of simplifying assumptions. Formula (6.3.68) yields dispersion relations for two bands. The energies $E_x^{\pm}(\underline{p})$ and $E_y^{\pm}(\underline{p})$ for the four other bands can be obtained from (6.3.68) using the cyclic permutation of indices.

To calculate the density of states corresponding to some vicinity of the point R it is necessary to introduce the cutoff of the momenta corresponding to the dimensions of the region selected. Integration over the region with a maximum radius q leads to the following compact expression [6.249]

$$N_R(E) = 3D_0 \int_0^1 dx\, K\left(\frac{\lambda x}{\epsilon - x}\right) \quad , \tag{6.3.69}$$

where K(x) is a first-order elliptic inegral,

$$\epsilon = \frac{E}{T^* q^2} \quad , \quad D_0 = \frac{q^2}{\upsilon T^* a^3} \tag{6.3.70}$$

and q is the momentum cutoff which is assumed to be small. The factor 3 in front of the integral in (6.3.69) arises from three pairs of bands making an equal contribution to N(E).

Expression (6.3.69) describes a peak in N(E) in the vicinity of $E = 0$, i.e., near the point R, whose width increases as λ deviates from 1. For $\lambda = 1$, (6.3.69) exhibits a logarithmic singularity $N(E) \sim \ln(1/E)$. This singularity was found by GOR'KOV [6.243] in model II since in the region of low momenta the electron spectrum (6.3.50) of his model coincides with (6.3.68). In the Birman-Lee-Williamson model which we have considered above N(E) does not contain any singularity, except for the case $\lambda = 1$, but involves only a local increase in the density of states in the vicinity of $E = 0$; it is this increase which is directly responsible for the physical anomalies occurring in A-15 systems. In [6.249] a set of lattice and magnetic properties of crystals having the electron spectrum (6.3.68) has been calculated.

For calculations of the magnetic susceptibility $\chi(T)$ one should use the general formula for the free energy of the ideal Fermi gas (6.3.38), by extending the summation over i to all the six bands and by including the magnetic field by the replacement: $E_i(\underline{p}) \to E_i(\underline{p}) - \mu_B H$. This gives

$$\chi(T) = -\left(\frac{\partial^2 \Omega}{\partial H^2}\right)_{H=0} = \chi_c + \frac{\mu_B^2}{kT} \int dE N_R(E) \exp\left(-\frac{E-\mu}{kT}\right) \Bigg/ \left[1 + \exp\left(\frac{E-\mu}{kT}\right)\right]^2 , \tag{6.3.71}$$

which must be computed numerically. Here χ_c is a temperature-independent term in the susceptibility and also includes the orbital contribution.

The elastic moduli may be calculated from the same general expression for the electron contribution to the free energy by adding to (6.3.68) deformation dependent terms. Using the method of invariants the electron spectrum in the presence of deformations is given by [6.249]

$$E_z^{\pm}(\underline{p})/T^* = p_x^2 + p_y^2 + \frac{d_3}{T^*}\left(\varepsilon_{xx} + \varepsilon_{yy} + \varepsilon_{zz}\right) + \frac{d_2}{T^*}\varepsilon_{zz}$$
$$\pm\sqrt{\upsilon^2 p_z^2 + \left[\lambda\left(p_x^2 - p_y^2\right) + \frac{d_1}{T^*}\left(\varepsilon_{xx} - \varepsilon_{yy}\right)\right]^2} \quad . \tag{6.3.72}$$

Next consider a tetragonal deformation like (6.3.63). Expanding the free energy in powers of deformations and comparing the result with (6.3.41) we find the shear modulus

$$c_s(T) = c_s^0 - \frac{3}{2}D_0 d_1^2\left[\left(1 + \frac{d_2^2}{3d_1^2}\right)\frac{\chi(T) - \chi_c}{\chi_c} + J(\mu,T)\right] \quad , \tag{6.3.73}$$

where $\chi(T)$ is determined by (6.3.71) and $J(\mu,T)$ was given in [6.249].

In order to calculate the spontaneous deformation during the first-order transition it is necessary to expand the free energy in a power series up to terms of $\sim \varepsilon_0^4$. This expansion may be written in the following form:

$$\Omega(\varepsilon,T) = \Omega_0(T) + \frac{1}{2}A\varepsilon_0^2 + \frac{1}{3}B\varepsilon_0^3 + \frac{1}{4}D\varepsilon_0^4 \quad , \tag{6.3.74}$$

where

$$A(T) = \frac{4}{3}c_s(T) \quad , \tag{6.3.75}$$

$$B(T) = B^0 + D_0 d_1^2 d_2^2 \frac{\partial}{\partial\mu}\left[J(\mu,T) + \frac{1 - d_2^2}{9d_1^2}\frac{\chi(T) - \chi_c}{\chi_c}\right] \tag{6.3.76}$$

and $D(T)$ is not given because it is rather cumbersome. The transition temperature T_c as well as the spontaneous deformation ε_m at the point of transition can be found from the condition of thermodynamic equilibrium of the cubic and tetragonal phases

$$\Omega(\varepsilon_m,T_m) = \Omega(0,T_m) \quad , \quad \left[\frac{\partial\Omega(\varepsilon_0,T)}{\partial\varepsilon_0}\right]_{\varepsilon_m,T_m} = 0 \quad . \tag{6.3.77}$$

These equations are equivalent to the two following expressions

$$\varepsilon_m = -3A(T_m)/B(T_m) \quad , \quad A(T_m) = 2B^2(T_m)/9D(T_m) \quad . \tag{6.3.78}$$

Since $A(T_m) \sim c_s(T_m)$ is not zero, the tetragonal deformation is finite, and the phase transition is of first order. This already follows from the presence of a cubic term in the expansion (6.3.74). The sign of the spontaneous deformation is determined by the sign of $B(T_m)$.

The effects of the magnetic field on the lattice properties follow from the thermodynamic properties of a system whose electron spectrum (6.3.72) contains the Zeeman energy. Thus, the velocity of sound, υ_s, in a magnetic field H is enhanced by $\Delta\upsilon_x$ [6.249]:

$$\frac{\upsilon_s \Delta\upsilon_s}{H^2} = \frac{\mu_B^2}{4\rho} \frac{\partial^2 c_s(T)}{\partial\mu^2} \quad , \tag{6.3.79}$$

where ρ is the density of the crystal. The magnetic transition point T_m is shifted in the magnetic field by a quantity ΔT_m according to [6.249]

$$\frac{\Delta T_m}{T_m} = -\alpha\left(\frac{\mu_B H}{T_m}\right)^2 \quad , \quad \alpha \approx \frac{1}{2}\left(T^2 \frac{\partial^2 c_s}{\partial\mu^2} \Big/ T \frac{\partial c_s}{\partial T}\right)_{T_m} \quad . \tag{6.3.80}$$

In this fashion, a number of experimentally measured properties such as $\chi(T)$, $c_s(T)$, ε_m, T_m, $\Delta\upsilon_s$ and ΔT_m can be calculated according to (6.3.71,73, 78-80). They are expressed through several phenomenological parameters: the T^*, υ and λ characteristics of the band spectrum, the cutoff parameter q and the chemical potential μ. The lattice properties are expressed in terms of one more couple of parameters, d_1 and d_2, which characterize the deformation potential.

The magnetic susceptibility $\chi(T)$ and the shear modulus $c_s(T)$ were measured over a wide temperature range for Nb_3Sn and V_3Si (Fig.6.36). The measurements of both properties for Nb_3Sn were made on one specimen. By fitting the theoretical expressions (6.3.71,73) to these experimental data the following values of the parameters have been found:

$$\begin{aligned} &T^* = 7.33\cdot10^{-2}\ \text{eV} \ , \quad \lambda = 0.7 \ , \quad \upsilon = 0.427 \\ &\mu = 0.401\cdot10^{-2}\ \text{eV} \ , \quad q = 0.427 \ , \quad d_1 = 2.11\ \text{eV} \ , \quad d_2 = 0 \quad , \end{aligned} \tag{6.3.81}$$

for Nb_3Sn and

$$\left.\begin{aligned} &T^* = 6.00\cdot10^{-2}\ \text{eV}, \quad \lambda = 0.6, \quad \upsilon = 0.291 \\ &\mu = 0, \quad q = 0.415, \quad d_2 = d_1 = 1.32\ \text{eV} \end{aligned}\right\} \tag{6.3.82}$$

for V_3Si.

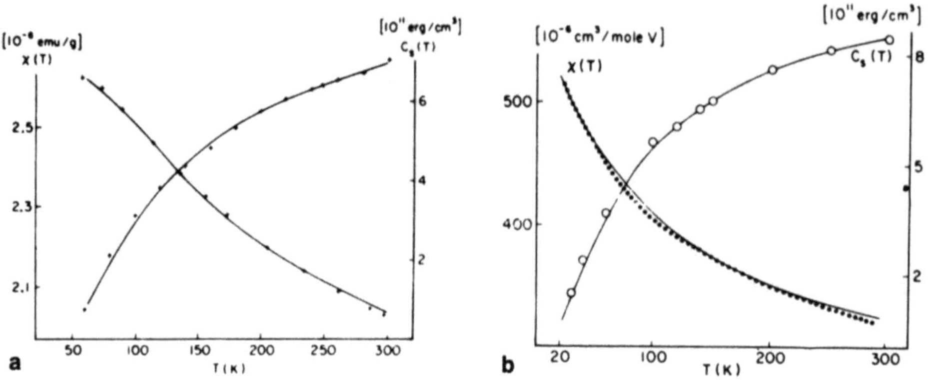

Fig.6.36. The magnetic susceptibility $\chi(T)$ and the shear modulus $c_S(T)$ as a function of temperature for (a) Nb_3Sn [6.158] and (b) V_3Si [6.139,156]

The parameters thus determined for V_3Si enable one to calculate the spontaneous deformation ε_m and the coefficient α according to (6.3.78,80): $\varepsilon_m = -2.10^{-3}$, $\alpha = 0.18$. These values and also the magnitude of $\Delta\upsilon_S$ are in satisfactory agreement with experiments [6.249]. Thus, the Birman-Lee-Williamson model which we have set forth here gives good agreement between theoretical predictions and experiments. The most substantial difference from other models describing the electron structure of A-15 compounds is that it is three-dimensional. Note that the anomalous behaviour of the physical properties is not ascribed to some singularity in the density of states (root singularity in the Labbé-Friedel model or logarithmic singularity in Gor'kov's model II), but to a local increase in N(E) in the region of energies corresponding to the position of the Fermi surface near the point R of the Brillouin zone.

The Birman-Lee-Williamson model was subjected to criticism by BHATT and MATTHEISS [6.250] who attempted to fit the band spectrum parameters of [6.187,204,251] for V_3Si and Nb_3Sn to the results of numerical calculation of the spectrum of these compounds according to the APW method. One might think that $T^* \sim 10^{-2}$ eV determines the width of the band, the point R of which is described by (6.3.68). However, that width turns out to be two orders of magnitude smaller than the one obtained in the calculations [6.187,204,251]. As correctly explained by LEE et al. [6.252], the parameter T^* has no direct relationship to the width of the entire zone; it characterizes only the dispersion curves in the vicinity of the point R. Moreover, the accuracy of band calculations is at present 3-30 mRy and thus permits no energy resolution in the interval of 0.3-1 mRy required for the theory of A-15 compounds with anomalous physical properties. At the same time LEE, BIRMAN and WILLIAMSON [6.252] have thoroughly analyzed a number of other characteristics of their model, particularly the sensitivity of the results to the cutoff parameter q, to the simplifying assumptions ($D = 0$, $F = 0$), etc. They have also shown that the results of calculations for different physical properties are highly stable.

The reader may wonder at the abundance of models which successfully account for the anomalous physical properties of a number of A-15 compounds. It should be borne in mind that each of the models described above contains a large number of fitting parameters enabling the scientists to obtain a satisfactory fit to a set of experimental facts. The available experimental data accumulated by now are inadequate and not reliable enough to permit an unambiguous choice of the appropriate model. But one thing is certain, the anomalous properties of A-15 compounds are explicable on the basis of some fine structure of the density of states in the vicinity of the Fermi level over an interval of about 100 K.

6.3.6 Bhatt's Model

Using the APW method, MATTHEISS performed band-structure calculations for a large number of A-15 compounds and compared the results with the calculations according to the LCAO method employing the strong-coupling approximation. This comparison yields the band-spectrum parameters, i.e., the matrix elements between different orbitals. As already noted in Sect.6.2, these results did not confirm the quasi-one-dimensional behaviour of the energy bands near the Fermi surface. It turned out that the density of states near the Fermi level is formed mainly by (ca.50%) atomic orbitals $\delta_1(x^2 - y^2)$ centred on the atoms of the transition metal in chains containing (ca. 25%) $\pi(xz, yz)$ orbitals. In all the calculated compounds, two relatively isolated bands, formed by these orbitals which may be assumed to be responsible for their anomalous properties, can be singled out near the Fermi surface.

Based on these data, BHATT [6.253,254] constructed a model of A-15 compounds by approximating the band structure near the Fermi surface by two bands formed by orbitals on the six atoms of the chain which are contained in a unit cell. The matrix elements between the nearest atoms (in the chains) and the next-nearest neighbours were chosen as the fitting parameters. For the interchain coupling parameters which amount in magnitude to less than 15% of the intrachain parameter, the density of states has a sharp peak from two saddle points in the bands, one of which (point M) lies in the lower band and the other (in the ΓX direction) in the upper band. If the Fermi level lies near this peak the electronic system exhibits a Jahn-Teller instability with respect to the tetragonal deformation associated with the removal of the three-fold degeneracy of states in the point M and a Peierls instability due to the contribution ΓX of the saddle points (for more details see Sect.6.5).

Using the experimental data on the structural transition temperature T_m and on the temperature behaviour of the elastic modulus $(C_{11} - C_{12})$ in the cubic phase, BHATT determined the electron-lattice coupling constant and the elastic lattice moduli A_{11} and A_{12}. These parameters were employed to calculate the magnetic susceptibility, tetragonal distortion below T_m and the jump in heat capacity at the transition point. The results are in good agreement with experiments for V_3Si and in a somewhat poorer agreement with Nb_3Sn.

Now we sum up the theoretical investigations into A-15 compounds. Several models were proposed to explain the anomalous properties of a number of these compounds: (i) Labbé-Friedel, (ii) Gor'kov I, (iii) Birman-Lee-Williamson, (iv) Bhatt. Models (i,ii) are quasi-one-dimensional, and (iii,iv) are three-dimensional. In spite of this substantial difference between them, all the models have some features in common. They all assume that the Fermi surface passes near a special point of the Brillouin zone where the density of states on the Fermi surface has a clearly marked nonmonotonicity such as a peak. This is the point Γ in the Labbé-Friedel model, the point X in Gor'kov's model I, the point R in the Birman-Lee-Williamson model, and saddle points (M and others) in Bhatt's model. If it is assumed that the energy scale of this peculiarity is small (about tens or hundreds of degrees), a satisfactory explanation may be offered for a large set of anomalous properties of these compounds: lattice, magnetic, elastically magnetic and a number of other superconducting properties.

However, one should not expect too much from the above models as for the anomalous properties since each of them introduces a great deal of fitting parameters. Thus, the main logical difficulty in explaining A-15 compounds is not lack of physical ideas but rather their abundance and the consequent uncertainty in the choice of the dominating idea. The present status in experiments does not enable one to choose between the existing models. At the same time, none of these models is microscopic in the direct sense of the word since each of them uses a number of undeterminate band-structure parameters. Incidentally, band-structure calculations from first principles do not agree with any of the models described above, except with Bhatt's model which is based explicitly on the results of these calculations.

We have considered the general properties of the models. A more detailed treatment of the structural transition on the basis of these models will be given in the two subsequent chapters, where the status of the problem of superconductivity in A-15 compounds will be discussed, too.

6.4 Analysis of the Electron-Phonon Interaction and T_c

6.4.1 Experimental Data on the Electron-Phonon Coupling Constant λ

It has already been noted in Sect.4.4 that the electron-phonon interaction constant in superconductors can be found either from McMillan's formula (2.3.27) by substituting experimental data on T_c and θ_D and making assumptions for the magnitude of μ^*, or from independent measurements of the physical properties related to λ by an explicit equation.

However, when determining λ from McMillan's formula (2.3.27) one runs into difficulties with the estimation of θ_D from heat capacity data. The difficulty of singling out the electron contribution with C_V/T extrapolated to 0 K, especially in compounds exhibiting structural instability, leads to a situation where, for example, for V_3Si (Fig.6.37) it is possible to suggest two methods of extrapolating C_V/T to 0 K and to obtain two values of γ and θ_D: $\gamma = 17.1$ mj/(mol K^2), $\theta_D = 700$ K and $\gamma = 15.1$ mj/(mol K^2), $\theta_D = 400$ K. As we see, the difference in the values of θ_D amounts to 300 K in this case. This causes a certain spread in the values of λ calculated from (2.3.27) by different

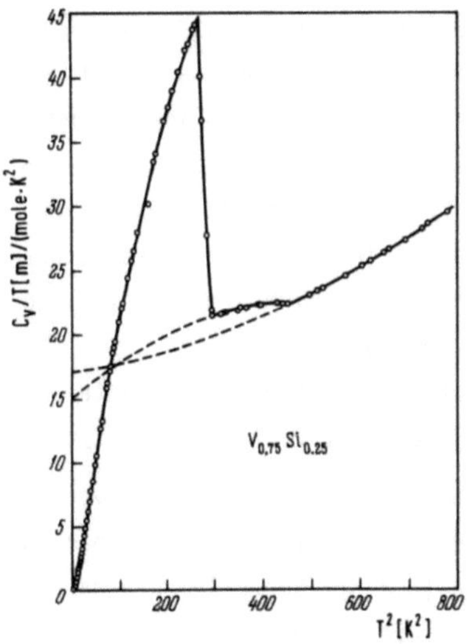

Fig.6.37. Temperature dependence of the heat capacity in V_3Si [6.19]

authors for the same compounds, depending on the choice of the values for γ and θ_D [6.127,255,256].

In Table 6.9 we cite the values obtained in [6.191] (according to McMillan) for the parameter λ in A-15 compounds. The values of $\langle\omega^2\rangle$ were estimated from measured phonon spectra [6.167]. In the same table we quote the values of λ determined by other methods, i.e., NMR spectra [6.257-260], resistivity, heat capacity and tunnel and optical measurements [6.261-264].

Table 6.9. Electron-phonon coupling constant λ in A-15 compounds

Material	λ according to McMillan	NMR	Optics	Resistivity	Electronic heat capacity	Tunneling
Nb_3Ge	1.80		0.08	0.22		1.55
Nb_3Sn	1.44		0.46	0.52	1.17	
Nb_3Al	1.60				1.07	
V_3Si	1.12	0.76	0.06	0.06	0.86	
V_3Ga	1.17	0.86			0.91	
V_3Sn	0.52				0.56	

Note: λ (according to McMillan) corresponds to λ_{exp} in Table 6.5

6.4.2 Analysis of λ and T_c

The highest $T_c \sim 20$ K occurring in the compounds Nb_3Ge, Nb_3Sn, V_3Si, etc., require high values of $\lambda \sim 1,2$, as follows from McMillan's formula with typical values of the Debye frequency $\omega_D \approx 300$ K. Unfortunately, as is evident from Table 6.9, these values of λ are far from always agreeing with estimates obtained from different experimental data, particularly on optical absorption and resistivity which lead to surprisingly small λ [6.262]. These estimates give values of the electron-phonon interaction $\lambda < 0.6$ which are too small to offer an explanation of such high T_c in these compounds.

The coupling constant λ is characteristic of the electron-phonon interaction whose energy structure is described by the spectral function $\alpha^2(\omega)F(\omega)$. Straightforward information concerning this characteristic can be derived from tunnel effect measurements with the subsequent solution of Eliashberg's equations following the McMillan-Rowell scheme [6.265]. Recently very careful tunnel effect measurements for Nb_3Sn have been made [6.266] which have shown an anomalous behaviour of $\alpha^2(\omega)F(\omega)$. It turns out that the values of interest calculated from this function according to (2.4.15) are $\lambda = 0.76$ and $\mu^* = -0.1$, and the form of the function $\alpha^2(\omega)F(\omega)$ differs appreciably from the density of states in the phonon spectrum $F(\omega)$ obtained from neutron diffraction data [6.167]. The value of λ proves to be less than that required to explain the high value of T_c, and, on the other hand, the nonphysical negative value of the Coulomb pseudopotential μ^* may point to the fact that in Nb_3Sn some new pairing mechanism operates [6.267] in addition to the normal pairing of electrons via phonons.

This should be kept in mind when analyzing T_c and λ for the entire class of A-15 compounds, the problem being in explaining not only the high T_c in a number of compounds but also the low T_c in their analogs.

According to BCS theory, T_c is determined by (1.2.11) in which $\lambda = VN(0)$. It might be assumed for the class of compounds having the same crystal and electron structure that the pairing interaction matrix element V varies weakly from substance to substance, and λ and, accordingly, T_c are determined mainly by the density of states on the Fermi surface N(0) [6.268]. However, the comparison of data on T_c and N(0) for a large number of compounds, presented in Fig.6.38, shows that this is far from being so in the class of A-15 compounds. The figure illustrates the densities of states calculated by the APW method [6.192] and also those found from electron heat capacity and magnetic susceptibility measurements. Although there is a considerable spread in the value of N(0) for every single compound, Fig.6.38 [6.267] demonstrates conclusively that T_c is not determined solely by the quantity N(0) at all, and the observed T_c does not fall on the dotted curve describing the functional dependence of T_c on λ [according to (1.2.11) of BCS theory]. The values of N(0) corresponding to the observed T_c turn out to be well below those expected in BCS theory, the values of N(0) proving to be very low for the superconductors Nb_3Ge, Nb_3Al and Nb_3Sn whose superconducting transition temperatures are the highest.

Special attention should be given to comparing the compounds Nb_3Ge and V_3Ge. Both of them have the same electron concentration which satisfies Matthias' rule for high T_c, and close values of N(0). Nevertheless, their T_c is quite different: Nb_3Ge has $T_c = 23$ K, and V_3Ge has $T_c = 6$ K.

The above discrepancies in the quantity λ for a number of high-temperature A-15 superconductors and the inability of the existing theory to explain the difference in T_c in such pairs as V_3Ge and Nb_3Ge, and also the absence of the isotopic effect in this class of compounds [6.182] call for a critical reconsideration of the existing theory of electron pairing via phonons in these compounds.

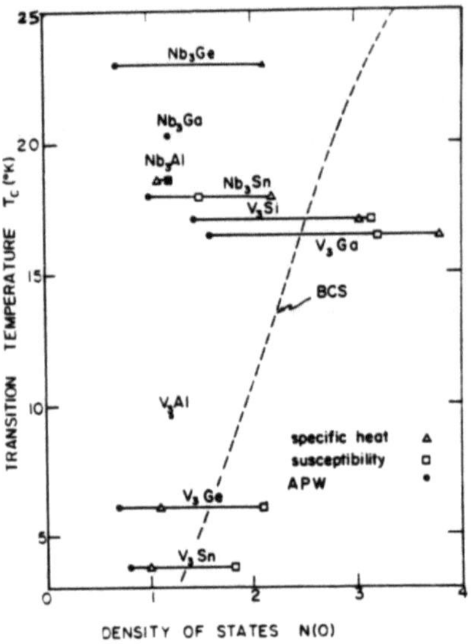

Fig.6.38. The superconducting transition temperature T_c as a function of N(0), the density of states on the Fermi surface in A-15 compounds [6.267]

6.4.3 Pairing of Electrons via Acoustic Plasmons

One of the possible mechanisms of electron pairing near the Fermi surface is the exchange of acoustic plasmons which was first considered by FRÖHLICH [6.269]. Subsequent papers [6.270,271] gave a general theoretical analysis of the influence which this mechanism might exert on the production of a superconducting state. Acoustic plasmons may arise in a two-component electron system as a collective branch of charge density oscillations similar to acoustic phonons occurring in a system of heavy ions and electrons. Much in the same way as the interaction of ions is screened by electrons, in a two-component electron system the lighter electrons screen the heavier ones, and this leads to a branch of plasma oscillations characterized by a frequency proportional to the wave vector.

The possibility of observing acoustic plasmons as well as their role in the pairing interaction of the electrons of the metal is determined by the decay of these plasmons. The latter is due to the disintegration of the collective oscillation into an electron-hole pair when the plasmon energy falls within the continuum of electron-hole excitations. First and foremost we discuss the problem concerning their decay in a simple two-component electron system. We shall call the electrons having a smaller effective mass s-electrons, and those with a greater mass d-electrons, although these conventional symbols may well not denote the relevant angular momenta. What is essential is only that s-electrons screen d-states so that acoustic plasmons will arise as oscillations of the density of localized d-states. To begin with, we shall give the fundamental equations of the theory of plasma oscillations of a two-component electron system [6.267].

The Hamiltonian of the electron system may be written in the form

$$H = \sum_{\underline{p}\sigma} E_s(\underline{p}) C^+_{\underline{p}\sigma} C_{\underline{p}\sigma} + \sum_{\underline{p}\sigma} E_d(\underline{p}) a^+_{\underline{p}\sigma} a_{\underline{p}\sigma} + H_{int} \quad , \tag{6.4.1}$$

where H_{int} is the Coulomb interaction

$$H_{int} = \sum_{\underline{q}\underline{K}} \frac{4\pi e^2}{q^2} \rho_{\underline{q}} \rho_{\underline{K}-\underline{q}} \tag{6.4.2}$$

and $\rho_{\underline{q}}$ is the Fourier component of the electron density

$$\rho_{\underline{K}} = \sum_{\underline{p}\sigma} C^+_{\underline{p}+\underline{q},\sigma} C_{\underline{p}\sigma} + \sum_{\underline{p}\sigma} a^+_{\underline{p}+\underline{q},\sigma} a_{\underline{p}\sigma} \quad . \tag{6.4.3}$$

The collective excitations of this system are determined by the dielectric function $\varepsilon(\underline{q},\omega)$ which in the random phase approximation is additive with respect to the s and d contributions

$$\varepsilon(\underline{q},\omega) = 1 + \frac{4\pi e^2}{q^2}\left[\Pi_s(\underline{q},\omega) + \Pi_d(\underline{q},\omega)\right] \quad , \tag{6.4.4}$$

where

$$\Pi_\nu(\underline{q},\omega) = \sum_{\underline{p}} \frac{f_{\underline{p}+\underline{q},\nu} - f_{\underline{p}\nu}}{\omega - E_\nu(\underline{p}+\underline{q}) + E_\nu(\underline{p}) + i\delta} \tag{6.4.5}$$

is the polarization operator for a system of electrons with $\nu = s$, d, and $f_{\underline{p}\nu}$ is a Fermi distribution function.

The optical plasmon branch in the long-wave limit has the frequency

$$\omega^2_{pl} = \omega^2_s + \omega^2_d \quad , \tag{6.4.6}$$

where

$$\omega^2_s = \frac{4\pi n_s e^2}{m^*_s} \quad , \tag{6.4.7}$$

and

$$\omega^2_d = \frac{4\pi n_d e^2}{m^*_d} \tag{6.4.8}$$

is the plasma frequency of the system of s and d electrons. The acoustic plasmon branch in the limit of a large difference in effective masses ($m_d \gg m_s$) was investigated by PINES [6.272] who obtained the following estimate for its frequency

$$\Omega_{ac} = sq \approx \frac{\omega_d}{\omega_s} \upsilon_{Fs} q \quad , \tag{6.4.9}$$

where υ_{Fs} is the velocity of s-electrons on the Fermi surface. This estimate is good on the assumption that $\omega_d \ll \omega_s$ so that the velocity of plasmons turns out to be considerably less than the Fermi velocity of s-electrons. However, for the frequency Ω_{ac} to leave the continuum of electron hole excitations of the d-system the velocity of s-electrons should exceed υ_{Fd}, but the latter is usually much higher than the typical phonon velocities. Thus, one obtains the following criterion for the existence of undamped acoustic plasmons [6.267]:

$$q\upsilon_{Fs} \gg \Omega_{ac} > q\upsilon_{Fd} \quad . \tag{6.4.10}$$

The last two estimates apply to a quadratic dispersion relation for s and d-electrons. In connection with a discussion relating to the situation for A-15 compounds, an investigation [6.273] was made into acoustic plasmons in the case of quasi-one-dimensional electron spectrum. Assume that the system of d-electrons is specified by a one-dimensional spectrum in the tight-binding approximation, i.e., $E_d(\underline{p})$ is given by (5.4.1), while s-electrons form a three-dimensional electron gas with a quadratic dispersion relation. Calculation of the polarization operator (6.4.5) for d-electrons in this case leads to a singular expression of the form

$$\Pi_d(\underline{q},\omega) \sim \frac{1}{\sqrt{\omega + 2E_m \cos\left(\frac{qa}{2}\right)}} \quad . \tag{6.4.11}$$

A consequence of this singularity is the fact that the full dielectric funfunction (6.4.4) always has zeroes beyond the electron-hole continuum of d-electrons. Thus, in a one-dimensional system acoustic plasmons will always exist beyond the electron-hole continuum when the screened Coulomb potential of d-electrons is arbitrarily low. This result is promising for the existence of acoustic plasmons over a fairly broad range of parameters in A-15 compounds with a quasi-one-dimensional spectrum of d-electrons.

TUTTO et al. [6.267] have calculated the dielectric function of d-electrons with a spectrum corresponding to the tight-binding approximation and with the symmetry of point Γ of the Brillouin zone in A-15 compounds. The spectrum given in the form

$$E_d(\underline{p}) = -A(\cos ap_x + \cos ap_y + \cos ap_z)$$
$$+ T\left(\cos\frac{ap_x}{2}\cos\frac{ap_y}{2} + \cos\frac{ap_x}{2}\cos\frac{ap_z}{2} + \cos\frac{ap_y}{2}\cos\frac{ap_z}{2}\right) \quad (6.4.12)$$

with parameters A and T corresponding to the best approach to the band-structure calculations for A-15 compounds. Numerical calculations have permitted determination of the regions (on the energy-wave vector plane) where nondecaying acoustic plasmons occur in definite positions of the Fermi energy in the d-band. It turns out that the Fermi level should cross some central part of the band.

When comparing the relationships between the Fermi level and the region of the occurrence of acoustic plasmons with the superconducting transition temperature, a remarkable correlation was disclosed for many A-15 compounds: T_c is high for those compounds whose Fermi level lies in the region admitting the appearance of acoustic plasmons. This is illustrated in Fig.6.39 borrowed from [6.267], where the hatched portion corresponds to the positions of the Fermi energy which admit acoustic plasmons. In the figure T_c is also plotted, depending on the position of the Fermi level which is known for the series of compounds under investigation from band-structure calculations [6.191,192,204]. As is obvious from the figure, all high-temperature superconductors fall within the region of the occurrence of acoustic plasmons. On the other hand, compounds with low T_c such as, for example, Nb_3Sn have a Fermi level crossing the d-band outside this region. Thus, the correlation between T_c and E_F, which is shown in Fig.6.39, is by no means brought about by the density of states on

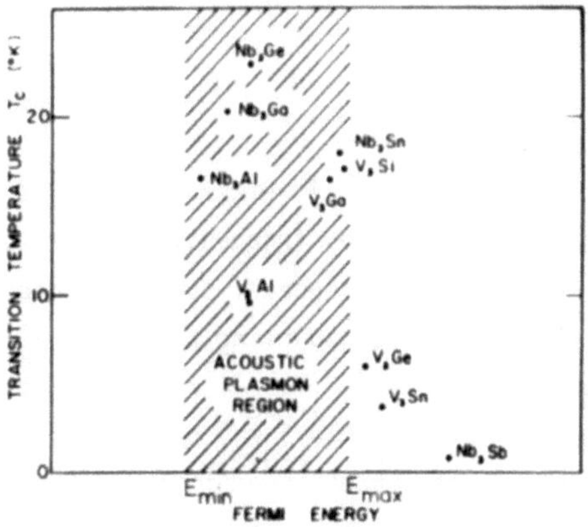

Fig.6.39. The correlation between T_c and the position of the Fermi level relative to the region of the existence of acoustic plasmons [6.267]

the Fermi level. This can readily be seen from the example of two compounds Nb_3Ge and V_3Ge which have close values of N(0) but have highly different T_c (23 K and 6 K, respectively). Thus, the anomaly involved is readily resolvable within the framework of the hypothesis concerning the contribution of the plasmon mechanism of pairing interaction in A-15 compounds.

TUTTO and RUVALDS [6.264] have discussed the problem concerning the possibility of singling out the plasmon contribution to the pairing interaction of electrons with the help of tunnel investigations. Their analysis is based on the strong coupling method described in Chapter 2, when in addition to the usual electron-phonon term H_{e-ph} in the Hamiltonian, it is necessary to take into account the electron-plasmon interaction term H_{e-pl}:

$$H_{e-pl} = \sum_{\underline{q}\underline{K}} g_{pl}\, \rho_{\underline{q}} \left(\beta^{+}_{\underline{K}} + \beta_{\underline{K}}\right) \quad , \tag{6.4.13}$$

where $\beta_{\underline{K}}$ and $\beta^{+}_{\underline{K}}$ are annihilation and creation Bose operators for an acoustic plasmon with a wave vector $\underline{K}$, and g_{pl} is the electron-plasmon interaction matrix element. In analogy with the electron-phonon interaction, TUTTO and RUVALDS [6.264] have obtained a formula for T_c which is a generalization of (2.3.27) of the strong-coupling theory

$$T_c = \frac{\theta_D}{1.45}\left(\frac{\theta_{pl}}{\theta_D}\right)^{\frac{\lambda_{pl}}{\lambda}} \exp\left(-\frac{1+\mu}{\lambda-\mu^*}\right) \quad . \tag{6.4.14}$$

Here $\lambda = \lambda_{ph} + \lambda_{pl}$ is the full electron-boson coupling constant consisting of the usual electron-phonon coupling constant λ_{ph}, (2.4.15), and of the electron-plasmon coupling constant for which there is an estimate [6.264] in the spirit of McMillan's formula (2.4.30) for λ_{ph}

$$\lambda_{pl} = \frac{g^2_{pl} N(0)}{2\langle\Omega_{ac}\rangle}\left(\frac{q_{max}}{P_F}\right)^2 \quad , \tag{6.4.15}$$

where $\langle\Omega_{ac}\rangle$ stands for the averaged frequency of acoustic plasmons, q_{max} is their cutoff momentum and $\theta_{pl} = sq_{max}$ labels the limited frequency (temperature).

Formula (6.4.14) shows the possibility of the existence of superconductors with high T_c even in the case of weak electron-phonon coupling λ_{ph} since θ_{pl} exceeds θ_D by about an order of magnitude. Evaluation of λ_{ph} for Nb_3Sn from optical data and resistivity yields a small value $\lambda_{ph} \approx 0.5$ which is not sufficient (as has already been noted above) for the appearance of the high value of $T_c = 18$ K. If we admit the electron-plasmon coupling mechanism it is possible to estimate the plasmon contribution to λ from (6.4.14): $\lambda_{pl} \approx 0.2$. Direct observation of this contribution could be obtained in tunnel-effect experiments. Theory [6.264] predicted that the spectral function $\alpha^2(\omega)F(\omega)$ determining the full coupling constant should be substantially different in form from the phonon contribution $\alpha^2_{ph}(\omega)F_{ph}(\omega)$. This is due to a decrease in the phonon maxima and to the transfer of the density towards higher energies (resulting from the fact that $\theta_{pl} \approx 10 \cdot \theta_D$). This behaviour of the observed spectral density and $\alpha^2(\omega)F(\omega)$ were found in tunnel experiments [6.266] on Nb_3Sn. This indicates that there is an additional electron pairing mechanism operating in this compounds, which may well be the virtual exchange of acoustic plasmons.

The results of the analysis of superconducting transition temperatures performed for A-15 compounds may be stated as follows: the picture of the observed T_c cannot be explained on the basis of the electron-phonon interaction alone because if the existing electron structure models for these compounds are used, the electron-phonon coupling constants prove to be anomalously small for a number of substances with high T_c. At the same time, the scheme of electron pairing via acoustic plasmons which has been proposed by RUVALDS and his collaborators accounts for many facts and anomalies in the distribution of T_c in this class of compounds. It must be said, in addition to this, that in terms of the plasmon mechanism it also becomes clear why T_c decreases anomalously as a function of the disorder, caused by impurities or radiation effects in the system (Chapter 9). This may arise from the fact that the decay of acoustic plasmons is very sensitive to this type of disorder in the system or that on reaching its critical value the suppressing mechanism in the pairing interaction of electrons becomes inoperative.

6.5 The Theory of Structural Transition in A-15 Compounds

6.5.1 The Character of the Phase Transition

In a number of A-15 compounds the structural transition from a cubic to a tetragonal phase results from the lattice-modulus $(c_{11} - c_{12})$ softening which reflects the instability of the cubic lattice with respect to shear strains.

Experiments showed [6.274] that in the tetragonal phase the crystal has the symmetry group D^9_{4h}. The symmetry analysis described in Sect.8.1 shows that the structural transition $O^3_h \rightarrow D^9_{4h}$ should proceed according to the irreducible representation Γ_{12} of the point symmetry O_h of the initial group. In accordance with Landau's concept of second-order phase transitions (or first-order transitions close to phase transitions of the second kind, as is the case with A-15), the transition to a low-symmetry phase goes through a soft mode which, at a temperature below the transition point, "freezes" into the initial structure, thus lowering it to the symmetry of this mode. In this case one might expect that the optical branch of lattice oscillations with $\underline{q} = 0$ and with symmetry Γ_{12} in Nb_3Sn and V_3Si would be unstable (Fig.6.40, Table 6.10). Instead of this, a softening of the transverse acoustic mode with $\underline{q} = [110]$ and with the polarization vector $[1\bar{1}0]$ occurs, whose velocity is proportional to the shear modulus $(c_{11} - c_{12})$. As yet, neutron diffraction experiments have not revealed any softening of the optical modes in V_3Si [6.245].

SHIRANE and AXE [6.274] assumed on the basis of those results that in the A-15 structure some relationship between the spontaneous displacements of atoms in the tetragonal phase described by the symmetry Γ_{12} and the tetragonal

Table 6.10. Polarization vectors of phonon modes in point Γ for the A-15 structure (according to SHAM [6.275])

κ	1	2	3	4	5	6	7	8
Γ_2	100	$\bar{1}00$	010	$0\bar{1}0$	001	$00\bar{1}$		
Γ_{12}	200	$\bar{2}00$	$0\bar{1}0$	010	$00\bar{1}$	001		
			010	$0\bar{1}0$	$00\bar{1}$	001		
Γ_{15}			001	$00\bar{1}$	010	010		
Γ_{25}			001	$00\bar{1}$	$0\bar{1}0$	010		
Γ_{25}			100	100	$\bar{1}00$	$\bar{1}00$		
Γ_{25}							100	$\bar{1}00$
Γ_{15}	100	100	100	100	100	100	A00	A00
Γ_{15}	100	100	100	100	100	100	B00	B00
Γ_{15}	$\bar{2}00$	$\bar{2}00$	100	100	100	100		

Note: The set of three digits denotes three projections of the displacement vector of an atom. For the three-dimensional representations Γ_{15}, Γ'_{15}, Γ_{25} and Γ'_{25}, only one mode is specified.
For the two-dimensional representation Γ_{12}, both modes are written out. Representations Γ_{15} are contained in the mechanical representation thrice, and Γ_{25} twice; therefore the tabulated basis functions of these representations are no polarization vectors. The quantities A and B are equal to $\sqrt{M_B/M_A}$ and $-3\sqrt{M_A/M_B}$, respectively. The digits 1-8 number the atoms (Fig.6.40)

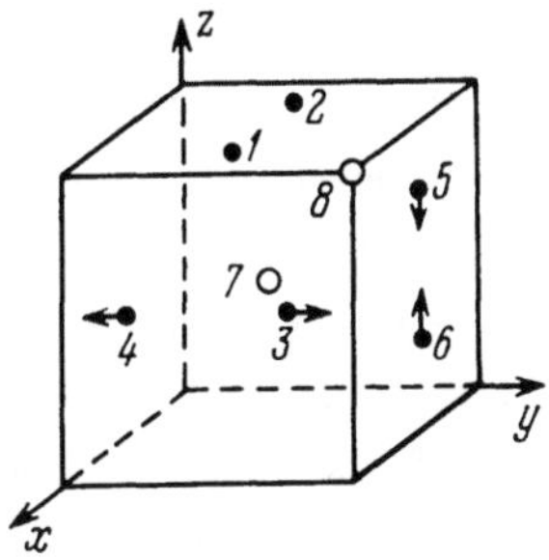

Fig.6.40. The displacement of atoms in A-15 structures due to the optical vibration of the symmetry Γ_{12}. Shown in the figure is one of the two modes which corresponds to the two-dimensional representations Γ_{12} and to the second line of the representation given in Table 6.10

deformation which is due to the instability of a shear-type acoustic wave should exist. Due to this relationship, both phenomena (spontaneous tetragonal deformation and Γ_{12} atomic displacements) occur simultaneously. The SHIRANE and AXE assumption is based on the experimental fact that the intensity of superstructural neutron diffraction peaks, which is proportional to the square of atomic displacements d in the tetragonal phase, and the value of spontaneous deformation $\delta = c/a - 1$ obtained from X-ray measurements, have the same temperature dependence (Fig.6.41) so that $d^2 \sim (c/a - 1)^2$.

This assumption was confirmed theoretically by SHAM [6.276,277] in the strong-coupling model. He took a comprehensive account of the contribution of d-electrons to the dynamic matrix of the A-15 crystal and calculated the elastic lattice moduli using the long-wave method [6.278]. The result obtained may be explained from the schematic formula

$$c_{11} - c_{12} = A_{11} - A_{12} + D - \frac{D'}{\tilde{\Omega}^2(\Gamma_{12})} \quad , \tag{6.5.1}$$

in which the term $(A_{11} - A_{12})$ denotes, as before, the contribution from the ion-ion interaction screened by s and p-electrons; the second term on the

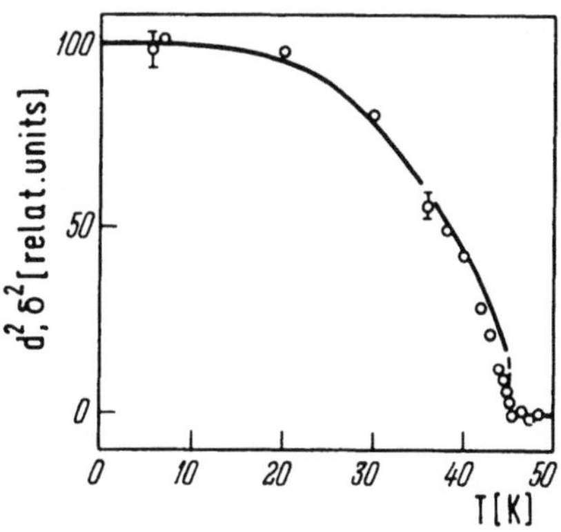

Fig.6.41. The temperature dependence of the displacement amplitude of d-sublattices in the tetragonal phase Nb_3Sn and of the spontaneous deformation parameter δ. The experimental points correspond to (300) neutron diffraction reflections, the solid line pertains to the parameter $(c/a - 1)^2$ plotted in relative scale

right-hand side of it describes the interaction of ions via d-electrons, and the third term arises from the dependence of the d-electron energy on the atomic displacement in the optical mode. $\tilde{\Omega}(\Gamma_{12})$ is the frequency of the optical mode when $q=0$. The electron activity of the optical oscillation Γ_{12} is due to the nature of the A-15 structure where, in the case of this type of oscillations, transition metal atoms move along the chains towards each other (Fig.6.40). This motion leads to a change in the overlap of the wave functions of d-states, resulting in an appreciable variation of the electron energy. In other optical oscillations with the symmetries Γ'_{15}, Γ'_{25} and Γ_{25}, the variation of electron energy is weaker and the coupling with these oscillations may be neglected. That is why the Γ_{12} mode which enters (6.4.1) because of the strong coupling of these oscillations with d-electrons produces some temperature dependence of the frequency $\tilde{\Omega}(\Gamma_{12})$; the frequency $\tilde{\Omega}(\Gamma_{12})$ diminishes as T decreases. This leads to an increase in the last term in (6.4.1) which makes a negative contribution to the elastic moduli so that at some temperature T the positive contribution to the shear modulus may be compensated for by the ion-ion interaction.

Thus, the optical mode Γ_{12} per se does not become unstable, but it promotes the softening of the acoustic shear mode whose instability comes earlier. As a result we face a situation typical of a first-order phase transition when the order parameter (the value of spontaneous deformation in this case) depends on the state of the other system (of optical modes) characterized by the amplitude of atomic displacements of the sublattices.

Using the Raman-spectroscopy method a direct investigation of the optical phonon mode with $q=0$ of symmetry Γ_{12} has recently [6.279] been carried out for the first time on A-15 compounds. The experiment has been made on a V_3Si single crystal in the temperature range between 50 and 340 K. It has turned out that the phonon frequency exhibits an inappreciable softening with decreasing temperature, thus indicating that the electron system is directly related to the Γ_{12}-symmetry lattice displacement. The width of the phonon peaks increases substantially when the temperatures decrease by a factor of two as in the temperature interval specified above. This effect also indicates a strong electron-lattice interaction in this compound [6.279].

6.5.2 The Noolandi and Sham Theory [6.280]

The frequencies and the polarization vector of the harmonic lattice are determinable from the well-known equation, see (2.4.4),

$$\omega_j^2(\underline{q})e_j^{\alpha}(\underline{q},\kappa) = \sum_{\kappa'\alpha'} C_{\alpha\alpha'}(\underline{q},\kappa\kappa')e_j^{\alpha'}(\underline{q},\kappa') \quad , \tag{6.5.2}$$

in which the dynamic matrix is expressed in terms of the matrix of force constants through the equation

$$C_{\alpha\alpha'}(\underline{q},\kappa\kappa') = \frac{\exp[-i\underline{q}(\underline{\rho}_\kappa - \underline{\rho}_{\kappa'})]}{\sqrt{M_\kappa M_{\kappa'}}} \sum_{\underline{l}-\underline{l}'} \phi^{\alpha\alpha'}_{\underline{l}\kappa,\underline{l}'\kappa'} \exp[-i\underline{q}(\underline{l}-\underline{l}')] \quad .$$

To calculate the lattice moduli there is no need to solve the complicated system of equations (6.5.2); it suffices to use the long-wave limit in the expansion of the dynamic matrix

$$C_{\alpha\alpha'}(\underline{q},\kappa\kappa') = C^{(0)}_{\alpha\alpha'}(\kappa\kappa') + iq_\beta C^{(1)}_{\alpha\alpha'\beta}(\kappa\kappa') + \frac{1}{2}q_\beta q_{\beta'} C^{(2)}_{\alpha\alpha'\beta\beta'}(\kappa\kappa') + \dots \; , \tag{6.5.3}$$

where the quantities $C^{(0)}$, $C^{(1)}$ and $C^{(2)}$ are the zeroth, first and second moments of the force matrix (the summation is over the recurring indices).

Following [6.280], we apply the homogeneous deformation method according to which the displacements of an arbitrary atom in the case of homogeneous deformation $\varepsilon_{\alpha\beta}$ may be represented as

$$u^\alpha_{l\kappa} = d^\alpha_\kappa + \varepsilon_{\alpha\beta}R^\beta_{l\kappa} \quad , \tag{6.5.4}$$

where d^α_κ is the displacement of the κ sublattice taken as a whole.

With aid of the above equations it is easy to represent the potential energy of the lattice in the harmonic approximation as

$$\Omega_0 = \frac{1}{2}\sum_j \tilde{\Omega}_j^2 d_j^2 + \sum_j P_{\alpha\beta}(j)d_j\varepsilon_{\alpha\beta} + \frac{1}{2}Q_{\alpha\alpha'\beta\beta'}\varepsilon_{\alpha\beta}\varepsilon_{\alpha'\beta'} \quad , \tag{6.5.5}$$

where $\tilde{\Omega}_j$ stands for the frequency of an optical phonon mode witn $\underline{q} = 0$; d_j is the degree of superposition of sublattice displacements which corresponds to this mode,

$$d_j = \sum_\kappa \sqrt{M_\kappa}\,\underline{e}_j(0,\kappa)\underline{d}_\kappa \quad , \tag{6.5.6}$$

and the quantities P and Q are expressed through the first moments of the force matrix

$$P_{\alpha\beta}(j) = \sum_{\kappa\kappa'} \sqrt{M_{\kappa'}}\, e_j^\alpha(0,\kappa)C^{(1)}_{\alpha\alpha'\beta}(\kappa\kappa') \quad , \tag{6.5.7}$$

$$Q_{\alpha\alpha'\beta\beta'} = \frac{1}{2}\sum_{\kappa\kappa'} \sqrt{M_\kappa M_{\kappa'}}\, C^{(2)}_{\alpha\alpha'\beta\beta'}(\kappa\kappa') \quad . \tag{6.5.8}$$

The first term in (6.5.5) is obtained by eliminating the dynamic matrix $\underset{-}{q}=0$ from the equation of motion (6.5.2).

We now investigate the symmetry of (6.5.7,8). The polarization vectors for the modes at $q=0$ are presented in Table 6.10, the use of which may show that for the A-15 structure the quantities $P_{\alpha\beta}(j)$ relating deformations to the displacements of sublattices are nonzero only for oscillations belonging to the irreducible representations Γ_{12} and Γ'_{25}. It is only the off-diagonal elements which are nonzero in $P_{\alpha\beta}(\Gamma'_{25})$ and therefore these quantities can contribute only to moduli associated with ε_{xy} type deformations. We are interested in shear strains of the Γ_{12}-symmetry oscillations. Calculation of the quantities $P_{\alpha\beta}(j)$ according to (6.5.7) involving the polarization vectors given in Table 6.10 and the symmetry of the components of the dynamic matrix of the A-15 structure, leads to the following form of the matrices $P_{\alpha\beta}(\Gamma_{12})$ corresponding to the $\Gamma_{12}^{(1)}$ and $\Gamma_{12}^{(2)}$ modes specified in Table 6.10:

$$P_{\alpha\beta}\left(\Gamma_{12}^{(1)}\right)=\begin{pmatrix}0 & 0 & 0\\ 0 & \sqrt{3}P & 0\\ 0 & 0 & -\sqrt{3}P\end{pmatrix}, \quad P_{\alpha\beta}\left(\Gamma_{12}^{(2)}\right)=\begin{pmatrix}-2P & 0 & 0\\ 0 & P & 0\\ 0 & 0 & P\end{pmatrix}, \tag{6.5.9}$$

where P is some constant which should be regarded as a phenomenological parameter here.

The coefficients $Q_{\alpha\alpha'\beta\beta'}$ have the same symmetry properties as the elastic moduli; therefore they comprise three independent constants which we shall designate in the same way as the elastic moduli in the cubic lattice were denoted Q_{11}, Q_{12} and Q_{44}.

Let us now minimize the energy (6.5.5) with respect to the quantities d_j. It follows from above that only the values $j=\Gamma_{12}^{(1)}$ and $\Gamma_{12}^{(2)}$ should be taken into account in the second term of (6.5.5). Allowing for the form of the matrices $P_{\alpha\beta}$ for these modes we obtain two equations relating the displacement of the sublattices to deformations:

$$\begin{aligned}\tilde{\Omega}^2\left(\Gamma_{12}^{(1)}\right)d\left(\Gamma_{12}^{(1)}\right) &= \sqrt{3}P(\varepsilon_{zz}-\varepsilon_{yy}) \quad ,\\ \tilde{\Omega}^2\left(\Gamma_{12}^{(2)}\right)d\left(\Gamma_{12}^{(2)}\right) &= P(2\varepsilon_{xx}-\varepsilon_{yy}-\varepsilon_{zz}) \quad .\end{aligned} \tag{6.5.10}$$

In A-15 compounds, the structural transition is related to a tetragonal deformation of the form

$$\varepsilon_{xx}=-\varepsilon_0 \quad , \quad \varepsilon_{yy}=\varepsilon_{zz}=\varepsilon_0/2 \tag{6.5.11}$$

(following [6.280], we have oriented the tetragonal axis along the x-axis). It then follows from (6.5.10) that

$$d\left(\Gamma_{12}^{(1)}\right) = 0 \quad , \quad d\left(\Gamma_{12}^{(2)}\right) \equiv d \quad . \tag{6.5.12}$$

Thus, tetragonal deformation brings about the displacement of the sublattices corresponding to mode $\Gamma_{12}^{(2)}$.

It will be assumed further that the dynamic matrix $C_{\alpha\alpha'}(\underline{q},\kappa\kappa')$ comprises only ion-ion interactions screened by s and p-electrons, and $\tilde{\Omega}_j$ represents the optical frequency arising from these interactions. To take into account the contribution of d-electrons to the energy of the crystal it is necessary to prescribe the model of the electron spectrum. In a chain model the electron energy of the deformed crystal may be written as [6.280]

$$H_e = \sum_{\lambda\lambda'} \sum_{\underline{p}} E_{\lambda\lambda'}(\underline{p}) a^+_{\underline{p}\lambda} a_{\underline{p}\lambda'} \quad , \tag{6.5.13}$$

where

$$E_{\lambda\lambda'}(\underline{p}) = \varepsilon_{\lambda\underline{p}}\delta_{\lambda\lambda'} + \sum_j \sigma_{\lambda\lambda'}(j)d_j + \xi_{\lambda\lambda'\alpha\beta}\varepsilon_{\alpha\beta} \quad . \tag{6.5.14}$$

$\varepsilon_{\lambda p}$ here stands for the energy of a d-electron in a chain oriented in the λ direction ($\lambda = 1,2,3$), the second term in (6.5.14) takes into account the variation of electron energy due to the displacement of the sublattices, and the third term comes from the deformation of the crystal. We assume that near the bottom of the d-band where the Fermi level is situated, the dependence of the deformation potentials ξ and of σ on momentum is not essential.

It may be shown that the coefficients $\sigma_{\lambda\lambda'}(j)$ are nonzero for the two modes Γ_{12} only. For example,

$$\sigma_{\lambda\lambda'}\left(\Gamma_{12}^{(2)}\right) = \begin{pmatrix} 2\sigma & -\sigma' & -\sigma' \\ -\sigma' & -\sigma & 2\sigma' \\ -\sigma' & 2\sigma' & -\sigma \end{pmatrix} \quad . \tag{6.5.15}$$

Finally, the coefficients $\xi_{\lambda\lambda'\alpha\alpha'}$ with λ, $\lambda' = 1,2,3$ and α, $\alpha' = x,y,z$ have the independent components $\xi_{11xx} \equiv \xi_{1x}$, $\xi_{11yy} \equiv \xi_{1y}$, ξ_{12xx}, and ξ_{12zz}.

In the model of noninteracting chains we neglect all the off-diagonal elements with respect to λ and λ' involved in (6.5.15). Then lattice deformation may only lead to energy variations within one chain and not to interchain transitions, so that

$$E_\lambda(\underline{p}) = \varepsilon_{\lambda\underline{p}} + \sum_j \sigma_{\lambda\lambda}(j)d_j + \xi_{\lambda\lambda\alpha\beta}\varepsilon_{\alpha\beta} \quad . \tag{6.5.16}$$

The general thermodynamic potential of the system Ω, as calculated for one cell, may be written as a sum of the lattice (Ω_0) and d-electron (Ω_e) contributions, where

$$\Omega_e = -kT\frac{1}{N}\sum_{\lambda \underline{p}} \ln\left[1+\exp\left(\frac{\mu - E_\lambda(\underline{p})}{kT}\right)\right] \quad . \tag{6.5.17}$$

Now write the full thermodynamic potential of the tetragonally deformed crystal calculated for one unit cell. Allowing for the symmetry of all the matrices involved in (6.5.5,16) we have

$$\Omega = \frac{1}{2}\tilde{\Omega}^2 d^2 + 3Pd\varepsilon_0 + \frac{3}{4}(Q_{11} - Q_{12})\varepsilon_0^2 - kT\frac{1}{N}\sum_{\underline{p}\lambda} \ln\left[1+\exp\left(\frac{\mu - E_\lambda - \varepsilon_{\lambda\underline{p}}}{kT}\right)\right] \quad . \tag{6.5.18}$$

Here $\tilde{\Omega}$ is the optical frequency of mode $\Gamma_{12}^{(2)}$, and d is the displacement of a sublattice in this mode. The quantity E_λ stands for the shift of the edge of the one-dimensional band $\varepsilon_{\lambda\underline{p}}$ of the distorted chain, with

$$E_1 = -2\Delta \quad , \quad E_2 = E_3 = \Delta \quad , \tag{6.5.19}$$

where Δ, depending on the quantities ε_0 and d, is

$$\Delta = -\sigma d + \frac{1}{2}(\xi_{1x} - \xi_{1y})\varepsilon_0 \quad . \tag{6.5.20}$$

Minimizing the energy (6.5.18) with respect to the parameters ε_0 and d, we obtain the value of spontaneous deformation and the displacement of the sublattices

$$\varepsilon_0 = \frac{U}{12A}(\nu_1 - \nu_2) \quad , \tag{6.5.21}$$

$$d = -3\varepsilon_0\left(P + \frac{8\sigma A}{U}\right)\frac{1}{\tilde{\Omega}^2} \quad . \tag{6.5.22}$$

Here

$$\nu_\lambda = \frac{1}{N}\sum_{\underline{p}} f_{\lambda\underline{p}} \tag{6.5.23}$$

is the number of d-electrons in the λ chain ($f_{p\lambda}$ stands for the Fermi distribution function for the energy $E_\lambda + \varepsilon_{\lambda\underline{p}}$), and A and U are combinations of constants:

$$U = \xi_{1x} - \xi_{1y} + \frac{6P\sigma}{\tilde{\Omega}^2} \quad , \quad 8A = Q_{11} - Q_{12} - \frac{6P^2}{\tilde{\Omega}^2} \quad . \tag{6.5.24}$$

We now calculate the shear moduli. When the crystal is deformed the sublattices should shift in it. Then the elastic moduli are determined by the equation

$$c_{\alpha\beta\alpha'\beta'} = \frac{\partial^2\Omega}{\partial\varepsilon_{\alpha\beta}\partial\varepsilon_{\alpha'\beta'}} - \sum_j \frac{\partial^2\Omega}{\partial\varepsilon_{\alpha\beta}\partial d_j}\,\frac{\partial^2\Omega}{\partial d_j\partial\varepsilon_{\alpha'\beta'}}\bigg/\frac{\partial^2\Omega}{\partial d_j^2} \tag{6.5.25}$$

from which one can obtain, for example, an expression for the shear modulus in the tetragonal phase

$$c_{11} - c_{12} = Q_{11} - Q_{12} - (\xi_{1x} - \xi_{1y})^2\,\frac{1}{3}\left(2\frac{\partial\nu_1}{\partial\mu} + \frac{\partial\nu_2}{\partial\mu}\right)$$
$$- 6\left[P + \sigma(\xi_{1x} - \xi_{1y})\frac{1}{3}\left(2\frac{\partial\nu_1}{\partial\mu} + \frac{\partial\nu_2}{\partial\mu}\right)\right]^2\bigg/\left[\tilde{\Omega}^2 - \sigma^2\left(4\frac{\partial\nu_1}{\partial\mu} + 2\frac{\partial\nu_2}{\partial\mu}\right)\right] \quad . \tag{6.5.26}$$

The corresponding expressions for the cubic phase can be obtained if $\nu_\lambda = \nu$ is taken to be equal to

$$c_{11} - c_{12} = 8A - \frac{\partial\nu}{\partial\mu}U^2\frac{\tilde{\Omega}^2}{\tilde{\Omega}^2 - 6\sigma^2\partial\nu/\partial\mu} \quad . \tag{6.5.27}$$

Note that if the constant σ defining the variation of energy of an electron due to the displacements of the sublattices is set equal to zero, (6.5.27) becomes similar to (6.5.20) obtained by REHWALD [6.235]. The temperature dependence of the shear modulus is determined by the factor $\partial\nu/\partial\mu$. Let us adopt the model of constant density of states for d-electrons, described in the previous section. In this model the density of states per unit cell for d band with which we are concerned is given by the expression

$$N_\lambda(E) = \frac{8}{3}N_0\theta(E - E_\lambda) \quad , \tag{6.5.28}$$

where N_0 is the total density of states of d-electrons for one crystal atom. The quantity ν_λ to be determined from (6.5.23) is equal to

$$\nu_\lambda = \frac{8}{3}N_0 kT\ln\left[1 + \exp\left(-\frac{E_\lambda - \mu}{kT}\right)\right] \quad . \tag{6.5.29}$$

According to [6.234], for the cubic phase we have

$$\frac{\partial\nu}{\partial\mu} = \frac{8}{3}N_0 F\left(\frac{T}{T_0}\right) \quad , \quad F\left(\frac{T}{T_0}\right) = 1 - \exp\left(-\frac{T_0}{T}\right) \tag{6.5.30}$$

and (6.5.27) for the shear modulus may be rewritten as

$$c_{11} - c_{12} = \frac{A + BF(T/T_0)}{1 + CF(T/T_0)} \quad . \tag{6.5.31}$$

The new constants B and C are

$$C = -\frac{16N_0\sigma^2}{\tilde{\Omega}^2} \quad , \quad B - AC = -\frac{1}{3}N_0U^2 \quad . \tag{6.5.32}$$

Equation (6.5.21) for the spontaneous deformation $\bar{\varepsilon}(T) = \varepsilon_0(T)/\varepsilon_0(0)$ in the tetragonal phase becomes

$$\bar{\varepsilon}(T) = \frac{A}{3B}\frac{T}{T_0}\ln\left(\frac{\exp\left[(1-\bar{\varepsilon})\frac{T_0}{T}\right]-1}{\exp\left[(1+2\bar{\varepsilon})\frac{T_0}{T}\right]-1}\right) \quad . \tag{6.5.33}$$

As can be easily shown, this $\bar{\varepsilon}(T)$ has a temperature dependence typical of first-order phase transitions. The spontaneous deformation at zero temperature is, in view of (6.5.21), given by

$$\varepsilon_0(0) = \frac{2}{3}N_0kT_0\frac{U}{A} \quad . \tag{6.5.34}$$

Combining the second of the expressions (6.5.32) with this formula we can express the density of states N_0 in terms of $\varepsilon_0(0)$ and the constants A, B and C:

$$N_0 = \frac{3}{4}\left(\frac{\varepsilon_0(0)}{kT_0}\right)^2 \frac{A^2}{AC - B} \quad . \tag{6.5.35}$$

Equations (6.5.31,34,35) permit the following comparison between theory and experiment. The above theory contains five phenomenological constants characterizing the ion-lattice and the electron-lattice interaction: $\tilde{\Omega}$, P, $Q_{11} - Q_{12}$, σ and $\xi_{1x} - \xi_{1y}$, plus two parameters of a purely electronic nature, viz., the density of states on the Fermi surface, N_0, and the Fermi temperature T_0. Instead of the former five constants, the final result involves three new constants: A, B and C. By comparing the measured temperature dependence of the shear modulus with (6.5.31) one can determine the three constants A, B and C if several values are tried for the quantity T_0. If the value of spontaneous deformation $\varepsilon_0(0)$ has been measured at zero temperature, it is also possible to determine the density of states from (6.5.35) by choosing some value of T_0.

A similar program was carried out by NOOLANDI and SHAM [6.280] for Nb_3Sn. They used experimental data for the shear modulus in a temperature interval from 50 to 300 K. The optimum value $T_0 = 195$ K was selected such as to get the experimentally observed structural-transformation temperature ($T_m = 45$ K). Accordingly, the density of states on the Fermi surface turned out to be close to $N_0 = 3.2$ states/(eV · atom). These values are close to those ($T_0 = 230$ K and $N_0 = 2.36$) obtained by REHWALD et al. [6.158] when comparing theory with magnetic susceptibility measurements. The value obtained for N_0 is close to that found from electron heat capacity measurements: $N(0) \approx 2.7$ states/(eV·atom). These data thus lead to a frequency $\Omega = 21$ meV for the optical mode Γ_{12}.

This value is in reasonable agreement with the limiting phonon frequency of 28 meV (measured in energy) in Nb_3Sn obtained from tunnel measurements. Sublattice displacements were not measured for V_3Si, and the agreement with experiment is not as good.

Compare now the NOOLANDI and SHAM major result (6.5.27) with SHAM's formula (6.5.1) obtained from microscopic calculations from first principles. Equation (6.5.27) may be written in the form of (6.5.1) noting that the quantity in the denominator of (6.5.27) is the square of the optical mode (Γ_{12}) frequency corrected for the electron-lattice interaction, i.e.,

$$\tilde{\Omega}^2(\Gamma_{12}) = \Omega^2 - 6\sigma^2 \frac{\partial\nu}{\partial\mu} \quad . \tag{6.5.36}$$

According to (6.5.27), the temperature dependence of the shear modulus is determined by the factor $\partial\nu/\partial\mu$. In SHAM's theory [6.275] the temperature dependence enters through parameters characterizing the screening of d-electrons. The latter is determined by the polarization operator which, in the case of a crystal lattice, has the usual RPA form

$$\chi(\underline{q}+\underline{G},\ \underline{q}+\underline{G}',\ \omega) = \sum_{ss'} \sum_{\underline{p}} \left\langle s\underline{p}\middle|\exp[-i(\underline{q}+\underline{G})\underline{r}]\middle|s'\ \underline{p}+\underline{q}\right\rangle$$
$$\times \left\langle s'\ \underline{p}+\underline{q}\middle|\exp[i(\underline{q}+\underline{G}')\underline{r}]\middle|s\underline{p}\right\rangle \frac{f_{\underline{p}+\underline{q}s'} - f_{\underline{p}s}}{\varepsilon_{\underline{p}+\underline{q}s'} - \varepsilon_{\underline{p}s} - \omega} \tag{6.5.37}$$

where s and s' stand for the indices of the bands and $\underline{G}$ and $\underline{G}'$ are the reciprocal-lattice vectors. In the chains model for A-15 compounds it is convenient to write the indices s and $\lambda\nu$ where λ is the number of a chain, and $\nu = 1,2$ is the number of an atom in a given chain within a unit cell.

In the tight-binding approximation the d-electron wave function may be written as a Bloch sum,

$$\psi_{\lambda\underline{p}}(\underline{r}) = \frac{1}{\sqrt{2N}} \sum_{l\nu} \exp(i\underline{p}R_{l\lambda\nu})\varphi_\lambda(\underline{r} - \underline{R}_{l\lambda\nu}) \quad . \tag{6.5.38}$$

To lowest order in the overlap of atomic functions $\varphi_\lambda(\underline{r})$ we have

$$\langle \lambda\underline{p}|\exp[-i(\underline{q}+\underline{G})r]|\lambda'\underline{p}+\underline{q}\rangle \approx \delta_{\lambda\lambda'}S_\lambda(\underline{G})\Psi_\lambda(\underline{q}+\underline{G}) \quad , \tag{6.5.39}$$

where S_λ and Ψ_λ are, respectively, the structural factor of the λ chain and the form factor of the atomic d function

$$S_\lambda(\underline{G}) = \frac{1}{2}\sum_\nu \exp(-i\underline{G}\underline{R}_{1\nu}) \quad , \quad \Psi_\lambda(\underline{q}+\underline{G}) = \int d\underline{r}|\varphi_\lambda(\underline{r})|^2\exp[-i(\underline{q}+\underline{G})\underline{r}] \quad .$$

In the chains model, for the static case (6.5.37) is approximated in the following manner:

$$\chi(\underline{q}+\underline{G},\ \underline{q}+\underline{G}') = \sum_\lambda \theta_\lambda(\underline{q})S_\lambda^*(\underline{G})S_\lambda(\underline{G}')\Psi_\lambda^*(\underline{q}+\underline{G})\Psi_\lambda(\underline{q}+\underline{G}') \quad , \tag{6.5.40}$$

$$\theta_\lambda(\underline{q}) = \sum_{\underline{p}} \frac{f_{\underline{p}+\underline{q}\lambda} - f_{\underline{p}\lambda}}{\varepsilon_{\underline{p}+\underline{q}\lambda} - \varepsilon_{\underline{p}\lambda}} \quad . \tag{6.5.41}$$

In the long-wave limit the lattice moduli depend on $\theta_\lambda(\underline{q})$ with $\underline{q}\to 0$. From (6.5.41) we have

$$\theta_\lambda(0) = \sum_{\underline{p}} \frac{\partial f_{\underline{p}\lambda}}{\partial\varepsilon_{\underline{p}}} = -\frac{\partial\nu_\lambda}{\partial_\mu} \quad , \tag{6.5.42}$$

where ν_λ is determined by (6.5.23). Thus, the quantity $\partial\nu/\partial\mu$ which, according to (6.5.27), determines the temperature dependence of the lattice moduli essentially is a polarization operator for the system of d-electrons. From this it may be said that in the model of noninteracting chains the temperature anomalies of a number of A-15 compounds result from the strong temperature dependence of the d-electron screening effect.

Thus, we have seen that the softening of the shear modulus $(c_{11}-c_{12})$ occurs in the model of noninteracting chains. SHAM's calculations have shown that the modulus c_{44} exhibits a weak temperature dependence. However, acoustic measurements in Nb_3Sn [6.158,166] have disclosed that c_{44} decreases by 30% when the temperature is decreased from 300 to 20 K. Inelastic neutron scattering [6.281] has also revealed the softening of acoustic phonons with the wave vector $\underline{q} = [\xi 00]$ and the polarization $\underline{e} = [001]$ throughout the Brillouin zone, i.e., at all ξ. These phonons are characterized just by c_{44} in the long-wave limit. Recently, HORSCH [6.282] has shown that allowing for interchain transitions of electrons interacting with phonons, the matrix element of which is a two-centre integral leads to the main features observed in experiments, viz., phonons with the wave vector $\underline{q} = [\xi 00]$. In addition, the transverse magnetization becomes a temperature-dependent correction from the interaction with d-electrons, the correction being independent of $\underline{q}$. This corresponds to their softening throughout the Brillouin zone. The calculated dependence of the modulus c_{44} on temperature represents a good fit to the experimental curves when the value of the two-centre matrix element of the crystal potential equals 690 meV [6.282].

6.5.3 Effect of Pressure on Structural Phase Transition

The microscopic phase-transition theory analyzed above uses the noninteracting chains model. It is also reasonable to apply Landau's phenomenological phase-transition theory, particularly to the problem of pressure [6.283]. Taking into account the smallness of the spontaneous deformation $|c-a|/a \sim 10^{-3}$ we expand the free energy of the crystal in a power series of the deformation tensor $\varepsilon_{\alpha\beta}$. In A-15 compounds the structural transition from a cubic to a tetragonal phase is known to proceed according to the irreducible representation Γ_{12}. Therefore it is convenient to use the basis functions η_1 and η_2 of this two-dimensional representation, composed of the deformation tensor components

$$\eta_1 = \frac{1}{\sqrt{6}}(2\varepsilon_{zz} - \varepsilon_{xx} - \varepsilon_{yy}) \quad , \quad \eta_2 = \frac{1}{\sqrt{2}}(\varepsilon_{xx} - \varepsilon_{yy}) \quad . \tag{6.5.43}$$

Subject to the condition of conservation of volume $(\varepsilon_{xx} + \varepsilon_{yy} + \varepsilon_{zz} = 0)$ the function η_1 describes a tetragonal $\sim (c-a)/a$ compressive or tensile strain along the z axis, and η_2 the shift in the (110) plane in the $[1\bar{1}0]$ direction.

Using these basis functions it is possible to make up invariants and to write the expansion of the free energy of a tetragonally deformed cubic crystal in the following form [6.283-285]

$$\Omega(\eta_1, \eta_2) = \frac{1}{2}A\left(\eta_1^2 + \eta_2^2\right) + \frac{1}{3}B\eta_1\left(\eta_1^2 - 3\eta_2^2\right) + \frac{1}{4}C\left(\eta_1^2 + \eta_2^2\right)^2 - \eta_1\sigma_1 - \eta_2\sigma_2 \quad . \tag{6.5.44}$$

The presence of the cubic invariant, first written for the A-15 structure by ANDERSON and BLOUNT [6.133], reflects the nonequivalence of the deformations ε and $-\varepsilon$ corresponding to tension and compression. The cubic invariant arises if the transition obeys a Γ_{12} representation which, according to Landau's theory, is passive. From this theory it follows that the presence of a cubic term in the expansion of the energy leads to a first-order phase transition, to be demonstrated below.

The last two terms in (6.5.44) take into account the dependence of the energy on temperature; σ_1 and σ_2 involved therein are given by expressions similar to (6.5.43), with the replacement of the strain tensor $\varepsilon_{\alpha\beta}$ by the stress tensor $\sigma_{\alpha\beta}$. The coefficient A, as is easily seen, is a shear modulus,

$$A = c_{11} - c_{12} \quad . \tag{6.5.45}$$

In transforming A-15 compounds it goes to zero at the point of the phase transition T_0 so that in the vicinity of this point it may be expected to have a temperature dependence of the form $A = A'(T - T_0)$. The coefficients B and C are assumed to be weakly dependent on T.

Using the form (6.5.44) of the thermodynamic potential for the tetragonal lattice we first find the set of possible states below the transition point in the absence of pressure. The equilibrium values of the spontaneous deformation parameter, η_1 and η_2, are determined by the two equations $\partial\Omega/\partial\eta_i = 0$ $(i = 1,2)$ having the following explicit form

$$\eta_2\left[A - 2B\eta_1 + C\left(\eta_1^2 + \eta_2^2\right)\right] = 0 \quad , \quad A\eta_1 + B\eta_1^2 - B\eta_2^2 + C\left(\eta_1^2 + \eta_2^2\right)\eta_1 = 0 \quad . \tag{6.5.46}$$

Its solution

$$\eta_1 = 0 \quad , \quad \eta_2 = 0 \quad , \tag{6.5.47}$$

corresponds to the cubic phase; the three tetragonal solutions are

$$\eta_1 = \eta \quad , \quad \eta_2 = 0 \quad ; \quad \eta_1 = -\frac{1}{2}\eta \quad , \quad \eta_2 = \pm\frac{\sqrt{3}}{2}\eta \quad , \tag{6.5.48}$$

where η is determined from the quadratic equation

$$A + B\eta + C\eta^2 = 0 \quad . \tag{6.5.49}$$

The solution (6.5.47) corresponds to the minimum of energy when $A > 0$, i.e., at a temperature $T > T_0$. The condition for the existence of the solution (6.5.48) is $A < B^2/4C$. Thus, at temperatures when

$$0 < A < B^2/4C \quad , \tag{6.5.50}$$

both the cubic solution (6.5.47) and the tetragonal solution (6.5.48) are possible. It can be easily verified that the three types of tetragonal solutions (6.5.48) have the same energy. This degeneracy arises from the fact that all the free states (6.5.48) pass into each other at the cyclic permutation of the coordinates and correspond to three domains oriented along z, x and y axes.

Equating the energies of the solutions (6.5.47,48) we find the condition $A = 2B^2/9C$ determining the phase-transition temperature

$$T_m = T_o + \frac{2B^2}{9A'C} \quad . \tag{6.5.51}$$

At the transition point T_m itself, the spontaneous deformation η_o to be determined from the conditions (6.5.49) is

$$\eta_o = -\frac{2B}{3C} \tag{6.5.52}$$

so that the phase transition is of first order. In accordance with Landau's theory we assume that $C > 0$, then the sign of η_o is determined by the sign of B. When $B > 0$, $\eta_o < 0$, hence $c/a < 1$, and this case corresponds to Nb_3Sn. When $B < 0$, $c/a > 1$, which corresponds to V_3Si.

We consider now the influence of pressure on the phase transition. In the case of uniaxial stress σ applied along the [001] axis, $\sigma_1 = 2\sigma/\sqrt{6}$ and $\sigma_2 = 0$. We find that the $(\eta_1, 0)$ solution is determined from the equation

$$A\eta_1 + B\eta_1^2 + C\eta_1^3 = \sigma_1 \quad , \tag{6.5.53}$$

and the (η_1, η_2) solution from the two equations

$$A - 2B\eta_1 + C\left(4\eta_1^2 - \frac{\sigma_1}{B}\right) = 0 \quad , \quad \eta_2^2 = 3\eta_1^2 - \frac{\sigma_1}{B} \quad . \tag{6.5.54}$$

Thus, the three solutions (6.5.48) are no longer degenerate. The behaviour of the deformation parameters in the $(\eta_1, 0)$ and (η_1, η_2) states as a function of stress and temperature is depicted in Fig.6.42. It is seen from Fig.6.42a, that with stress applied the deformations remain finite at any temperature. From Fig.6.42b it follows that in the state (η_1, η_2) at some stress the phase transition changes its character as a function of temperature. With increasing σ the first-order transition is replaced by a second-order transition. The phase diagram of states at uniaxial stress is shown in Fig.6.43. On the curve separating the regions of the $(\eta_1, 0)$ and (η_1, η_2) states, the conversion to the second-order transition occurs at the tricritical point t.

An elaborate comparison of the theoretical predictions with experimental data for V_3Si and Nb_3Sn has been drawn by PIETRASS [6.283]. This comparison has made it possible to determine the parameters A, B and C of the phenomenological expansion of the thermodynamic potential, which turned out to be satisfactorily self-consistent. However, it has also been found that the be-

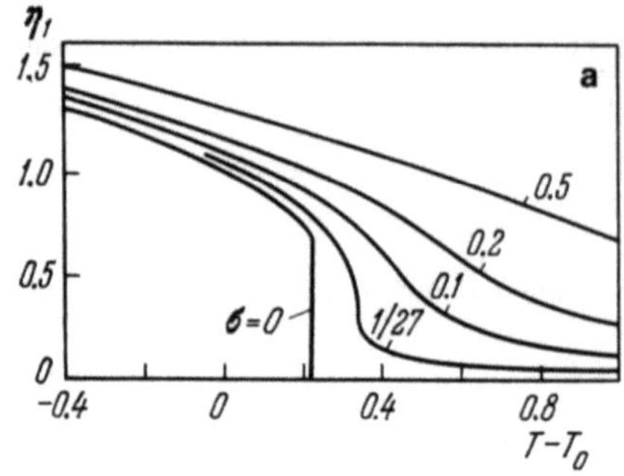

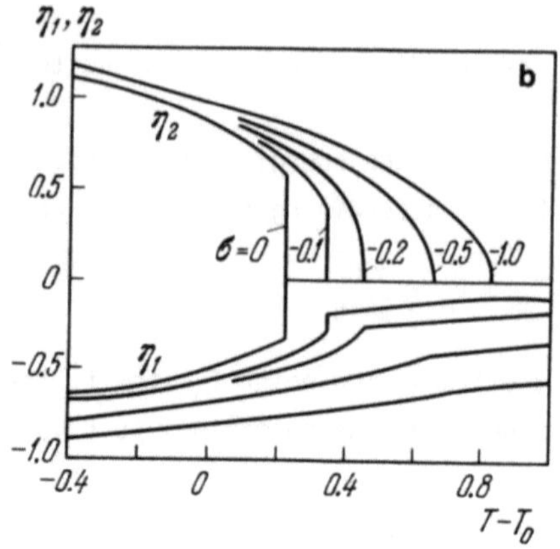

Fig.6.42. The deformations η_1 and η_2 as a function of temperature at different uniaxial stresses along the [001] axis. The deformations have been calculated from equations (a) (6.5.53) and (b) (6.5.54). η_1 and η_2 are expressed in the units B/C, σ through B^3/C^2, and T through $B^2/A'C$

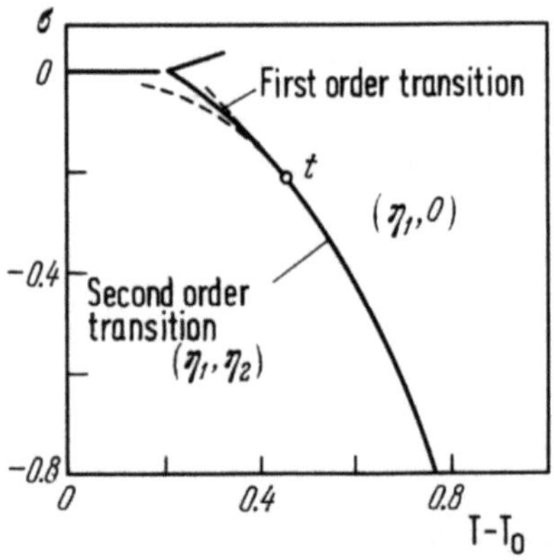

Fig.6.43. The phase-rule diagram of a tetragonally distorted crystal at uniaxial stress along the [001] axis

haviour at uniaxial deformation contradicts the phase diagram portrayed in Fig.6.43.

The phase diagram given in Fig.6.43 corresponds to uniaxial stress at which $\sigma_2 = 0$. If we investigate the general case of stress we obtain the phase diagram shown in Fig.6.44 [6.284,286]. The truncation by the plane containing the points q, t_1 and the axis of the temperature is nothing but the phase diagram represented in Fig.6.43, which corresponds to uniaxial deformation. The symmetry (Fig.6.44) with respect to the rotation by 120° about the T axis arises due to the invariance of the cubic system under cyclic permutation of the coordinates. A surface of the type qt_1t_3 is the interface of two phases: tetragonal and cubic. The notion "the cubic phase" should here be understood with the following restriction. In the presence of stress the crystal is cer-

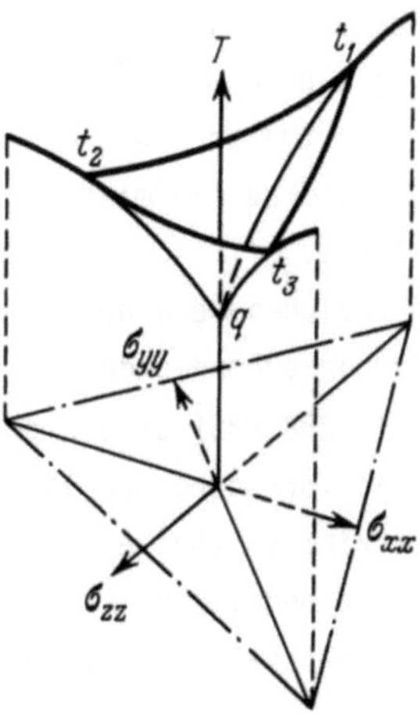

Fig.6.44. The phase diagram for an A-15 structure at an arbitrary tetragonal deformation. q is a quadrupolar point where four phases coexist: a cubic phase and three tetragonal phases; t_1, t_2 and t_3 are three critical points

tainly not cubic, but if upon removal of stress it returns to its initial form we shall call this stressed state cubic.

It is seen from Fig.6.44 that in the absence of stresses the phase-transition temperature T_m is a quadrupolar point q. If there are internal stresses in the A-15 crystal the phase-transition temperature increases. In accordance with the assumption [6.284], the variation of T_m in different Nb_3Sn specimens from 35 to 45 K and in V_3Si from 18 to 23 K may be explained by internal stresses in the crystal.

As already stated above, this theory does not explain the measurements for V_3Si. We shall mention two weak points of this phenomenological theory. Firstly, the entire analysis of the effect of stresses on the martensitic transformation temperature has been performed on the basis of expanding the energy in powers of order parameters keeping the first invariants only. As shown in a number of papers (see, for example, [6.287]), taking the full series or using the whole rational basis of invariants can appreciably change the results of the first analysis of LANDAU. However, this remark is not specific to the A-15 structure.

As we have already seen in the beginning of this section, the deformations in A-15 compounds are accompanied by atomic displacements of sublattices which themselves may play the role of an order parameter. At the same time, the energy expansion (6.5.44) involves only deformations as order parameters. Apparently one should write more general expressions for the energy, taking two systems of basis functions of the irreducible representation Γ_{12} for each case: one system (6.5.43) composed of deformation-tensor components and the other consisting of atomic displacements of sublattices, and make up invariants of a mixed kind. As second-order invariants one should take an expression similar to (6.5.5) used by SHAM, with the optical frequency being a phenomenological parameter. The temperature dependence of this parameter and the presence of additional variables, with respect to which the energy should be minimized, may alter part of the results obtained in the above analysis.

Nevertheless, this theory gives the main features of the phenomena observed in experiments, particularly the existence of first-order or second-order transitions under pressure. Recently KLOSE and SCHUSTER [6.288,289] predicted, on the basis of a chain model, that in the case of uniaxial pressure the lattice transition in Nb_3Sn should transform from the first kind to the second. Since the same result has been obtained in the general phenomenological theory of PIETRASS above, the observation of this phenomenon may not yet serve as a verification of the chain model.

6.5.4 Structural Transition as a Result of a Peierls Instability in Chains

BHATT and McMILLAN [6.290] developed a phenomenological theory of structural phase transition in A-15 compounds using LANDAU's idea of expanding the free energy in powers of the order parameters. They have assumed that what is responsible for the structural transition is the electronic system. This system is unstable with respect to the formation of charge density waves corresponding to the observed spontaneous displacement of transition-element atoms in linear chains, described by the symmetry Γ_{12}. All this is accompanied by a tetragonal lattice deformation. Thus, the system should be characterized by the following order parameters: the charge density wave amplitude Φ_μ (corresponding to the doubling of the period of a separate chain), the displacement Q_μ of an atom along a chain in mode Γ_{12} and the diagonal deformation tensor components $\varepsilon_{\mu\mu}$ ($\mu = x,y,z$) which form, all in all, 9 quantities.

The spontaneous atomic displacements of the Γ_{12} symmetry, which reduce to a doubling of the period along orthogonal chains, may be regarded as a Peierls instability of one-dimensional electron systems. In GOR'KOV's model I for A-15 compounds one assumes that the Fermi surface passes in the vicinity of the point X of the Brillouin zone since the dispersion relation in the vicinity of this point satisfies the symmetry $\varepsilon_{p+2p_F} = -\varepsilon_p$, required for the Peierls instability. Thus, adopting Gor'kov's model BHATT and McMILLAN [6.290] considered the martensitic transition in A-15 compounds as a Peierls transition of interacting one-dimensional chains forming a three-dimensional structure. Because the system is three-dimensional the fluctuations play no catastrophic role, as in purely one-dimensional systems, and use can be made of the simple Landau phase transition theory.

The expansion of the free energy in the vicinity of the structural-transition point has the form

$$\Omega = \frac{1}{2}\sum_{\mu} a_{\mu}\phi_{\mu}^{2} + \frac{1}{2}\sum_{\mu}(K_{o}+G)Q_{\mu}^{2} + \frac{1}{2}\sum_{\mu\neq\nu} GQ_{\mu}Q_{\nu}$$
$$+ \sum_{\mu} f_{\mu}Q_{\mu}\phi_{\mu} + \frac{1}{2}\sum_{\mu} A_{11}\varepsilon_{\mu\mu}^{2} + \frac{1}{2}\sum_{\mu\neq\nu}(A_{12}\varepsilon_{\mu\mu}\varepsilon_{\nu\nu} + 2A_{44}\varepsilon_{\mu\nu}^{2})$$
$$+ h[\varepsilon_{xx}(Q_{y}-Q_{z}) + \varepsilon_{yy}(Q_{z}-Q_{x}) + \varepsilon_{zz}(Q_{x}-Q_{y})] \quad . \qquad (6.5.55)$$

The first row of this expression describes the energy of the charge density waves interacting with each other in the chains, and of lattice oscillations with optical symmetry Γ_{12}. The second row gives the elastic energy of the crystal in the presence of deformations, and the third describes the interaction of the optical and acoustic lattice modes, which is due to the coupling of either modes with conduction electrons. Thus, all the degrees of freedom of the system turn out to be coupled to each other. The peculiar form of the last term is due to the Γ_{12} symmetry of the optical lattice oscillation.

BHATT and McMILLAN argued that structural transitions in A-15 compounds arise from the instability of the electronic system with respect to the formation of charge density waves so that the coefficient a_{μ} should change sign at the transition point, whereas the other coefficients in (6.5.55) may be treated as weakly varying with temperature. Suppose a_{μ} is linearly dependent on T in the vicinity of the transition temperature

$$a_{\mu} = a'(T - T_{mo}^{\mu}) \quad . \qquad (6.5.56)$$

In addition, the gain in electronic energy is largest when the Fermi surface passes through the point X of the Brillouin zone, which has the energy $E_{X\mu}$. In the vicinity of this maximum we have

$$T_{mo}^{\mu} = T_{m}^{max} - B(E_{F} - E_{X\mu})^{2} \quad . \qquad (6.5.57)$$

In the deformed crystal the energy level shifts, i.e.,

$$E_{X\mu} = E_{X}^{o} + \lambda_{1}\varepsilon_{\mu\mu} + \lambda_{2}\sum_{\nu}\varepsilon_{\nu\nu} \quad , \qquad (6.5.58)$$

so that we have the following expression of the coefficient a_{μ} in the vicinity of T_{m}^{o} in terms of the deformation-tensor component

$$a_\mu = a'(T - T_m^o) + \Lambda_1 \varepsilon_{\mu\mu} + \Lambda_2 \sum_\nu \varepsilon_{\nu\nu} \quad , \tag{6.5.59}$$

where the parameters T_m^o, Λ_1 and Λ_2 are $(i = 1,2)$

$$T_m^o = T_m^{max} - B(E_F - E_X^o)^2 \quad , \quad \Lambda_i = 2B\lambda_i a'(E_X^o - E_F) \quad . \tag{6.5.60}$$

Instead of the nine variables Φ_μ, Q_μ, $\varepsilon_{\mu\mu}$ $(\mu = x,y,z)$ we introduce nine other variables Φ_i, Q_i, ε_i $(i = 1,2,3)$ which are linear combinations of them and are transformed according to the irreducible representations Γ_1 and Γ_{12} [cf. (6.5.43)]:

$$\varepsilon_1 = \frac{1}{\sqrt{3}}(\varepsilon_{xx} + \varepsilon_{yy} + \varepsilon_{zz}) \quad , \quad \varepsilon_2 = \frac{1}{\sqrt{2}}(\varepsilon_{xx} - \varepsilon_{yy}) \quad ,$$

$$\varepsilon_3 = \frac{1}{\sqrt{6}}(\varepsilon_{xx} + \varepsilon_{yy} - 2\varepsilon_{zz}) \tag{6.5.61}$$

(similarly for Φ_i and Q_i). It can be readily shown that the 9×9 matrix of the quadratic form (6.5.55) on these basis functions breaks up into 3×3 blocks M_i so that (6.5.55) may be written as

$$\Omega = \frac{1}{2} \sum_{i=1}^{3} \tilde{V}_i M_i V_i \quad , \tag{6.5.62}$$

where

$$V_1 = \begin{pmatrix} \Phi_1 \\ Q_1 \\ \varepsilon_1 \end{pmatrix} \quad , \quad V_2 = \begin{pmatrix} \Phi_2 \\ Q_2 \\ \varepsilon_3 \end{pmatrix} \quad , \quad V_3 = \begin{pmatrix} \Phi_3 \\ Q_3 \\ \varepsilon_2 \end{pmatrix} \quad .$$

The transition temperature is determined by the vanishing of one of the eigenvalues of the matrix M_i. Thus, we have the following three critical temperatures

$$T_1 = T_m^o + \frac{f^2}{(K_o + 3G)a'} \quad , \quad T_2 = T_3 = T_m^o + \frac{f^2}{\left(K_o - \dfrac{3h^2}{A_{11} - A_{12}}\right)a'} \equiv T_m^* \; . \tag{6.5.63}$$

It is easy to show that the first eigenvalue belongs to the eigenstate V_1 when $\varepsilon_1 = 0$, i.e., there is no tetragonal deformation. This state corresponds to the Peierls doubling of atoms of each of the crystal chains (owing to this doubling the crystal remains in the cubic phase). Doubling should occur when $T_1 > T_2 = T_3$, i.e., when $G < -3h^2/(A_{11} - A_{12})$. Otherwise, the state V_2 or V_3

with spontaneous tetragonal or orthorhombic deformation occurs. The degeneracy of these states with respect to the transition temperature ($T_2 = T_3$) arises from the harmonic approximation (6.5.55). It should be removed when taking into account the highest terms of the free-energy expansion.

From (6.5.55) we easily get the shear moduli. It turns out that the moduli $c_{11} + 2c_{12}$ and c_{44} are not renormalized as compared with the initial values of $(A_{11} + 2A_{12})$ and A_{44} and are therefore temperature independent. But the shear modulus becomes equal to

$$c_{11} - c_{12} = \left(c_{11}^{o} - c_{12}^{o}\right)\left[1 - \frac{1}{1 + \frac{K_o^2\left(c_{11}^{o} - c_{12}^{o}\right)a_r}{3h^2f^2}}\right] . \tag{6.5.64}$$

In this expression

$$c_{11}^{o} - c_{12}^{o} = A_{11} - A_{12} - \frac{3h^2}{K_o} \tag{6.5.65}$$

is the limiting value of the modulus far from the transition point, i.e., with $a_r \to \infty$, where a_r is the renormalized value of the parameter a_μ for which, instead of (6.5.56), we have to write $a_r = a'(T - T_m^*)$ or

$$a_r = a - \frac{f^2}{K_o - \frac{3h^2}{A_{11} - A_{12}}} . \tag{6.5.66}$$

Expression (6.5.64) may be brought into the following form

$$c_{11} - c_{12} = \left(c_{11}^{o} - c_{12}^{o}\right)\left(1 - \frac{1}{1 + \beta\theta}\right) , \tag{6.5.67}$$

where $\theta = (T - T_m^*)/T_m^*$ and β is a dimensionless parameter of electron-phonon coupling. Thus, the harmonic approximation leads to a shear-modulus softening which, however, obeys not a logarithmic law, as is the case in the Gor'kov model, but a power law.

Now consider the effects produced by higher-order terms in the free-energy expansion. Since the structural transition is assumed to be caused by electron factors we have to add the fourth-order term $b\left(\sum_\mu \phi_\mu^2\right)^2/2$ to (6.5.55). Additional third-order terms occur if, using (6.5.59), we take into account the dependence of the electron energy on deformation in the expression for the coefficient a_μ. If we now write the full free energy in terms of (6.5.61) it can be easily verified that only the quantities ε_2 and ε_3 are coupled with the electronic parameters Φ_2 and Φ_3. Therefore, setting the variables Φ_1, Q_1 and ε_1 equal to zero (they are not essential to the phase transition), we have the following expression for the free energy of the system of electrons and optical phonons interacting with each other and with acoustic phonons

$$\Omega = \frac{1}{2}a\left(\Phi_2^2+\Phi_3^2\right) + \frac{b}{4}\left(\Phi_2^2+\Phi_3^2\right)^2 + \frac{1}{2}K_o\left(Q_2^2+Q_3^2\right) + f\left(\Phi_2Q_2+\Phi_3Q_3\right)$$
$$+\frac{\Lambda_1}{12}\left[\varepsilon_3\left(\Phi_2^2-\Phi_3^2\right)+2\varepsilon_2\Phi_2\Phi_3\right] + \frac{1}{2}\left(A_{11}-A_{12}\right)\left(\varepsilon_2^2+\varepsilon_3^2\right) + \sqrt{3}h\left(\varepsilon_2Q_3-\varepsilon_3Q_2\right) \quad . \tag{6.5.68}$$

After minimization with respect to the variables Φ_i and Q_i we can express the free energy in terms of deformations alone

$$\Omega = \alpha_2\left(\varepsilon_2^2+\varepsilon_3^2\right) + \alpha_3\varepsilon_3\left(\varepsilon_3^2-3\varepsilon_2^2\right) + \alpha_4\left(\varepsilon_2^2+\varepsilon_3^2\right)^2 \quad , \tag{6.5.69}$$

where the coefficients α_i are

$$\alpha_2 = \frac{a_r}{2}K \quad , \quad \alpha_3 = \frac{\Lambda_1}{2\sqrt{6}}K \quad , \quad \alpha_4 = \frac{b}{4}K^2 \quad ; \quad K \equiv \frac{K_o^2\left(c_{11}^o-c_{12}^o\right)^2}{3h^2f^2} \tag{6.5.70}$$

The resulting expression (6.5.69) coincides with the earlier phenomenological expression (6.5.44) in which the quantities ε_2 and ε_3 were designated as η_1 and η_2, but in contrast to (6.5.44), the parameters in the free-energy expansion with respect to the deformations are expressed through a set of coefficients of the expansion (6.5.68) in powers of all the order parameters. In view of the coincidence of these expressions, all the results of the preceding section, which are based on the form of the free energy (6.5.69), remain the same. In particular, the structural transition in the system under consideration is a first-order phase transition having the following parameters: the transition temperature

$$T_m = T_m^* + \frac{\alpha_3^2}{4\alpha_2'\alpha_4} \tag{6.5.71}$$

and the jump in spontaneous deformation at the point of transition

$$\varepsilon_3(T_m) = -\frac{\alpha_3}{2\alpha_4} \quad . \tag{6.5.72}$$

Since $\alpha_3 \sim \Lambda_1$, the sign of spontaneous deformation is determined by the sign of the difference $(E_F - E_x^o)$.

Thus, based on the mechanism of the formation of charge density waves and on the assumption that the Fermi level passes near the point X, Landau's simple phenomenological theory for the structural transition in A-15 compounds describes all the main features of this transition. Moreover, BHATT and McMILLAN [6.290] generalized the free-energy expansion for the inhomogeneous case permitting consideration of long-wave phonons.

For the transverse acoustic phonon with wave vector $\underline{q} \sim [110]$ and with polarization $[1\bar{1}0]$, the frequency $\omega_T(\underline{q})$ turns out to be equal to

$$\omega_T(\underline{q}) = \frac{c_{11}^0 - c_{12}^0}{2m} q^2 \left(1 - \frac{1}{1 + \beta\theta + \eta q^2}\right) \quad . \tag{6.5.73}$$

This phenomenological expression with the only fitting parameter η satisfactorily describes the frequency of the soft mode in a broad interval of wave vectors (up to the middle of the Brillouin zone). The long-wave limit of this expression corresponds to (6.5.67) which describes the softening of the shear modulus. Another important result of the theory is the temperature dependence of the optical frequency of Γ_{12} mode. With decreasing temperature this frequency diminishes, although when $T \to T_m$ it remains finite, just as in the SHAM and NOOLANDI theory.

The results obtained for the temperature dependence of the velocity of sound as well as for the frequency of the optical mode prove to be in good agreement with experiments, and also with the central peak observed in neutron scattering. For details, the reader is referred to the original paper [6.290].

Within the framework of the Gor'kov model I, BILBRO and McMILLAN [6.291] have considered the relationship between the martensitic and the superconducting transition in A-15 compounds, and found: If the Fermi level passes through the point X during the martensitic transition arising from the Peierls instability in the chains, a gap appears on the sections of the Fermi surface which are adjacent to point X. As a result, the density of states on the Fermi surface decreases and the superconducting transition temperature should decrease. However, since the sections of the Fermi surface distorted by the Peierls transition are small, the resultant variation of T_c is also small (0.3 K for V_3Si).

Another aspect of the problem is the influence of the superconducting transition on the temperature behaviour of the lattice. The model predicts the inhibition of the growth of the Peierls gap with decreasing temperature and of the tetragonality parameter in accordance with experiments. This agrees qualitatively with the theoretical results of BIRMAN et al. [6.241,292] who have shown that the softening of the shear modulus in a nontransforming crystal stops at the superconducting transition.

The relationship between T_c and T_m in A-15 compounds requires a more thorough investigation in light of the work by KOPAEV [6.293].

6.5.5 A Comparison of Different Theories of Martensitic Transformations

The BHATT-McMILLAN theory described here is a phenomenological version of GOR'KOV's theory I. In the latter the mechanism of the lattice phase transition consists of an electronic instability. The order parameter in the electronic system is associated with the optical mode Γ_{12} leading to a dimerization of transition metal atoms in the chains. As a result of the coupling of the optical mode to crystal deformations, an indirect coupling of electrons to deformations takes place. On the other hand, a number of models (Labbé-Friedel, Bhatt, etc.) assume that the electronic system is directly related to deformations, and the martensitic transition resulting from this relationship is treated as a Jahn-Teller effect. These models can be phenomenologically described if a Landau functional is given. The easiest thing to do is to resort to the Labbé-Friedel noninteracting chains model.

An important starting point in the discussion of the structural phase transition in this model is the three-fold degeneracy of the electron spectrum of d-states due to the existence of three mutually perpendicular systems of chains in the $\mu = x,y,z$ directions. In the cubic phase, the filling of all three bands associated with these chains is the same. In the tetragonal phase the three-fold degeneracy is removed, and the band spectrum of the chains aligned with the tetragonal axis is populated in a manner different from that for perpendicular chains. Due to the removal of the degeneracy, the electron energy decreases. If this decrease exceeds the gain in the elastic energy of the deformed crystal, the latter undergoes a spontaneous tetragonal distortion. It is clear that the order parameter in the system is the degree of filling of the three bands (to be more exact, the difference between these quantities in the tetragonal and cubic phases). Denote this filling as φ_μ. Taking into account the *direct* relationship of the electron energy to the field of deformations, we write the Landau expansion as

$$\Omega = \frac{1}{2}\sum_{\mu}\left[A_\mu\varphi_\mu^2 + \left(K_o + G\right)Q_\mu^2 + A_{11}\varepsilon_{\mu\mu}^2 + 2F\varepsilon_{\mu\mu}\left(\varphi_\mu - \frac{1}{3}\sum_{\nu}\varphi_\nu\right)\right]$$

$$+ \frac{1}{2}\sum_{\mu\neq\nu}\left\{A_{12}\varepsilon_{\mu\mu}\varepsilon_{\nu\nu} + 2A_{44}\varepsilon_{\mu\nu}^2 + GQ_\mu Q_\nu\right.$$

$$+ h\left[\varepsilon_{xx}\left(Q_y - Q_z\right) + \varepsilon_{yy}\left(Q_z - Q_x\right) + \varepsilon_{zz}\left(Q_x - Q_y\right)\right]\Big\} \quad , \tag{6.5.74}$$

where

$$A_\mu = A'(T - T_{mo}) \quad . \tag{6.5.75}$$

Proceeding to the variables ε_i, φ_i and Q_i according to (6.5.61), we note that the off-diagonal components of the strain tensor as well as ε_1 and the symmetric optical mode Q_1 turn out to be uncoupled. In the second order of perturbation the related portion of the free energy is of the form

$$\Omega = \frac{1}{2}A\left(\varphi_2^2 - \varphi_3^2\right) + \frac{1}{2}\left(A_{11} - A_{12}\right)\left(\varepsilon_2^2 + \varepsilon_3^2\right) + \frac{1}{2}K_o\left(Q_2^2 + Q_3^2\right)$$

$$+ F(\varphi_2\varepsilon_2 + \varphi_3\varepsilon_3) + \sqrt{3}\,h\,(\varepsilon_2 Q_3 - \varepsilon_3 Q_2) \quad . \tag{6.7.76}$$

The condensation of two coupled modes $(\Phi_2, Q_3, \varepsilon_2)$ and $(\Phi_3, Q_2, \varepsilon_3)$ occurs at a temperature

$$T_m^* = T_{mo} + \frac{F^2/A'}{A_{11} - A_{12} - 3h^2/K_o} \quad . \tag{6.5.77}$$

Above this temperature the functional (6.5.76) leads to the following formula for the shear modulus

$$c_{11}(T) - c_{12}(T) = \left(c_{11}^o - c_{12}^o\right)\left(1 - \frac{1}{1 + B\theta}\right) \quad , \tag{6.5.78}$$

where $(c_{11}^o - c_{12}^o)$ is given by (6.5.65), and B is the dimensionless parameter of the electron-deformation coupling. It also follows from (6.5.76) that the moduli $c_{44}(T)$ and $c_{11}(T) + 2c_{12}(T)$ are not renormalized and therefore are temperature-independent.

Comparison of (6.5.78) and (6.5.67) shows that the shear moduli have the same temperature dependence in both models. Apart from this, the frequency of the transverse acoustic mode with $\underline{q} \sim [110]$ and with the polarization $\sim [1\bar{1}0]$ is described by the previous formula (6.5.73), but with a different coefficient, N instead of η.

The temperature dependence of the lattice moduli and of the phonon freqencies arises from the quantities a_μ and A_μ in (6.5.55) and (6.5.74) which are the only temperature parameters of the models. This leads to the same behaviour of these quantities with temperature, but with parameters different in their microscopic sense and magnitude.

Thus, the behaviour of the moduli and of the above phonon frequencies cannot be separated in Landau's phenomenological theory for the Gor'kov and Labbé-Friedel models. Differences arise in the behaviour of the frequency of the optical mode Γ_{12}. Calculation shows that in the Jahn-Teller phase-transition mechanism it is temperature-independent, whereas in the Peierls mechanism some softening at temperatures approaching T_m is predicted. As already stated above, experiments of Raman light scattering in V_3Si have shown that the mode Γ_{12} is weakly softened. It is unlikely that these data may serve as a basis for our choice between the two theories under consideration.

Thus, at the moment we are not in a position to say definitely which of the two order parameters describing the structural transition is the more fundamental one, viz., ε or Q. The SHIRANE and AXE [6.274] experiments showed that a linear relationship exists between them, and at a temperature below T_m both parameters appear simultaneously. These parameters arise from the instability in the electronic system described by the parameters Φ or φ. The microscopic foundation of the two phenomenological (Bhatt-McMillan and Bhatt) theories is constituted by the Labbé-Friedel and Gor'kov models, which are both one-dimensional.

Three-dimensional models of the electron structure of A-15 compounds may also have a phenomenological interpretation. This applies to Bhatt's model which assumes that the atomic states of symmetry $(x^2 - y^2)$ play a dominant role on the Fermi surface. This model, too, leads to sharp peaks in the density of states owing to the contribution made by the vicinity of the saddle points M and other points lying on the ΓX line, which have an energy close to that of the saddle points. Both points are three-fold degenerate, and this degeneracy is removed in the tetragonal phase for one and two-fold degenerate states. Their relative filling, just as is the case in the one-dimensional Labbé-Friedel model, may serve as the order parameter Φ_μ governing the structural phase transition. The behaviour of lattice moduli in such a model turns out to be the same as that in one-dimensional models.

Some information concerning the applicability of a particular model can be obtained from a study of the correlation length. Expressions (6.5.73,78) for the frequency of transverse acoustic phonons (the soft mode) in two different models at temperatures which ar not too far from T_m may be written in the form

$$\omega_{\underline{q}}^2 \sim q^2\left(\theta + \xi_0^2 q^2\right) \quad . \tag{6.5.79}$$

The parameter ξ_0 is the correlation at zero temperature. It is associated with the fundamental concept of the correlation length of order-parameter fluctuations which appears in Landau's phenomenological theory. If the Landau expansion is written in terms of the order parameter

$$F = a\psi^2 + b\psi^4 + c(\underline{\nabla}\psi)^2 \quad , \quad a = a'(T - T_m) \quad , \tag{6.5.80}$$

then the correlation length $\xi(T)$ at a given temperature T is given by the equation

$$\xi^2(T) = \frac{c}{a} = \frac{c}{a'(T - T_m)} = \frac{\xi_0^2}{\theta} \quad ,$$

where $\xi_0^2 \equiv c/a'T_m$. In Landau's theory, a functional such as (6.5.80) leads to an expression for the soft mode in the form of (6.5.79). Thus, (6.5.73,78) for the soft mode frequency have the correct limiting form of the phenomenological theory.

The ξ_0 depends on the microscopic parameters of the system and is a characteristic of the model. Neutron diffraction experiments show a considerable softening of the transverse mode throughout almost all the Brillouin zone. This indicates that ξ_0 is of the order of magnitude of the a parameter of the lattice. BHATT and LEE [6.294] have computed ξ_0 for various models of A-15 compounds. It turned out that in the case of Gor'kov's model I in which the quasi-one-dimensional electron band develops a Peierls instability,

$$\xi_0/a \sim E_F/T_m >> 1 \quad , \tag{6.5.81}$$

the Labbé-Friedel model leads to an appreciably smaller value of the correlation length

$$\xi_0/a \sim \sqrt{B/T_m} \tag{6.5.82}$$

(B stands for the band width) which is nevertheless much greater than a. Bhatt's three-dimensional model relating the anomalous behaviour of A-15 compounds to the contribution to N(E) of the singularities from the saddle points leads to a sufficiently short correlation length $\xi_0 \sim a$, observed in experiments. This gives additional arguments in favour of the three-dimensional model of the band structure of A-15 compounds.

Let it be noted in conclusion that all the theories of structural phase transitions in V_3Si and Nb_3Sn which we have considered here are based on a mean-field approximation or on the expansion of the free energy in powers of order parameters. An order parameter is a basis function of the two-dimensional irreducible representation Γ_{12} for $\underline{q}=0$. This representation admits third-order terms in the free energy expansion and therefore, according to the general theory of Landau, describes the first-order phase transition observed in experiments. In principle, transitions according to other irreducible representations might also be observed in A-15 compounds. A symmetry description of the possible transitions from group theory according to irreducible representations with $\underline{q}=0$ (point Γ) on the basis of Landau's theory has been made in [6.134,295].

It is known that allowing fluctuations of the order parameters may change the second-order phase transition to a first-order transition. On the basis of the renormalization group method and of the ε-expansion [6.296], JARIC and BIRMAN [6.297] have investigated the character of the phase transition for all the active representations of the point Γ, which in Landau's theory corresponds to a second-order transition. They have shown that taking into account the fluctuations does not change the character of this transition. A similar analysis has also been performed for the phase transitions in the irreducible representations of the point R and of the point X. Although no such phase transitions have been detected in A-15 compounds, the above analysis may be helpful if these should be discovered. In this connection we would like to draw the reader's attention to [6.298,299] containing evidence that in Nb_3Sn there is a structural transition which occurs with the increase of the crystal cell.

7. Other Compounds Based on Transition Metals

This chapter is devoted to the superconducting properties of intermetallics (compounds with A-12 structure, B-2, Laves phases and σ-phases and of chemical compounds based on transition elements (compounds with B-1 structure, hydrides, sesquicarbides, ternary oxides and borides, Chevrel phases). The influence of alloying on the superconducting characteristics of these compounds are discussed and results for their electronic and lattice properties are given.

7.1 Intermetallics

7.1.1 Laves Phases

Laves phases constitute a numerous group of metallic compounds: about 270 binary compounds of this type are now known [7.1].

These compounds have a common chemical formula AB_2 and may belong to three different structural types, viz., $MgNi_2$ (C-14 structure), $MgZn_2$ (C-15 structure) and $MgCu_2$ (C-36 structure). Most of the binary Laves phases (ca. 140) pertain to the structural type C-15. The C-15 structure has a cubic symmetry and belongs to the space group O_h^7 - Fd3m. The other group which ranks second in the number of representatives belongs to the structural type C-36 (112 compounds) [7.1]. This structure possesses a hexagonal symmetry and pertains to the space group D_{6h}^4 - $P6_3$/mmc. The remaining compounds have a C-14 structure which, just as C-36, pertains to hexagonal symmetry (space group D_{6h}^4 - $P6_3$/mmc) but, in contrast to the C-36 structure, contains not 12 but 24 atoms in the unit cell.

The three structural types mentioned above have one common element: tetrahedra of B atoms forming a spatial skeleton in whose voids A atoms are located (Fig.7.1 shows a C-15 type structure).

All the metals in subgroups of the periodic system (except for mercury), and also Al, Pb and Bi appear as the atomic components of these compounds.

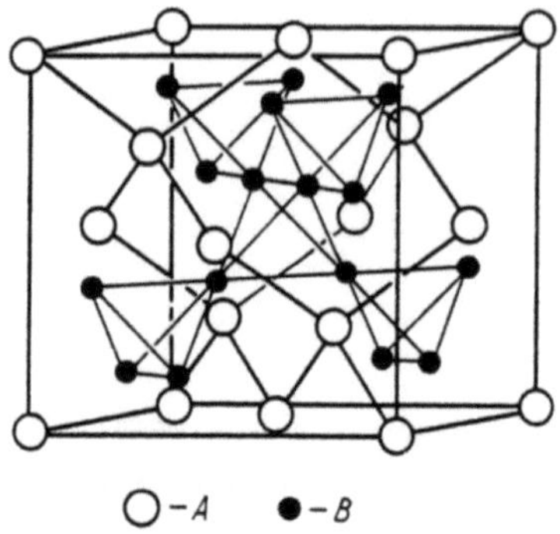

Fig.7.1. The C-15 crystal structure

The distribution of A and B components among the metals specified is not governed by any law - both A and B components occur in all the subgroups (except for IIIa and IVa).

The superconducting properties have been studied only in some of the compounds. Of the 100 compounds investigated, superconductivity has been detected in 75 compounds (Table 7.1) [7.2-8].

Among the binary C-15 structure compounds, the following compounds possess maximum T_c: ZrV_2 ($T_c = 9.62$ K [7.9]), HfV_2 ($T_c = 9.4$ K [7.10]) and $LaOs_2$ ($T_c = 8.9$ K [7.11]). In two of the three examples given above, the B partner in these compounds is vanadium. It was assumed that the third vanadium-based Laves phase should also have a high T_c [7.12]. However, as shown by measurements [7.13], $T_c = 3.6$ K in the C-15 compound TaV_2. According to [7.10,14], TaV_2 has a high-temperature hexagonal modification with $T_c = 10$ K. This value is the highest for C-14 structure compounds whose critical temperatures are, as a rule, lower than those with C-15 structure. Almost all of the superconducting Laves phases belong to two structural types, C-15 and C-14. So far only one superconducting compound has been found with C-36 structure: $HfMo_2$ ($T_c = 0.7065$ to 0.082 K ·[7.5]).

Most interesting are the properties of the intermetallics ZrV_2 and HfV_2 and of the alloys on their basis. Apart from sufficiently high T_c, these materials possess high H_{c2} ($T_c = 10.1$ K, $H_{c2} = 240$ kOe for $(Zr_{0.5}Hf_{0.5})V_2$ [7.15]). It should be noted that these materials have low hardness and are less brittle compared to other high-field superconductors and thus are technologically more interesting. On the basis of $(Zr_xHf_{1-x})V_2$ compounds, composites possessing high critical parameters have already been obtained (T_c to 10.2 K, H_{c2} to 208 kOe, J_c to $4 \cdot 10^5$ A/cm^2), some of which are higher than those in V_3Ga and Nb_3Sn based composites used now [7.16-18].

Alloying by a third component affects the T_c of ZrV_2 and HfV_2 in different ways. In ternary systems such as $ZrV_2 - ZrX_2$ and $HfV_2 - HfX_2$ where X stands for

Table 7.1. T_c in Laves-phase compounds

A \ B_2	Be	Mg	Al	Zn	Ge	La	Lu	Bi	Hf	V	Te	W	Re	Fe	Ru	Os	Co	Rh	Ir	Pd	Pt	Cu	Au	Mo	Tc
Na																							<1.0		
K								3.6																	
Rb								4.2																	
Cs								4.7																	
Be					<1.7																				
Mg						<1.0																<1.0			
Ca		<1.0	<1.0															6.4	6.1	<1.0	<1.0				
Sr																		6.2	5.7		<1.0				
Ba																		6.0		<1.0	<1.0				
Al							<1.0																		
Pb																							1.2		
Bi																							1.8		
Sc			<1.0										4.2		2.2	4.6		6.2	2.5		0.7				
																			-5.7		1.6				
Y			<1.0										1.8		2.4	4.7		<0.3	2.1		0.54				
La			3.2												4.4	8.9		<1.0	0.48						

Table 7.1 (cont.)

A \ B_2	Be	Mg	Al	Zn	Ge	La	Lu	Bi	Hf	V	Te	W	Re	Fe	Ru	Os	Co	Rh	Ir	Pd	Pt	Cu	Au	Mo	Tc
Ce															6.2	x	1.5	<0.3	<0.4		<1.0				
Pr															x	x		x	x		x				
Nd															x	x		x	x		x				
Cd													$\underline{<4.2}$		<1.2			x	x		<1.0				
Tb													$\underline{<4.2}$												
Ho													$\underline{<4.2}$												
Er													$\underline{<4.2}$												
Lu															$\underline{0.9}$	$\underline{3.5}$		1.27	2.89						
Th													$\underline{5.0}$		3.5	<1.0			6.5						$\underline{5.3}$
U		<1.0												1.0		<0.4			<0.3						
Ti	<1.0			<1.0																					
Zr		$\underline{<0.3}$		$\underline{<0.1}$						9.6	$\underline{7.6}$	<0.03	$\underline{6.8}$		$\underline{1.8}$	$\underline{3.0}$	<1.0		4.1					0.125	
Hf										9.4	$\underline{5.6}$	<0.03	$\underline{5.6}$			$\underline{2.7}$								0.07	
Nb	2.15																								
Ta										3.6															
										$\underline{10}$															
Mo	$\underline{<1.7}$																								
W	$\underline{<1.7}$																								
Re	$\underline{<1.7}$																								
Rh									1.9																

Note: The underlined values of T_c belong to Laves phases with C-14 structure, x marks the Laves phases in which the superconducting properties were not investigated.

elements of the subgroups V_a - VII_a of the periodic table, the situation is as follows. As the X component content increases to 10 or 15 at.%, T_c rises somewhat and then drops [7.19]. The T_c decreases in the case of alloying by aluminum [7.20], iron [7.21] and hydrogen [7.22]. It is worth noting that, in contrast to binary compounds, the introduction of hydrogen into $(Zr_{0.5}Hf_{0.5})V_2$ leads to an increase of the T_c to 11.8 K [7.23].

Electron heat capacity measurements in HfV_2, ZrV_2 and $(Zr_{0.5}Hf_{0.5})V_2$ [7.15] have established the same correlation between T_c and γ which occurred in pure transition metals and in disordered alloys. Table 7.2 contains values of T_c, γ, θ_D, λ and N(0) for a number of Laves-phase compounds. It is seen from this table that at such large N(0) and λ the values of T_c in these compounds are not high enough apparently because of the low θ_D.

Table 7.2. Superconducting and electronic properties of a number of Laves phases

Compound	T_c [K]	γ $\left[\frac{mJ}{mole \cdot K^2}\right]$	θ_D [K]	λ	N(0) $\left[\frac{\text{number of states}}{eV \cdot atom}\right]$
HfV_2	9.2	21.7	187	1.00	2.30
$(Hf_{0.5}Zr_{0.5})V_2$	10.1	28.3	197	1.02	2.97
ZrV_2	8.5	16.5	219	0.88	1.86

It is worth noting that the lattice properties of the compounds ZrV_2, HfV_2 and TaV_2 turned out to be particularly interesting. According to X-ray microscopy data [7.24], the crystal structure of these materials is characterized by the presence of plane defects: twins and stacking faults. As the temperature decreases, a structural transformation takes place in ZrV_2 and HfV_2. Early X-ray structure investigations of powders of these materials revealed the existence of a transformation from a cubic (C-15 type) to a body-centred orthorhombic lattice [space group $Imm2\left(C_{2v}^{20}\right)$] at temperatures below 130 K [7.4,25]. However, these data differ from subsequent neutron diffraction and X-ray measurements according to which the structure of the low-temperature phase ZrV_2 is rhombohedral [7.26-28] and that of HfV_2 is tetragonal [7.28,29]. The above investigations show that the variation of the lattice parameters during structural transformation is of a spasmodic character and, as well as this, there are hysteresis effects and a two-phase region of the "martensitic" and cubic phases. According to [7.28], reduction of a specimen or reduction of the mean size of a grain in a polycrystal leads to a broadening of the two-phase region and to a lowering of the transformation temperature, which is one of the properties inherent in martensitic transformation. For the compound TaV_2, no structural transformation at temperatures as low as 4.2 K was registered.

LAWSON et al. [7.30,31] have discovered the low-temperature structural transformation in one more compound with a C-15 structure. It was $LaRu_2$ which at a temperature close to 30 K exhibited a tetragonal distortion of the initial cubic structure with the lattice parameters $\sqrt{2}a = 7.663$ Å and $c = 7.773$ Å. This transformation occurs without change in crystal volume.

In the temperature range where the lattice symmetry of ZrV_2, HfV_2 and $LaRu_2$ decreases, these materials were found to exhibit an anomalous behaviour in a number of physical properties. Figure 7.2 presents results of specific heat measurements, which show a sharp peak in HfV_2 at temperatures between 105 and 128 K [7.32]. An anomalous decrease of the magnetic susceptibility in ZrV_2 (about 10 to 15%) occurs at temperatures below 150 K [7.33-35]. In [7.36,37], the above effect was attributed to the spin contribution of d-electrons, on the assumption that the Fermi level passes near the peak in the density of d-states. The calculation of the electron structure of Laves phases [7.38], and also the results of NMR measurements [7.39] and investigations of X-ray emission spectra indicate that the Fermi level in the cubic phase of the com-

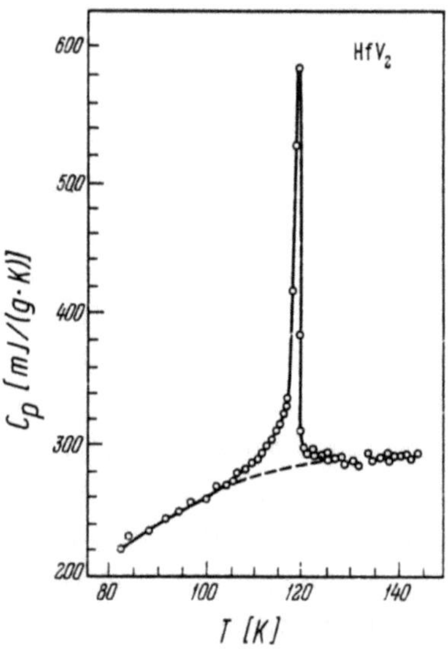

Fig.7.2. Heat capacity in the compound HfV_2

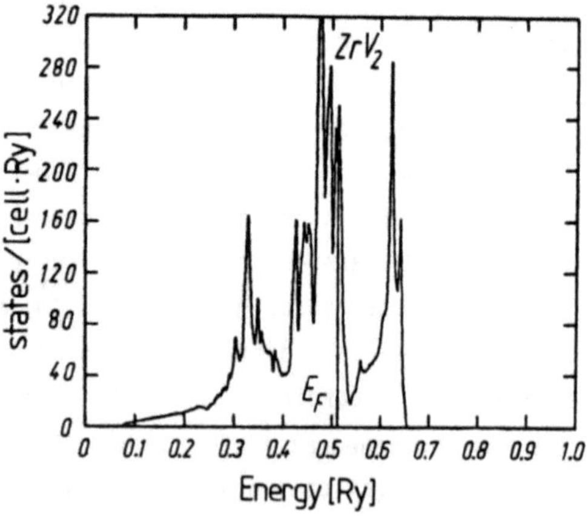

Fig. 7.3. Density of electron states in HfV_2 [7.38]

pounds ZrV_2 and HfV_2 does indeed pass in the vicinity of the peak in the density of d-electron states (Fig.7.3). With the temperature decreasing from 300 to 120-150 K, the resistivity of annealed ZrV_2 and HfV_2 alloys decreases monotonically almost in a linear fashion, whereas at lower temperatures it increases anomalously. An investigation of ZrV_2 specimens with a varying character of structural transformation revealed that the temperature intervals, within which the structural transformation and the anomaly in resistivity occur, coincide [7.27]. The fact that the anomalies have a form common to all specimens, and the presence of anomalies in the temperature dependences of resistivity in specimens where the structural transformation is absent, enabled LEVINSON et al. [7.27] to assume that structural and electronic transitions exist "independently". The change of the character of structural transformation from specimen to specimen is ascribed to the difference in the microstructure of specimens. According to the data obtained in [7.15], martensitic originates on defects whose number in the single crystallite undergoing no structural transformation is inadequate to initiate a new phase.

PAN et al. [7.20,40-46] assumed that, in addition to the change of the crystal structure, the electron spectrum becomes partially dielectrized during the phase transition in ZrV_2 and HfV_2. Dielectrization is accompanied by a freezing-out of part of the carriers, which motivates the observed anomalies in the temperature dependences of resistivity and magnetic susceptibility. Very interesting results have recently been obtained in experiments on ultrasound in polycrystalline HfV_2, ZrV_2 and TaV_2 specimens [7.47]. These data are given in Fig.7.4. It is seen that the velocities of longitudinal waves in HfV_2 and ZrV_2 decrease initially when cooled from 300 K but then increase sharply at a temperature of about 115 K. At the same time, TaV_2 shows an enhancement of lattice rigidity with cooling, which is typical of metals. These data indicate conclusively that in ZrV_2 and HfV_2 the lattice moduli become softened in the temperature range close to the structural-transition temperature. Thus, comparatively high (for this family of compounds) superconducting transition temperatures were observed in compounds exhibiting a softening of lattice moduli at low temperatures and having a structural transition. This indicates that high-temperature superconductivity is related to lattice instability.

In the premartensitic region of temperatures, the temperature dependences of the intensity of Bragg X-ray reflections deviates from the behaviour calculated by a harmonic approximation. This shows that the contribution to the intensity of diffuse scattering due to the distortion of the crystal lattice

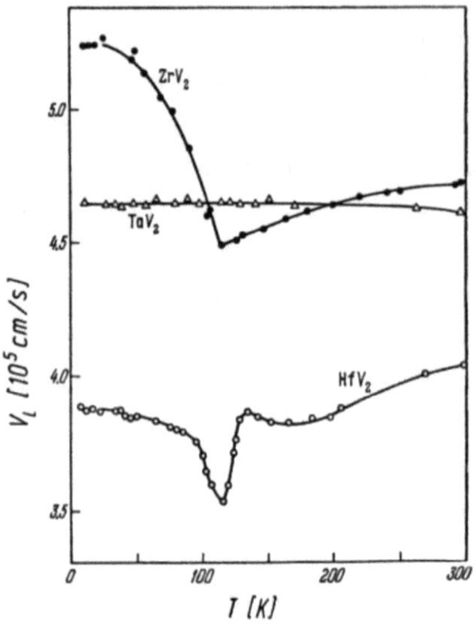

Fig.7.4. Temperature dependence of the velocity of longitudinal sound waves (10 MHz) in ZrV_2, HfV_2 and TaV_2

increases as the point of transition is approached [7.28]. The presence of an anomalous diffuse scattering was revealed in electron diffraction patterns of ZrV_2 and TaV_2 [7.24]. In TaV_2 it is practically absent. Analysis of the diffuse scattering permitted construction of the spectrum of atomic displacement, being essentially anisotropic. The displacement waves are mainly shear waves; both short-wave and long-wave components are present and the amplitudes increase as the temperature decreases. The maximum displacements occur in the {111} directions lying in the <111> planes.

Since anomalies in resistivity, in the velocity of longitudinal sound, and the structural transformation in ZrV_2 and HfV_2 occur at the same temperature, no structural measurements were made in the subsequent study of ternary compounds such as C-15 structure Laves phases based on ZrV_2 and HfV_2. The presence of a phase transition was judged by the anomaly in resistivity or in the velocity of propagation of ultrasound. Figure 7.5 displays T_c vs. composition curves for alloys of the ZrV_2 - HfV_2 system on the basis of the data obtained in [7.48,49]. Also shown here is the variation of the structural transformation temperature T_m determined from the temperature dependences of the resistivity (Fig.7.5a) and the velocity of ultrasound (Fig.7.5b). In both cases a minimum in T_m corresponds to the maximum value of T_c. Thus, T_c and T_m have a clearly marked tendency to come closer to each other in high-temperature superconductors such as Laves phases.

The substitution of tantalum for hafnium in the compound HfV_2 quickly destructs the anomaly in resistivity. With the Ta content being 20%, the resistivity vs. temperature curve has practically no peculiarities [7.50]. X-ray

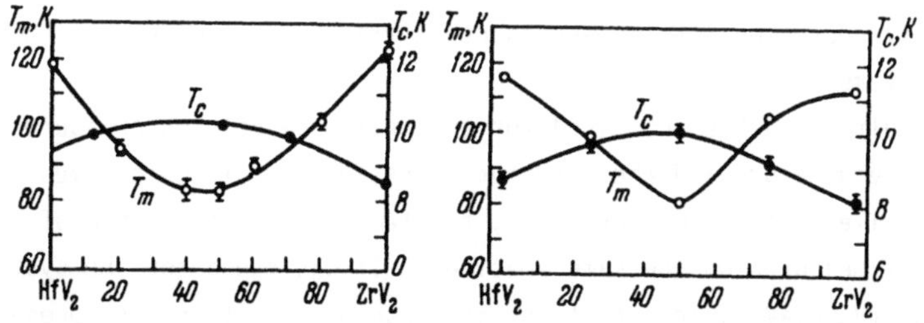

Fig.7.5. The dependence of the superconducting transition temperature and of the onset of lattice instability in compounds based on HfV_2 and ZrV_2

measurements have shown that the disappearance of the anomaly in resistivity is accompanied by a decrease in the degree of lattice distortion which tends to zero for about 15% of Ta. The suppression of the structural instability proceeds still more quickly when molybdenum is substituted for vanadium [7.50]. It is worth noting that in both cases no substantial change in T_c occurs in this range of compositions.

From a combined analysis of $C_p(T)$ and $\chi(T)$ HAFTSTROM et al. [7.51,52] obtained the temperature dependence of the mean geometric frequency of the phonon spectrum $\langle\omega\rangle$ in $(Ta_xHf_{1-x})V_2$ compounds. Long before the structural transformation, a decrease of $\langle\omega\rangle$ comes about, which indicates softening of the phonon spectrum. In alloying, the degree of softening becomes lower, but it does not vanish completely even for specimens undergoing no structural transformation.

Interesting information concerning the behaviour of the density of electron states on the Fermi level with concentration and temperature in $Zr(Nb_xV_{2-x})$ and $(Nb_xZr_{1-x})V_2$ compounds was derived by KIMURA [7.53] from a combined analysis of the temperature dependences of magnetic susceptibility and specific electrical resistivity. It turned out that N(0) decreased with increasing zirconium content of the high-temperature cubic phase. Such tendency is also specific to $Hf(Nb_xV_{2-x})$ and $(Zr_xHf_{1-x})V_2$, as follows from analyses of the NMR spectra [7.39,54]. Comparison of the concentration dependences of the density of states determined before and after the structural transformation [7.54], of the latent heat of transformation [7.55] and of the lattice parameters at 4 K [7.55] in the system $(Zr_xHf_{1-x})V_2$ shows that the most substantial readjustment of the structure and the electron spectrum during martensitic transformation occurs in binary compounds.

The problem concerning the influence of alloying on the superconducting properties and lattice instability in HfV_2 and ZrV_2 was considered in detail by INOUE and TACHIKAWA [7.4]. In all the cases a reverse relationship was observed between T_c and T_m. In other words, whereever the temperature of the lattice instability decreased, the superconducting transition temperature increased, and vice versa.

Interesting results have been obtained for the influence of high pressure on the transition temperature and the structural instability of cubic Laves phases [7.56]. Parallel with T_c measurements, the temperature dependence of the resistivity was measured on the same specimens. It was found for the system of ZrV_2 - HfV_2 and HfV_2 - TaV_2 alloys that the increase of T_c with rising pressure is accompanied by a decrease in the temperature at which the lattice becomes instable.

Furthermore, it has turned out that, although T_c and T_m show strong pressure dependences for HfV_2 and for alloys rich in HfV_2, these dependences weaken appreciably when the compound HfV_2 is alloyed by ZrV_2 (Figs.7.6,7). Thus, despite the fact that both ZrV_2 and HfV_2 exhibit a structural transformation at low temperatures and show a softening of lattice moduli, the stability of the cubic phase of ZrV_2 is less sensitive to application of pressure than that for HfV_2. Possibly, this difference in the behaviour of the compounds concerned arises from a difference in the character of structural instability. Proceeding from experiments on pressure it may be assumed that when ZrV_2 undergoes a structural transformation the lattice volume undergoes no appreciable changes. This agrees with neutron-diffraction measurements. According to these data, what occurs in ZrV_2 is not a transition from a cubic to an orthorhombic lattice but a transition from a cubic to an octahedral lattice. This transition is, indeed, not accompanied by a change in volume [7.26].

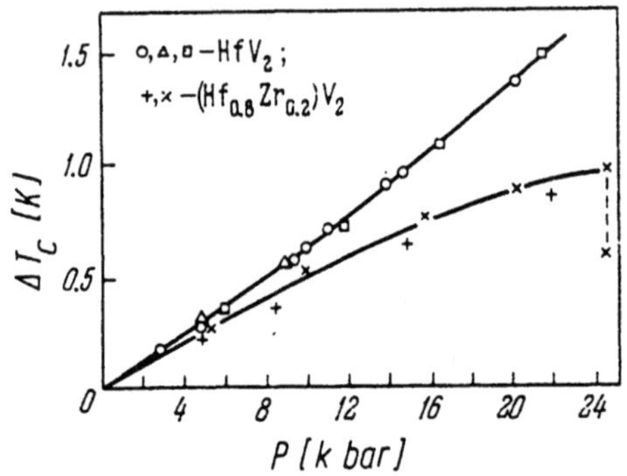

Fig.7.6. Variation of the superconducting transition temperature with pressure in HfV_2 and $(Hf_{0.8}Zr_{0.2})V_2$

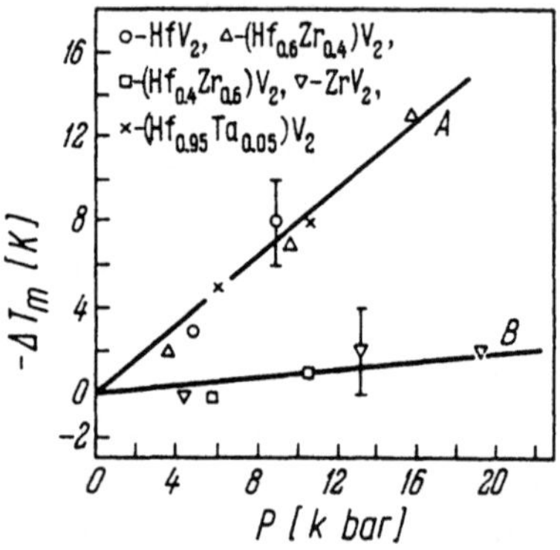

Fig.7.7. Variation of the temperature of lattice instability in HfV_2 and ZrV_2-based alloys

7.1.2 Compounds with CsCl (B-2) Structure

Until quite recently, only three superconductors were detected among the transition-metal compounds with CsCl (B-2) structure (Fig.7.8). Here belong TiRu ($T_c = 1.07$ K), TiOs ($T_c = 0.46$ K) [7.2] and VRu ($T_c = 5$ K) [7.57]. In 1976 MATTHIAS et al. [7.58] reported the existence of four other binary and three other ternary compounds of this type: ScCu ($T_c = 0.47$ K), ScAu ($T_c = 2.0$ K), LuAu ($T_c = 0.37$ K), AuZn ($T_c = 0.4$ K), $ScAlAu_2$ ($T_c = 4.4$ K), $ScInAu_2$ ($T_c = 3.02$ K) and $YAlAu_2$ ($T_c = 1.9$ K). VRu is the most interesting of them. According to [7.59,60], in a comparatively low concentration interval (in the vicinity of the equiatomic composition) V and Ru form alloys possessing bcc, CsCl (cubic), CsCl (tetragonal) and hcp structures. MAREZIO et al. [7.61] have revealed a structural transformation from CsCl (cubic) structure to CsCl (tetragonal) structure for $V_{0.51}Ru_{0.49}$ and $V_{0.50}Ru_{0.50}$ specimens in the temperature range 360 - 110 K. In $V_{0.50}Ru_{0.50}$ specimens, approximately one third of the material had a tetragonal structure at room temperature. With decreasing temperature the tetragonal Bragg reflections increased in intensity, whereas the cubic reflections weakened. Measurements have shown that the transition from the cubic phase to the tetragonal one in ordered V - Ru alloys is accompanied by an increase in lattice volume. For $V_{0.50}Ru_{0.50}$ specimens, this increase grew from 0% to 1.4% at transition from 360 K to 3000 K and did not vary further at temperatures down to 110 K. Apparently, the volume variation does not depend on composition since this variation of volume amounted to about 1.5% at 110 K in $V_{0.51}Ru_{0.49}$.

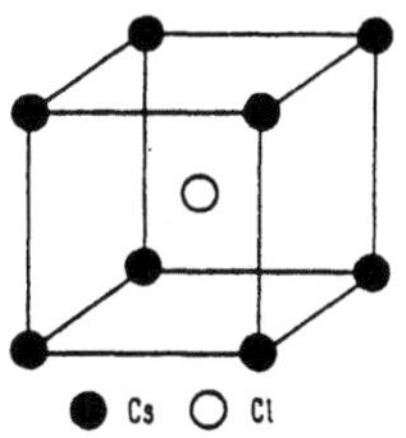

Fig.7.8. The B-2 (CsCl) crystal structure

The instability of the CsCl structure in the equiatomic VRu alloy also manifests itself in the behaviour of different physical properties. When the VRu alloy had been cooled from 400 K, the resistivity rose quickly, the susceptibility dropped sharply and the Knight shift on the vanadium nuclei increased [7.57]. It is significant that the temperature region in which anomalies occur in the physical properties coincides with the region where the structure has been found to display a structural distortion [7.61]. It was shown in [7.57] that the temperature range in which the above anomalies are observed is very sensitive to alloy composition. When the ruthenium concentration of the alloy varies from 51% to 46%, i.e., by 5% only, the temperature of the observed jump in resistivity drops from 300 K to 50 K (Fig.7.9). The changes in the above-mentioned electronic properties were attributed to the variation of the density of states at the Fermi level on the assumption that the states near the Fermi surface have a well-defined d-character and are rather related genetically with vanadium atoms than with ruthenium atoms [7.25,62]. The calculations of the electron structure of the compound VRu have confirmed these qualitative considerations and established that the V3d-states do, indeed, prevail in the energy region near the Fermi energy [7.63]. The results of this non-self-consistent calculation have turned out to be in full agreement with spectral measurements data [7.64]. Investigations of the X-ray photoelectron valence spectrum have shown that the basic structure of the theoretically calculated energy bands of the compound VRu is reproduced over the entire interval of occupied states (Fig.7.10), and the position of the X-ray VLα emission band (2p - 3d4s transition) has confirmed that the V3d-states are concentrated near the Fermi level.

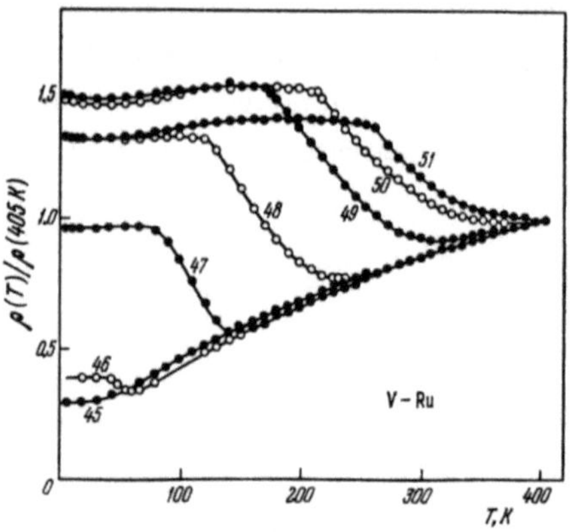

Fig.7.9. Temperature dependence of the resistivity in V-Ru alloys. The numbers at each curve indicate the Ru concentration of the alloy

Of interest are the results of the velocity of sound measurements for $V_{0.60}Ru_{0.40}$ ($T_c < 1$ K), $V_{0.548}Ru_{0.452}$ ($T_c \approx 4$ K) and $V_{0.51}Ru_{0.49}$ ($T_c < 0.4$ K) in a temperature range from 1 to 120 K [7.65]. For $V_{0.60}Ru_{0.40}$ and $V_{0.51}Ru_{0.49}$ (which have been found to exhibit no superconductivity), the variation of the velocity of longitudinal waves with temperature is usual in character, displaying a monotonic increase by 0.5% and 3%, respectively, for both specimens when cooled. The results of a measurement of the velocity of longitudinal sound waves for $V_{0.548}Ru_{0.452}$ ($T_c = 4$ K) are shown in Fig.7.11.

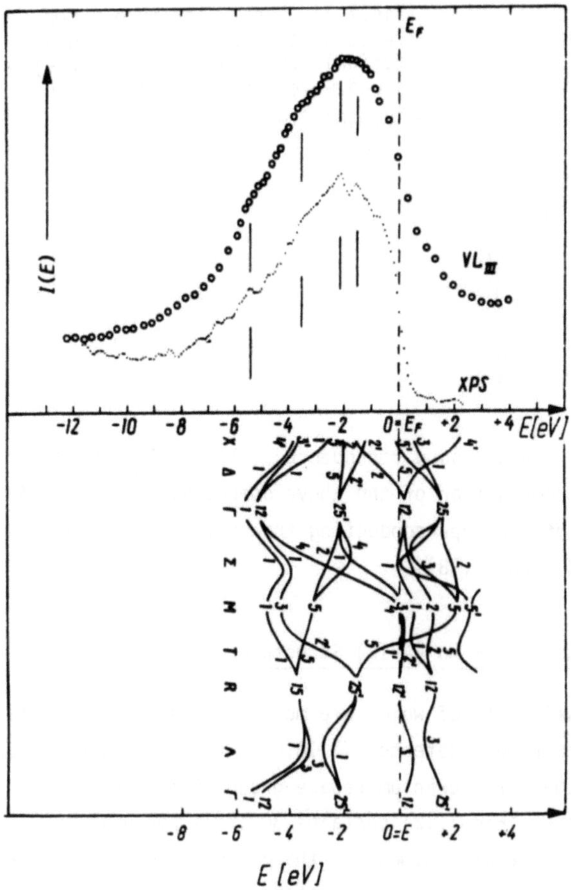

Fig.7.10. The X-ray photoelectron and emission spectra of the alloy $V_{0.55}Ru_{0.45}$ (CsCl cub.) and their comparison with the band structure calculation [7.64]

The data presented here demonstrate that at temperatures close to 55 K, an anomaly in the propagation both of longitudinal and of transverse waves occurs, which indicates that the lattice is unstable.

Interesting data have recently been obtained concerning the behaviour of T_c and T_m in V - Ru compounds near the equiatomic composition with exposure to hydrostatic pressure up to 21 kbar [7.66,67]. For the compound $V_{0.54}Ru_{0.46}$, it has been found that the structural transformation is suppressed and the superconducting transition temperature rises under pressure. Let us recall that a similar situation took place for the compound V_3Si (Fig.6.14). The complete suppression of the structural transformation occurred at some critical pressure P = 14 kbar. It is worth noting that the decrease of $\partial T_c/\partial P$

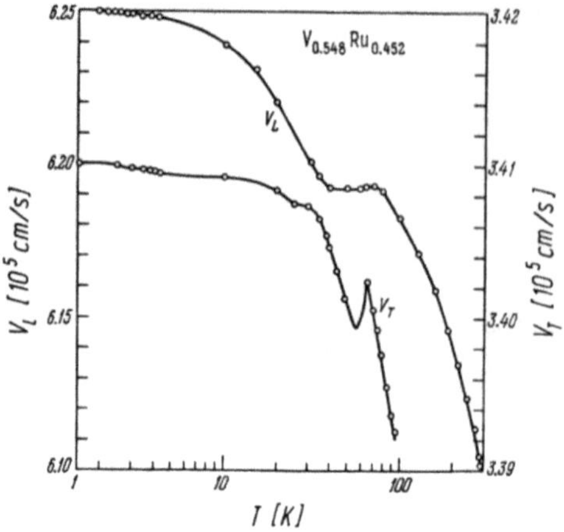

Fig.7.11. The temperature dependence of the velocity of longitudinal (V_L) and transverse (V_T) sound waves in a polycrystalline $V_{0.548}Ru_{0.452}$ specimen

occurs at the same pressure. The pressure dependence of T_c and T_m turned out to be very sensitive to the composition of the above compounds [7.66-68]. A similar pressure dependence of the superconducting transition temperatures was revealed in the compound AuZn [7.58].

7.1.3 σ Phases and Compounds with α-Mn (A-12) Structure

In fifty binary systems constituents of which are mostly transition elements, compounds such as σ phases have been detected [7.10,69]. The structure of the σ phases is isomorphous to that of β uranium (space group P4nm). The tetragonal lattice of this structure (Fig.7.12) contains 30 atoms per unit cell [7.70]. A feature peculiar to the arrangement of the atoms in this structure is the presence of quasi-hexagonal grids (illustrated in Fig.7.12) positioned over each other and turned by 90° relative to each other. These grids are centred by atomic chains parallelly to the tetragonal axis [001]. The distance between the atoms in a chain is equal to half the lattice parameter (a/2).

As a rule, the A component of an AB compound (such as σ phases) can only be represented by elements of the groups situated left of the manganese

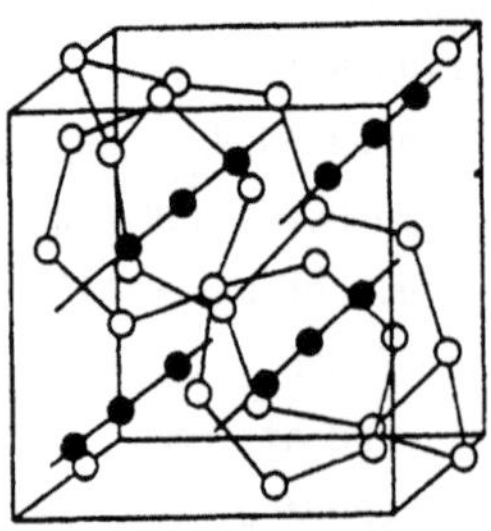

Fig.7.12. The β-U structure (σ phase)

groups in the periodic system. Most of the σ phases have been revealed in alloys whose A component is represented by Nb or where the B component is rhenium. The ranges of compositions at which the σ phases have been revealed may vary over wide limits from A_4B to AB_4 [7.69]. For a number of compounds of the type under consideration it has been reported that the atoms of the components have an ordered arrangement in alloys such as (Cr - Fe) [7.71], (Cr - Ru), (Cr - Os) [7.72], (Nb - Al) [7.73], (Ta - Al) [7.74] and others.

According to the data given in [7.2,10,75,76], 33 compounds of the σ phase type were investigated for superconductivity and 25 of them turned out to be superconducting ones. What is worth noting is that in two of them, $Mo_{0.3}Tc_{0.7}$ and $Mo_{0.38}Re_{0.62}$, T_c does not exceed 10 K (Table 7.3).

The highest value of $T_c = 14.6$ K was obtained for the σ phase in the Mo - Re system. It must be noted that the compound with maximum T_c, $Mo_{0.38}Re_{0.62}$, was synthesized by sputtering on a sapphire substrate heated to a temperature of 1250°C [7.28]. In a bulk material of the same composition prepared by conventional arc melting, T_c did not exceed 6-7.5 K. The superconducting properties of a $Mo_{0.38}Re_{0.62}$ specimen obtained by sputtering turned out to be very sensitive to the precipitation temperature which was regulated by heating the substrate. Figure 7.13 shows that T_c vs. the precipitation temperature exhibits a sharp maximum at 1250°C for this material.

The increase of T_c in film specimens of the compound $Mo_{0.38}Re_{0.62}$ is attributed to the structural instability of the σ phase in the Mo - Re system [7.77]. The reason for the instability is that at the precipitation temperature of 1250°C, the σ phase of the above composition is near the boundary with a phase having a different crystalline structure (such as α - Mn). It is characteristic that the σ phase thus obtained has a unit cell which differs in size from the cell in a bulk material [7.77].

About 20 transition-metal compounds crystallize in an α-Mn type structure (space group T_3^d - $I\bar{4}3m$). The A component of such compounds usually consists of transition metals standing on the left-hand side of the period (groups III - V), whereas the B component is formed by transition metals standing at the end of the period, which is mostly the 3d-period. The composition intervals in which the compounds with α-Mn (A-12) structure occur are rather large: from 50 to 90 at.% of component B.

The α-Mn type structure derives from the β-U structure (Fig.7.12). The cubic body-centred cell of α-Mn is obtained if two cells of β-U are sandwiched so that the point (000) of the second cell and the point (1/2,1/2,1) of the first cell coincide; two atoms coincide in this case so that α-Mn contains 58 atoms in a cell [7.70].

Table 7.3. Superconducting transition temperature in σ phases[1]

A \ B	Al	Mn	Tc	Re	Fe	Ru	Os
V		x		6.26	x		
Nb	<1.3	x		2.5	x		1.85
Ta	<1.02			1.4			<1.2
Cr		x	x	2.50	x	2.10	x
Mo			12	14.6	x	7.0	5.65
W			7.9	4.9		5.12	4.4
Mn				<4.2			
Tc		x			x		
Re		<1.0			6.55		
Zr				<4.2			
Hf				<4.2			

A \ B	Co	Rh	Ir	Ni	Pd	Pt	Au
V	x			x			
Nb		4.04	9.8		2.0	4.01	
Ta		2.35	1.2		x	1.0	<1.2
Cr	x						
Mo	x		6.8				
W			4.46				
Mn							
Tc							
Re							
Zr							<1.02
Hf							

[1] σ phases exist in a wide range of compositions in each binary system and T_c differs for different compositions. The table gives the highest T_c obtained in the σ phase in each binary system. x stands for σ phases in which the superconductivity was not investigated.

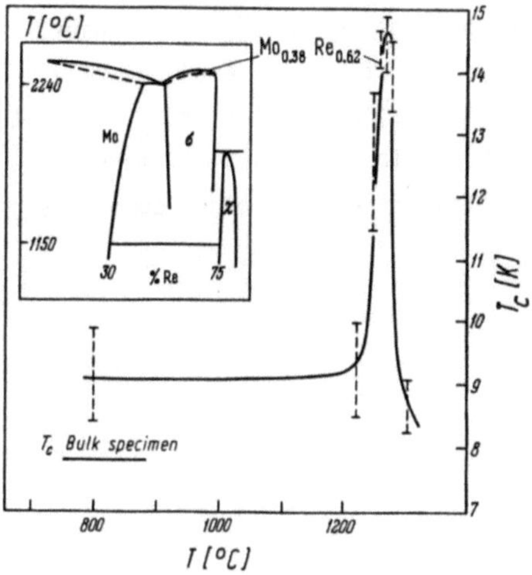

Fig.7.13. The dependence of T_c on the precipitation temperature for the alloy $Mo_{0.38}Re_{0.62}$. At the top part of the Mo-Re phase diagram is shown

The superconducting properties were investigated in 14 binary transition-metal compounds with α-Mn structure (A-12). All of these compounds turned out to be superconducting (Table 7.4 [7.3,78]). The highest T_c ($T_c = 9 - 10.5$ K) found in compounds whose A component is Nb, Mo and W, and in which the B component is Tc and Re. There a practically no papers dealing with the physical properties of the above compounds.

Table 7.4. T_c in compounds with α-Mn structure

A \ B	Tc	Re	Os	Pd	A \ B	Tc	Re	Os	Pd
Sc	x	2.2			Nb	10.5	9.7	2.86	2.47
Ti	x	6.6			Ta	x	6.78	1.95	
Zr	9.7	7.40			Mo		9.89		
Hf	x	5.86			W		9.0		
					Al		3.35		

7.1.4 C-40 Structure Compounds

Interest in transition-metal compounds with C-40 structure showed up quite recently following the detection of $T_c \sim 16$ K in film specimens of the compound $NbGe_2$ [7.79]. For nearly 25 years before that time it had been assumed that T_c would not exceed 1.2 K in this class of compounds [7.80]. The C-40 structure of compounds with the chemical formula AB_2, where A stands for a transition metal of the fourth, fifth or sixth group and B is represented by

Si, Ge and Sn, has a hexagonal symmetry and belongs to the $D^4_{6h} - P6^{22}_2$ space group [7.81] (Fig.7.14). By now seven binary C-40 structure compounds have been synthesized (Table 7.5). Their bulk specimens, except for $CrSi_2$, exhibit metallic conductivity at room temperature. Magnetic susceptibility measurements show that VSi_2 and $CrSi_2$ are paramagnets, whereas the other compounds are diamagnetic [7.82].

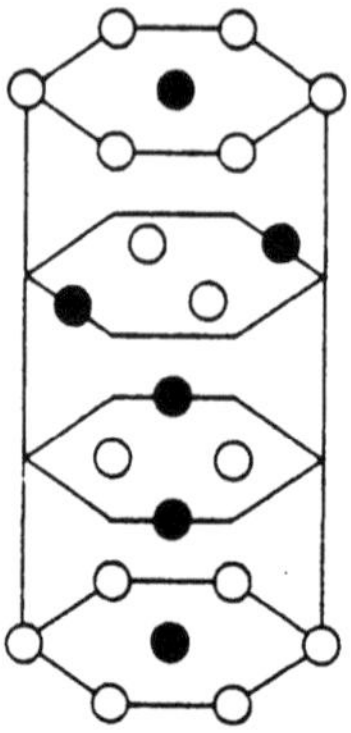

Fig.7.14. The crystal structure of AB_2 compounds such as C-40

The superconducting properties of the compounds with C-40 structure turned out to be very sensitive to heat treatment and to conditions of synthesis. Variation of annealing conditions for bulk specimens of the compound $NbGe_2$ led to variations of T_c from 2.3 to 7.09 K [7.79]. A still larger spread in the values of T_c was registered in film specimens of the compound $NbGe_2$: from 7.8 to 16 K when the temperature and the substrate material were varied. In bulk specimens of $NbSi_2$ and $TaSi_2$, no superconductivity was revealed at temperatures down to 1.2 K. But the films of the same compounds turned out to be superconducting at $T_c \sim 3$ to 4 K. Higher values of T_c (compared to the bulk material) were also observed in film specimens of the compound $TaGe_2$ (Table 7.5). The above experimental findings show that, just as in the case of A-15 compounds (Nb_3Ge), in film specimens of C-40 structure it becomes possible in a number of cases to reveal for the first time, or to substantially increase, the values of T_c in comparison with those in bulk specimens. This enhancement of superconductivity in such materials remains to be explained. To this end, it is necessary to perform detailed investigations into the physical properties and the structural state of the film specimens of the above-mentioned compounds.

Table 7.5. Physical properties and superconductivity of transition-metal compounds with C-40 structure

Compound	Resistivity[a)] μΩ cm	Magnetic susceptibility[a)] $10^{-6} cm^3/mole$	T_c [K][a)]	T_c [K][b)]
$CrSi_2$	1420	+41	<1.2	<1.2
VSi_2	9.5	+161	<1.2	<1.2
$NbGe_2$	-	-	2.34 -7.09	7.8 -16.0
$NbSi_2$	6.3	-37	<1.2	2.2 -3.75
$TaGe_2$	-	-	1.93	2.0 -2.7
$TaSi_2$	8.5	-40	<1.2	3.2 -4.4
$HfSn_2$	-	-	<1.2	<1.2

a) bulk specimens b) film specimens

7.2 Compounds with NaCl Structure

There is a large class of compounds comprising borides, carbides, nitrides, and transition-metal oxides which crystallize in the NaCl structure. Among them there are quite a number of compounds exhibiting high superconducting transition temperatures which are becoming useful for various technical devices [7.83].

7.2.1 Crystal Chemistry

The crystal structure of this class of compounds with the chemical formula AB exhibits the following features. Metallic A atoms form a cubic face-centred lattice, and nonmetallic B atoms occupy all the octahedral interstices. This results in a structure identical to that of NaCl (B-1), i.e., two cubic face-centred lattices are inserted into each other (Fig.7.15). Each atom of the metal is surrounded by an octahedron composed of atoms of a different sort, and vice versa, i.e., the coordination of atoms of different sorts is alike.

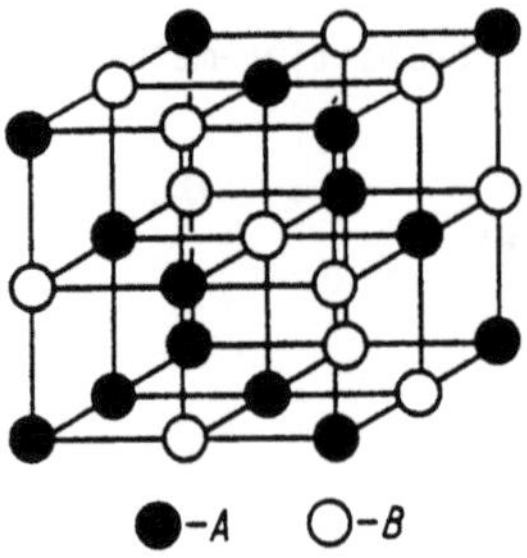

Fig.7.15. The B-1 (NaCl) structure

Usually the transition metals of groups III, IV, V and VI of the periodic system of elements are the A atoms, and the nontransition elements of the same groups the B atoms. Most of the compounds being considered are compounds of variable composition, i.e., deviations from stoichiometric composition in either direction may be observed within one and the same crystal structure (of the NaCl type in our case). For example, a compound such as TiO_x has the region of homogeneity $x = 0.7$ to 1.25. Hereafter, for specifying a compound of variable composition we shall use a notation which more accurately reflects the degree of completeness of the metallic or metalloid sublattice. For example, we write $Ti_{0.8}O_{1.0}$ when the sublattice of a metalloid is completely occupied by oxygen and titanium atoms in the metallic sublattice occupy about 80% sites.

The physical properties of NaCl structure compounds cannot be understood if one disregards the fact that they always have a certain amount of vacancies. Indeed, comparison of the pycnometric and the estimated (with due regard for the lattice parameter) densities has shown that both metallic and nonmetallic atoms may be simultaneously absent over a broad range of compositions even in a number of stoichiometric compounds of the above-mentioned class (see Table 7.6 drawn up according to the data obtained in [7.84,85]). It turned out that what corresponds to equiatomic composition is not the formula $Ti_{1.0}O_{1.0}$ but $Ti_{0.85}O_{0.85}$; or that the formula $Nb_{0.75}O_{0.75}$ corresponds to NbO. This actually means that the vacancies present in both sublattices amount to 15% and 25%, respectively.

As a rule, the vacancies are distributed in a random way throughout the lattice. However, structural investigations show that at certain compositions of these compounds the vacancies tend to become ordered and to produce a superstructure both in the metalloid ($Ti_{1.0}O_{0.7}$ [7.86], $V_{1.0}C_{0.84}$ [7.87-91], $V_{1.0}N_{0.75}$ [7.92], $Nb_{1.0}C_{0.75}$, $Ta_{1.0}C_{0.76}$ [7.93]) and in the metallic ($V_{0.763}O_{1.0}$ [7.94,95]) sublattices. The ordering of vacancies in both sublattices has been detected in $Ti_{0.85}O_{0.85}$ and $Nb_{0.75}O_{0.75}$ [7.96].

Table 7.6. Vacancy concentration of compounds of stoichiometric composition with NaCl structure (concentration in each of the sublattices)

Compound	TiC	ZrC	HfC	VC	NbC	TaC	TiN	ZrN	NV
Vacancy concentration, %	2.0	3.5	4.0	8.5	0.5 -3.0	0.5	4.0	3.5	1.0

Compound	NbN	TaN	TiO	VO	NbO		ZrS	ZrSe	YS
Vacancy concentration, %	1.3	2.0	15	11-15	25		20	16	10

Note: The exact compositions of the compounds are not reported in [7.84,85] according to which data this table has been drawn up.

7.2.2 Superconductivity

We now turn our attention to the discussion of the superconducting properties of NaCl structure transition-metal compounds. Among the 40 compounds investigated, only 26 compounds have turned out to be superconducting [7.97-120]. In six of them, T_c exceeds 10 K (Table 7.7). These high-temperature superconductors are formed by transition metals of groups IV, V and VI with the compounds of the 4d and 5d-metals (Zr, Nb, Mo, Ta, W) having higher T_c than the compounds of the 3d-metals (Ti, V, Cr). On the other hand, it must be noted that the partners of the metals concerned which are favourable for high-temperature superconductivity, actually reduce to two elements, viz., carbon

Table 7.7. T_c in compounds with NaCl structure

B \ A	Sc	Y	La	Ti	Zr	Hf	V	Nb	Ta	Cr	Mo	W	Re
B					3.4	3.1							
C	<1.38	<1.38		3.42	<0.3	<1.20	0.03 3.2*	12	10.35		14.3	10.0	3.4
N	<1.38	<1.4	1.35	5.49	10.7	8.83	8.5	17.3	6.5	<1.28	5.0	<1.38	
P			<1.68										
Sb		<1.02	<1.02										
O				2.0			<0.3	1.39					
S	<0.33	1.9	0.87		3.3								
Se	<0.33	2.5	1.02										
Te		2.05	1.48										

* $T_c = 3.2$ K was registered in vanadium carbide after implantation of C^+ ions [7.122]

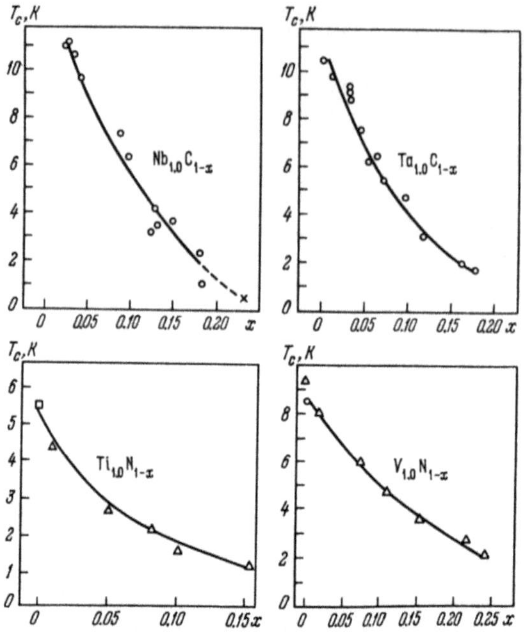

Fig.7.16. The influence of deviations from stoichiometry on the superconducting properties of transition-metal compounds with NaCl structure

and nitrogen. In the search for new high-temperature superconductors this circumstance drastically restricts the number of possible combinations.

Just as in A-15 structures, the superconducting properties of NaCl structure compounds have turned out to be very sensitive to deviations from stoichiometric composition. Figure 7.16 presents experimental data on T_c in Nb and Ta carbides and in Ti and V nitrides [7.116-120] which convincingly demonstrate the fact that the temperature of transition to the superconducting state drops with decreasing carbon/metal or nitrogen/metal ratios. This decrease is accompanied by an increase in the number of vacancies in the nonmetallic sublattice. On the other hand, it has been found that the formation of vacancies in the metallic sublattice also decreases the T_c in compounds of this type [7.121].

It is shown in [7.122,123] that implantation of C^+ ions into nonstoichiometric transition-metal carbides allows to obtain a composition closer to the stoichiometric one. Using the above technique it has been possible to raise T_c from 3.7 to 7.0 K in nonstoichiometric NbC_x [7.123] and to bring T_c in vanadium carbide to 3.2 K [7.122].

First attempts to alloy transition-metal compounds with A-15 structure yielded encouraging results. In 1953, MATTHIAS obtained a superconducting transition temperature of 17.8 K in the NbC - NbN system [7.118]. During the subsequent 14 years this was one of the highest T_c among all the superconductors investigated.

In the alloying of an AB compound by an A'B' compound, three cases are possible:

i) substitution of A' atoms for A atoms, with the B atoms remaining invariable, B(A - A') (Fig.7.17);

ii) substitution of B' atoms for B atoms, with the A atoms remaining invariable, A(B - B') (Fig.7.18);

iii) substitution of atoms of both sorts, (A - A')(B - B') (Fig.7.18).

Consideration of the concentration dependence of the relative transition temperature $T_c/T_{co} = t$ (where T_{co} corresponds to the initial state) in all the three cases shown above [7.98,116,119,124,125] leads to the following conclusions.

Substitution of a 5d-metal (A') for a 4d-metal (A) has little effect on the magnitude of relative superconducting transition temperatures in the

Fig.7.17

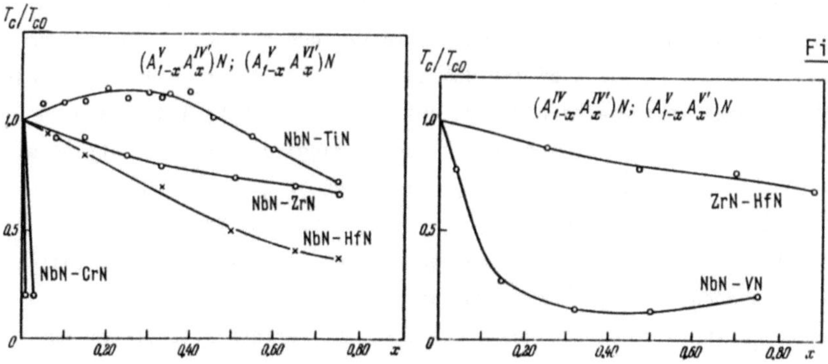

Fig.7.18

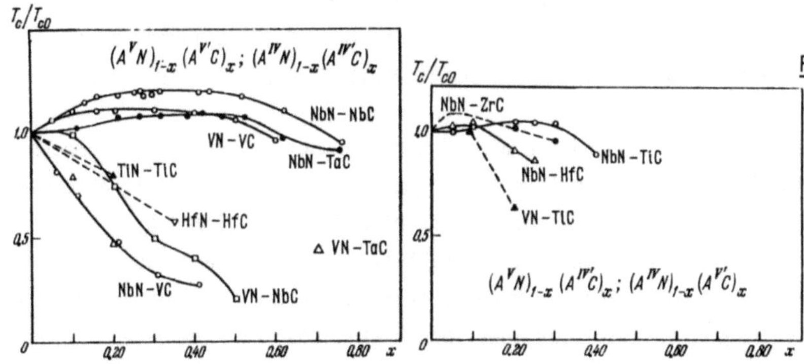

Fig.7.17. The composition dependence of T_c/T_{co} for the alloys $(A_{1-x}A'_x)N$. In the left figure A and A' are transition metals from different groups of the periodic table, in the right figure these are transition metals from one group

Fig.7.18. The composition dependence of T_c/T_{co} for the alloys $(AN)_{1-x}(A'C)_x$. In the left figure A and A' are transition metals from one group, in the right figure these are transition metals from different groups

(A - A')B system. Vice versa, when a 4d-metal is replaced by a 3d-metal, the transition temperature drops drastically, causing an abrupt decrease in the values of T_c/T_{co}. When B' atoms are substituted for B atoms in the A(B - B') system the situation is favourable for compounds of transition metals of the fifth group: V(C - N), Nb(C - N) where $t > 1$ for a large region of concentrations. For compounds of transition metals of group IV: Ti(C - N) and Hf(C - N), conversely, replacement of nonmetallic atoms brings about a substantial decrease of T_c, i.e., t drops sharply.

For (A - A')(B - B') alloys, substitution of transition-metal atoms has a strong impact. Just as in the case of (A - A')B alloys, substitution of 3d-metals for 4d-metals drastically deteriorates the superconducting properties of the material.

Concluding our consideration of the superconducting properties of NaCl structure compounds of transition metals and their alloys, it may be noted that the compounds in which high T_c were attained due to their alloying were obtained chiefly on the basis of the NbN compound. The superconducting transition temperatures in these ternary systems (17.8 to 18.0 K) and in a pure NbN compound (17.0 to 17.3 K) are very close in magnitude. Further attempts to alloy the "luckiest" ternary alloy $(NbN)_{0.75}(NbC)_{0.25}$ with $T_c = 17.8$ K by introducing different additives (TiN, TiC, HfN, HfC, ZrN, VN, VC) [7.119] did not yield positive results. On the other hand, it is worth noting that $T_c = 17.3$ K has been obtained for a composition of NbN which is not strictly stoichiometric. In this connection we believe that further attempts at improving the superconducting properties of NaCl structure compounds should not be confined to alloying alone. Until compounds of stoichiometric composition have been obtained, the potential of pure compounds may not be regarded as exhausted.

If we construct the dependence of the superconducting transition temperature on the electron concentration of a compound it will be readily observed that there are two values of electron concentration $N_e = 4.5$ and $N_e = 5.0$ e/a which are the most favourable ones for high-temperature superconductivity. The first value is satisfied by group IV transition-metal nitrides (ZrN, HfN) and group V transition-metal carbides (NbC, TaC). Most of them have a superconducting transition temperature of 8 to 12 K. To the second value correspond nitrides of transition metals of the fifth group (VN, NbN) and carbides of transition metals of the fourth group (MoC, WC) with the values $T_c = 8.5 \div 17.3$ K.

Consideration of other physical properties (electron heat capacity, paramagnetic susceptibility [7.106,110,126]) as a function of electron concentration, shows that high values of the superconducting transition temperature are not necessarily accompanied by higher values of γ and χ. For example, the value of γ in the compound TiO exceeds that in TiN, whereas the superconducting transition temperature in TiN, on the contrary, is 2.5 times larger than that in TiO. Again, just like in A-15 structure compounds, a substantial difference in the behaviour of γ and χ may be noted for vanadium and niobium compounds. For example, the NbN compound has comparatively low values of electron heat capacity and magnetic susceptibility and high T_c ($\sim$17 K), whereas the reverse is true of VN ($T_c = 8.5$ K).

7.2.3 Band Structure

A number of band structure calculations using the APW method have been carried out for compounds of transition metals with NaCl structure. The APW method requires knowledge of the one-electron potential which is spherically symmetric within the spheres centred on atomic sites and which is constant between them. This potential is determined by the given charge distribution in the crystal. Therefore the choice of the starting potential in calculations of chemical compounds requires the assumption of a certain ionicity which causes charge transfer between spheres of different types.

In accordance with the various initial assumptions concerning the configuration of electrons in the atomic components of the same compounds [7.127-131], electron energy spectra were obtained in which the sequence of bands and the energy distances between the bands differed appreciably. Which of the calculations performed should be preferred? A most reliable criterion for the correctness of a calculation may be the degree of agreement with the data obtained from X-ray and X-ray photoelectron spectroscopy methods.

As an example, we shall quote the results obtained for the compound TiC. Figure 7.19 presents the density of electron states curves obtained in [7.132, 133]. Curve 1 [7.132] corresponds to the configurations of the atomic components: C - $(2s)^2(2p)^{3/4}$, Ti - $(3d)^4(4s)^{3/4}(4p)^{1/2}$, i.e., it is assumed that 1.25 electrons will make a transition from carbon atoms to titanium atoms. Accordingly, the band constituted by 2p-states of carbon lies higher than the 3d-band and is in the region of the free states of the energy spectrum.

Curve 2 has resulted from a self-consistent calculation and corresponds to an electron configuration of the atomic components close to C - $(2s)^2(2p)^4$, Ti - $(3d)^2(4s)^0$, i.e., it is assumed that two electrons travel in the reverse sense - from titanium atoms to carbon atoms. The band formed by 2p-states of carbon is located in the region of occupied states and strongly overlaps with the 3d-band.

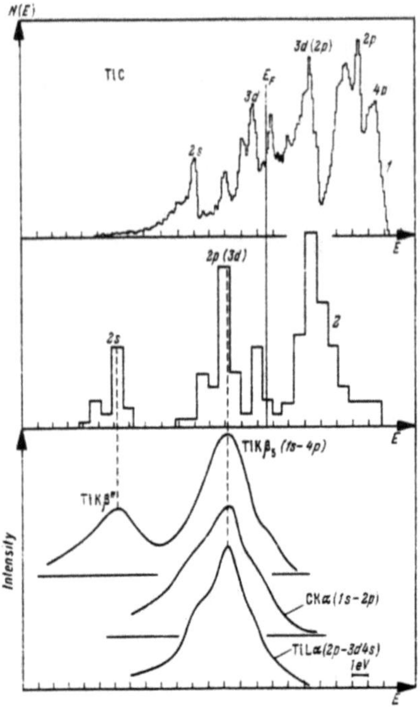

Fig.7.19. Comparison of calculations of the electron energy spectrum for the compound TiC with X-ray emission spectra

So, what has the experiment shown? In the lower part of the figure there are given the X-ray emission $TiK\beta_5$ (1s - 4p transition), TiLα (2p - 3d4s transition) and CKα (1s - 2p transition) spectra are given from [7.134,135]. These spectra give an insight into the distribution of Ti4p, Ti3d4s and C2p states, respectively, in the occupied portion of the conduction band. Thus, the relative position of these spectra in the energy scale should determine the sequence of bands in the TiC compound exactly. A rigorous comparison of these spectra accounting for the difference in the relative energies of the inner levels of titanium and carbon atoms in TiC [7.136] has shown that the intensity maxima of the CKα and TiLα-bands are close in energy. This suggests that the 2p-states of carbon and the 3d-states of titanium lie in the same spectrum region. Thus, the results of the experiment contradict the energy band structure shown in Curve 1 (Fig.7.19) and agree fully with the data given in Curve 2. This gives reason to believe that in TiC there is no transition of electrons from carbon atoms to titanium atoms.

Comparison of different electron spectrum calculations for other compounds has shown that self-consistent APW calculations give a closer fit to experimental facts in all the cases (see, for example, [7.137]). Self-consistent calculations have been carried out for the following NaCl structure compounds of transition metals: TiC [7.133], VC [7.138], NbC [7.139], NbN [7.140], TiN, TiO, VN, VO [7.141], HfC, TaC [7.142]. Considering the results of these investigations, the following conclusions may be drawn concerning the electron structure of these compounds. Regardless of the type of atoms of the metal (Ti, V, Nb) and the metalloid (C, N, O), three major bands may be separated in the occupied portion of the energy spectrum. In the order of increasing energy, these bands are populated by 2s, 2p and nd-electrons. The energy difference of the 2s and 2p-bands coincides with that of the 2s and 2p-levels of the free atoms of the nontransition element [7.143,144]. The highest-energy band is the nd-band whose electrons are apparently responsible for the superconducting properties of the compounds concerned. In all the cases enumerated above self-consistent calculations have shown that a certain transfer of electrons from metallic atoms to nonmetallic ones occurs.

All of the above electron energy spectrum calculations apply to compounds of stoichiometric composition with a complete filling of atoms in both sublattices. Meanwhile, the results given in Table 7.6 show that even the compounds with stoichiometric composition always contain a certain amount of vacancies.

In the case of deviation from stoichiometry the number of vacancies increases substantially. The question arises whether the conclusions about the electron structure of complete compounds of stoichiometric composition will be applicable to actual existing stoichiometric and nonstoichiometric compounds with vacancies.

GELD et al. [7.145,146] proposed that in nonstoichiometric carbides of transition metals the vacancies in the carbon sublattice may be regarded as a perturbation of the electronic state of the crystal which leads to the formation of local energy levels. At $\sim$25% vacancy concentrations, vacancies appear spaced at a distance of the order of the lattice parameter and in this case, according to GELD et al., it may be expected that they will become ordered and the local levels will transform into an energy band. However, no spectrum calculations have actually been made in this model.

7.2.4 Phonon Spectra

Let us now proceed to the discussion of the lattice properties of transition-metal compounds with NaCl structure. First of all, consider the results of neutron diffraction investigations of the phonon spectra of ZrC, $Nb_{1.0}C_{0.97}$, $Nb_{1.0}C_{0.76}$, HfC and TaC compounds which have been obtained by SMITH [7.147, 148]. Figure 7.20 presents dispersion curves of the phonon spectrum $\omega(\underline{q})$ for

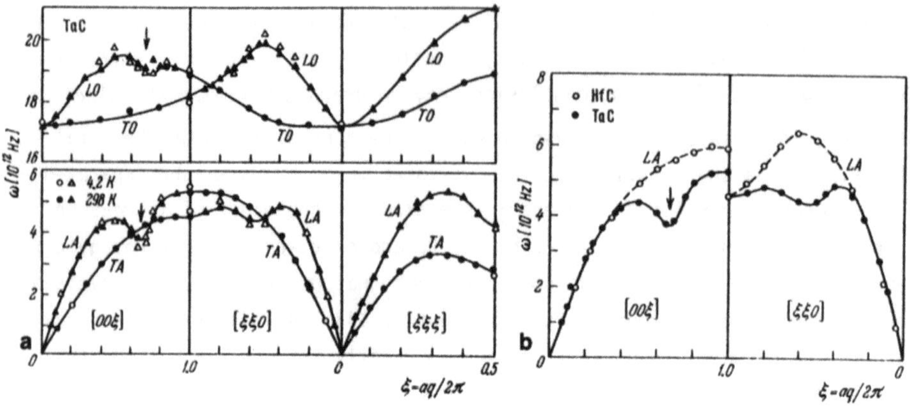

Fig.7.20. The dispersion curves of the phonon spectrum for the compounds TaC and HfC. The designation of the branches is the same as that used in Fig.4.22

stoichiometric compounds TaC and HfC which differ drastically in their superconducting properties (TaC: $T_c \approx 10$ K; HfC: $T_c < 1$ K). What is most interesting is that the high-temperature superconductor TaC exhibits anomalies in longitudinal-acoustic and optical modes near the boundary of the band (marked by an arrow). At the same time, the above anomalies are absent in the non-superconducting HfC. Anomalies in the phonon spectra were also detected in TiN ($T_c \sim 5.5$ K) [7.149], ZrN ($T_c \sim 10.7$ K [7.150] and VN ($T_c \sim 8.5$ K) [7.151].

Several attempts have been undertaken to account for the above anomalies in a number of carbides on the basis of the shell model [7.152] and pseudopotential calculations [7.153]. Possessing high ionicity and covalency, NaCl structure compounds may be described by the model of shells in which the dip on the dispersion curve in TaC is attributed to the resonance excitation of the oscillation of the mobile electron shell. On the other hand, since carbides are high-conductivity phases, an attempt has been made to describe their phonon spectra on the basis of pseudopotential calculations [7.153]. These, however, give a poorer fit to experimental data. A good agreement for two compounds (TaC and HfC) has also been obtained with the help of calculations according to a modified shell model [7.154] which takes into account the long-range three-particle forces. This means that for carbides, the approach based on the Heitler-London approximation is just as good as the allowance made for the screening in the free-electron approximation used in [7.152].

Another approach to accounting for the anomalies in the phonon spectrum in the Nb carbide has been developed by SINHA and HARMON [7.155]. In contrast to WEBER's [7.152] approach assuming the existence of an interaction between the shells such as attraction, SINHA and HARMON [7.155] proceeded from considering the dielectric permeability of a system of highly localized d-electrons. In the case of a high density of states on the Fermi surface, an instability with respect to the formation of a charge density wave arises. However, due to the electron-phonon coupling the instability does not occur, but it leads to anomalies in the phonon spectrum which take place in a number of carbides.

GUPTA and FREEMAN [7.156-158] have developed the concept that the phonon anomalies in transition-metal carbides arise from the anomalous increase of the conduction-electron response function and that this increase results in a strong screening of the relevant phonon modes. Numerical band-structure calculations were performed for the compounds NbC and TaC (having $T_c > 10$ K and high phonon anomalies), and also for ZrC and HfC (free from phonon anomalies and with $T_c < 1$ K). In addition, the form of the Fermi surface was obtained for several intersecting conduction bands. The data thus obtained permitted calculation of the generalized electron-electron susceptibility as a function of the wave vector $\chi(\underline{q})$. It has turned out that in TaC and NbC, $\chi(\underline{q})$ has maxima in those points of the Brillouin zone where one experimentally observes dips of the phonon spectrum. It is worth noting that the size and depth of the dips correlate with the width and height of the maxima of $\chi(\underline{q})$. For the compounds HfC and ZrC, $\chi(\underline{q})$ falls off monotonically outside the centre of the Brillouin zone.

The spectra shown in Fig.7.20 have been obtained at room temperature. In order to correlate the detected anomalies in the phonon spectra with the superconducting properties it was of interest to perform low-temperature measurements. For the compound TaC, such measurements were made at a temperature of 4.2 K. In Fig.7.20, the data obtained from low-temperature measurements are given along with the spectra obtained at room temperature. It is seen that the transition to low temperatures is not accompanied by any appreciable changes in the phonon spectra.

The influence of deviations from stoichiometry on the structure of phonon spectra was considered by SMITH for the compound NbC [7.148]. Like TaC, the niobium carbide $Nb_{1.0}C_{0.97}$ (whose composition is almost stoichiometric) and vanadium nitride VN [7.151] showed a softening of longitudinal acoustic and optical modes. At a considerable deviation from stoichiometry ($Nb_{1.0}C_{0.76}$), these anomalies vanished and the spectrum became similar to that of the compounds HfC and ZrC [7.148]. It is remarkable that $Nb_{1.0}C_{0.97}$ as well as TaC is a high-temperature superconductor with $T_c \sim 11$ K, and $Nb_{1.0}C_{0.76}$, just as HfC and ZrC, exhibits no superconductivity at temperatures down to 1 K. Thus, the relationship between high-temperature superconductivity and the softening of phonon modes and, consequently, the lattice instability has received further confirmation.

Interesting information on the distribution of the phonon density of states in NbC and TaC has been obtained from tunneling experiments [7.159] which have revealed a strong nonmonotonic dependence of the second derivative of the tunneling characteristic d^2I/dV^2 in the region of acoustic phonons. At the same time, this dependence is slight in the region of optical phonons. The conclusion drawn from this fact is that the main contribution to the electron-phonon interaction parameter is made by acoustic phonons [7.159].

It has been shown experimentally [7.161-164] that the vacancies cause first-order Raman scattering in non-stoichiometric transition-metal carbides and nitrides. The Raman spectrum is approximately proportional to the phonon density of states, and it has been established that for TiC [7.164], TiN and NbC [7.161,162] the peaks of the Raman spectrum correspond to the maxima of the phonon density of states determined from inelastic neutron scattering.

7.2.5 Electron-Phonon Interactions

An attempt to account for the enormous difference of T_c in NbC and TaC on the one hand, and in ZrC and HfC on the other, as well as to ascertain the role of the lattice anomalies described above was undertaken by WEBER [7.152] on the basis of McMillan's formula (2.3.28). The calculations performed by WEBER according to McMillan's formulae [(2.3.28 and (2.4.30)] for several carbides are given in Table 7.8. The quantity λ was determined according to (2.3.27) from the data on T_c and the Debye temperature; N(0) was determined from electron heat capacity measurements (with corrections for the electron-phonon interaction), and $<\omega^2>$ was calculated from the phonon spectrum dispersion curves. Then, using (2.4.30), the values of $<J^2>$ can be determined. As is seen from the table, the quantity $N(0)<J^2>$ is not constant for carbides, contrary to the McMillan-Hopfield rule. Its values for carbides with 8 and 9 valence electrons per cell differ by a factor of two. One of the reasons for this, as WEBER noted, may be the overlap of the d-functions of neighbouring atoms which is completely neglected in Hopfield's formula (2.4.39).

Table 7.8. Values of $<J^2>$ for carbides calculated by WEBER [7.152] from McMillan's formulae

Compound	T_c [K]	λ	N(0) $[\mathrm{ev\cdot atom}]^{-1}$	$\sqrt{<\omega^2>}$ [K]	(J^2) $[\mathrm{ev/\AA}]^2$	$N(0)<J^2>$, $[\mathrm{eV/(atom\cdot \AA^2)}]$
ZrC	0.25	0.3	0.14	295	30	4.2
HfC	0.25	0.3	0.14	220	33	4.6
NbC	11.1	0.7	0.40	268	21	8.4
TaC	10.3	0.7	0.40	196	22	8.8

An important conclusion following from these calculations is that the big difference in T_c for the two groups of carbides cannot be explained on the basis of phonon-spectrum softening in TaC and NbC in the short-wave region. As is seen from the data presented in Table 7.8, the difference in $\langle\omega^2\rangle$ of the compounds TaC and NbC exhibiting a phonon anomaly, and of HfC and ZrC which are without it, does not exceed 25%.

KLEIN and PAPACONSTANTOPOULOS [7.165] arrived at a similar conclusion. Using the APW method they have calculated, in a straightforward manner, the matrix elements $\langle \underline{J}^2\rangle$ for the atomic components in the carbides NbC, TaC and HfC and then employed PHILLIP's formula for λ in the two-component system

$$\lambda = \lambda_A + \lambda_B = \frac{[N(0)\langle \underline{J}^2\rangle]_A}{M_A\langle\omega^2\rangle_A} + \frac{[N(0)\langle \underline{J}^2\rangle]_B}{M_B\langle\omega^2\rangle_B} \quad ,$$

where A and B stand for a metallic and a nonmetallic atom in a compound having the chemical formula AB. Such separation of λ is valid only when the difference in masses is large, which is the case in carbides ($M_A \gg M_B$). In virtue of this inequality, the heavy metallic atom mainly oscillates in the acoustic frequency range, and the light atom of carbon in the optical range. In the same approximation it turns out that $\langle\omega^2\rangle_A \sim 1/M_A$ and $\langle\omega^2\rangle_B \sim 1/M_B$ so that the denominators in the expression for λ are weakly dependent on mass and are of the same order of magnitude. Numerical calculation according to the APW method has shown that the first term, which is due to the oscillation of the metallic atom, prevails in the expression for λ. This is in agreement with ZELLER's [7.159] measurements which have shown that λ is primarily determined by the interaction of electrons with acoustic phonons. Calculation of λ from first principles has led to values of T_c for NbC and TaC which give an extraordinarily good fit to experiments (within 10%).

These results have enabled HAUFE et al. [7.166] to perform more simplified calculations of λ and T_c in a large number of carbides and nitrides with NaCl structure. These calculations neglect the contribution made by the interaction of electrons with optical phonons to the electron-phonon coupling constant, i.e., it is assumed that

$$\frac{[N(0)\langle \underline{J}^2\rangle]_A}{M_A\langle\omega^2\rangle_A} \gg \frac{[N(0)\langle \underline{J}^2\rangle]_B}{M_B\langle\omega^2\rangle_B} \quad .$$

The agreement obtained between the calculated and experimental values of T_c confirms this assumption (the smallness of the contribution of λ_B to λ).

It thus turns out that high T_c in NbC and TaC are not associated with the mode softening mechanism but are rather phonon anomalies, and high T_c result from the strong electron-lattice interaction.

A calculation of the electron-phonon interaction parameter λ and of T_c in nonstoichiometric carbides NbC_x, TaC_x and HfC_x $(0.7 \le x \le 1.0)$ was carried out in [7.165]. The use of the CPA formalism developed to calculate the spectrum in disordered alloys (Sect.5.5) made it possible to take into account the s and p-orbitals of randomized carbon vacancies. The calculation has shown that, although accompanied by a certain increase in the values of N(0) and $N(0)\langle J^2 \rangle$, the deviation from stoichiometry in NbC and TaC leads, in the final analysis, to a decrease of λ and T_c due to the stiffening of the phonon spectrum.

Another attempt to account for the sharp decrease of T_c in NbC by deviations from stoichiometry was made by SCHWARZ et al. [7.167] by calculation of the electron structure of a cluster incorporating three coordination spheres with respect to the central carbon ($CNb_6C_{12}Nb_8$) atom. To model the nonstoichiometric niobium carbide the same cluster was considered, but the central carbon atom was replaced by a vacancy ($\square Nb_6C_{12}Nb_8$). The calculation has shown that the population of the free-electron p-type states in the vicinity of the Fermi level in a cluster with a vacancy is much less than that in a complete cluster. According to [7.167], a required condition for the presence of anomalies in the phonon spectrum and of high T_c in transition-metal carbides, is a high density of d-states in the t_{2g} band and a sufficient density of unoccupied p-states in the vicinity of the Fermi level. In the case of a cluster with a vacancy the number of free p-states turns out to be very small, which inhibits the d-p scattering and leads to a decrease of T_c in nonstoichiometric carbides.

7.3 Compounds of Palladium with Hydrogen and Deuterium

When palladium is saturated with hydrogen or deuterium, the latter occupies the octahedral interstices and forms an NaCl structure (Fig.7.15). Formally, we should have considered palladium hydrides in the previous section along with the other compounds with NaCl structure. However, on account of their specific features, and because of the mechanisms leading to the occurrence of a superconducting state, we will treat these compounds in a separate section.

7.3.1 The Superconductivity of PdH and PdD and the Reverse Isotopic Effect

Recently, considerable interest has been developed in compounds of palladium with hydrogen and deuterium due to the discovery of superconductivity in PdH by SKOWSKIEWICZ in 1972 [7.168]. SKOWSKIEWICZ found that T_c depends largely on the H/Pd ratio. At H/Pd = 0.73, no superconductivity in palladium hydride occurs at temperatures down to 1.25 K. When H/Pd = 0.81 this compound exhibits features of a superconducting transition at 2.5 K. When H/Pd = 0.87, a clear-cut transition to a superconducting state takes place at $T_c = 4$ K. In the same year, STRITZKER and BUCKEL [7.169] obtained $T_c \sim 9$ K for H/Pd ≈ 1 using the ion-implantation technique which makes it possible to obtain a higher saturation of a metal by hydrogen. They discovered the reverse isotopic effect due to saturation of Pd by deuterium, when superconductivity with $T_c \sim 11$ K was registered in PdD [7.169]. The detection of high-temperature superconductivity in Pd-H and Pd-D systems was a rather unexpected effect since there is no superconductivity in pure Pd and its alloys with noble metals. In view of this, the results obtained in [7.168,169] were put to a thorough experimental test [7.78,170-176,178,179] which showed that palladium hydrides synthesized with the help of different methods display superconductivity with $T_c = 2$ to 9 K as a function of the H/Pd ratio.

7.3.2 The Mechanisms of Superconductivity in Pd Hydrides

First attempts to explain such high T_c in palladium hydrides rested on the assumption that metallic hydrogen, in which T_c should be very high (about 100 K), according to the estimates of [7.180,181] has an independent sublattice in PdH. However, if metallic hydrogen really does make a decisive contribution to the superconductivity of palladium hydride, the value of T_c in compounds with other hydrogen isotopes should obey the relation $T_c \sim M^{-1/2}$ (M being the mass of an ion) and should be less in PdD than it is in PdH. But, as already stated above, experiments show the reverse.

The mechanisms leading to the occurrence of a superconducting state in PdH are connected with the absence of superconductivity in pure Pd. This problem, discussed in detail in Sect.4.3, will be briefly recalled here. Because of the high density of states on the Fermi level in metallic Pd its conduction electrons feel a strong Coulomb interaction and, as a consequence of this, the electronic subsystem is in the vicinity of the critical conditions for the occurrence of magnetic moments, as shown by the high magnetic susceptibility. Under these conditions anomalously large spin fluctuations should arise, as shown by BERK and SCHRIEFFER [7.182], lead to the repulsion of electrons on the Fermi surface and, at some value of this repulsion, to the annihilation of Cooper pairs.

The effect of spin fluctuations in T_c in strong-coupling superconductors can be understood from McMillan's formula written in the form of (3.8.10). The parameters entering the effective electron-electron attraction constant λ_m through spin fluctuations, were estimated by BENNEMANN and GARLAND [7.183]. From magnetic susceptibility and electron heat capacity data they got $\lambda_m(Pd) \gtrsim 0.1$ and $\lambda_m(PdH) \approx 0$. At these values of λ_m, estimates according to (3.8.10) give $T_c = 0$ K in pure Pd and $T_c = 8$ K for PdH, which gives a close fit to experimental values.

Thus, according to these concepts, the role of hydrogen in PdH consists of the suppression of spin fluctuations due to the filling of the narrow d-band of palladium by hydrogen electrons and due to the decrease of the high density of states on the Fermi level. In all other respects, the properties of a pure metal hardly vary and the electrons near the Fermi surface in PdH are in tightly bound d-states, just as in Pd [7.183].

However, experiments on photoemission in PdH and Pd [7.184] and the electron-energy-spectrum calculations carried out by SWITENDICK [7.185,186] have shown that, although the d-band of palladium is, indeed, partly filled with 1s-electrons when hydrogen is added to it, the rigid-band model is not suitable to describe the electron structure of PdH. It has been established that the Fermi level in the compound passes through a wide s-band with a low density of states, which thus disagrees with the assumption of the d-symmetry of electrons on the Fermi surface.

According to MAKSIMOV et al. [7.187], addition of H to Pd exerts a much stronger influence on the properties of this compound than a mere suppression of spin fluctuations, namely, it appreciably increases the electron-phonon coupling constant λ. In the majority of the existing theories [7.188-191], it is assumed that the direct contribution to the constant λ comes from the interaction of electrons with optical high-frequency oscillations of H or D atoms. Because of the difference in masses, the optical modes in the Pd-D system lie lower than those in the Pd-H system and, consequently, should make a greater contribution to the interelectron attraction constant λ. Thus, within the framework of these concepts one can qualitatively understand the existence of a reverse isotopic effect in the Pd-H system.

Recently, calculations of λ in Pd and in PdH have been performed [7.192] within the framework of the Gaspari-Gyorffy model proposed for transition metals and their carbides [7.193,194] (Sect.5.3). These calculations were based on neutron-diffraction investigations of the phonon spectrum [7.195,196] and on Pd, PdH and PdD electron-energy-spectrum calculations [7.192, 197-199]. The value obtained for pure Pd is $\lambda = 0.47$. Substituting this value into McMillan's formula (2.3.27) we obtain the result that in the absence of superconductivity in Pd the value of the Coulomb pseudopotential should be equal to $\mu^* = 0.28$, i.e., much greater than that in superconducting transition metals (Table 4.1).

Calculations for PdD have given the value $\lambda = 0.72$. It has turned out that the main contribution to λ is made by the interaction of electrons with optical phonons; it is approximately three times larger than the interaction with acoustic phonons. With $\lambda = 0.72$ and $T_c = 11$ K we obtain $\mu^* = 0.22$. Thus, although addition of deuterium to palladium suppresses spin fluctuations and decreases the Coulomb interaction somewhat (μ^* diminishes from 0.28 in pure Pd to 0.22 in PdD), the main cause of the occurrence of superconductivity lies in a substantial increase of the electron-phonon interaction constant due to the coupling of electrons with optical phonons (λ increases from 0.47 in Pd to 0.72 in PdD).

Another point of view concerning the nature of superconductivity and of the reverse isotopic effect is based on the consideration of differences in the PdH and PdD lattice parameters [7.200]. According to [7.201], the $PdH_{0.58}$ and $PdD_{0.56}$ lattice parameters are equal to 4.025 Å and 4.027 Å, respectively, which implies that the introduction of deuterium into palladium extends its lattice to a somewhat greater degree than the introduction of hydrogen does. As a result, compared to PdH, the overlap of the wave functions of the component atoms in PdD is weaker and the number of 1s-electrons which form bonds with Pd is less. Consequently, the amount of electrons available on the Fermi surface in PdD is somewhat larger than that in PdH and, accordingly, the values of N(0), λ and T_c are higher. The theory proposed explains the negative sign of the derivative $\partial T_c/\partial P = -3 \cdot 10^{-5}$ K/bar [7.202] by the decrease in volume during compression and by the relevant changes in the overlap of the wave functions, of the component atoms and in the number of Me - H bonds leading to a decrease in the values of N(0), λ and T_c.

However, more elaborate X-ray structure investigations of PdH_x and PdD_x have revealed [7.203], in contrast to measurements [7.201], that the lattice parameter in PdD_x is not larger but, on the contrary, 0.1% less than that in PdH_x in the entire region of the occurrence of the β-phase (with NaCl structure) in these compounds.

In view of this, we are more inclined to prefer the first of the above theories, which relates the occurrence of superconductivity in PdH to the interaction of electrons with optical phonons. It may be noted that the possibility of explaining the reverse isotopic effect in this model will come only after λ has been calculated from first principles, but such a calculation has not yet been carried out for lack of information concerning the phonon spectrum of this compound.

7.3.3 Other Systems Based on Palladium

Recently experimental investigations have been carried out on the saturation of metallic Pd by other light elements: Li, B, C and N [7.204,205]. It turns out that Pd becomes superconducting upon implantation of boron and carbon at low temperatures with $T_c \sim 1.3$ to 3.8 K. For the other cases no superconductivity was observed at temperatures down to 0.1 or 0.2 K. It is obvious from the data given in Table 7.9 that the ion/palladium ratio at which the T_c is maximum for this system, decreases if passing from Pd-B to Pd-C. The mechanism of the superconductivity in such systems has not yet been ascertained. It is somewhat surprising that superconductivity is absent in experiments with Li, although in its properties lithium, as compared to other elements, is closer to hydrogen. This is possibly due to the fact that lithium atoms are of larger size than C and N atoms and are less capable of implantation in the octahedral and tetrahedral interstices of Pd.

Table 7.9. Superconductivity in palladium saturated with light elements

Ion	T_c^{max} [K]	Optimal ion-palladium ratio	Ion	T_c^{max} [K]	Optimal ion-palladium ratio
$^{7}Li^{+}$	<0.1	>1.3	$^{14}N_2^{+}$	<0.2	>2.0
$^{10}B^{+}$	3.8	1.5	$^{1}H_2^{+}$	8.8	1.0
$^{12}C^{+}$	1.3	0.6	$^{2}D_2^{+}$	10.7	1.0

To conclude, it may be noted that the discovery of superconductivity in the binary Pd-H and Pd-D systems gave a powerful impetus to search for superconductivity in more complicated Pd alloys by saturating them with hydrogen and deuterium. The T_c obtained for the ternary Pd-Ag-H, Pd-Cu-H and Pd-Au-H systems was 13 to 17 K [7.205-207], which is some of the highest T_c in the alloys and compounds based on transition metals. The results of these investigations, presented in Fig.7.21, show that the maximum T_c is reached at a definite

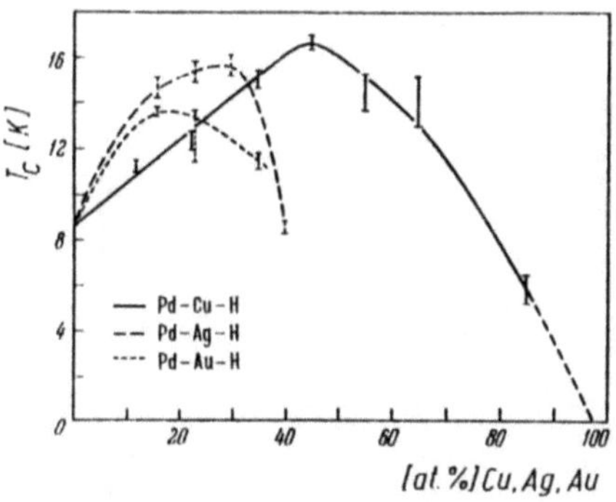

Fig.7.21. The superconducting transition temperature of the ternary systems Pd-Me-H

concentration of the noble metal (the higher the atomic number the less the concentration). The optimum value of saturation of these alloys with hydrogen varies: in the Pd-Cu-H system $T_c = 16.6$ K is reached at $H/Pd_{0.55}Cu_{0.45} \approx 0.7$, in the Pd-Ag-H system $T_c = 15.6$ K is reached at $H/Pd_{0.7}Ag_{0.3} = 0.8$, and in the Pd-Au-H system $T_c = 13.6$ K is obtained when $H/Pd_{0.84}Au_{0.16} = 0.9$.

An increase in T_c was also observed when hydrogen was introduced into other Pd alloys in the Nb-Pd-H, Nb-Pd-Mo-H and Nb-Pd-W-H systems [7.208], Pd-Al-H [7.209], and Pd-Si-H systems [7.210].

Electron structures were calculated for ternary alloys $Pd_{1-y}Ag_yH_x$ and $Pd_{1-y}Rh_yH_x$, with λ and T_c calculated for different values of x and y [7.211]. Use was made of the hybridized APW-CPA method, or of the APW-VCA method, in which the partial substitution of Pd for Ag/Rh atoms is taken into account using the APW technique. The disorder in the H sublattice was described in the coherent potential approximation [7.197-199] or in the virtual crystal approximation [7.212,213]. Figure 7.22 presents curves of the density of states for RhH, PdH, AgH and the ternary alloy $Pd_{0.7}Ag_{0.3}H$. One can see a certain similarity in the electron structure of all the compounds shown above. The basic differences are due to the location of the Fermi level, depending on the electron concentration. The quantities λ and T_c, respectively, were calculated by Phillips' formula [7.214], and by McMillan's formula (2.3.27) with corrections by ALLEN and DYNES [7.215]. Calculations have shown that λ and T_c in ternary hydrides of noble metals, as well as in the case of PdH, are determined by the interaction of electrons with optical phonons. The variation of T_c with component concentration depends largely on the density of hydrogen s-states on the Fermi level.

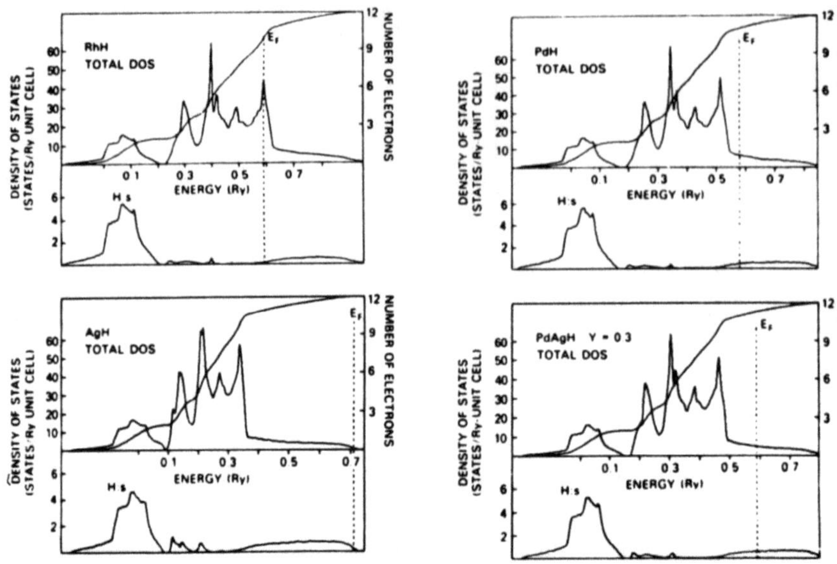

Fig.7.22. The full and partial (Hs-type) densities of electron states in binary and ternary hydrides of noble metals according to [7.211]

7.4 Sesquicarbides and Ternary Borides

In this section we consider a group of carbides with the chemical formula A_2C_3 in which the A-component is represented by the transition metal Y: lanthanides La, Ce, Pr, Nd, Sm, Gd, Tb, Dy and Ho, and actinides Th, U, Np and Pu.

7.4.1 Crystal Chemistry

The above compounds have a structure of the type Pu_2C_3 and belong to the space group I43d-T_d^6 [7.216,217]. The A_2C_3 compounds with the Pu_2C_3 structure possess a complicated unit cell incorporating 16 A atoms and 24 carbon atoms. In the sublattice which is constituted of A atoms (Fig.7.23a) we can single out a polyhedron composed of 6 tetrahedra. Translation of this polyhedron using three operations gives the position of all the atoms of the metal in the Pu_2C_3 structure. The carbon atoms are arranged pairwise and are separated by a gap as small as two tenths of the lattice parameter (Fig.7.23b). It is significant that the pairs consisting of carbon atoms are oriented parallel to the coordinate axes. It is only by convention that a structure such as Pu_2C_3

may be regarded as a bcc structure. According to SCHUBERT [7.70], of the eight metallic atoms occupying the cube vertices, seven atoms are displaced by 0.05 of the lattice parameter and actually belong to other unit cells (Fig.7.23a).

Most of the carbides under consideration form when A elements and graphite undergo ordinary arc melting. However, the sesquicarbides Y, and Th and Pu_2C_3 structure ternary compounds based on them cannot be obtained using the standard synthesis technique. Their preparation requires high-temperature (1200 to 1500°C) and high pressure (15 to 30 kbar) [7.218-222].

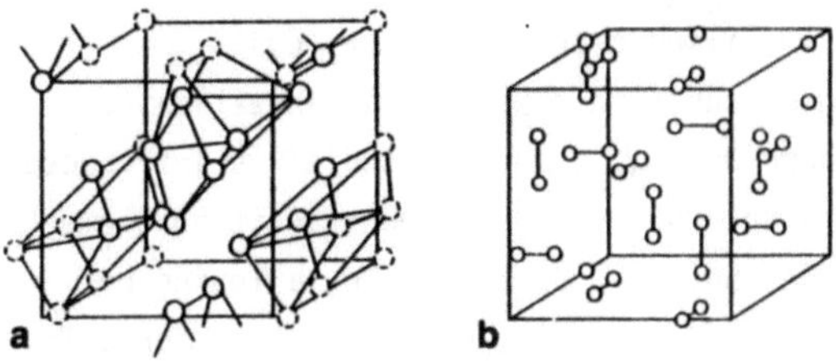

Fig.7.23a,b. The Pu_2C_3 structure (a) sublattice composed of metallic atoms (b) carbon sublattice

7.4.2 Superconductivity

In spite of the circumstance that the class of sesquicarbides is highly representative, the superconducting properties were investigated only in a limited number of pure compounds, viz., Y_2C_3, Th_2C_3 and La_2C_3 [7.219,222,223]. In two of these compounds T_c turned out to be above 10 K (Table 7.10). This immediately stimulated interest in studying the superconducting properties of sesquicarbides, and soon high-temperature superconductivity ($T_c \sim 14$ to 17 K) was revealed in ternary compounds based on Y_2C_3 and La_2C_3 [7.220,224]. Subsequent investigations showed that, just as in NaCl structure carbides of transition metals, the superconducting properties of sesquicarbides were very

Table 7.10. T_c in sesquicarbides

Compound	$Y_2C_{2.68}$	$La_2C_{2.70}$	Th_2C_3	$(Y_{0.7}Th_{0.3})_2C_{3.1}$
T_c[K]	11.5*	11.0	4.1*	17.0*

* These compounds were synthesized using the high-pressure technique

sensitive to deviations from stoichiometry. Maximum T_c, as a rule, occur close to stoichiometric compositions [7.219]. The electron concentration in the compounds Y_2C_3, La_2C_3 and Th_2C_3 amounts to 3.6 e/a, i.e., it is less than that in transition-metal carbides with NaCl structure.

In contrast to A-15 and B-1 structure compounds, the T_c in the alloys of sesquicarbides substantially exceed those in the basic compounds (Table 7.10). On account of this, attempts were made to investigate the influence of alloying on the superconducting properties of the sesquicarbides Y and Th in detail [7.220,222]. Analysis of these data leads us to the conclusion that the superconducting properties of the compound Th_2C_3 are improved when the alloying elements have an unfilled or a nearly filled f-shell (La, Er, Lu).

The alloying by transition and nontransition elements was investigated in [7.220]. The initial compound there was Y_2C_3, and the additives introduced were transition d-metals (Ti, V, Cr, Nb, Mo, Ru, W) and nontransition elements primarily of the fourth group (Si, Ge, Sn, Bi). On the whole, alloying by these elements gives a positive effect (for example, when W is used for alloying in the ratio $W/Y = 1/9$, T_c increases by a factor of 2.5 approximately). However, in a number of cases (when In and Zn are added) the ternary alloy becomes nonsuperconducting even at temperatures down to 4 K.

The influence of hydrostatic pressure on T_c was studied in the compounds Y_2C_3 and Th_2C_3 and in solid solutions of $(Th_xY_{1-x})_2C_3$ ($x = 0.3$; 0.35) in the interval between 0 and 10 kbar [7.225]. In the compounds Y_2C_3 and Th_2C_3, T_c was found to drop with increasing pressure at a rate of $\partial T_c/\partial P = -2.5 \cdot 10^{-5}$ K/bar. In ternary compounds $(Th_xY_{1-x})_2C_3$, conversely, an increase of T_c due to compression ($\partial T_c/\partial P = 2\text{-}3 \cdot 10^{-5}$ K/bar) was registered.

The results of an investigation of the solid solutions of the sesquicarbides Y_2C_3 and Th_2C_3 are very interesting. We have already noted above (Table 7.10) that among these ternary alloys of the Y_2C_3-Th_2C_3 system the superconducting sesquicarbides with highest transition temperatures were found. A body-centred cubic structure (such as Pu_2C_3) favourable to superconductivity in these compounds may arise due to combined application of high pressure and high temperatures. When synthesizing specimens of this type by ordinary arc melting it was found that the ternary Y_2C_3-Th_2C_3 alloys had a body-centred tetragonal lattice and did not exhibit superconductivity even at temperatures down to 4 K. Thus, although one can, under certain conditions of synthesis, obtain either bcc or bct structures, no direct transition from one structure to another is observed in the above compounds. Nevertheless, consideration of the geometric relations between tetragonal and cubic carbide phases shows [7.226] that such a transition may occur.

The information given here shows that sesquicarbides are very interesting. Unfortunately, the physical properties of these compounds have practically not been studied. It is of special interest to measure the magnetic susceptibiliy in alloys with the addition of rare earths, and to determine the localized magnetic moments in those cases when alloying decreases T_c. On the other hand, investigations of the structural transformation from a bcc to a bct structure and of the phonon spectrum in the above compounds are required.

As a rule, binary borides of transition metals do not exhibit high superconducting transition temperatures. The highest value of T_c in these materials does not exceed 8 K [7.227]. Nevertheless, in ternary borides such as ARh_4B_4 (where A stands for Y, Nd, Sm, Er, Tm, Lu, Th) and in their solid solutions, MATTHIAS et al. [7.228] registered higher T_c, which in a number of cases exceed 11 K. According to X-ray structural [7.229] and neutron-diffraction [7.230] investigations, these compounds have a tetragonal structure (the corresponding space group is $P4_2/nmc$). In this structure (Fig.7.24), the Rh atoms form tetrahedra constituting sheets perpendicular to the c axis, the sheets being interconnected by A atoms. The boron atoms are arranged pairwise parallel to the a axis and have five A atoms as their nearest neighbours. The A atom may be represented by the above-mentioned rare-earth elements whose metallic radii range between 1.70 and 1.82 Å.

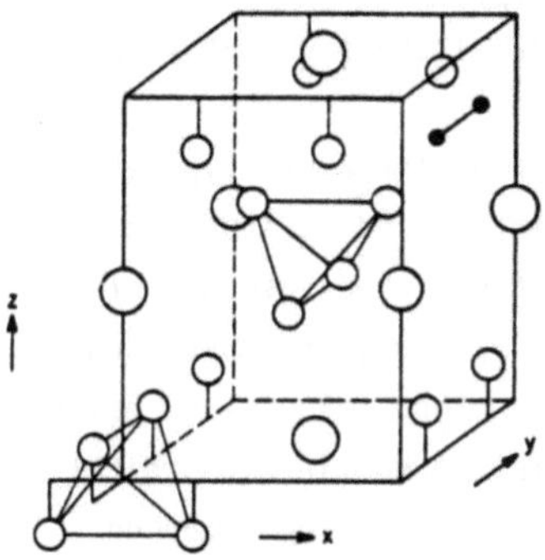

Fig.7.24. Crystal structure of YRh_4B_4

Like in the case of ternary molybdenum chalcogenides (Sect.7.5), the presence of a comparatively large concentration of rare-earth magnetic atoms does not lead to suppression of superconductivity in the compounds concerned (Table 7.11). The problem concerning the coexistence of magnetism and superconductivity in ternary borides was considered in Chapter 3. Here we present results of the search for superconductivity in other structural systems and in more complicated quaternary compounds, and also consider some theoretical data.

Table 7.11. T_c in ternary borides

Compound	T_c [K]	Compound	T_c [K]
YRh_4B_4	11.34	$CeRu_4B_4$	<1.5
$NdRh_4B_4$	5.36	$PrRu_4B_4$	<1.5
$SmRh_4B_4$	2.51	$SmRu_4B_4$	<1.5
$ErRh_4B_4$	8.55	$EuRu_4B_4$	<1.5
$TmRh_4B_4$	9.86	$TmRu_4B_4$	<1.5
$LuRh_4B_4$	11.76	$YbRu_4B_4$	<1.5
$ThRh_4B_4$	4.34	$LuRu_4B_4$	1.77
		YRu_4B_4	<1.5
		$ThRu_4B_4$	<1.5

It was found in [7.231] that compounds such as ARu_4B_4 crystallize in the same structure as ARh_4B_4, and also that the A component is represented by rare-earth elements (A = Ce, Pr, Sm, Eu, Tm, Yb, Lu, Y, Th). An investigation of these compounds showed that only one of them, $LuRu_4B_4$, was a superconductor with $T_c = 1.77$ K. For all the other compounds of this type, no superconductivity was revealed at temperatures down to 1.5 K. The search for superconductivity in quaternary systems such as $(A_{0.75}Th_{0.25})Rh_4B_4$ [7.228] and $A(Rh_{0.85}Ru_{0.15})_4B_4$ [7.231] was most fruitful. The T_c registered for the compounds $Lu_{0.75}Th_{0.25}Rh_4B_4$ and $Y(Rh_{0.85}Ru_{0.15})_4B_4$ were 11.93 and 9.56 K, respectively.

The present state of investigations of compounds of the above type may be characterized as follows. So far attention has been centred on studying the superconducting properties due to alloying and on the various aspects of the coexistence of superconductivity and magnetism [7.232-239], the most-studied compound being $ErRh_4B_4$ [7.233-239]. An electron-structure calculation has recently been made for this compound using the linearized muffin-tin orbitals (LMTO) method [7.235]. Analysis of these data shows that in the compound $ErRh_4B_4$, the Fermi energy falls within the density-of-states peak to the formation of which the Rh4d-states make the greatest contribution. The value of the electron-phonon coupling parameter was estimated from a comparison of band calculation results and electron heat capacity measurements [7.228] and turned out to be $\lambda = 0.96$. The contribution of the Er5d-states to the formation of a peak near the Fermi level in $ErRh_4B_4$ was approximately 1/3 less than that in pure rare-earth metals.

There is a new class of superconducting ternary borides of transition metals such as $A_{0.67}Pt_3B_2$ (A = Ga, Sr, Ba) with $T_c \sim 1.5$ to 5.6 K [7.240].

These compounds crystallize in the hexagonal structure (the space group is $P6_3/mmc$). The stability of this structure requires the atoms of the A components to have comparatively large dimensions. The number of isostructural compounds of this type is, therefore, rather limited.

7.5 Chalcogenides

7.5.1 Sulphides

One of the most striking discoveries made in recent years in the field of the superconductivity of materials is the detection of high-temperature superconductivity in ternary transition-metal chalcogenides. In 1972, BARZ et al. [7.241] synthesized a compound $Li_xTi_{1.1}S_2$ ($0.1 \le x \le 0.3$) with a hexagonal structure such as Ti_3S_4 and with $T_c = 10$ to 13 K. This was the first compound of a class of high-temperature noncubic superconductors. An interesting feature of this ternary compound is that none of the binary compounds of the components in the Li-S and Li-Ti or Ti-S systems is superconducting at temperatures higher than 1 K.

The Ti_3S_4 structure belongs to the space group $P6_3/mmc$ [7.242]. The sulphur atoms in this structure (Fig.7.25a) alternately occupy two nonequivalent sets of positions: in the alternate hexagonal and cubic (S_2) close packing. Titanium atoms are located at the centre of regular octahedra composed of sulphur atoms ($3S_1 + 3S_2$) (Fig.7.25b). If the Ti_1 type sites are completely populated, the Ti_2 type sites are only partially occupied, but not more than by half.

Thus, one of the structural features peculiar to the compound $Li_xTi_{1.1}S_2$ is the presence of layers of metallic atoms. However, this compounds may, nevertheless, not be classed a layered compound. There is practically no inter-

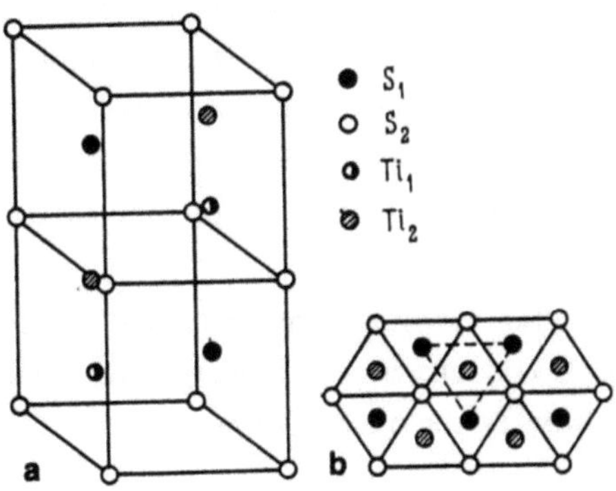

Fig.7.25a,b. The Ti_3S_4 structure

action between the layers in layered compounds, while in Ti_3S_4-structure compounds the Ti_2 type metallic atoms from the partially occupied layers serve as "bridges" and completely bind the filled layers of Ti_1 atoms and thereby ensure their interaction.

One of the important features of the compound $Li_xTi_{1.1}S_2$ is its instability in the temperature range from 77 to 600 K. Initially, upon synthesizing the above compound, measurements showed high values of T_c ($T_c \sim 10$ to 13 K), but then, after the specimens had been held at room temperature for several hours, T_c dropped by several degrees. After having been kept for a week, the specimen's superconducting properties deteriorated to a still greater extent. With the specimen subjected to repeated annealing at 1020 K, the high-temperature superconductivity was recovered completely. On the other hand, if upon synthesis the specimen was immediately cooled to helium temperature, the high-temperature superconductivity remained for a long time. It has not been ascertained yet whether the detected instability is of a structural character, but the essence of the detected effect cannot be overestimated. In MATTHIAS' opinion, one should reconsider the large number of compounds which did not exhibit good superconducting properties and check whether they became instable between the synthesis and time of measurement [7.243].

7.5.2 Chevrel Phases

Another representative of the superconducting ternary chalcogenides of transition metals is a large class of compounds with the general formula $A_xMo_6X_8$, where $X = S$, Se and Te, and A stands for practically any chemical element: an alkali, alkali-earth, simple, transition, noble, or rare-earth metal [7.244,245]. These compounds were first synthesized in 1971 by CHEVREL et al. [7.246], and thus they are conventionally referred to as Chevrel phases. Soon after it was found that a number of ternary molybdenum sulfides and selenides exhibited comparatively high superconducting transition temperatures [7.245, 247,248] (Table 7.12 [7.244,249]) and extraordinarily high critical fields [7.179,250-252] (Table 7.13 [7.244,253]). The latter circumstance gave an impetus to the study of the superconducting properties as well as of the various characteristics of the electron and phonon spectra in ternary molybdenum chalcogenides.

The considerable interest expressed in the investigation of Chevrel phases is also due to the fact that superconductivity and magnetic order were found to coexist in these compounds (this problem is considered in detail in Chapter 3).

Table 7.12. T_c in ternary molybdenum sulphides and selenides

Compound	T_c [K]	Compound	T_c [K]	Compound	T_c [K]
$LiMo_6S_8$	4.4-3.7	$Er_{1.2}Mo_6S_8$	1.85-2.19	$V_{1.2}Mo_6Se_8$	-
$NaMo_6S_8$	9.0-8.1	$Tm_{1.2}Mo_6S_8$	2.1	$Mn_{1.2}Mo_6Se_8$	-
KMo_6S_8	3.1-2.7	$Yb_{1.2}Mo_6S_8$	9.1-8.2	$Fe_{1.2}Mo_6Se_8$	-
$Rb_2Mo_6S_8$	<0.6	$Lu_{1.2}Mo_6S_8$	2.0	$Co_{1.2}Mo_6Se_8$	-
$Cs_2Mo_6S_8$	<0.6	$Mn_{1.04}Mo_{6.6}S_8$	-	$Ni_{1.6}Mo_6Se_8$	-
$Cu_2Mo_6S_8$	11.3-10.0	$Fe_{1.3}Mo_{6.6}S_8$	-	$Cu_2Mo_6Se_8$	5.9
$AgMo_6S_8$	9.1-8.3	$Co_2Mo_6S_8$	-	$Ag_{1.2}Mo_6Se_8$	5.9
$BeMo_6S_8$	<1.5	$Ni_2Mo_6S_8$	-	$Zn_{1.2}Mo_6Se_8$	-
$MgMo_6S_8$	3.5-2.5	$Pd_{1.6}Mo_6S_8$	-	$Cd_{1.2}Mo_6Se_8$	-
$ZnMo_6S_8$	3.6-3.5	$Zn_{1.1}Mo_{6.6}S_8$	3.6	$Sn_{1.2}Mo_6Se_8$	6.8
$CdMo_6S_8$	3.5-3.2	$CdMo_{6.6}S_8$	3.5	$Pb_{1.2}Mo_6Se_8$	6.7-3.9
$ScMo_6S_8$	3.6-2.6	Mo_6S_8	1.6	$In_3Mo_{15}Se_{19}$	4.3
YMo_6S_8	3.0-2.6	$Y_{1.2}Mo_6Se_8$	6.2	Mo_6Se_8	6.4-6.3
CMo_6S_8	4.1-3.5	$LaMo_6Se_8$	12-11.4	$Mo_6S_6Br_2$	13.8
$Sn_{1.2}Mo_6S_8$	14.2-10.8	$CeMo_6Se_8$	<1.1	$Mo_6S_6I_2$	14.0
$PbMo_{6.4}S_8$	15.2-14.0	$PrMo_6Se_8$	9.2	Mo_6Se_7Br	7.1
VMo_6S_8	8.2-7.3	$NdMo_6Se_8$	8.5-8.2	Mo_6Se_7I	7.6
$NbMo_6S_8$	3.5-3.2	$Sm_{1.2}Mo_6Se_8$	6.8	$Mo_6Te_6I_2$	2.6
$LaMo_6S_8$	7.1-6.0	$Eu_{1.2}Mo_6Se_8$	<1.1	$Mo_4Re_2Te_8$	3.5
$CeMo_6S_8$	<0.6	$Gd_{1.2}Mo_6Se_8$	5.6	$Mo_2Re_4Se_8$	-
$PrMo_6S_8$	4.0	$Tb_{1.2}Mo_6Se_8$	5.7	$Mo_2Re_4S_8$	-
$NdMo_6S_8$	3.5	$Dy_{1.2}Mo_6Se_8$	5.8	$Mo_6Rb_2S_6$	<0.6
$SmMo_6S_8$	<0.6	$Ho_{1.2}Mo_6Se_8$	6.1	$Mo_6Cs_2S_6$	<0.6
$Sm_{1.2}Mo_6S_8$	2.9	$Er_{1.2}Mo_6Se_8$	6.2	$Mo_6S_{4.8}Te_{3.2}$	2.69-2.26
$Gd_2Mo_6S_8$	2.2	$Tm_{1.2}Mo_6Se_8$	6.3		
$Tb_{1.2}Mo_6S_8$	1.65	$Yb_{1.2}Mo_6Se_8$	5.8		
$Dy_{1.2}Mo_6S_8$	2.1	$Lu_{1.2}Mo_6Se_8$	6.2		
$Ho_{1.2}Mo_6S_8$	2.0	$Ti_{1.2}Mo_6Se_8$	-		

A feature peculiar to the crystal structure of Chevrel phases (an Mo_3Se_4 type hexagonal-rhombohedral structure, the space group C^2_{3i}-$R\bar{3}$) is the presence of Mo_6X_8 blocks which are slightly deformed cubes whose apices accommodate sulphur atoms centred by the octahedron of molybdenum atoms. These blocks are rotated through 25° about the <111> axis and thus form the channels in which the A atoms are located. The large cations (Ag, In, Sn, Pb,...) are

Table 7.13. Superconducting parameters and physical properties of some ternary molybdenum chalcogenides

Compounds	T_c [K]	γ [mJ/K^2-mole]	χ (300 K) [10^{-6} cm^3/mole]	λ	$H_{c2}(0)$ [kGauss]
$SnMo_6S_8$	11.8	105	32.8	1.14	340
$PbMo_6S_8$	12.6	79	35.6	1.20	500
$Gd_{0.2}PbMo_6S_8$	14.3				600
$LaMo_6S_8$	5.8-7.2	38	12.9	0.83-0.85	54
YMo_6S_8	2.1		8.4	0.53	
Mo_6Se_8	6.3	21-44		0.74-0.79	110
$PbMo_6Se_8$	3.8	28	23.5	0.64	38
$LaMo_6Se_8$	10.8-11.2	53-79	15.7	1.01-1.06	445
$Lu_{1.2}Mo_6Se_8$	5.84	31		0.77	
YMo_6Se_7	6.26	34		0.79	
$In_3Mo_{15}Se_{19}$	4.3				212
$PbGa_{0.4}W_{0.4}Mo_6S_8$	14.9				580
$PbGa_{0.4}Re_{0.4}Mo_6S_8$	14.4				520

situated at the centre between eight Mo_6X_8 blocks so that one deals with a distorted CsCl type structure (Fig.7.26). The small-sized A cations (Fe, Co, Ni, Cu,...) occupy up to four of the twelve tetrahedral interstices available (Fig.7.27) [7.254-258].

Among the superconducting ternary Mo chalcogenides a new compound $In_3Mo_{15}Se_{19}$ with a high critical magnetic field has recently been found

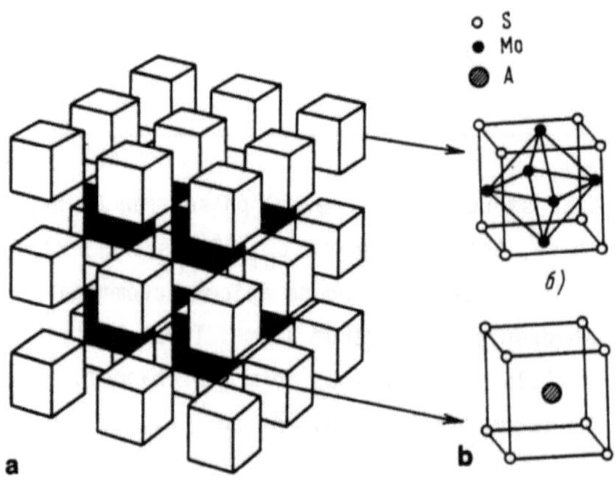

Fig.7.26. Structure of compounds of the type $A_xMo_6X_8$

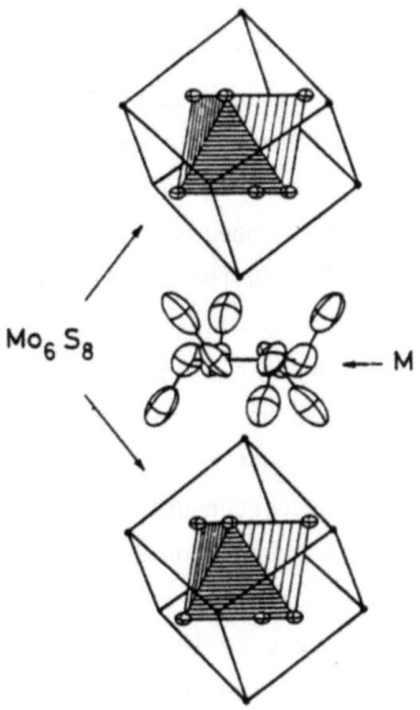

Fig.7.27. Crystal structure of the Chevrel phases $A_xMo_6X_8$ (A = Fe, Co, Ni, Cu)

[7.259], which possesses a hexagonal structure. As distinct from rhombohedral compounds such as $AxMo_6X_8$, two types of structural blocks, Mo_6Se_8 and Mo_9Se_{11}, may be singled out in the compound $In_3Mo_{15}Se_{19}$. Accordingly, the atoms of In in the structure of this compound occupy two nonequivalent positions.

It was found that binary compounds such as Mo_6Se_8 and Mo_6Te_8 [7.260,261] can also exist individually. In a number of cases it is possible to substitute part of the component atoms in these binary phases: rhenium for molybdenum to form compounds such as $(Mo_2Re_4)X_8$ (X = S, Se, Te) [7.262], and group VII A elements (Cl, Br, I) for the X element to form $Mo_6(S_6Br_2)$, $Mo_6(S_6I_2)$ [7.263], $Mo_6(Se_7Br)$ and $Mo_6(Te_6I_2)$ [7.264].

Structural investigations have shown that in ternary molybdenum chalcogenides such as $A_xMo_6S_8$ the distance between the atoms of Mo in the Mo_6 octahedron (cluster) amounts to 2.6-2.8 Å and depends little on the sort of the A component. At the same time, the atoms of molybdenum in the various clusters are bound much more weakly; the corresponding distance ranges between 3.1 and 3.6 Å and varies from compound to compound [7.256,265-269]. In IVON and PAOLI's opinion [7.270], the highest compression of the Mo_6 cluster when the A component is added results from a charge transfer from the A element to the Mo_6 cluster.

The superconducting properties of ternary molybdenum sulphides and selenides have received a most detailed study. This is so because it is among these compounds that superconductors with the highest temperatures and fields

have been found. Except in the compounds $(Mo_4Re_2)Te_8$ and $Mo_6(S_{6\pm x}Te_{2\pm y})$, no superconductivity has been detected in ternary molybdenum tellurides. Analysis of the data given in Tables 7.12 and 13 permits the following conclusions.
i) No clear correlation exists between the sort of A atom and its position in the periodic table of elements and the superconducting properties of the compounds. The highest values of T_c and H_{c_2} are shown by compounds whose A components are both the nontransition elements Sn and Pb and the noble metals Cu and Ag, or the rare-earth elements La, Pr and Yb.
ii) The magnetic state of the A component exerts no appreciable influence on T_c of these compounds when the A component is represented by rare-earth elements and suppresses the superconductivity in the case of A transition metals.

The above differences noted in the behaviour of the superconducting properties of compounds whose A components are rare-earth and transition metals is, according to FISCHER [7.271], due to their different interaction with Mo atoms, which follows from the peculiarities of the crystal structure. The rare-earth A atoms which are large in size, have only X atoms as their nearest neighbours and are remote from Mo atoms (the nearest Mo atoms are 4.2 Å away). This leads to a weak d-f exchange interaction. On the other hand, the small-sized Fe or Ni atoms occupying the tetrahedral positions have Mo atoms as their nearest neighbours, which ensures a sufficiently strong d-d exchange interaction and the suppression of superconductivity in compounds of this type.

We consider some empirical correlations between T_c and the various parameters of ternary molybdenum chalcogenides. FISCHER [7.271] gave the dependence of T_c on the concentration of valence electrons available for each atom in the Mo_6 cluster for ternary molybdenum sulphides and their solid solutions. This parameter is calculated with due regard for the charge transfer from the A atom to the Mo_6 cluster, on the assumption that in the Mo_6S_8 cell, sulphur has the valence (2-). Despite the considerable scatter in the data presented, one can note a general tendency for a decrease of T_c when the limiting electron concentration in the cluster becomes equal to 4 electrons. This may point to the fact that at this electron concentration the band is filled and the Fermi level passes near the maximum in the density of states. Figure 7.28 illustrates the dependence of T_c on the volume of the hexagonal cell for a series of solid solutions: $A_xPb_{1-x}Mo_6S_8$ (where A = La, Gd, Lu, Y) [7.272]. It is seen that T_c falls off gradually with decreasing volume. However, in this case the substitution of a trivalent rare-earth element for the divalent Pb occurs along with a decrease in volume, as a result of which the charge transfer and the electron concentration increase. A more straightforward way of studying the influence of volume on the superconducting transition temperature

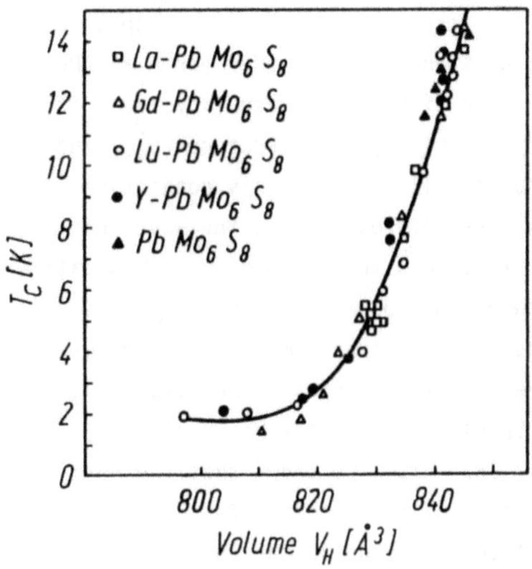

Fig.7.28. Dependence of T_c on the hexagonal-cell volume for solid solutions such as $(RE)_xPb_{1-x}Mo_6S_8$, where RE = La, Gd, Lu, Y [7.272]

is to measure T_c under pressure. Such measurements [7.273-275] have shown that the values of T_c decrease substantially with increasing pressure. This is in qualitative agreement with the data on the variation of T_c as a function of cell volume due to allyoing (Fig.7.28).

Although there is no explicit correlation between the sort of A atom, its position in the periodic table and the superconducting properties of Chevrel phases, many attempts have been made to study the role which the A component plays in the formation of the properties of the above materials. According to FISCHER [7.244], the A atoms in the Chevrel phases manifest themselves as monovalent (Na, Cu), divalent (Pb, Sn, Yb) and trivalent ions (In, La, Lu, Y) and donate electrons to the Mo_6 octahedron. The difference in T_c of the compounds being considered is attributed to the variation of the electron concentration in the Mo_6 cluster. This concentration determines the position of the Fermi level and the magnitude of the density of states on the Fermi level.

On the other hand, analyzing the measurements of the Mössbauer spectra and heat capacity in $SnMo_6S_8$, one arrives at the conclusion that the A component in the Chevrel phases forms additional soft lattice modes whose presence may explain the higher T_c in ternary Chevrel phases as compared to binary Mo_6X_8 compounds [7.276,277].

In view of the above statements it is of interest to consider the direct investigations of the electron and phonon spectra in ternary molybdenum chalcogenides.

At present we know four calculations of the electron structure of Chevrel phases [7.278-281]. Despite the crudeness of the models and approximations used, they give us a rough idea about the structure of the energy bands in these materials. A straightforward use of standard techniques (such as APW or KKR) to calculate the electron structure of Chevrel phases involves difficulties arising from the low symmetry of the crystal structure (C_{3i}^2) and from the large number of atoms in the unit cell (15). In this connection, MATTHEISS [7.279] has calculated, as a first approximation, the wave functions and the energy levels of a cubic Mo_6S_8 cluster. According to his data the Pb6s levels form a narrow band between the S3s and S3p orbitals, and the centre of gravity of the Pb6p-bands is above the Fermi level. The position of the Pb6p-bands in the region of free states presupposes the divalent state of Pb in $PbMo_6S_8$ and the transport of 6p electrons to the Mo_6S_8 cluster.

The calculation of JARLBORG et al. [7.281] has also shown that the Fermi level in $SnMo_6S_8$ is located in the vicinity of the density-of-states peak, with the Mo4d-states making the largest contribution to the formation of this maximum (Fig.7.29). The main result of the calculation of ANDERSON et al. [7.278] is that the effect of the intercluster d-d interaction in $PbMo_6S_8$ has been found to be comparatively weak. It has turned out that the Mo4d-bands determined from the calculation of the Mo_6 cluster widen only slightly due to the intercluster interaction.

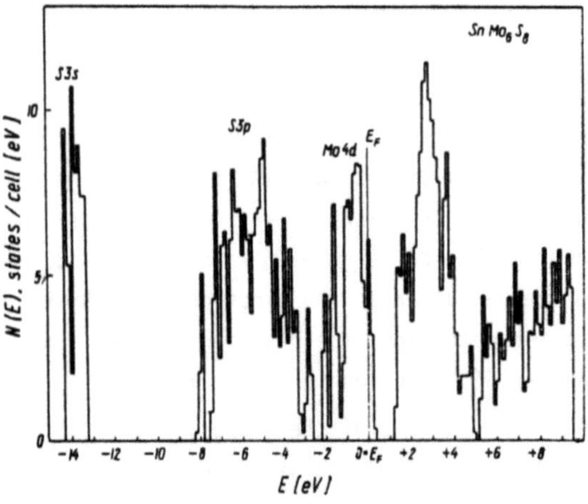

Fig.7.29. Density of states in $SnMo_6S$ according to the calculation [7.281]

The results of cluster calculations [7.278,279] agree in that the electrons near the Fermi level in these clusters have primarily $d_{x^2-y^2}$ symmetry. However, the relative position of the doubly degenerate level $\Gamma_{12}(E_g)$ and of the triply degenerate level $\Gamma_{25}(T_{2u})$ has turned out to be different in these calculations. According to ANDERSON et al., the last occupied level has a $\Gamma_{12}(E_g)$ symmetry, and the band consisting of these cluster states localized in the

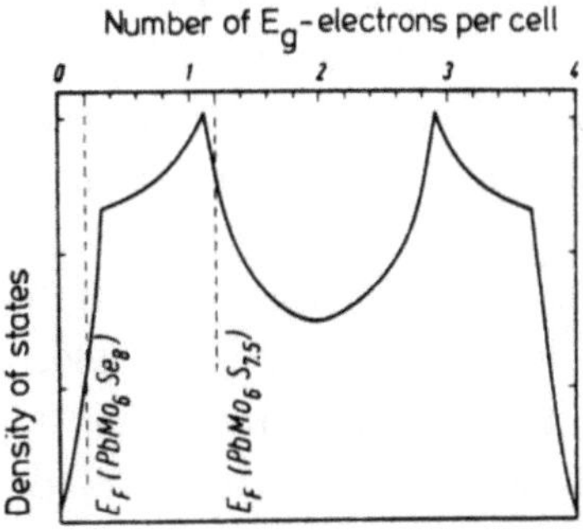

Fig.7.30. Density of states in the E_g-band of Chevrel phases in the cluster approximation [7.278]

Mo_6 octahedron plays an important role in the analysis of the superconductivity in Chevrel phases. Figure 7.30 presents a curve for the density of states in the E_g-band and shows the position of the Fermi level for $PbMo_6S_{7.5}$ and $PbMo_6Se_8$. The location of the Fermi level near the van Hove singularity in $PbMo_6S_{7.5}$ ensures, according to [7.278], a high T_c and an anomalously large dependence of T_c on pressure in this compound. Using the GASPARI-GYORFFY [7.282] approximation, ANDERSON et al. [7.278] calculated the electronion interaction parameter $\langle J^2 \rangle$ and estimated λ according to McMILLAN's formula [7.283] in a number of ternary molybdenum chalcogenides. It has turned out that the quantity $\langle J^2 \rangle$ in Chevrel phases is approximately 2.5 times less than the corresponding values in transition metals, including molybdenum. On the other hand, the weakness of the intercluster interactions in ternary molybdenum chalcogenides ensures higher values of N(0) (which agrees with the experimental electron heat capacity measurements made in [7.276,284,285]) and, as a result, leads to values of $\eta = N(0) \langle J^2 \rangle$ close to those in transition metals ($\eta_{PbMo_6S_{7.5}} = 8$ eV/Å^2, $\eta_{Nb} = 11$ eV/Å^2, $\eta_{Mo} = 5$ eV/Å^2). However, λ and T_c in $PbMo_6S_{7.5}$ exceed those in pure Mo on account of the lower phonon frequencies in Chevrel phases.

The intercluster interactions and their influence on the electron structure of Chevrel phases have been investigated in detail by BULLETT [7.280], calculating the electron densities of states for Mo_6S_8, Mo_6Se_8, $PbMo_6S_8$ and $PbMo_6Se_8$. In accordance with the data of previous calculations [7.278,279], BULLETT showed that the removal of Pb atoms from the structure of $PbMo_6S_8$ does not introduce any substantial change in the spectrum of the energy levels of ternary molybdenum chalcogenides. On the other hand, a comparison of the curves of the density of states for binary and ternary Chevrel phases shows that the addition of the A component does not reduce the variation of the electron concentration and the shift of the Fermi level (in the rigid-band approximation). BULLETT [7.280] believed that the appreciable change in the value (correlating with the variation of T_c) of the density of states on the Fermi level at transition from Mo_6S_8 to $PbMo_6S_8$ or from Mo_6Se_8 to $PbMo_6Se_8$, arises from the distortion of the Mo_6X_8 blocks and from their rotation about the rhombohedral axis in ternary compounds, which leads to a modification of the intercluster interactions.

An experimental investigation of the X-ray photoelectron spectra (XPS) in the compounds $Cu_xMo_6S_8$, $PbMo_6S_8$ and $SnMo_6S_8$ has been carried out in [7.160, 177] (Fig.7.31). Identification of the spectral bands was performed with due regard for the data on the cross section of the photoionization of electrons of the component atoms and for the comparison of X-ray emission spectra of different series. An example of such an analysis is given in Fig.7.32 which presents CuLα(2p-3d4s transition), SKβ_1(1s-3p transition), $SL_{II,III}$(2p-3s transition) and MoLβ_2(2p-4d5s transition) emission bands together with the XPS valent spectrum of the compound $Cu_{2.8}Mo_6S_8$. It is obvious that near the bottom of the valence band there is a concentration of S3s-states which weakly intermix with states of other types. The next band has a complicated (hybridized) character. It may, nevertheless, be assumed that S3p-states prevail at the bottom of this band, Cu3d-states at the centre and Mo4d-states at the top, in the immediate vicinity of the Fermi level.

On the whole, the experimental results confirm the sequence (determined in theoretical calculations [7.278-281]) of the energy levels (bands) in

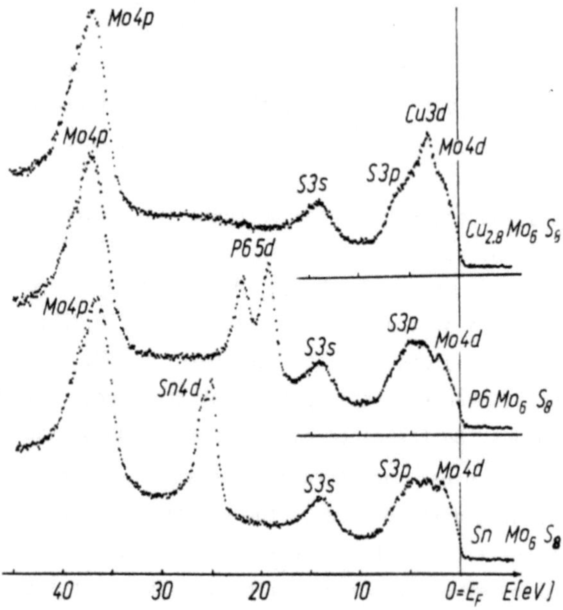

Fig.7.31. The X-ray photoelectron spectra of Chevrel phases [7.177]

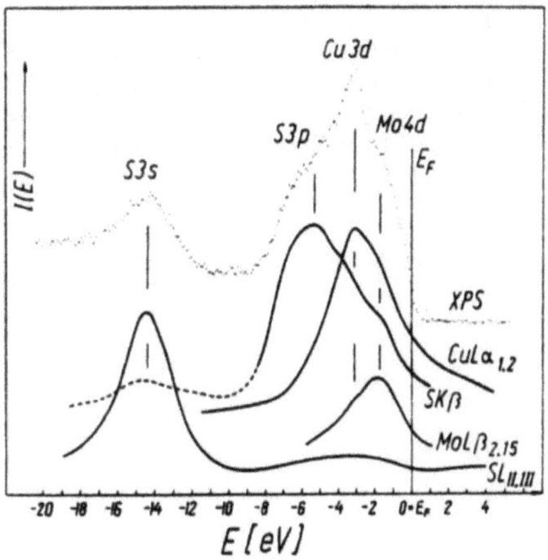

Fig.7.32. Comparison of the X-ray emission and photoelectron spectra in $Cu_{2.8}Mo_6S_8$ [7.177]

Chevrel phases. Experiments show that when the third (A) component of the compound is an sp-metal its valence electrons weakly participate in chemical bonds and Pb5d, Sn4d levels lie rather deep with respect to the Fermi level. When a 3d-element (Cu) is introduced into a compound the d-states substantially modify the spectrum and, although the contribution of Mo4d-states still dominates at the Fermi level, the participation of 3d-states in the formation of an external valence band arouses no doubt. Another important result of X-ray electron investigations of Chevrel phases is the absence of the narrow Pb6s or Sn5s-band, predicted between the S3p and S3s-bands in the experimental spectra of $SnMo_6S_8$ or $PbMo_6S_8$ [7.278,279]. This indicates that either the Pb6s and Sn5s-states are considerably delocalized or that the valence of the sp-element in ternary Chevrel phases is (4-) and not (2-). This problem requires further exploration.

Let us now turn our attention to the characteristics of the phonon spectrum. The experimental heat capacity measurement results for $PbMo_{5.1}S_6$ and $SnMo_5S_6$ (Fig.7.33) were treated on the basis of a simple-phonon density of states model [7.286]. This model considers Chevrel phases as molecular crystals consisting of A atoms binding the Mo_6X_8 quasimolecular blocks. Then the 45 modes of the unit cell of ternary molybdenum chalcogenides, which comprises 15 atoms per formula unit, may be grouped into 9 soft "external" modes and 36 rigid "internal" modes. The "external" ones contain (three acoustic and three optical) modes associated with the translational motion of the X atom and of the quasi-rigid Mo_6X_8 blocks, and also incorporate three torsional modes of the Mo_6X_8 blocks. The corresponding expression for the phonon density of states may be written in the form

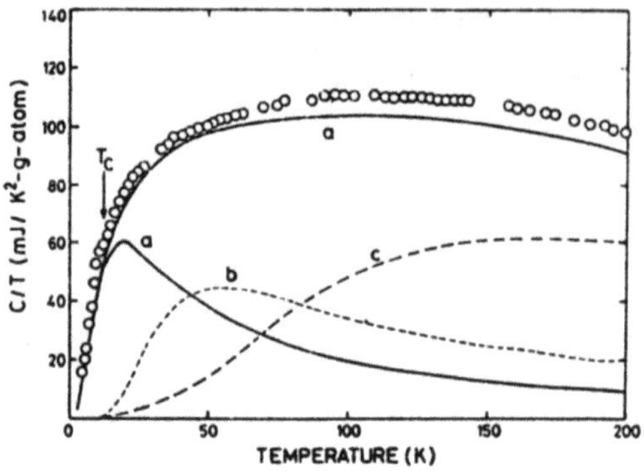

Fig.7.33. The measured (circles) and calculated (curve d) heat capacities in the compound $PbMo_6S_8$ [7.286]. [Curves a,b, and c correspond to the contributions by three terms in (7.5.1)]

$$F(\omega) = 3F_a(\omega) + 6F_b(\omega) + (3n-9)F_c(\omega) \quad (7.5.1)$$

where n is the number of atoms in a unit cell, F_a is the contribution of acoustic modes, F_b is the contribution of optical and torsional modes, and F_c is the contribution of rigid "internal" modes.

Within the framework of the model described above the heat capacity in $PbMo_6S_8$ was calculated. The calculation is in reasonable agreement with the results of experimental measurements [7.286] over the entire temperature range (Fig.7.33).

Subsequent measurements of the inelastic scattering of neutrons in binary and ternary molybdenum chalcogenides [7.287-289] and tunnel measurements [7.290] made it possible to obtain the experimental distribution of the phonon density of states and to test the validity of the model proposed. Analysis of neutron scattering data (Fig.7.34) disclosed that the phonon spectra of Chevrel phases may be divided into two parts, consisting of external low-frequency and internal high-frequency modes with the interface occurring at 18 meV and 14 meV for sulphides and selenides, respectively. A general feature of the structure of the external modes is the presence of a peak at 15 meV (for sulphides) and 12 meV (for selenides). BADER et al. [7.286] ascribed this peak to torsional oscillations of the Mo_6X_8 blocks near their axis of inertia. The narrowest low-energy peak (at 4.3 meV for $PbMo_6S_8$, $SnMo_6S_8$ and $PbMo_6Se_8$ or at 6 meV for $Cu_2Mo_6S_8$) is present only in the spectrum of ternary molybdenum chalcogenides (Fig.7.34). Its position correlates with the mass of the A atom, and for this reason it is attributed to the optical mode associated with the oscillations of weakly bound A atoms in the open channels between adjacent Mo_6X_8 blocks. The wider second peak at 10 meV is ascribed to acoustic oscillations of the unit cell. The structure of the internal modes is determined by three wide maxima, the most intensive of these peaks in the spectra of sulphides is due to the oscillations of light sulphur atoms.

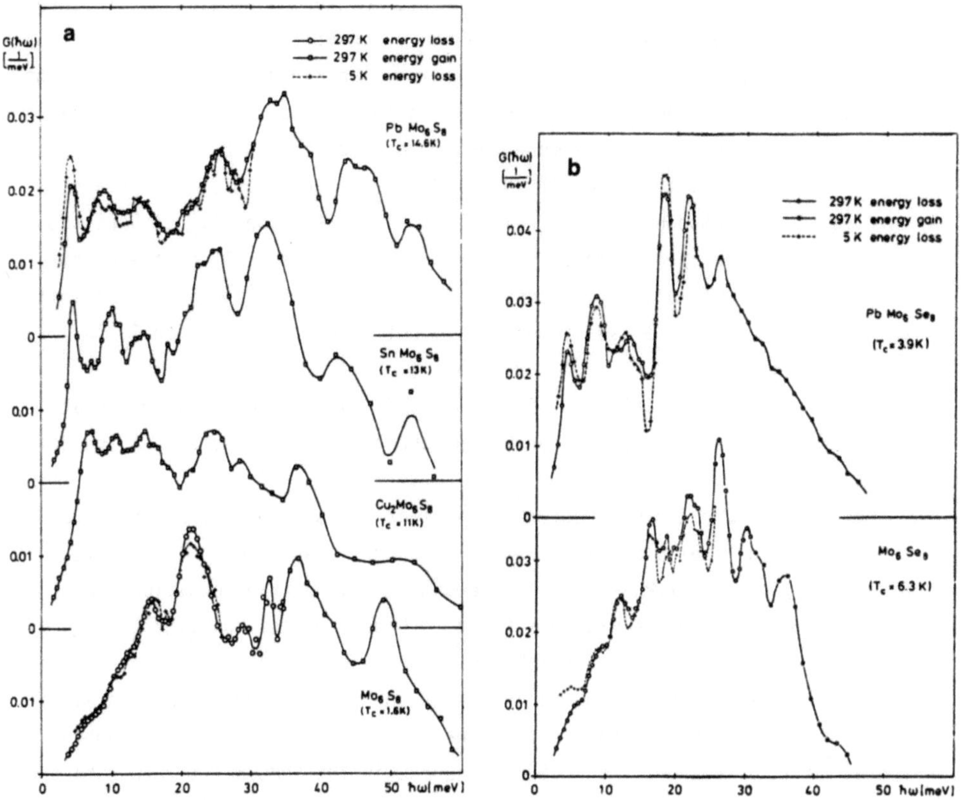

Fig.7.34. The phonon densities of states, $F(\omega)$, in Chevrel phases (a) in sulphides, (b) in selenides [7.289]

Tunnel measurements made for the compound $Cu_{1.8}Mo_6S_8$ [7.290] permitted investigation of a comparatively small section of the phonon spectrum. All the elements of the structure of the curve d^2I/dV^2 (except for the elements of the structure at 2 and 3.5 meV) have found their reflection in the phonon spectrum of the compound $Cu_2Mo_6S_8$ constructed from neutron diffraction measurements.

Thus, it turned out to be possible to reasonably interpret the experimentally measured phonon spectra of compounds such as Chevrel phases within the framework of a molecular model. While the general applicability of the above model to the lattice dynamics of ternary molybdenum chalcogenides does not pose special problems (this has recently been confirmed by more rigorous calculations [7.291]), the role of a particular type of phonon mode played in the formation of superconductivity in these compounds still remains open to discussion.

Temperature measurements of the inelastic scattering of neutrons in ternary Chevrel phases revealed an increase in the intensity of the low-frequency peak at 4 meV when passing from 297 to 5 K. This is ascribed to the appearance of a soft optical mode. The importance of taking into account the acoustic translational and external optical modes has been demonstrated by measurements of the isotopic effect in Mo_6Se_8 [7.292]. However, comparison of the phonon spectra of the compounds $PbMo_6S_8$ ($T_c = 14.6$ K) and $PbMo_6Se_8$ ($T_c = 3.9$ K) shows that although the optical mode is softened in both cases, T_c of these compounds differ substantially. According to [7.289], the contribution of soft phonon modes to $\langle\omega^2\rangle$ per se is not sufficient to explain the high T_c in Chevrel phases.

What is more essential, as shown by [7.286], is the torsional modes associated with the displacement of Mo_6 octahedra in the structure of the compounds concerned.

Low-temperature X-ray structure investigations showed that a structural transformation takes place in a number of Chevrel phases with small-sized A cations (Cu, Co) [7.293-297]. At the same time, no transformation for temperatures down to 6 K was registered in binary phases Mo_3S_4, Mo_3Se_4, Mo_3Te_4 as well as in ternary sulphides with large A cations (Sn, Pb, Ag) [7.254,293,294]. It was reported in one of the first papers [7.298] dealing with investigations into this effect that a transformation from a rhombohedral to an orthorhombic structure occurred in the $CuMo_4S_5$ ($T_c = 10.5$ K) and $CoMo_6S_8$ compounds (T_c was not measured) at temperatures of 260 and 220 K, respectively. A further, more elaborate study of the structural transformation in $Cu_xMo_6S_8$ has revealed that the structure of the low-temperature phase is very complicated [7.295,296]. But this structure has not yet been identified unambiguously. It is anticipated that in the case of a structural transformation the A cations in the tetrahedral 6f sites become ordered and that this ordering causes a triclinic distortion of the rhombohedral lattice, as was observed for the compound $Ni_{0.66}Mo_6Se_8$ [7.299].

The high critical fields at comparatively high T_c, revealed in ternary molybdenum chalcogenides, make them most promising materials for possible technological applications. Figure 7.35 compares the values of H_{c_2} in $PbMo_6S_8$ and technical superconductors Nb-Ti and Nb_3Sn, and also in Nb_3Ge [7.244]. This comparison clearly demonstrates the advantages of $PbMo_6S_8$ over other materials in the range of helium temperatures.

At present there are two obstacles impeding the extensive use of Chevrel phases in various superconducting devices. The rather low critical currents which normally do not exceed 2 to $5 \cdot 10^2$ A/cm^2 primarily belong here [7.300,

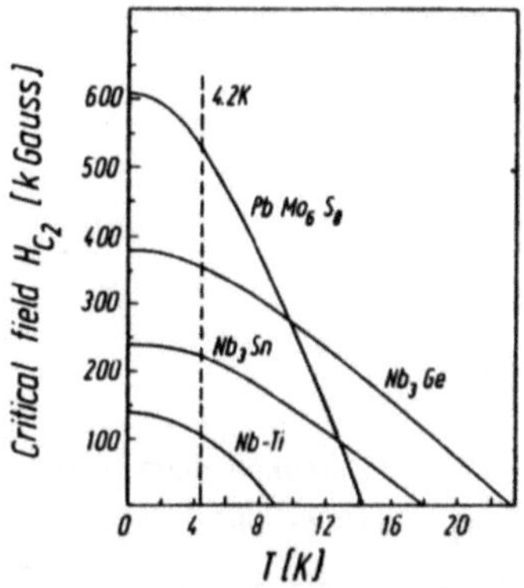

Fig.7.35. Comparison of the critical fields in $PbMo_6S_8$, Nb_3Ge, Nb_3Sn and NbTi [7.244]

301]. The superconducting properties of Chevrel phases are also sensitive to hydrostatic pressure. As a consequence, superconductivity is expected to deteriorate in wires under load.

ALEKSEEVSKII et al. [7.302] have recently found that the magnitude of critical current in ternary molybdenum chalcogenides depends largely on the specimen preparation conditions. By making use of an optimum procedure they managed to get values of critical currents of about $2 \cdot 10^4$ A/cm^2 in $SnGa_{0.25}Mo_6S_8$ in an 80 kGauss field at 4.2 K, which is close to the values obtained for V_3Ga and Nb_3Sn [7.303] (Fig.7.36).

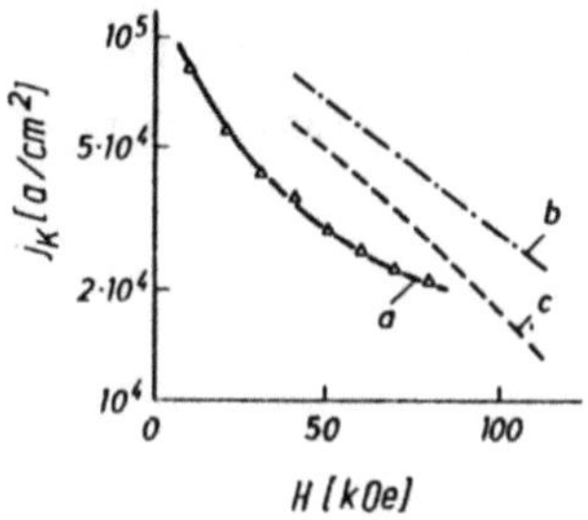

Fig.7.36. Comparison of the critical current densities in (a) $SnGa_{0.25}Mo_6S_8$, (b) V_3Ga and (c) Nb_3Sn [7.302]

7.5.3 Oxides

There are some oxides that fall into the group of chalcogenides. Oxygen compounds of transition metals normally exhibit no high T_c. Until 1973, the highest T_c reported for this class of materials was 6 to 7 K for $Rb_{0.2}WO_3$ [7.304]. In 1973, JOHNSTON et al. [7.305,306] synthesized the compound $Li_xTi_{3-x}O_4$ $(0.8 \le x \le 1.33)$ which had $T_c = 7 \div 13.7$ K, and thus was a first oxygen compound in the class of high-temperature superconductors.

The compound $Li_xT_{3-x}O_4$ $(0.8 \le x \le 1.33)$ crystallizes with a structure such as the $MgAl_2O_4$ spinel (with space group $H1_1$-Fd3m). The spinel structure of

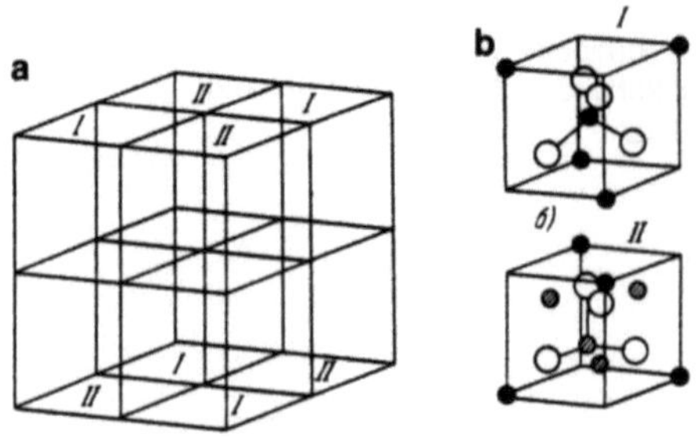

Fig.7.37. The spinel structure

the compound AB_2O_4 is very complicated, and it is normally broken down into eight simple structural elements, octants (Fig.7.37a), half of which (octants of type I, Fig.7.37b) contain A cations in a tetrahedral coordination. As for the other octants (type II, Fig.7.37b), they are in an octahedral coordination although the octahedra formed by oxygen atoms go beyond one octant.

It is worth noting that up till now an fcc structure of the spinel type has been considered to be unfavourable for superconductivity. Only three of the 200 known spinel compounds turned out to possess superconducting properties: $CuRh_2S_4$, CuV_2S_4 and $CuRh_2Se_4$ with $T_c = 4.8$, 4.5 and 3.5, respectively [7.307].

One of the peculiarities of the spinel phase in the Li-Ti-O system is that its region of homogeneity contains both dielectric compositions $Li_{\sim 1.3}T_{1.7}O_4$ and compositions with metallic conductivity $Li_{1.0}Ti_2O_4$. Thus, the compound $Li_{1.0}Ti_2O_4$, for which T_c is the highest registered (13.7 K), is in the vicinity of compositions at which the metal-insulator transition takes place.

8. High Temperature Superconductors and Lattice Instability of Compounds

In considering the superconducting properties of intermetallides and chemical compounds of transition metals in the preceding chapters we have focussed our attention on a number of empirical correlations between T_c and various physical properties. In the present chapter we investigate which of these correlations are universal for all types of compounds. We compare the superconductivity in different crystal structures, give a crystallochemical analysis of them, and consider the influence of alloying, of deviations from stoichiometry and of the variation of electron concentration on the formation of the superconducting properties.

Special attention is given to the various types of lattice instability in transition-metal compounds: structural transition, softening of lattice moduli with decreasing temperature, coexistence of various structural phases near a given composition, etc. It turns out that these anomalies of lattice properties are accompanied by high values of T_c. Numerous examples permit formulation of the concept advanced, for example, in [8.1-7]: a high superconducting temperature is accompanied by an instable lattice.

8.1 Comparison of Superconductivity in Various Crystal Structures

8.1.1 Symmetry

For a long time it was widely believed that cubic symmetry of the crystal lattice is the most favourable one for high-temperature superconductivity. However, recent studies show that this conclusion was premature. Indeed, if we analyze the superconducting properties of transition-metal compounds which have been discussed so far, we shall inevitably come to the conclusion that high T_c (>10 K) may be detected not only in a cubic system (A-15, B-1), but also in rhombohedral, tetragonal and hexagonal systems. It follows from the data presented in Table 8.1 that symmetry plays no decisive role in the

Table 8.1. Comparison of T_c in different structures [8.8]

Type of structure	Symmetry	Maximum T_c[K]	T_c of components of compounds [K]
A-15	cubic	Nb_3Ge:23.2	Nb:9.2
C-15	cubic	$LaOs_2$:8.9	La:6.0; Os:0.66
C-14	hexagonal	$ZrRe_2$:6.8	Zr:0.61; Re:1.7
B-2(CsCl)	cubic	VRu:5.0	V:5.4; Ru:0.49
A-12(α-Mn)	cubic	$NbTc_3$:10.5	Nb:9.2; Tc:7.8
σ-phase (β-U)	tetragonal	$Mo_{0.38}Re_{0.62}$:14.6	Mo:0.92; Re:1.7
B-1(NaCl)	cubic	NbN:17.3	Nb:9.2
		Pd-Ag-D:16	Pd<0.01; Ag<0.01
Pu_2C_3	cubic	$(Y_{0.7}Th_{0.3})_2C_{3.1}$:17.0	Th:1.38; Y<0.03
Ti_3S_4	hexagonal	$Li_{0.1-0.3}Ti_{1.1}S_2$:10-13	Li<0.08; Ti:0.40
Mo_3Se_4	rhombohedral	$Pb_{0.92}Mo_6S_{7.5}$:15.2	Pb:7.2; Mo:0.92
Spinel	cubic	$Li_xTi_{3-x}O_4$:7-13	Li<0.08; Ti:0.40
	orthorhombic	(x = 0.8-1.3)	
C-49	orthorhombic	$ZrGe_2$:8.0	Zr:0.61

formation of superconducting properties of transition-metal compounds.

The table also contains the T_c-values of components of the compounds involved. Analysis of these data allows us to conclude that the components of a compound may have low T_c or even be no superconductors, whereas the compound displays good superconducting properties. Of interest in this respect are the examples Pd-Ag-D [8.9], Pd-Cu-H [8.10] or Li-Ti-S [8.11], exhibiting high-temperature superconductivity with components the T_c of which does not exceed 0.5 K.

8.1.2 Crystallochemistry

The crystallochemical analysis of the compounds under consideration includes determining the type of chemical bound, coordination numbers, the way in which the atoms are packed in the crystal, etc. The following conclusions may be drawn:

The presence of wide homogeneous regions (an indication of metallic bond) and metallic conductivity in compounds with A-15 structure go hand in hand with considerable brittleness inherent in covalent crystals. An indication of the covalent bond is the low coordination number in the atoms of the metal with A-15 structure.

The conclusion as to the high degree of covalency in A-15 compounds has also been drawn from hardness measurements [8.12]. It is shown in [8.13] that the hardness (H) is proportional to the shear modulus C_{44} for several classes of solids with a cubic structure. The ratio H/C_{44} has a rather high value ($\sim$0.1) for covalent crystals with a diamond or sphalerite lattice and is very small ($\sim$0.006) for fcc metals. In V_3Si, V_3Ge and Nb_3Sn compounds, the value of H/C_{44} is equal to 0.11-0.14, i.e., close in magnitude to the values of this parameter in the covalent crystals Si (0.13) and Ge (0.12). X-ray structure measurements of the distribution of electron density in V_3Si and Cr_3Si [8.14-16] confirmed the conclusion about the strong covalent bond between metallic atoms in those compounds.

In B-1 structure compounds (NaCl) there is one more indication, in addition to the above features (the presence of a region of homogeneity, metallic conductivity, brittleness). Charge transfer from metallic atoms to nonmetallic atoms has been established in experiments on X-ray electron spectroscopy and in electron-energy spectrum calculations. This indicates the presence of an ionic component of the bonding forces. Thus, in NaCl structure compounds three types of bond exist: metallic, covalent and ionic.

Compounds of the Laves phase type possess metallic conductivity, but other "metallic" characteristics thereof are not so clearly marked (insignificant regions of homogeneity or complete absence of them). Brittleness and high melting temperatures characterize the presence of a covalent component of bonding forces. There is evidence of a similarity in the geometric arrangement of atoms in Laves phases and in the diamond structure of which rigid covalent bonds are typical [8.17]. This enables one to group the above compounds among those which preferentially have a covalent-metallic character of bond.

Chalcogenides and sesquicarbides have been studied less than the above compounds. Nevertheless, it may be assumed that the presence of a region of homogeneity in chalcogenides indicates the existence of a metallic bond, and the charge transfer from a metallic atom to a nonmetallic atom (as suggested by some preliminary results of X-ray spectral investigations) points to an ionic contribution to the bonding forces.

Both components of σ phases, and of α-Mn and B-2 compounds are metals with a small difference in electronegativity, therefore ionic-bond effects are hardly to be expected here. Wide regions of homogeneity, high plasticity and metallic conductivity give good reason to believe that the type of bond occurring in these compounds is preferentially metallic.

The above discussion of the character of interatomic interaction in transition-metal compounds does not go beyond the scope of qualitative conclusions. A quantitative relationship between different types of bond may be obtained from a quantum-mechanical treatment. Particularly convenient for investigating the problem of the chemical bond is the method of linear combination of atomic orbitals (LCAO) which in the case of crystals is referred to as the tight-binding method (for a discussion of the LCAO method see [8.18]). By determining the population of the atomic orbitals one can find the effective charges of the atomic components in a compound and the relationship between the purely covalent and purely ionic states of the system [8.19]. For transition-metal compounds with NaCl structure (TiC, TiN, VC, VN, TiO, VO), such a treatment has recently been performed in [8.20,21]. No similar quantitative estimates have been made for other compounds.

Most of the superconducting transition-metal compounds have structures with a fairly tight atomic packing (A-15, C-15, etc.). In a number of cases this enables one to determine the effective radii of the atomic components and to obtain information on the crystallochemical stability of the compounds. Consider this with reference to the example of A_3B, i.e., a compound with A-15 structure, for which such an analysis was recently performed [8.22,23].

In the A-15 structure the distance between the neighbouring A atoms forming linear chains in $R_{AA} = a/2$, a being the lattice parameter. If one assumes that the A atoms touch each other, one can determine the effective radius of the A atom, $R_A^{eff} = R_{AA}/2 = a/4$. Comparison of the effective radii of A compononents in A_3B compounds (thus calculated) indicates that R_A^{eff} is close to covalent radius and this, in its turn, reveals the presence of additional covalent bonds in the chain.

Proceeding from similar assumptions that atoms A and B touch each other and knowing that the distance between them amounts to $R_{AB} = \sqrt{5}a/4$, one finds the effective radius of the B atom

$$R_B^{eff} = R_{AB} - R_A^{eff} = \frac{\sqrt{5}}{4}a - \frac{1}{4}a = \frac{\sqrt{5}-1}{4}a \quad .$$

The condition for the crystallochemical stability of the A-15 lattice, when the B atom simultaneously contacts all the twelve A atoms surrounding it, is

$$R_B^{eff}/R_A^{eff} \geq \sqrt{5} - 1 = 1.23 \quad .$$

It follows from the data given in [8.22] that none of the known superconducting compounds with an A-15 structure satisfies this condition. In all cases the relationship $R_B^{eff}/R_A^{eff} < 1.23$, which attests that none of the compounds specified above is crystallochemically stable. This follows from the fact that A atoms are somewhat compressed, whereas B atoms are extended in the A-15 structure. The degree of compression of the A sublattice, or the degree of extension of the B sublattice, may serve as a certain measure of crystallochemical stability of the A-15 structure. The degree of extension of the B sublattice may be expressed by the parameter $\varepsilon = R_B^{eff}/R_B^c$, ($R_B^c$ being the covalent radius of a B element). If one now introduces the values of T_c into the function $\varepsilon = R_B^{eff}/R_B^c$ one finds that T_c tends to grow as the crystallochemical instability of the A-15 structure increases (Fig.8.1). The estimates made in [8.22] show that high T_c should be expected to occur in Nb_3Si (which was confirmed in [8.24,25] where $T_c = 19$ K was found for Nb_3Si) and also in V_3C, Nb_3C and Ta_3C compounds which have not been synthesized yet.

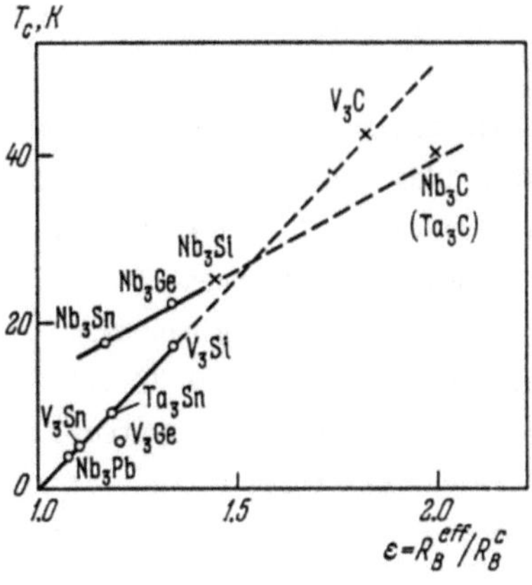

Fig.8.1. The influence of the degree of crystallochemical nonequilibrium of an A 15 structure on T_c

8.1.3 Stoichiometry

A general property of superconducting compounds of transition metals is the high sensitivity of T_c to deviations from stoichiometry. As a rule, the highest values of T_c are reached in stoichiometric compounds with A-15 structure, in carbides and nitrides with B-1 structure and in palladium hydride. It may be noted that the T_c values (23.2 K and 17.3 K, respectively) obtained for Nb_3Ge and NbN are closest to stoichiometry and are the highest for the entire class of A-15 and B-1 (NaCl) structure compounds.

Analysis of the phase equilibrium diagrams for A_3B compounds with A-15 structure and for AB compounds with B-1 (NaCl) structure shows [8.26] that the regions of homogeneity, as a rule, are restricted by stoichiometric composition on the one hand, but the regions extended due to an increase in the

A component content on the other hand. For A-15 compounds this means that at deviations from stoichiometry the A element replaces the B element in almost all cases, i.e., the integrity of the chain remains unbroken. This suggests that the A-A interaction is decisive in the A-15 structure.

8.1.4 Alloying

If one considers the influence of alloying on the superconducting properties of transition-metal compounds one may arrive at the conclusion that it is different for various families of compounds.

In compounds of A-15 and B-1 structures, alloying yields practically no inrease of T_c. A much greater effect is reached here as stoichiometry is approached. This has recently been demonstrated brilliantly on the example of an Nb_3Ge compound where a gradual approach to stoichiometry enabled T_c to be increased from 6 K to 17 K and finally to 23.2 K, i.e., to the highest T_c known at present.

On the other hand, in sesquicarbides, in chalcogenides, in compounds such as Laves phases, in hydrides and palladium deuterides alloying has produced a positive effect. For example, if in palladium deuteride $T_c = 11$ K [8.27], then in the ternary Pd-Ag-D system it has been possible to increase T_c to 16 K [8.9]. In binary Ti-S and Li-S systems no superconductivity has been detected at all, and in the ternary Ti-Li-S system T_c reaches 17 K [8.11].

No clear-cut criteria have been established so far as to which systems exhibit an improvement in the superconducting characteristics of transition-metal compounds owing to alloying. The large body of information accumulated in recent years concerning the alloying of superconductors is still empirical in character.

8.1.5 Electron Concentration

Much time has passed since MATTHIAS paid attention to the relationship of T_c on the electron concentration (the number of valence electrons available for one atom). For transition metals of the first, second and third periods, the values of T_c group around two maxima corresponding to the electron concentrations of 5 and 7 e/a [8.28] (Fig.4.2). When the values of T_c for alloys of transition metals were plotted as functions of the electron concentration it turned out that T_c for alloys gave a good fit to the functional dependence calculated for pure metals. The positions of the main maxima of alloys formed by transition metals of the periods I, II and III, viz., 4.7 and 6.5 e/a, differ inappreciably from those of the corresponding maxima of pure metals (5 and 7 e/a) [8.29].

It seemd at first that the electron-concentration dependence of T_c constructed for transition metals and their alloys would be universal and would also apply to other superconductors based on transition metals. Indeed, analysis of the superconducting properties of A_3B compounds with A-15 structure has shown that the highest values of T_c group around two electron-concentration intervals, 4.50 to 4.75 and 6.25 to 6.60 e/a, which actually coincide with the regions of "favourable" concentrations found for transition metals and their alloys.

In other superconducting compounds whose components are transition metals: σ phases, Laves phases, compounds with B-2 (CsCl) and α-Mn structures, the correlation which has been described above also occurs, i.e., high values of T_c account for the electron-concentration regions of 4.6 to 4.7 and 6.5 to 6.75 e/a. However, to high values of T_c in B-1 compounds (NaCl) correspond electron concentrations of 3.6 e/a (for sesquicarbides) and 4.5 and 5.0 e/a (for carbides and nitrides with B-1 structure).

Thus, a new value of 3.6 e/a arises, which does not agree with MATTHIAS' rule, although in most of the compounds as well as in alloys, stable values approximately equal to 4.5 and 6.5 e/a remain surprisingly favourable.

8.2 Types of Lattice Instability

The experimental data given in the previous section show that the superconducting compounds of transition metals exhibit lattice instability to some degree. The manifestation of this instability may be quite varied, e.g., as direct structural transformation (such as weak distortion of the initial lattice), softening of lattice moduli, softening of short-wave branches of the phonon spectrum, etc. Below we consider the types of lattice instability occurring in the compounds under examination.

8.2.1 Structural Transition

Low-temperature structural transformation was observed in compounds with A-15 structure, in Laves phases, compounds of B-2 (CsCl) structure, and in

Table 8.2. Superconducting and structural characteristics of compounds undergoing lattice transformation

Compound	Initial structure (cubic)	Structure after transformation	T_m [K]	T_c [K]
V_3Si	A-15	tetragonal	21-23	17
Nb_3Sn	A-15	tetragonal	43-45	18
HfV_2	C-15	body-centred, orthorhombic	120	8.7
ZrV_2	C-15	rhombohedral	90-117	8.8
$LaRu_2$	C-15	tetragonal	30	4.45
VRu	B-2	tetragonal	110-360	5
La_3S_4	Th_3P_4	tetragonal	90	8.2
La_3Se_4	Th_3P_4	tetragonal	60	7.6

chalcogenides. Data on the basic structure of these compounds and on structures upon transformation are summarized in Table 8.2 together with values of T_c and T_m [8.8,30]. In six of the eight compounds the character of transformation is the same: from a cubic structure to a tetragonal one. It is interesting that in three compounds of the Laves-phase type with an initial C-15 structure, which show transformations, the low-temperature lattice distortion is different in all the three cases: tetragonal, rhombohedral and orthorhombic.

It is worth noting that in all cases temperatures of structural transitions exceed those of superconducting transitions. However, while the difference in the values of T_m and T_c in compounds with A-15 (V_3Si, Nb_3Sn) and C-15 ($LaRu_2$) structures amounts from 6 to 27 K only, this quantity in other compounds is much larger and reaches hundreds of degrees. It is not yet clear whether a relationship exists between the proximity of the temperature regions of superconducting and structural transitions and the quantity T_c. The experimental findings available are inadequate for analysis.

8.2.2 "Virtual" Structural Transition

Ternary alloys of the Y_2C_3-Th_2C_3 system undergo no structural transformation with varying temperatures. However, by varying the conditions of synthesis it is possible to produce these alloys in a bcc structure (such as Pu_2C_3) or in a body-centred tetragonal structure. By convention, this situation is called "virtual" structural transition. It is of interest that T_c of a compound with cubic structure (~17 K) is the highest among the entire class of sesquicarbides, whereas a tetragonal-lattice compound exhibits no superconductivity at temperatures down to 4 K.

8.2.3 Softening of Long-Wave Lattice Moduli with Decreasing Temperature

As already discussed in Chapter 6, investigations into the temperature dependence of the velocity of sound have shown that this parameter diminishes anomalously with decreasing temperature in compounds with A-15 (V_3Si, Nb_3Sn, Nb_3Al, V_3Ga) and C-15 (ZrV_2, HfV_2) structures, and that there is a peculiarity in compounds of B-2 structure (VRu). These data indicate softening of the lattice moduli in the above compounds, the temperature region in which maximum softening occurs coincides with T_m.

In some compounds (V_3Si, Nb_3Sn) this softening begins at room temperature, i.e., well above the structural transformation, and is the "forerunner of the phase transition". It is worth noting that there may well be no phase transition as such (for example, in V_3Si and Nb_3Sn it disappears at low deviations from stoichiometry or at small impurity concentrations).

8.2.4 Softening of Short-Wave Phonons

The softening of lattice moduli observed in A-15, C-15 and B-2 compounds points to a developing instability of acoustic long-wave phonons. As we have seen in compounds with B-1 structure, the dispersion curves of acoustic phonons have anomalies in the short-wave region. These anomalies are insensitive to temperature, but their position and depth are determined by the interatomic and interelectron interaction parameters. By varying these we can, in principle, obtain instable phonon modes with a finite wave vector. In this sense the occurring anomalies also indicate a developing lattice instability (with respect to the interaction parameters). It is significant that such anomalies have been detected only in high-temperature superconductors of the B-1 structure: TaC and NbC.

8.2.5 The Coexistence of Different Structural Phases in the Vicinity of a Given Composition

When synthesizing a number of high-temperature superconductors with A-15 and B-1 structures, it was found that in the vicinity of stoichiometric compositions of Nb_3Si, MoC and NbN different structural phases coexist. In other words, by slightly varying the composition, temperature and other conditions of synthesis one can obtain different structural modifications of one and the same compound. For example, the compound Nb_3Si may have three modifications: with a Cu_3Au structure ($T_c = 1.5$ K) [8.31], with a Ti_3P-type structure ($T_c = 5.3 \div 9$ K) [8.4,32-34] and with A-15 structure ($T_c = 6.9$, 9.2, $18 \div 19$ K) [8.24,25,33,35,36].

The compound NbN may also be synthesized with a cubic B-1 type (NaCl) structure ($T_c = 17.3$ K) [8.37,38], with a tetragonally distorted NaCl-type structure ($T_c = 8.9$ K) [8.39] and with a hexagonal TiP-type structure ($T_c <$ 1.94 K) [8.40]. Molybdenum carbide MoC has two structural modifications: a cubic one (such as B-1) with $T_c = 12.1$ to 14.3 K [8.41,42] and a hexagonal one with $T_c = 7.1$ K [8.42].

The σ phase in the Mo-Re system at a temperature of about 1250°C is near the boundary with a phase having an α-Mn type crystal structure [8.43]. Synthesis of this compound (of the σ phase) by sputtering on a substrate heated up to the same temperature resulted in the production of a σ phase, whose unit-cell dimensions differ from those in a bulk material and lead to an increase of T_c by a factor of two [8.43]. A similar effect was reported in [8.38] where sputtering of an NbN compound onto a substrate heated to 700°C made it possible to increase the lattice parameter and to raise T_c from 13.5 to 17.3 K.

Using the temperature of the substrate as an adjustable parameter, TESTARDI et al. [8.4] recently undertook a search for new superconducting phases in compounds and alloys based on transition metals. As a result, they found a large number of new metastable phases in which the increase of T_c was achieved at those precipitation temperatures which corresponded to temperatures of phase transformations in these materials. Thus, it is quite probable that the proximity of a given composition to phase-equilibrium boundaries may characterize one type of lattice instability which leads (in one of the phases) to high T_c. In order to substantiate this one would have to show that some phonon modes are unstable in one of the phases of the system. The difficulty in reaching a stoichiometric composition can be considered as an indication of that. For example, in the synthesis of A-15 and B-1 compounds one has not been able to obtain perfect stoichiometry. This was usually the case with compounds having high T_c(Nb_3Ge, Nb_3Ga, MoC, NbN). The phase-transition boundary at low temperatures (<200 K) in these compounds does not encompass the stoichiometric composition, but is shifted toward a higher transition-metal content. Therefore, to faciliate a maximum possible approach to stoichiometry and to obtain higher T_c, special conditions of synthesis involving the use of rapid quenching from high temperatures at which the phase equilibrium boundary shifts toward stoichiometric composition, are required.

8.3 Phase Transition in Structures from the Point of View of Symmetry Theory

8.3.1 Characteristics of the Phase Transitions

As we have seen, one of the manifestations of the lattice instability occurring in high-temperature superconductors is a structural transition. In this section we investigate such transitions in A-15, C-15 and B-2 structures per se, without reference to the phenomenon of superconductivity and solely on the basis of symmetry theory. This will permit a reevaluation of the phonon-spectrum modes responsible for the transition itself and allow us to map out ways for further experimental study of such transitions. The analysis performed below rests on the papers by IZYUMOV et al. [8.44,45].

Prior to transformation, all the three structures are cubic and when below the transformation temperature T_m the A-15 and B-2 compounds become tetragonal, and the C-15 compound (according to different data) is rhombic [8.46] or rhombohedral [8.47]. The distortion parameter is as low as 10^{-3}, but the study of these transitions is very important for ascertaining the physical nature of the lattice instability and of the high-temperature superconductivity in the compounds being considered.

A number of features suggest a second-order transition, however, in the point T_m, apparently producing a finite spontaneous deformation. This indicates a first-order transition which likely is close to a transition of the second kind. To predict the symmetry group of a new phase in A-15 compounds, BIRMAN and PEREL [8.48,49] used Landau's second-order phase transition theory. As noted above, the transitions observed are more like first-order phase transitions. Thus, it is necessary to reconsider the existing results regarding A-15 compounds, and also to make predictions for C-15 and B-2 compounds. One should not use though, a power series expansion of the free energy in terms of the order parameter which in this case describes the spontaneous atomic displacement. Those displacements are finite at the transition point T_m and the free energy is not an analytic function of these parameters. Anyway, it is useful to retain, for this general case, Landau's constructive idea of second-order phase transition theory about the relationship of an undistorted structure to the irreducible representation of the group of the initial symmetric phase. What is typical of all these structures is that the transition occurs without a change in the number of atoms in the unit cell, therefore the instable modes of the cubic phase in point Γ of the Brillouin zone must be responsible for this transition. Thus, first and foremost, we must investigate the phonon spectrum of our crystals at point Γ.

The A-15, C-15 and B-2 structures of interest belong to the same crystal class O_h and space groups O_h^3, O_h^7 and O_h^1, respectively. Since the representations of the space groups in the point Γ reduce to point group representations, the phonon modes of all the three lattices should be classified according to the irreducible representations of the point group O_h. This point group is known to have ten irreducible representations: four one-dimensional representations Γ_1, Γ_1', Γ_2, Γ_2'; two two-dimensional representations Γ_{12}, Γ_{12}' and four three-dimensional representations Γ_{25}, Γ_{25}', Γ_{15}, Γ_{15}'. The matrices of these representations may be found in [8.50]. Although the matrices of the two-dimensional representations Γ_{12} and Γ_{12}' are given in complex form, these representations are reducible to a real form (the identification in the general case is given by Herring's criterion). For our purposes, this reduction is faciliated through the use of the unitary matrix.

$$U = \frac{1}{2}\begin{pmatrix} 1+i & -1+i \\ 1-i & -1-i \end{pmatrix} \quad . \tag{8.3.1}$$

From the matrices of the irreducible representations of the group O_h thus obtained we construct a mechanical representation which describes the phonon modes of the lattices at the point Γ.

As far as A-15 compounds are concerned, a number of symmetry properties have been noted in Chapt.6. Since it is reasonable to analyze the structures of class O_h in parallel, we repeat some material pertaining to A-15 compounds. Below, the various data concerning A-15, C-15 and B-2 structures will be given concurrently. For example, the coordinates of the atoms in a cubic primitive cell:

$$\begin{aligned}
\text{A-15:}\ & B_1(0\,0\,0),\ B_2\left(\tfrac{1}{2}\tfrac{1}{2}\tfrac{1}{2}\right),\ \left(A_3\ \tfrac{3}{4}\tfrac{1}{2}0\right),\ A_4\left(\tfrac{1}{4}\tfrac{1}{2}0\right),\\
& A_5\left(0\tfrac{3}{4}\tfrac{1}{2}\right),\ A_6\left(0\tfrac{1}{4}\tfrac{1}{2}\right),\ A_7\left(\tfrac{1}{2}0\tfrac{3}{4}\right),\ A_8\left(\tfrac{1}{2}0\tfrac{1}{4}\right) \quad ;\\
\text{C-15:}\ & B_1(0\,0\,0),\ B_2\left(\tfrac{1}{4}\tfrac{1}{4}\tfrac{1}{4}\right),\\
& A_3\left(\tfrac{5}{8}\tfrac{5}{8}\tfrac{5}{8}\right),\ A_4\left(\tfrac{5}{8}\tfrac{7}{8}\tfrac{7}{8}\right),\ A_5\left(\tfrac{7}{8}\tfrac{5}{8}\tfrac{7}{8}\right),\ A_6\left(\tfrac{7}{8}\tfrac{7}{8}\tfrac{5}{8}\right) \quad ;\\
\text{B-2:}\ & B_1(0\,0\,0),\ A_2\left(\tfrac{1}{2}\tfrac{1}{2}\tfrac{1}{2}\right)
\end{aligned} \tag{8.3.2}$$

(the chemical formulae of these compounds are A_3B, A_2B and AB, respectively).

8.3.2 Construction of the Mechanical Representation and Analysis of Phonon Modes at the Point Γ

The mechanical representation D_m of the wave vector group $G_{\underline{p}}$ of the crystal reflects not only its spatial symmetry but also the proper structure, i.e., the equivalent positions of atoms. This is reason why it is convenient in investigating the phonon spectrum of the crystal. For our purposes we should find the phonon modes at point Γ, i.e., basis functions of the mechanical representation transforming according to the irreducible representations of the wave vector group at point Γ. The method for finding the mechanical representation is described in a number of papers (see, for example, [8.51]). We shall present a very convenient version of this technique [8.52] by considering the general case of an arbitrary wave vector $\underline{p}$ in the Brillouin zone.

The mechanical representation of the space group of the crystal may be constructed in 3sN dimensional space (s being the number of atoms in a unit cell and N the number of cells) of the eigenfunctions of the translation operator $\psi_{\underline{p}}^{i\alpha}$

$$\psi_{\underline{p}}^{i\alpha} \quad \sum_{\underline{a}}^{\oplus} \varepsilon^{i\alpha}\, e^{i\underline{p}\underline{a}} \quad , \tag{8.3.3}$$

where $\sum^{\oplus}$ stands for the direct sum over all translation vectors and $\varepsilon^{i\alpha}$ is a vector in 3s-dimensional space describing a single deviation of the ith atom in α^{th} direction in the zeroth cell of the crystal (where the origin of coordinates is taken to be).

We find the wave vector group representation $G_{\underline{p}}$ with the basis functions $\psi_{\underline{p}}^{i\alpha}$. For each element the $g \in G_{\underline{p}}$ operation of symmetry $\hat{T}(g)$ converts $\psi_{\underline{p}}^{j\beta}$ into a linear combination of other vectors $\psi_{\underline{p}}^{i\alpha}$ with the same wave vector $\underline{p}$:

$$\hat{T}(g)\psi_{\underline{p}}^{j\beta} = \exp(-i\underline{p}\underline{a}_p) \sum_{a} R_{\alpha\beta}\, \psi_{\underline{p}}^{i\alpha} \quad , \tag{8.3.4}$$

where $R_{\alpha\beta}$ is the matrix of rotation in a three-dimensional space, which corresponds to the element of the space group $G_{\underline{p}}$

$$g = \{R|\underline{\tau}_R + \underline{a}_p\} \tag{8.3.5}$$

($\underline{\tau}_R$ being a fractional translation and $\underline{a}_p$ a vector of translation to the unit cell into which atom number j passes from the zero cell by the action of element g). In the new cell it already has the number i so that the element g performs the atomic transition which is symbolically represented as

$$(j0) \rightarrow (i\underline{a}_p) \quad . \tag{8.3.6}$$

Equation (8.3.4) may be written in the standard form

$$\hat{T}(g)\psi_{\underline{p}}^{j\beta} = \sum_{i\alpha} \{D_p(g)\}_{i\alpha,j\beta}\, \psi_{\underline{p}}^{i\alpha} \quad , \tag{8.3.7}$$

where the $3s \times 3s$ matrices

$$\{D_p(g)\}_{i\alpha,j\beta} = \exp(-i\underline{p}\underline{a}_p)\; R_{\alpha\beta}\; \delta_{g(j0),(i\underline{a}_p)} \tag{8.3.8}$$

perform the mechanical representation of the $G_{\underline{p}}$ group. The Kronecker symbol in (8.3.8) instruments condition (8.3.6). This representation is generally reducible and may be expanded into irreducible representations of the G_p wave vector group. In turns they are expressable in terms of projective irreducible representations of the point group $G^o_{\underline{p}}$. The multiplicity of the νth irreducible representation in the mechanical representation is determined according to the formula

$$n_\nu = \frac{1}{n(G^o_{\underline{p}})} \sum_{R \in G^o_{\underline{p}}} \exp(i\underline{p}\underline{\tau}_R)\, \chi_{\underline{p}}(\{R|\underline{\tau}_R\})\, \chi^{\nu *}_{\underline{p}}(R) \quad . \tag{8.3.9}$$

Here $n(G^o_{\underline{p}})$ is the order of the group $G^o_{\underline{p}}$, the summation is taken over the elements of this group, $\chi^\nu_{\underline{p}}(R)$ is the character of the νth irreducible representation of group $G^o_{\underline{p}}$, and $\chi_{\underline{p}}$ is the character of the mechanical representation which, by definition, may be calculated from

$$\chi_{\underline{p}}(g) = \mathrm{Sp}\, R \sum_j \exp(-i\underline{p}\underline{a}_p)\, \delta_{g(j0),(j\underline{a}_p)} \quad . \tag{8.3.10}$$

Thus, to perform calculations according to (8.3.9,10) it is necessary to draw up a table of transitions (8.3.6) for a given structure and to take the characters $\chi^\nu_{\underline{p}}(R)$ from the literature (for example, [8.50]).

In the following we need the linear combinations

$$\varphi^{(\nu)}_{\underline{p}\xi} = \sum_{i\alpha} C^{(\nu)\xi}_{i\alpha}\, \psi^{i\alpha}_{\underline{p}} \tag{8.3.11}$$

of basis functions in the mechanical representation transforming according to the νth irreducible representation of the group $G_{\underline{p}}$. The coefficients $C^{(\nu)\xi}_{i\alpha}$ should be matrix elements of the relevant projection operator

$$C^{(\nu)\xi}_{i\alpha} \equiv P^{(\nu)\xi\delta}_{i\alpha,j\beta} \quad , \tag{8.3.12}$$

where ξ specifies the basis function in the νth representation. The additional indices δ and $j\beta$ are fixed. The passage from this set δ, j, β to another one generates other sets of basis functions of this irreducible representation.

We express the projection-operator matrix in terms of the matrix of the νth irreducible projective representation of the group $G^o_{\underline{p}}$ and write the mechanical representation matrix (8.3.8) as

$$P^{(\nu)\xi\delta}_{i\alpha,j\beta} = \sum_{R\in G^o_{\underline{p}}} \{D^\nu_p(R)\}^*_{\xi\delta} \exp(i\underline{p}\underline{\tau}_R)\{D_p(R|\underline{\tau}_R)\}_{i\alpha,j\beta} \quad . \tag{8.3.13}$$

In final form the 3sN dimensional vector $\varphi^{(\nu)}_{\underline{p}\xi}$ may be represented as a double direct sum of three-dimensional vectors $\underline{e}\left(\begin{smallmatrix}\nu\\ i\end{smallmatrix}\middle|\xi\right)$, i.e.,

$$\varphi^{(\nu)}_{\underline{p}\xi} = \sum_{\underline{a}}{}^{\oplus}\left[\sum_{i=1}^{s}{}^{\oplus}\underline{e}\left(\begin{matrix}\nu\\ i\end{matrix}\middle|\xi\right)\right]e^{i\underline{p}\underline{a}} \quad , \tag{8.3.14}$$

with

$$\underline{e}\left(\begin{matrix}\nu\\ i\end{matrix}\middle|\xi\right) = \sum_{R\in G^o_{\underline{p}}} \{D^\nu_p(R)\}^*_{\xi\delta} \exp(i\underline{p}\underline{\tau}_R) \exp(-i\underline{p}\underline{a}_p)\delta_{g(j0),(i\underline{a}_p)}\underline{R}_\beta \quad , \tag{8.3.15}$$

where $\underline{R}_\beta$ is a vector consisting of the components $R_{\alpha\beta}$, $R_{y\beta}$ and $R_{z\beta}$ of the rotational matrix. The vectors $\underline{e}\left(\begin{smallmatrix}\nu\\ i\end{smallmatrix}\middle|\xi\right)$ have close bearing on the known phonon polarization vectors. In particular, if a given representation enters the expansion of the mechanical representation only once, the vector $\underline{e}$ coincides with the polarization vector. However, if it is involved several times, the block of the dynamic matrix needs further diagonalization which is no longer connected with the space symmetry of the system.

Utilization of (8.3.9,10) for $\underline{p}=0$ leads to the following expansion of the mechanical representation at the point Γ in our structures:

$$\text{A-15:}\quad D_m = (\Gamma_2 + \Gamma_{12} + \Gamma'_{25} + \Gamma_{25} + \Gamma'_{15} + 2\Gamma_{15})_A + (\Gamma_{25} + \Gamma_{15})_B \quad ,$$

$$\text{C-15:}\quad D_m = (\Gamma'_2 + \Gamma'_{12} + \Gamma_{25} + 2\Gamma_{15})_A + (\Gamma'_{25} + \Gamma_{15})_B \quad ,$$

$$\text{B-2 :}\quad D_m = (\Gamma_{15})_A + (\Gamma_{15})_B \quad . \tag{8.3.16}$$

Here it is evident what sort of atoms participate in the displacements, described by a given irreducible representation. The sum of the dimensions of the irreducible representations is, of course, equal to that of the mechanical representation, i.e., 3s. The result obtained for the A-15 structure coincides with that given in [8.49].

Table 8.3. Phonon modes at the point Γ for A-15, C-15 and B-2 structures [8.44,45]

A-15

	1	2	3	4	5	6	7	8
Γ_2			100	$\bar{1}00$	010	$0\bar{1}0$	001	$00\bar{1}$
Γ_{12}			$u00$ $\bar{1}00$	$\bar{u}00$ 100	$0\bar{1}0$ $0u0$	010 $0\bar{u}0$	$00\bar{v}$ $00\bar{v}$	00 00
Γ'_{25}			$00\bar{1}$ 010	001 $0\bar{1}0$	001 $\bar{1}00$	$00\bar{1}$ 100	$0\bar{1}0$ 100	010 $\bar{1}00$
Γ_{25}	100 010 001	$\bar{1}00$ $0\bar{1}0$ $00\bar{1}$						
			$0\bar{1}0$ 001	$0\bar{1}0$ 001	100 $00\bar{1}$	100 $00\bar{1}$	$\bar{1}00$ 010	$\bar{1}00$ 010
Γ'_{15}			001 010	$00\bar{1}$ $0\bar{1}0$	001 100	$00\bar{1}$ $\bar{1}00$	010 100	$0\bar{1}0$ $\bar{1}00$
Γ_{15}	100 010 001	100 010 001	100 010 001	100 010 001	100 010 001	100 010 001	100 010 001	100 010 001
$\Gamma_{15}^{(1)}$	$\bar{3}00$ $0\bar{3}0$ $00\bar{3}$	$\bar{3}00$ $0\bar{3}0$ $00\bar{3}$	100 010 001	100 010 001	100 010 001	100 010 001	100 010 001	100 010 001
$\Gamma_{15}^{(2)}$			$\bar{2}00$ 010 001	$\bar{2}00$ 010 001	100 $0\bar{2}0$ 001	100 $0\bar{2}0$ 001	100 010 $00\bar{2}$	100 010 $00\bar{2}$

C-15

	1	2	3	4	5	6
Γ'_2			111	$1\bar{1}\bar{1}$	$\bar{1}1\bar{1}$	$\bar{1}\bar{1}1$
Γ'_{12}			$u\bar{1}\bar{v}$ $\bar{1}u\bar{v}$	$u1v$ $\bar{1}\bar{u}v$	$\bar{u}\bar{1}v$ $1uv$	$\bar{u}1\bar{v}$ $1\bar{u}\bar{v}$
Γ'_{25}	100 010 001	$\bar{1}00$ $0\bar{1}0$ $00\bar{1}$				
Γ_{25}			$01\bar{1}$ $\bar{1}01$ $1\bar{1}0$	$0\bar{1}1$ 101 $\bar{1}\bar{1}0$	$0\bar{1}\bar{1}$ $10\bar{1}$ 110	011 $\bar{1}0\bar{1}$ $\bar{1}10$
Γ_{15}	100 010 001	100 010 001	100 010 001	100 010 001	100 010 001	100 010 001
$\Gamma_{15}^{(1)}$	$\bar{2}00$ $0\bar{2}0$ $00\bar{2}$	$\bar{2}00$ $0\bar{2}0$ $00\bar{2}$	100 010 001	100 010 001	100 010 001	100 010 001
$\Gamma_{15}^{(2)}$			011 101 110	$0\bar{1}\bar{1}$ $\bar{1}01$ $\bar{1}10$	$0\bar{1}1$ $\bar{1}0\bar{1}$ $1\bar{1}0$	$01\bar{1}$ $10\bar{1}$ $\bar{1}\bar{1}0$

B-2

	1	2
Γ_{15}	100 010 001	100 010 001
	100 010 001	$\bar{1}00$ $0\bar{1}0$ $00\bar{1}$

$u = 2 - \sqrt{3}$

$v = 1 - \sqrt{3}$

The results of mode calculation according to (8.3.16) are given in Table 8.3 which specifies the composition of the mechanical representation for each of the A-15, C-15 and B-2 structures. Regarding the representation Γ_{15} for A-15 and C-15 structures which appears three times in D_m, it must be noted that by appropriate unitary transformation it is possible to single out a set of three basis functions corresponding to the acoustic branch. Then, the remaining sets of basis functions are orthogonal with respect to the acoustic set Γ_{15} and to each other, and are labeled in the table by the symbols $\Gamma_{15}^{(1)}$ and $\Gamma_{15}^{(2)}$.

8.3.3 Seeking the Group of the New Phase

We introduce the density function ρ of the crystal in the new phase. This density function was introduced by LANDAU to analyze second-order phase transitions, but we shall use it in the more general case of first-order phase transitions which are close to phase transitions of the second kind. We designate the density function in the initial phase by ρ_0. Since ρ differs from ρ_0 in some set of atomic displacements one can write

$$\rho = \rho_0 + \delta\rho \quad , \tag{8.3.17}$$

where the distortion of the structure $\delta\rho$ may be represented as a superposition of the basis function φ_ξ of some ν^{th} irreducible representation (ξ specifies the function) characterizing the instable phonon mode responsible for the phase transition. Thus,

$$\delta\rho = \sum_\xi C_\xi \varphi_\xi^\nu \quad . \tag{8.3.18}$$

The quantity $\delta\rho$ should be invariant under the symmetry group G_1 of the new phase. In our search for this phase we shall be guided by the following principles.

i) G_1 should be a subgroup G_0 of the initial phase.
ii) The restriction of the νth representation of a group on the subgroup G_1 should contain an identity representation (BIRMAN criterion [8.48]). This requirement results from the fact that $\delta\rho$ should be invariant under group G_1, and that an invariant transforms according to the identity representation.
iii) The composition of the mechanical representation of the crystal should be taken into account. In fact, the last two principles apply to some fixed irreducible representation of the group G_0. If one does not know in advance according to which representation the transition goes, it is necessary to consider all the representations involved in the decomposition of the mechanical representation one by one.

The above requirements, which are purely of a group theory nature, drastically limit the list of possible groups of the new phase.

As already noted above, the transition in compounds such as A-15, C-15 and B-2 occurs without any change in the primitive (unit) cell, i.e., with the conservation of the translational periodicity. All variations of symmetry are contained in rotational elements; accordingly, of all the subgroups G_1 of the space group G_0 one should only consider the subgroups in which the cell is conserved. In [8.53] one can find such subgroups. The number of them is the same

and equals 32 for each of the three groups O_h^3, O_h^7 and O_h^1. The reason for this is that the point group O_h has 32 subgroups with crystallographically non-equivalent orientations. Each of them will correspond to its own space group.
iv) The fourth principle may, by convention, be called the principle of the maximum supergroup[1] of the group of the new phase G_1. The essence of this principle is as follows. The superposition coefficients C_ξ in (8.3.18) for $\delta\rho$ can give different sets. Their numerical values are not important for the prediction of the new phase, what matters is only the type of mode mixing. For two-dimensional representations, for example, the following sets are possible: {C,0}; {C,C}; {C,-C}; $\{C_1,C_2\}$. Taking all the kinds of mode mixing separately, one can obtain all the types of real atomic displacements on the one hand, and only the groups G_1 admissible according to BIRMAN, on the other. For example, take such a group G for a phase transition according to some irreducible representation and see to which type of mode mixing it corresponds. The quantity $\delta\rho$ should be invariant under the group G_1. On the other hand, we know how the functions φ_ξ^ν should transform by the action of the group G_1 (to this end it suffices to take the limitation of the representation group G_0 on the subgroup G_1). This leads to specialization of the mixing coefficients.

If for a fixed irreducible representation and for a fixed type of mode mixing one obtains several groups which are in subgroup relations, it is necessary that the maximum group should be chosen as the group describing the transition.

8.3.4 Discussion of Results

A list of the possible symmetry groups of the new phase, produced following the method described above, is given in [8.44]. For a discussion of the results we should now reduce the list of groups using experimental data on the crystallographic system of the new phase. The final results will assume the form

$$\begin{aligned}&\text{A-15: } \underline{D_{4h}^9},\ D_{2d}^5,\ C_{4v}^7,\ C_{4h}^2,\\ &\text{C-15: } \underline{D_{2h}^{28}},\ C_{2v}^{22},\ D_2^7 \text{ or } \underline{D_{3d}^5},\ D_3^7,\ C_{3v}^5,\\ &\text{B-2 : } C_{4v}^1 \quad .\end{aligned} \tag{8.3.19}$$

[1] The principle discussed here has actually nothing to do with ASCHER's [8.54] maximum principle, encountered earlier in the literature, whose inconsistency was shown by SHUVALOV [8.55].

The results for the A-15 system can be compared to those obtained in [8.48,49]. In [8.48] only the groups D_{2d}^{5} and C_{4v}^{7} were determined, in [8.49] one more group is added to them, indicating that it is omitted in [8.48]. The results for the C-15 and B-2 systems were first obtained in [8.44]. Group D_{4h}^{9} for the A-15 system was deliberately excluded in [8.48,49] since the representation Γ_{12} corresponding to it is inactive (following the terminology of LANDAU's [8.56] second-order phase transition theory). To this representation corresponds a third-order invariant in a free-energy expansion in powers of the order parameter. In the summary of results (8.3.19) we have underlined the groups arising only at transitions according to passive representations which are excluded in second-order phase transition theory. Here these groups are retained in the list so that both second-order and first-order (close to second order) phase transitions can be united.

It is group D_{4h}^{9} which is detected in experiments with A-15 systems in the low-temperature phase. It may hence be inferred that some additional thermodynamic variable is not disclosed, either in the second-order transition or in the free-energy expansion in powers of the order parameter. As already stated in Sect.6.5, the latter consideration was developed by ANDERSON and BLOUNT [8.57]. They showed that the phase transition in this structure should also be characterized by the parameter of displacement of atomic sublattices relative to each other, which accompany the tetragonal deformation of the lattice. This has been verified by neutron diffraction experiments on Nb_3Sn [8.58].

As for the C-15 structure, experiments with HfV_2 and ZrV_2 indicate that the new phase has an orthorhombic crystal structure [8.46,59]. By using X-ray structure analysis it has been possible to determine even the space group C_{2v}^{20} [8.46]. At the same time, according to neutron diffraction data [8.47], this phase is rhombohedral. Therefore in (8.3.19) we have retained the groups of the two crystal systems involved. It is remarkable that the group C_{2v}^{20} is absent in (8.3.19). If we retrospectively trace its fate it appears to fall out since this group is not maximum supergroup at transition according to the representation Γ_{25}' [its maximum supergroup D_{2h}^{28} is included in (8.3.19)]. Since the results (8.3.19) barely apply to group theory, in order to finally determine the symmetry of the low-temperature phase in C-15 compounds one has to perform additional diffraction experiments. So far the experimental data concerning the structure of the low-temperature phase in VRu (type B-2) have only indicated the tetragonal crystal system [8.60], whereas its space group remains to be determined.

Thus, when searching for the group of the new phase in the structures under investigation, we can well do without the use of the free-energy expansion in powers of the order parameter in employing only group-theoretical considerations. The predictions of the groups of the new phase thus hold true for first-order transitions as well. Instead of the free-energy minimization procedure usually employed in Landau's theory, we have taken all the possible versions of mode mixing one by one. We thereby found all solutions to the minimization problem irrespective of which order the invariants would be taken in the expansion. In some way, the method used in [8.44,45] to find the new-phase group is the opposite of that applied in Landau's theory. Rather than minimize the energy, which yields the mode mixing coefficients, and then determine the group corresponding to this mixing, we set out from a list of all the subgroups of the initial group and select appropriate mixing coefficients for each of them. This abandons some model assumptions connected with allowance for first-order invariants only and, moreover, from the assumption that the free energy may be expanded in the vicinity of the phase-transition point.

It may also prove to be important for further studies of the new phases and their properties to reveal a possible domain structure associated with several ways of inserting the subgroup G_1 into the group G_0. These domains may be different in some experimentally observed properties. In particular, it is worth noting that among the admissible groups of the new phase there are also polar groups such as $C_{n\upsilon}$ so that the domains may be distinguished, just like segnetoelectrics, by some vecotr property. In the C-15 system the $C_{2\upsilon}^{20}$ group determined in experiments is polar, too.

8.4 Correlation Between High T_c and Lattice Instability

It follows from the preceding material that each class of superconducting compounds, based on transition metals, has high-temperature superconductors exhibiting a certain lattice instability. In order to ascertain whether the correlation between high T_c and lattice instability is universal one ought to study the lattice properties of all the compounds with high T_c within a given family. We shall now analyze how far such investigations have been carried out.

Lattice instability in the A-15 compounds V_3Si and Nb_3Sn was first detected as a structural phase transition. This transition is preceded by a softening of the lattice moduli which is the forerunner of the phase transition over a wide temperature range. Somewhat later a similar transition was found to occur in V_3Ga. What can be said with regard to other high-temperature superconductors of this class? Initial searches for structural transformation in Nb_3Al ($T_c = 18.8$ K) were not successful [8.61], and we should recall that in V_3Si and Nb_3Sn, too, this transition occurs not in all specimens but only in chemically pure samples of stoichiometric composition. KODESS [8.62] made attempts to find a transformation in Nb_3Al and other A-15 compounds. He performed an X-ray structure study at low temperatures and found a structural transformation in Nb_3Al at 80 K, and also in the alloys $Nb_3(Al_{0.75}Ge_{0.25})$ ($T_m = 105$ K) and $Nb_{3.1}(Al_{0.7}Ge_{0.3})$ ($T_m = 130$ K). Measurements of the temperature dependence of magnetic susceptibility on the same specimens [8.63] have shown anomalies in just the temperature range where the structural transition occurred (as in V_3Si). Somewhat earlier, ALEKSEEVSKII [8.64] observed a similar anomaly for $Nb_3(Al_{0.75}Ge_{0.25})$. There is other evidence that in Nb_3Al a structural transformation occurs. Thus, for example, when measuring the low-temperature heat capacity in Nb_3Al and V_3Ga, VISWANATHAN [8.65] registered two individual superconducting transitions, one of which is ascribed to the tetragonal phase. In addition, it should be pointed out that upon measurement of the elastic properties no structural transition has been revealed in Nb_3Ge. As for other A-15 compounds with high T_c such as Nb_3Ga, Nb_3Si, Nb_3Au, Nb_3Pt, Ta_3Au and V_3Al, this transition has not yet been searched for. Undoubtedly, further investigations are required; however, there is no doubt that structural transformation occurs not only in V_3Si and Nb_3Sn but also in other high-temperature superconductors with A-15 structure.

Investigations of superconductors with NaCl structure have shown that longitudinal acoustic modes soften in the compounds NbC and TaC ($T_c = 10$ to 12 K) and that there is no softening in HfC and ZrC ($T_c < 1$ K). Unfortunately, such measurements were not made for other compounds of this class with comparatively high T_c (ZrN, NbN, MoC). It is difficult to say anything definite about the universality of the correlation observed. There are, however, other experimental facts certifying that the anomalies of the lattice properties, revealed in NbC and TaC, are not fortuitous. An anomalously large variation of the thermal expansion coefficient over a temperature range of 60 to 70 K has been registered for NbC and TaC [8.66] resulting in an abrupt increase of the atomic displacement amplitude. If we compare these data with similar investigations for V_3Si [8.67], we can assure ourselves that an anomaly in the temperature dependence of the thermal coefficient was also observed for the V_3Si specimen undergoing structural transformation.

Regarding other classes of compounds we have already noted that structural transformation was found to occur in ternary molybdenum sulphides $CuMo_4S_5$

($T_c \approx 11K$) and $CoMo_3S_4$ (T_c was not measured) [8.68-70]. However, an attempt has been made in the work [8.68] to find a structural transformation in other compounds of this class, $SnMo_5S_6$ ($T_c = 11.3$ K), $AgMo_4S_5$ ($T_c = 8.9$ K) and $NiMo_3S_4$ (T_c was not measured), but it has not been detected. It is of great interest to gain deeper insight into this problem.

For compounds such as σ phase and α-Mn sesquicarbides and hydrides, there are only some indirect indications of lattice instability. An extensive study of acoustic properties and X-ray structure investigations are required here.

In summing up this discussion it may be noted that in spite of the scantiness of the available experimental data the material considered here lends support to the existence of a correlation between lattice instability and superconductivity in transition-metal compounds. It would be still premature to speak of a universal correlation since the experimental data on lattice properties of many of these compounds are far from being complete.

As far as the theoretical aspect of this correlation is concerned, the situation is much worse. Per se, lattice instability may promote high T_c, for example, through the phonon-spectrum softening mechanism, but the structural transformation temperature T_m is, as a rule, much higher than T_c. Thus, superconductivity starts in the low-temperature phase in which the lattice becomes more rigid. In cases where the lattice instability results from anomalies of short-wave length phonons (NbC, TaC) and is not associated with temperature, the corresponding softening of the high frequency region of the spectrum cannot give a sufficiently large increase of T_c according to the existing theories (McMILLAN's formula). Thus, the theoretical interpretation of the experimentally obtained correlation between high T_c and lattice instability is at present an unsolved problem. It is quite probable that in the future its solution may be as follows: by itself, a lattice instability is not responsible for high T_c, but both phenomena, i.e., high T_c and lattice instability, are manifestations of the effect of strong electron-lattice coupling in metallic compounds based on transition metals [8.77]. In such a case the solution of the problem (the correlation under discussion) should be preceded by the development of a theory for the electron-phonon interaction in these metallic compounds.

An approach to the electron-phonon interaction in transition metals and their compounds was advanced in [8.71,72] on the basis of the strong-coupling method, see also the reviews by ALLEN [8.73] and SINHA [8.74]. T_c may increase due to the pairing interaction of electrons via quasi-localized excitations of the lattice system in structurally instable crystals [8.75,76]. Such excitations probably arise in systems where transitions of light atoms between two closely spaced potential wells are likely to take place.

9. Radiation Effects on Superconductors

In connection with the potential applications of superconductors based on transition metals to devices to be employed in radiation fields, much attention has been given in recent years to experimental and theoretical investigation of the radiation effects. Of immediate practical interest are the actual data on radiation resistance in various groups of superconducting materials, which ultimately determines the service life of a superconducting device at different radiation doses. It is particularly important to gain insight into the mechanisms of degradation of superconductivity in irradiated materials. This is a necessary step in the creation of superconducting materials with high radiation resistance.

In this chapter we analyze data on radiation resistance of superconducting alloys and compounds based on transition metals (primarily the A-15 type compounds), and relate those data to the formation of different kinds of radiation defects.

9.1 Radiation Resistance of Superconducting Materials

Early investigations of the superconducting properties of materials exposed to radiation showed that there are at least two groups of substances for which the effect of a radiation field on the superconducting characteristics are different substantially [9.1]. The first group of superconductors incorporates pure transition metals [9.1-3], binary alloys Nb-Ti, Nb-Zr, Nb-Ta [9.1,3,4], Laves phases (HfV_2) [9.6,7], niobium carbides, nitrides and carbonitrides [9.8-11]. These materials are characterized by a relatively weak sensitivity of the superconducting transition temperature to the effect of radiation. The second group is represented by A-15 compounds [9.12-14] and Chevrel phases [9.6,15] the irradiation of which by ionizing radiation sources leads to a high degradation of T_c and H_{c_2}.

9.1.1 Compounds with A-15 Structure

Figure 9.1 shows experimental data for the dependence of T_c on the neutron radiation dose in A-15 compounds [9.7]. Except for Mo_3Os and Mo_3Ge [9.16], all the other compounds exhibit a very strong suppression of T_c (ca. 60-80%). The rate of degradation of T_c, as follows from Fig.9.1, is largest for low doses ($\sim 10^{18}$ cm^{-2}) and decreases continuously with increasing doses. It saturates to $T_c \sim 2$ to 4 K at a dose of $\sim 3 \cdot 10^{19}$ cm^{-2} for a neutron energy of $E > 1$ MeV. The compound Mo_3Os is fairly resistant to radiation. With the irradiation dose increasing to $\sim 1.5 \cdot 10^{20}$ cm^{-2}, the decrease in T_c is as low as $\sim 16\%$.

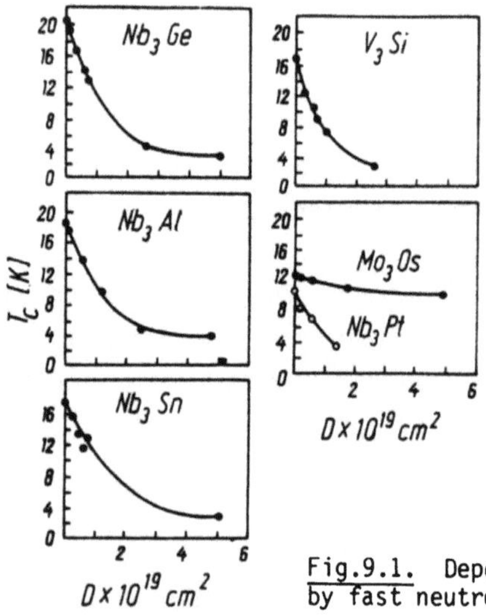

Fig.9.1. Dependence of T_c on the dose of irradiation by fast neutrons ($E > 1$ MeV) for A-15 compounds [9.7]

In Fig.9.1 the dependence of T_c on the irradiation dose is presented in a linear scale. In the literature such correlations are frequently constructed for a logarithmic dose scale and a relative superconducting transition temperature T_c/T_{co} (T_{co} being the transition temperature of a nonirradiated specimen). In this case it is customary to speak of some threshold dose at which T_c has relatively weak sensitivity to the radiation effect for certain compounds. It should be noted in this connection that the size of the degradation of T_c in superconductors also depends on the temperature of the irradiated specimen and on the purity of the basic material. Specimens were irradiated at comparatively high temperatures ($T \sim 340$ K) and the T_c degradation registered was insignificant even at high flux (in Nb_3Sn at an exposure dose of 1.5 to $2.7 \cdot 10^{18}$ cm^{-2}, $\Delta T_c \sim 0.2$ K was found when $E > 0.1$ MeV) [9.1,17]. Decreasing the temperature at which the specimens were irradiated from T = 340 K to 4.6 K, led to an increase of ΔT_c by a factor of four [9.18]. On the other hand, special investigations have shown that the rate of degradation of T_c in materials exposed to radiation depends on the degree of purity of the superconductors [9.19].

Figure 9.2 illustrates the dependence of T_c on the residual resistivity in films of Nb_3Sn and Nb_3Ge irradiated by α particles [9.19]. These data indicate why the degradation of T_c with the exposure dose in Nb_3Ge (with initial $\rho \sim 45$ μΩ cm) is more pronounced at low doses than that in Nb_3Sn (with initial $\rho < 30$ μΩ cm). It follows from the data given in Fig.9.2 that T_c rapidly degrades beyond some level given by a residual resistivity of ∼45-50 μΩ cm. In [9.19] the above effect was explained by the fact that the radiation damage of the specimen Nb_3Sn leads to an increase of the number of defects and thus of the number of scattering centres for conduction electrons. At some critical defect concentration, the degradation of T_c occurs equally rapidly for both materials (Fig.9.2).

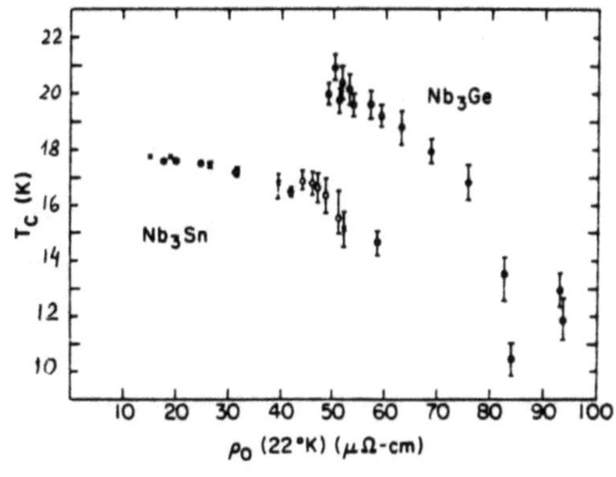

Fig.9.2. Dependence of T_c on the residual resistivity in Nb_3Ge (⊗,⊙) and Nb_3Sn (x, O, ●) irradiated by α particles [9.19]

The influence of deviations from stoichiometric composition in A-15 compounds on the degradation of the superconducting parameters by radiation was studied in [9.7,20]. The investigation was carried out for the compounds Nb_3Al (in the range of compositions from 18.7 to 27 at.% Al) and Nb_3Pt (from 22.1 to 28 at.% Pt). It was noticed for these two systems that the rate of degradation of T_c was higher when the compound was closer to the stoichiometric composition.

As compared to the degradation of T_c, the influence of the radiation field on the second critical field in compounds with A-15 structure has received the least attention. For very low doses of irradiation by fast neutrons (∿5 to $6 \cdot 10^{17}$ cm^{-2}), some increase in $H_{c_2}(0)$ (∿5 to 9%) was observed for Nb_3Al [9.21], Nb_3Sn [9.15] and V_3Ga [9.22]. For higher exposure doses, $H_{c_2}(0)$ is presented in Fig.9.3 [9.8,23,24].

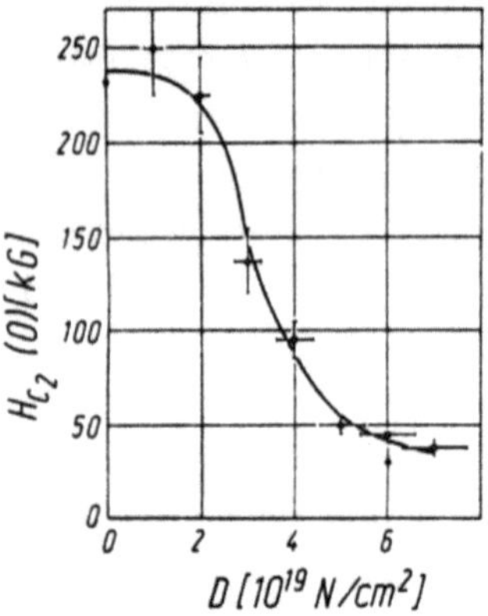

Fig.9.3. Dependence of the second critical field H_{c2} in Nb_3Sn on the dose of irradiation by fast neutrons (○[9.23], △[9.8])

9.1.2 Disordered Alloys of Transition Metals

For the transition-metal alloys, primary attention has been given to the investigation of the radiation resistance in NbTi and NbZr, which are widely used as technical superconductors, as well as in the alloy Nb + 30%Ta. According to the data given in [9.1,4,24-27], the radiation does not decrease T_c appreciably in these materials even when the doses are very high. Thus, for instance, at a dose of $\sim 2 \cdot 10^{19}$ cm^{-2} (E > 0.1 MeV) the decrease of T_c is only 5 to 6% [9.27]. A similar effect was revealed for pure transition metals [9.3,28,29]. However, in a number of investigations an increase of T_c in Nb and Pd foils and films was registered after irradiation by oxygen ions and α particles [9.30,31]. The effect of the radiation on Pd turned out to be particularly strong in cases where no superconductivity was registered up to 2 mK under normal conditions [9.31]. After low-temperature irradiation by α particles, Pd exhibited superconductivity with T_c = 3.2 K [9.31].

9.1.3 Laves Phases

The effect of irradiation on T_c of Laves phases turned out to be identical to that observed in NbTi and NbZr alloys [9.6,7]. With the C-15 compound HfV_2 irradiated by fast neutrons, T_c decreased from 9.4 K (for a nonirradiated specimen) to 8.6 and 8.4 K for doses of $6 \cdot 10^{18}$ cm^{-2} and $1 \cdot 10^{19}$ cm^{-2}, respectively [9.7].

9.1.4 Chevrel Phases

The effect of the radiation on T_c in Chevrel phases is comparable with that in compounds with A-15 structure or even exceeds it somewhat [9.6,7]. According to the data given in [9.7], irradiation of $PbMo_6S_8$ by fast neutrons decreases T_c from 12.8 K to <4.2 K at a dose of $1 \cdot 10^{19}$ cm^{-2}.

9.1.5 Carbides and Nitrides of Transition Metals

Investigations of the superconducting properties of niobium carbide, nitride and carbonitride irradiated by neutrons have shown that the T_c of these materials is comparatively insensitive to radiation [9.8-10]. Irradiation of niobium carbide by fast neutrons at a dose of $5 \cdot 10^{19}$ cm^{-2} at $T \sim 150^{\circ}C$ decreases T_c by 15%, but $H_{c_2}(0)$ increases from 26 kOe to 50 kOe [9.10].

Figure 9.4 presents data on the variation of the relative superconducting transition temperature as a function of the dose of fast neutrons for different groups of superconducting transition-metal alloys and compounds. These data are an illustrative example of their comparative radiation resistance. It is seen that at a dose of $1.6 \cdot 10^{19}$ cm^{-2}, the degradation of T_c is 5% for NbTi and HfV_2, 23% for Nb_3Al, 51% for $SnMo_5S_8$ and 61% for $PbMo_6S_7$ [9.6]. This indicates that today the disordered alloys of transition metals and the Laves phases are the most radiation-resistant superconductors.

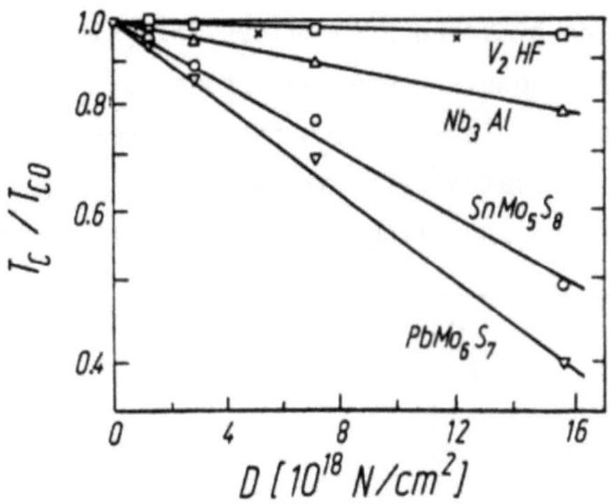

Fig.9.4. Dependence of the relative superconducting transition temperature on the dose of irradiation by fast neutrons in superconducting materials [9.6,27], x-NbTi [9.27]

The superconducting characteristics of irradiated materials may be recovered by subsequent annealing at comparatively moderate temperatures (ca. 500-700°C) [9.20,23,32-34]. Figure 9.5 illustrates results of isothermal annealing (at a constant temperature and with varying annealing time) in Nb_3Sn and Nb_3Ga compounds irradiated by fast neutrons at a dose of $5 \cdot 10^{19}$ cm^{-2}. It is evident that T_c is almost completely recovered in both materials, but compared to

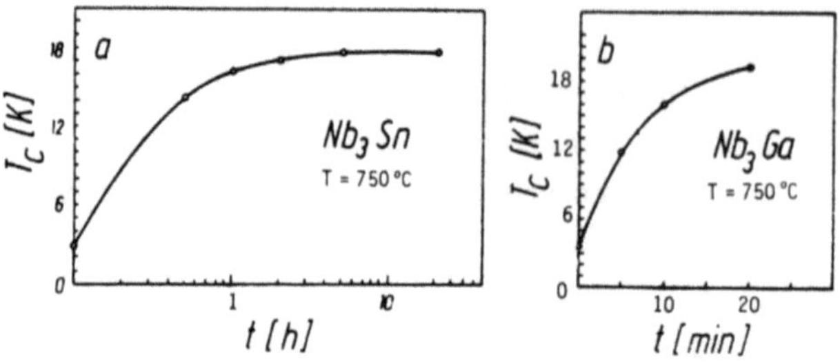

Fig.9.5. Isothermal annealing curves for the superconducting compounds of (a) Nb_3Sn and (b) Nb_3Ga irradiated by fast neutrons to a dose of $5 \cdot 10^{19}$ cm^{-2} [9.7]

Nb_3Ga, much more annealing time is required for the recovery of the initial value of T_c in Nb_3Sn. This effect was explained in [9.7] by the difference in the ratio of the atomic radii of Sn and Ga to those of Nb, which affects the rate of the mutual substitution of the A and B sites in the A-15 lattice in the process of annealing. Figure 9.6 shows the dependence of T_c on the temperature of isochronous annealing (during which the annealing time is kept constant and the temperature varies) in Nb_3Ge [9.33] and Nb_3Sn [9.23] for doses of irradiation of the above materials by fast neutrons. It is seen that in the temperature range from 100 to 400°C, T_c very weakly depends on the annealing time. At temperatures above 400°C, the value of T_c recovers quickly: up to 700°C for Nb_3Ge and 900°C for Nb_3Sn. At these temperatures T_c has recovered to ~90% of its initial value in a nonirradiated material. A similar result was obtained when annealing a series of irradiated alloys $Nb_{3\pm x}Al_{1\pm x}$ with different initial compositions [9.20]. However, when analogous $Nb_{3\pm x}Pt_{1\pm x}$ alloys were subjected to isochronous annealing it was found that the recovery of T_c was as low as 40% of T_c of the nonirradiated material [9.20]. No substantial recovery of T_c at comparatively high annealing temperatures of ~650°C

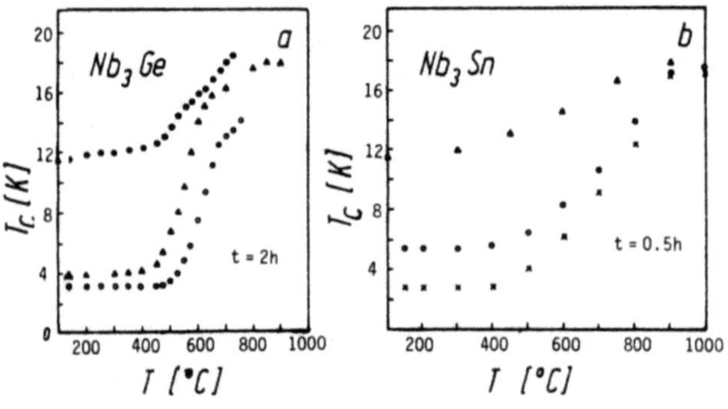

Fig.9.6. Isochronous annealing curves for (a) Nb_3Ge [9.33] and (b) Nb_3Sn [9.23] irradiated by fast neutrons {●-0.9 · 10^{19} cm^{-2}, Δ-2:2.5 · 10^{19} cm^{-2}, o-4÷5 · 10^{19} cm^{-2}, x-7 · 10^{19} cm^{-2}}

was revealed in the compound Mo_3Os, either [9.7]. An interesting result of isochronous annealing of a number of A-15 type compounds is not only the complete recovery of T_c but even a small excess of T_c above the initial value [9.7]. This excess amounts to about 1 K for nonstoichiometric compositions and is about 0.3 K for A-15 compounds close to stoichiometry. The abovementioned effect is explained in [9.7] by the fact that when an irradiated material is annealed, the diffusion processes proceed more quickly. This yields a higher degree of order than in the absence of defects produced by the radiation.

In summary, it may be noted that at present there is an extremely broad spectrum of sources of ionizing radiation (neutrons, protons, α particles, deutrons, electrons and charged particles) whose energy varies over a wide range. It has been shown in a number of papers that the character of the dependence of T_c on the dose does not change appreciably at equal doses of irradiation of a material by neutrons, α particles and other kinds of radiations [9.35]. In a number of cases a certain difference in the level of degradation of T_c has been noted which is somewhat lower when a superconductor is irradiated by α particles and electrons, as compared to neutrons [9.32,36].

9.2 Structural Investigations of Radiation Defects

Since irradiated A-15 compounds display a considerable degradation in their superconducting properties, structural investigations of radiation effects have been carried out mainly for these materials.

9.2.1 Lattice Parameter Measurements

It was found for all the compounds with A-15 structure exposed to radiation that the crystal lattice parameter increases with the exposure dose [9.7, 37-39]. The lattice expansion rate is highest at low doses ($<1\cdot10^{19}$ cm^{-2}) and decreases at higher doses. The lattice parameter continues to increase at exposure doses where T_c no longer varies. The increase in the lattice parameter is quite appreciable and amounts to 0.37% and 0.91% for Nb_3Al and Nb_3Ge irradiated at a dose of $5\cdot10^{19}$ cm^{-2} and is somewhat lower for Nb_3Pt and Mo_3Os (0.3% and 0.21% for a flux of $3\cdot10^{19}$ cm^{-2} and $1\cdot10^{20}$ cm^{-2}, respectively) [9.7]. X-ray and neutron diffraction patterns show that the crystal A-15 structure is conserved in the overwhelming majority of materials when the dose of neutrons increases up to the maximum attainable dose [9.7,20].

An exception is Nb_3Ge in which a considerable widening of high-angle reflections and an appreciable weakening of the intensity of Bragg peaks at doses of $5 \cdot 10^{19}$ cm^{-2} were registered, which indicates that the irradiated material becomes amorphous [9.7]. A similar effect was found in [9.40] when irradiating an Nb_3Ge film by α particles.

9.2.2 Determintation of the Degree of Long-Range Order

Investigations of the diffraction of X-rays and neutrons showed that the long-range order parameter (S) in A_3B compounds with A-15 structure decreases after irradiation and that the B atoms from the bcc position are substituted for part of the A atoms in the chain (the formation of antisite defects). This effect was registered in Nb_3Al [9.7,37], Nb_3Sn [9.23,38,41], V_3Si [9.27,42], Nb_3Ge [9.7,32], Mo_3Os [9.7] and Nb_3Pt [9.20]. Figure 9.7 presents T_c vs. S curves for Nb_3Al, Nb_3Ge and Nb_3Pt irradiated by fast neutrons. The rate of suppression of T_c with disorder in the case of Nb_3Ge with the Nb sites replaced by Ge atoms is 3.2 K/%. For Nb_3Al, with the Nb sites replaced by Al atoms it is 2.4 K/%. A much lower rate of degradation of T_c with disorder is observed for Mo_3Os (∿0.3 K/% for the Mo sites replaced by Os atoms) (Fig.9.8).

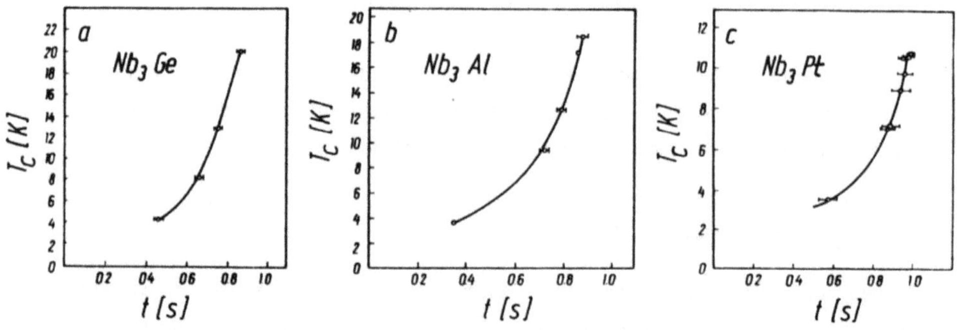

Fig.9.7a-c.

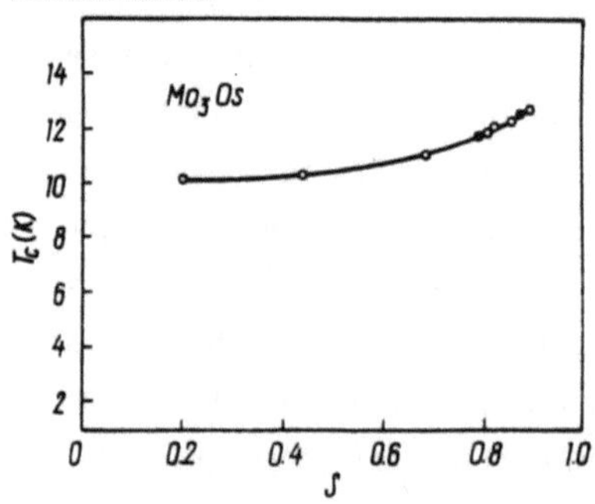

Fig.9.7. Dependence of T_c on the long-range order parameter S in (a) Nb_3Al, (b) Nb_3Pt irradiated by fast neutrons [9.7,20]{□corresponds to the disordering of Nb_3Pt due to heat treatment}

Fig.9.8. Dependence of T_c on S in Mo_3Os disordered by neutron irradiation (o) and heat treatment (●) [9.13]

Note some difference in the variation of the long-range order parameter in different A-15 superconductors upon exposure to radiation. For the compound Nb_3Sn irradiated by neutrons at a dose of $4 \cdot 10^{20}$ cm^{-2}, the decrease of T_c from 17.5 to 2.2 K was accompanied by a sharp drop in the degree of long-range order from $S = 0.98$ to $S \leq 0.1$ [9.23], while for V_3Si, with the same dose of irradiation by neutrons and with a close level of degradation of T_c (from 17 to 1.6 K), the decrease of S (from 0.95 to 0.8 ÷ 0.7) was rather moderate [9.27].

It is of interest to compare the dependence of T_c on the degree of long-range order when disordering a material by radiation and heat treatment. Such a comparison has been drawn for Nb_3Pt and Mo_3Os [9.7,20,43]. It revealed that the dependence of T_c on S is described by one smooth curve, irrespective of the method which has caused the disordering of a compound (Figs.9.7,8).

9.2.3 Investigation by Small-Angle Scattering of Neutrons

Figure 9.9 presents data on the angular dependence of diffuse neutron scattering in Nb_3Sn specimens irradiated by different doses. It is seen that in a nonirradiated Nb_3Sn which is an inhomogeneous medium for neutron scattering, no small-angle scattering occurs, in accordance with the predictions of diffraction theory [9.44]. But the same phenomenon should be observed if irradiation results in a homogenous disordering of the crystal. Meanwhile, experiment yielded diffuse small-angle scattering of thermal neutrons in irradiated Nb_3Sn specimens. This indicates that in the ordered matrix of this compound, a microregion of the disordered phase appears. An increase in the exposure dose leads to the disappearance of the small-angle scattering signal, which may be due to the overlap of microinhomogeneities at a large dose [9.23]. Estimates of the

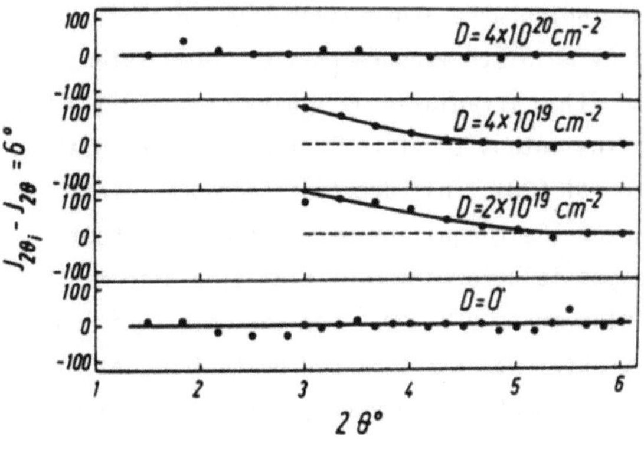

Fig.9.9. Angular dependence of the intensity of diffuse small-angle scattering of thermal neutrons in Nb_3Sn when irradiated by fast neutrons [9.23]

effective size of the microregions of the disordered phase by GUINIER's [9.45] approximation gave dimensions of the order of ~20 Å [9.23].

9.2.4 Electron-Microscopic Investigations

The presence of regions of a disordered phase in an ordered matrix had already been revealed in early electron-microscopic investigations of irradiated superconductors with A-15 structure [9.46]. Later, more detailed measurements using the diffraction contrast technique and involving both basic and superstructural reflections confirmed the existence of structural microinhomogeneities in the ordered phase of Nb_3Sn [9.47,48]. Those measurements established, firstly, that the size of the regions of the disordered phase amounts to 20-60 Å and, secondly, that the structural microinhomogeneities contain a high density of antisite defects and large stress fields. It follows from the experiments that with increasing irradiation doses the number of disordered regions increases, although their overlap and merging are not observed. This differs from the data on small-angle neutron scattering for Nb_3Sn with large irradiation doses.

9.2.5 Channelling of α Particles

Measurements of the channelling and backward scattering of α particles with an energy of $E \sim 1.9$ to 2.0 MeV in irradiated single crystals of V_3Si (the specimens were irradiated by α particles to doses of $2.5 \cdot 10^{17}$ cm^{-2} at $E \sim 1.9$ MeV [9.5]) showed that for the [100] and [110] channelling directions, there is an atomic displacement of 0.115 and 0.1 Å, respectively, perpendicularly to the chains. Normal thermal displacements of atoms in a nonirradiated specimen are 0.088 and 0.061 Å, respectively [9.5]. TESTARDI et al. [9.49] carried out similar experiments at different temperatures and also found additional displacements of the V atoms in the chains, which were approximately 0.1 Å. Recently, these results have been confirmed by absolute measurements of X-ray intensities in thin monocrystalline Nb_3Sn films irradiated by α particles ($E \sim 1.9$ MeV) [9.50]. A substantial decrease in the intensity of reflection with irradiation dose was registered (about a factor of 600). This decrease indicates static displacements of Nb and Sn atoms near their equilibrium positions of about 0.2 Å. The problem concerning the possible formation of radiation defects in A-15 compounds such as static displacements of atoms, and the resulting loss in translational symmetry of the crystal was first raised by TESTARDI et al. [9.40,51] on the basis of X-ray measurements of polycrystalline Nb_3Ge films irradiated by α particles. They found that at irradiation doses of $D \geq 10^{17}$ cm^{-2}, all the X-ray lines beyond the (211) reflection became very weak. Special investigations disclosed that the formation of simple antisite defects does not lead to the observed weakening of X-ray reflections. It was proposed that defects of an unknown nature exist whose presence leads to static displacements of atoms. This proposal was confirmed later by straightforward measurements of channelling and X-ray scattering on single crystals [9.5,49,50].

9.2.6 The Mössbauer Effect

An investigation of the γ radiation from ^{119}Sn in Nb_3Sn specimens irrradiated by protons ($E \sim 2$ MeV) and by oxygen ions ($E \sim 25$ MeV) was made in [9.52,53]. These experiments revealed no additional line, expected for antisite defects with an isomeric shift substantially different from the shift of the ^{119}Sn line in a nonirradiated Nb_3Sn. The line broadening observed in irradiated specimens was attributed to the displacement of Sn atoms from the bcc positions to the interstices available in a sufficient number (16) in the A-15 structure [9.52].

Thus, structural investigations of A_3B compounds with A-15 structure subjected to irradiation have revealed the presence of different types of radiation defects:

i) antisite defects;

ii) static displacements of atoms from equilibrium positions;

iii) the displacement of B atoms from bcc positions to interstices.

Antisite defects were, for the most part, detected when irradiating A-15 compounds by neutrons, while static displacements of atoms in linear chains were revealed after exposing the material to α particles. Evaluations of the mean size of the regions of collision cascades (R) and of the mean free path of the particles forming them (L) for Nb_3Sn have shown that the effective size of the damaged zone for α particles ($E = 2$ MeV) and oxygen ions ($E = 25$ MeV) amounts to $R \sim 20$ Å and $L \sim 10^3$ Å, for fast neutrons ($E \gtrsim 1$ MeV) $R \sim 70$ Å and $L \sim 10$ Å [9.54,55]. This big difference in R and L results in that when the material is irradiated by fast neutrons the free path length of the primarily hit electron becomes comparable with the interatomic distance and its energy is redistributed quickly among the atoms on the nearest lattice sites. As a result, in a small region of irradiated material (of radius ~ 50 Å) a sharp local increase in temperature takes place (up to the melting point), followed by a rapid crystallization with a random distribution of atoms in the lattice sites (the region of a disordered phase with a high density of antisite defects). But when the material is exposed to α particles or oxygen ions, the condition that the free path of the primarily hit metal and the interatomic distance of the material be similar, is not satisfied and no appreciable disordering occurs [9.54,55]. However, radiation defects such as vacancy interstices will occur. This usually leads to large internal stresses in the lattice and to static displacements of the atoms of the components [9.56].

9.3 Physical Properties of Irradiated Superconducting Materials

In order to clarify the degradation of the superconductivity parameters of transition-metal alloys and compounds, it is of interest to study the physical properties of irradiated materials. Of particular interest are measurements of a set of physical properties carried out on the same material exposed to one type of radiation to eliminate the distorting influence of metallurgical factors and the dependence of the properties on the radiation energy. Such measurements have recently been carried out on V_3Si single crystals [9.57] irradiated by fast neutrons to doses of 0.25, 1.66, 3.54 and $22.2 \cdot 10^{18}$ cm^{-2}. The superconducting transition temperature of a nonirradiated specimen [9.42, 58,59] was 17.4 K; upon irradiation by a maximum dose it reduced to 7.4 K. The measurements covered the velocity of sound, the magnetic susceptibility, the upper critical field [9.58], the heat capacity and resistivity [9.59] and elastic and inelastic neutron scattering [9.42]. The results of these measurements are outlined below.

9.3.1 Magnetic Susceptibility

Consider first the electronic properties of irradiated superconductors. Irradiation of V_3Si by a $22.2 \cdot 10^{18}$ cm^{-2} dose leads to a reduction of the susceptibility at 20 K by almost one half, and to an appreciable weakening of its temperature dependence (Fig.9.10). The temperature dependence of the magnetic susceptibility of irradiated V_3Si becomes similar to that in nonirradiated A-15 specimens which have close values of T_c ($\sim$7 K), for example, in V_3Ge (Fig.6.8). It is worth noting that the temperature dependence of the susceptibility of irradiated V_3Si disappears as the temperature decreases, starting from 17 K (which corresponds to T_c of a nonirradiated specimen) down to the new value of 7.4 K (see insert in Fig.9.10). This implies that the sus-

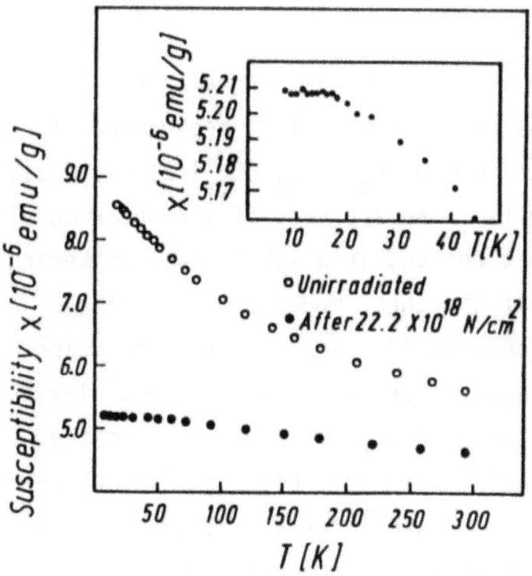

Fig.9.10. The influence of the irradiation by fast neutrons on the temperature dependence of the susceptibility in V_3Si [9.58]

ceptibility of irradiated V_3Si with a low T_c has some memory of the higher T_c of the nonirradiated material. An estimate, based on magnetic susceptibility data in [9.58], of the variation of the d density of states at the Fermi level when passing from nonirradiated to irradiated V_3Si has shown that $N_d(0)$ decreases by almost a factor of 2.5.

9.3.2 Heat Capacity

Similar conclusions have been drawn from heat capacity measurements in irradiated V_3Si specimens [9.59]. With the exposure doses being the same ($D = 22.2 \cdot 10^{18}$ cm^{-2}), a substantial decrease occurs in the values of the heat capacity and the density of states at the Fermi level from $N(0) = 2.40$ states/eV spin atom (for a nonirradiated specimen) to $N(0) = 0.93$ states/eV spin atom. It should be noted that some decrease also occurs in the height of the jump in heat capacity from a transition to superconducting state: from $\Delta c/\gamma T_c = 1.95$ (for a nonirradiated specimen) to $\Delta c/\gamma T_c = 1.40$ (for a specimen exposed at a dose of $22.2 \cdot 10^{18}$ cm^{-2}). This was associated with the presence of regions of the disordered phase with low T_c in the ordered matrix of the compound [9.59]. From heat capacity measurement data [9.59], estimates were made of the Debye temperature for a nonirradiated V_3Si specimen ($\theta_D = 410$ K) and for an irradiated specimen ($\theta_D = 455$ K) with a maximum exposure dose. This seems to indicate that the lattice of V_3Si compounds becomes more rigid when this material is irradiated by fast neutrons.

9.3.3 The Velocity of Sound

Measurements of the velocity of sound in irradiated V_3Si single crystals [9.58] have provided adequate information on the temperature dependence of the moduli C_{11}, C_{12}, C_{44} and of the shear modulus $C_s = (C_{11} - C_{12})/2$, Fig.9.11. It is seen that irradiation of V_3Si results in an increase of the shear moduli over the entire range of measured temperatures and in a substantial weakening of their temperature dependence. Note that the shear modulus of irradiated V_3Si continues varying with temperature between 17 K and the new transition temperature of $T_c = 7.4$ K. The fact that upon irradiation V_3Si becomes elastically isotropic (the elastic anisotropy factor C_s/C_{44} is close to 1 in the entire temperature range) is also worth noting. In this connection, such behaviour occurs extremely rarely in cubic-structure materials [9.60] and may serve as a sign of a partial amorphization of the irradiated compound.

The temperature dependences of the shear moduli of an irradiated V_3Si specimen were found to have minor anomalies at T = 20.5 K, which in [9.42] was ascribed to the beginning martensitic transition [9.61]. The weak display of these anomalies shows that only part of the specimen (∿10%) undergoes structural transformation. Upon irradiation by fast neutrons, these

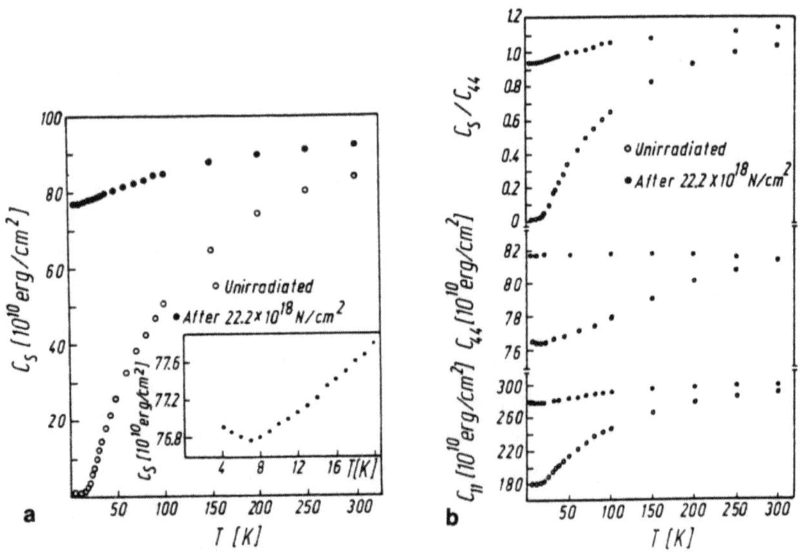

Fig.9.11. The influence of the irradiation by fast neutrons on the temperature dependence (a) of the shear modulus and (b) of the elastic moduli C_{11}, C_{44} and C_s/C_{44} in a single crystal V_3Si [9.58]

anomalies disappear even at small doses. Suppression of the structural transformation in an irradiated V_3Si also indicates a stiffening of the crystal lattice of this compound when subjected to radiation.

9.3.4 Inelastic and Elastic Scattering of Neutrons

Inelastic neutron-scattering measurements in an irradiated V_3Si have shown an increase in the lattice parameter from $a = 4.725_5$ Å (for a nonirradiated specimen) to $a = 4.735_5$ Å (for a specimen exposed to a dose of $22.2 \cdot 10^{18}$ cm^{-2}), which corresponds to a decrease in the theoretical X-ray density to 5.69 g/cm^3. Real-density measurements by a hydrostatic method for irradiated specimens yielded 5.67 g/cm^3. From this difference it was inferred in [9.42] that the concentration of isolated vacancies in an irradiated V_3Si does not exceed 0.3%. Such a small amount of vacancies attests that the concentration of radiation defects such as interstices vacancies in A-15 compounds is low.

Inelastic neutron scattering measurements have shown that irradiation of V_3Si by a dose of $22.2 \cdot 10^{18}$ cm^{-2} heavily suppresses the softening of phonons belonging to the transverse acoustic [110] branch with $[1\bar{1}0]$ polarization; this softening is sufficiently pronounced in A-15 compounds exhibiting structural transition [9.62].

9.3.5 Resistivity

In most of the papers dealing with a study of the superconducting characteristics of transition-metal alloys and compounds it is common practice to construct the dependence of T_c on residual resistivity or to give the ratio of resistivity at room temperature to that at low temperatures. In Sect.9.1 we noted the importance of this dependence for the analysis of differences in the rate of degradation of T_c in materials of different initial purity. On the other hand, these data appear to be useful in refining the concepts of radiation defects in superconducting materials. Figure 9.12 presents data (the hatched region) for the dependence of T_c on the ratio $\rho_{300\,K}/\rho_{25\,K}$ obtained for 130 Nb_3Ge films in which the resistivity and T_c were varied by changing the stoichiometry, the film thickness, the sputtering conditions and other factors [9.40,63]. Thus measured T_c and $\rho_{300\,K}/\rho_{25\,K}$ of irradiated Nb_3Ge films, which are also given in this figure, do not go beyond the above values. (A similar correlation was also found for other compounds of A-15 structure [9.51]. Figure 9.13 illustrates the temperature dependence of resistivity for specimens having different T_c (from $T_c = 23.2$ K to $T_c < 4.2$ K)

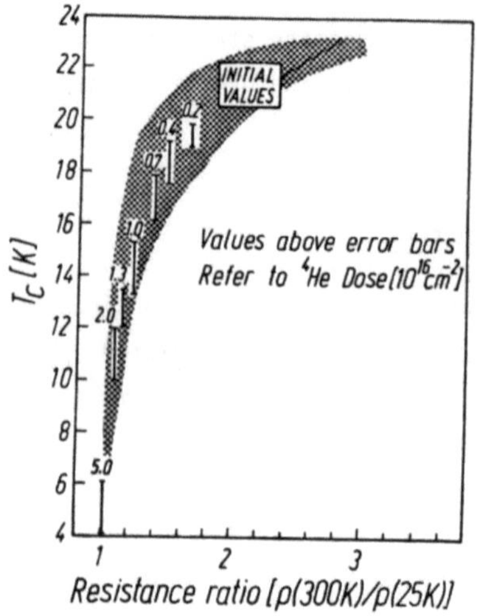

Fig.9.12. Correlation between T_c and the resistivity ratio $\rho_{300\ K}/\rho_{25\ K}$ for film specimens of Nb_3Ge irradiated by α particles. The figures indicate the exposure dose x 10^{16} cm^{-2}, the hatched region presents data on the dependence of T_c on $\rho_{300\ K}/\rho_{25\ K}$ for 130 nonirradiated Nb_3Ge films in which the resistivity and T_c varied due to deviations from stoichiometry, sputtering conditions and other metallurgical factors [9.40,63]

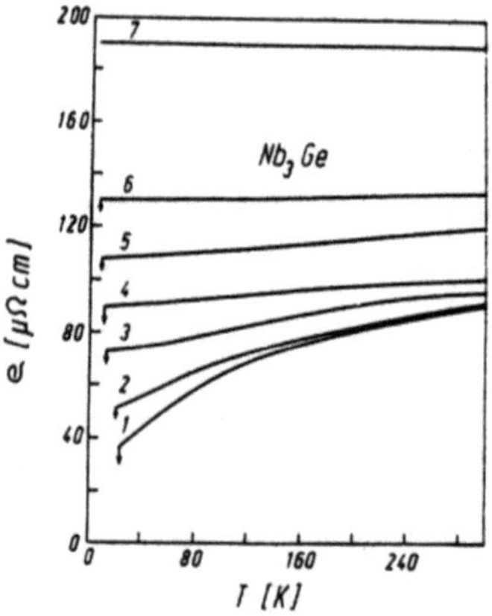

Fig.9.13. Temperature dependence of the resistivity in Nb_3Ge films having different values of T_c [9.51]

[9.51]. The anomalous temperature dependence of resistivity which is typical of A-15 compounds with high T_c [9.52,53], gradually disappears for Nb_3Ge specimens with lower T_c, the results turning out to be independent of whether the drop in T_c was attached due to irradiation or due to a variation of the metallurgical factors.

9.3.6 The Temperature Coefficient of the Upper Critical Field

As already noted in Chapter 4, measurements of the temperature dependence of H_{c_2} permit evaluation of the electronic portion of the heat capacity (γ) and hence the density of states on the Fermi level N(0). By this procedure the variation of the density of states on the Fermi level in transition metals under pressure (Sect.4.1) was determined in [9.64]. A similar approach has recently been used to investigate the variations of γ and N(0) in A-15 compounds subjected to a radiation field [9.16,65,66]. In contrast to [9.64], the analysis also comprised, in addition to the quantity $(dH_{c_2}/dT)_{T_c}$, the residual resistivity ρ_0. Figures 9.14,15 present the dependences of γ and N(0) on the T_c determined for Nb_3Sn and Nb_3Ge [9.66], irradiated by electrons up to a dose of $5.5 \cdot 10^{20}$ cm^{-2}, and by α particles (the dose was not specified). The solid line displays the calculation of the dependence of N(0) on T_c according to McMILLAN's formula on the assumption that $\langle J^2 \rangle / M \langle \omega^2 \rangle$ = const. [9.67,68]. Analysis of these data shows that, on the one hand, irradiation of superconductors with A-15 structure leads to a decrease in the values of T_c and N(0). On the other hand, the dependence of N(0) on T_c for Nb_3Sn is close to the theoretical one and the approximation $\lambda \sim N(0)$ presents a reasonable basis for an analysis of the superconducting properties in irradiated Nb_3Sn. No such correlation occurs for Nb_3Ge and it should also be noted that at higher T_c the values of N(0) for Nb_3Ge turned out to be much less than those for Nb_3Sn. In [9.66] this difference was explained by the fact that Nb_3Ge contained a considerable amount of the other phase.

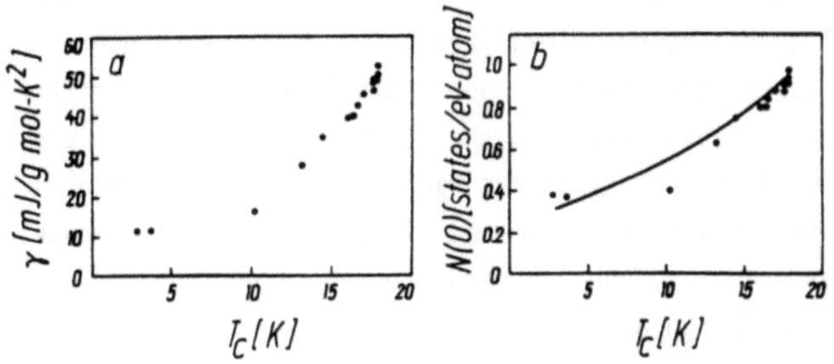

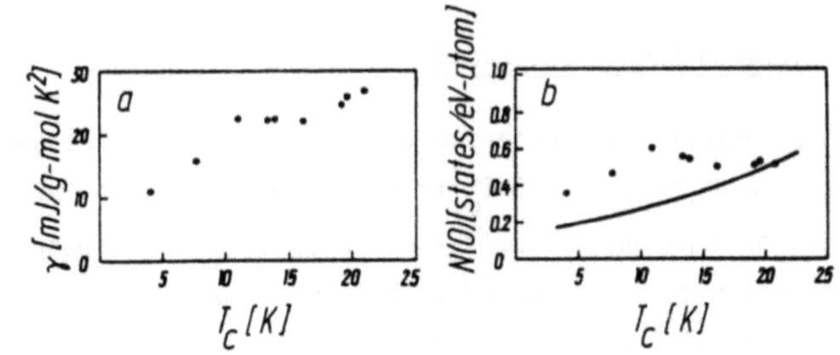

Fig.9.14. The T_c dependence of (a) γ and (b) N(0) in Nb_3Sn irradiated by α particles (o) and electrons (•) [9.66]

Fig.9.15. The T_c dependence of (a) γ and (b) N(0) in Nb_3Ge irradiated by α particles (o) and electrons (•) [9.66]

For an Mo_3Ge compound irradiated by α particles, measurements of $(dH_{c2}/dT)_{T_c}$ and ρ_0 have shown that N(0) increases by about 50% compared to the nonirradiated material. The radiation effect in this compound, as distinct from other compounds of A-15 structure, leads to an increase in T_c from 1.5 to 5 K [9.16], and therefore the increase of N(0) correlates with the increase of T_c in the irradiated material. One of the possible causes of this increase is that the Fermi level in the compound under investigation passes through a minimum in the density of states, thus separating two peaks with occupied and vacant states [9.69]. A smearing of the above peaks due to radiation defects may lead to an increase of the total density of states at the Fermi level in Mo_3Ge [9.16].

9.4 Theoretical Models for the Degradation of Superconducting Properties

The experimental evidence outlined in the previous sections concerning the influence of irradiation on T_c and other properties of superconductors must be analyzed and generalized by a theory. But such a theory does not yet exist. The proposals advanced in the literature are restricted to general considerations, part of which will be discussed below. These considerations concern the influence which factors such as the electron-phonon coupling constant or the density of states on the Fermi level exert on the T_c of irradiated materials. It should be noted that until the nature of the radiation defects in each particular class of compounds has been determined either by means of computer modelling or by special experiments, one can hardly expect a profound theoretical analysis of the degradation of T_c.

9.4.1 Variations of the Electron-Phonon Interaction

If one regards the radiation defects as scattering impurities, one can use the ANDERSON theory to analyze the degradation of T_c in irradiated materials [9.70]. Accordingly, the scattering on nonmagnetic impurities leads to some decrease of the effective electron-phonon interaction and to a lowering of T_c because the Fermi surface becomes isotropic. In a real superconductor, the electron-phonon interaction depends on the direction of the electron momentum relative to the crystal. The averaging of the anisotropy of the gap by the action of im-impurities (defects) decreases T_c, as compared to nonirradiated materials. According to the estimates made in [9.71] in the presence of impurities in the superconductor, the averaging of the anisotropy of the gap decreases T_c by 4% at the most. Using this approach, GURVITCH et al. [9.16] explained the initial region of the dose dependence of T_c in superconducting compounds with A-15 structure. But the extension of the above mechanism to the entire region of doses, proposed in [9.72,73], may not be considered to be sufficiently justified. The point is that when the defect concentration exceeds 1%, the mean free path of electrons becomes much less than the dimensions of a Cooper pair and the mechanism of the lowering of T_c due to the averaging of the anisotropy of the gap does not work any more.

The other approach, which also associates the variation of T_c in irradiated materials with the variation of the electron-phonon interaction, emphasizes the importance of the PIPPARD-ZIMAN condition [9.74]. Accordingly, the phonons with a wavelength exceeding the free path of an electron are noneffective scatterers. At a sufficient radiation-defect concentration the PIPPARD-ZIMAN condition restricts the contribution of phonons to the electron-phonon interaction and leads to a decrease of T_c. One of the possible explanations offered for the abrupt lowering of T_c as a function of defect concentration (or of the exposure dose in A-15 compounds) is associated with the hypothesis of the plasmon mechanism of the electron pairing interaction (Sect.6.4). This mechanism is very sensitive to the decay of electrons at the Fermi surface. At a critical value of decay, Γ_c, coinciding with the plasmon energy $\Omega_{ac}(\Gamma_c \sim \Omega_{ac})$, the plasmon mechanism should become insignificant and this will lead to a drop in T_c.

9.4.2 Smearing of the Peak in the Density of Electron States

Most of the theoretical papers treating the dose dependence of the superconducting characteristics associate the variation of T_c in irradiated materials with the variation of the density of electronic states at the Fermi level [9.73,75-82]. This approach rests on the assumption that the sharp peak in the density of states of the superconductor is smeared out by electron scattering on radiation defects (the fact that high T_c occur in nonirradiated compounds of A-15 structure is normally attributed to the location of the Fermi level in the vicinity of that peak).

In [9.75,76], the dependence of T_c on the exposure dose in A-15 compounds has been calculated by standard methods [9.83] proceeding from the assumption that the density of states in these materials is described by the following expression

$$N(E) = N_0\left(1 + \frac{A^2}{E^2 + a^2}\right)$$

where E is the energy of an electron, A and a define the height and width of the peak. By appropriate selection of the parameters A, a and Γ (Γ being the inverse lifetime of an electron), NAKAYAMA and TSUNETO [9.75,76] obtained good agreement with experimental results.

In addition, mention should be made of [9.73,77,80] which dealt with the modelling of antisite radiation defects in A_3B compounds with A-15 structure. The approach adopted is the same as that used by LABBE and VAN REUTH to describe the dependence of T_c as a function of the disorder due to heat treatment [9.84]. Accordingly, substitution of B atoms for A atoms breaks down the integrity of the A chains along which the one-dimensional motion of d-electron comes about. At a limiting antisite defect concentration the singularity of N(E) corresponding to a quasi-one-dimensional spectrum becomes smeared out, and this leads to a decrease of N(0) and T_c.

Each of the approximations described above gives a qualtitative explanation of the experimentally observed dependences of the superconducting parameters on irradiation dose. This indicates that each of these factors makes its own contribution to the formation of the above dependences. The next stage of the development of the theory is a quantitative estimation of these contributions. Of some interest in this respect is estimating the variation of the density of states on the Fermi level upon irradiation of the superconducting alloys and compounds Zr-Nb-Mo, Ti-V-Cr, V_3Si, Nb_3Sn and Nb_3Ge [9.82,9.86]. This estimation was made on the basis of calculations for the band structure of these substances and from an assessment of the lifetime of the electrons in the irradiated materials [9.81].

It remains an unfortunate fact that these quantitative estimates only take into account the integral effect of the total number of radiation defects on the density of electronic states and on the superconducting properties of transition-metal alloys and compounds. Meanwhile, to understand the mechanisms governing the degradation of T_c in the materials concerned it is essential to quantitatively estimate the influence of the various types of radiation defects on N(0) and T_c. Below we survey the results of such estimates, made on the basis of cluster calculations, for A-15 compounds.

9.4.3 Radiation Defects in the Cluster Approximation

One of the ways to overcome the difficulties arising from consideration of the electronic structure of crystals with a random distribution of defects is to use a cluster approximation. The latter is based on a small fragment of the crystal (a cluster) in which different types of defects are present and the electronic spectrum is calculated within the MO LCAO approximation [9.87-92]. In [9.93,94] this approximation was used to analyze the electron structure of nonstoichiometric carbides and oxides of transition metals. An analysis of the radiation defects using the cluster approximation was first made in [9.79] for A_3B compounds with A-15 structure (V_3Si, Mo_3Ge and Nb_3Sn). An $[A_3B_4]$ type cluster taking into account the nearest neighbours of the A atom and modelling the crystal well enough was selected from the A-15 structure [9.95]. The following types of radiation defects were introduced into the cluster:

i) antisite defects or substitution ($A \leftrightarrows B$) type defects;

ii) zigzag type static displacements of A atoms [9.96];

iii) defects such as B-type site vacancy-interstice.

Thereupon, the spectrum of the molecular orbitals of the clusters $[V_3Si_4]$, $[Mo_3Ge_4]$ and $[Nb_3Sn_4]$ was calculated and the value of the density of states at the Fermi level was estimated according to the method described in [9.90]. It turned out that the static displacement of A atoms in the chains exert almost no influence on N(0) in the clusters concerned. It is the antisite defects that produce a much greater effect in N(0). In the case of V_3Si and Nb_3Sn, the presence of a pair of antisite defects in the relevant clusters brings about a decrease of N(0) by 22 to 24%. The reverse is true for Mo_3Ge in which N(0) becomes twice as large. This difference in the variation of N(0), which correlates with the behaviour of T_c in the above materials, arises from the fact that in V_3Si and Nb_3Sn the E_F in complete clusters passes near the peak in N(E), whereas in Mo_3Ge it passes in the region of the minimum of the density of states. Introduction of radiation defects into a cluster smears out the peaks in N(E), which leads to a decrease of N(0) in V_3Si and Nb_3Sn and to an increase of N(0) in Mo_3Ge.

References

Chapter 1

1.1 B.W. Roberts: Nat. Bur. Stand. U.S. Tech. Note *983* (1978) Suppl.
1.2 E.I. Nikulin, N.V. Volkenshtein, V.E. Startsev: JETP Lett. *17*, 470 (1973)
1.3 B.T. Matthias: Phys. Today *24(8)*, 23 (1971)
1.4 J.L. Smith, R.G. Haire: Science *200*, 535 (1978)
1.5 C. Reale: Vacuum *27*, 3 (1977)
1.6 N. Jacobsen, C.G. Granqvist, T. Claeson: Z. Phys. B*25*, 265 (1976)
1.7 M. Belzons, R. Blanc, R. Rayan: C. R. Acad. Sci. B*283*, 241 (1976)
1.8 C. Reale: Appl. Phys. Lett. *27*, 157 (1975)
1.9 C. Reale: Phys. Lett. A*55*, 165 (1975)
1.10 P.H. Schmidt, R.N. Castellano, H. Barz, A.S. Cooper, E.G. Spencer: J. Appl. Phys. *44*, 1833 (1973)
1.11 C. Reale: Thin Solid Films *28*, L29 (1975)
1.12 B. Stritzker: J. Phys. Paris Lett. *39*, L397 (1978)
1.13 B. Stritzker: Phys. Rev. Lett. *42*, 1769 (1979)
1.14 R.F. Hoyt, A.S. Mota: Solid State Commun. *18*, 139 (1976)
1.15 V.P. Dmitriev, A.K. Milai, A.F. Orlov: Fiz. Nizk. Temp. *1*, 914 (1975)
1.16 H.H. Balster: Ber. Kernforschungsanlage Jülich *1126*, 639 (1974)
1.17 B.W. Roberts: Nat. Bur. Stand. U.S. Tech. Note *724* (1972) Suppl.; *482* (1969); in *Intermetallic Compounds*, ed. by J.H. Westbrook (Wiley, New York 1967) p.581
1.18 E.M. Savitskii, V.V. Baron, Yu.F. Efimov, M.I. Bychkov, L.F. Myzenkova: *Mjetallovjedjenije svjerkhprovodjachikh matjerialov* (Nauka, Moscow 1969)
1.19 S.V. Popova, L.N. Fomitcheva, N.I. Palnikova: JETP Lett. *20*, 648 (1974)
1.20 G. Hardy, J.K. Hulm: Phys. Rev. *89*, 884 (1953)
1.21 L.R. Testardi, J.H. Wernick, W.A. Royer: Solid State Commun. *15*, 1 (1974)
1.22 A.L. Robinson: Science *185*, 846, 848 (1974)
1.23 Yu.A. Izyumov, E.Z. Kurmaev: Usp. Fiz. Nauk *118*, 53 (1976)
1.24 J.K. Hulm, M. Ashkin, D.W. Deis, C.K. Jones: Prog. Low Temp. Phys. *6*, 205 (1970)
1.25 J.M. Vandenberg, B.T. Matthias: Mater. Res. Bull. *9*, 1085 (1974)
1.26 R.L. Greene, G.B. Street, L.J. Suter: Phys. Rev. Lett. *34*, 577 (1975)
1.27 D.C. Johnston, H. Prakash, W.H. Zachariasen, R. Viswanathan: Mater. Res. Bull. *8*, 777 (1973)
1.28 S. Foner, E.J. McNiff, E. Alexander: Phys. Lett. A*49*, 269 (1974)
1.29 H.E. Barz, A.S. Cooper, E. Corenzwit, M. Marezio, B.T. Matthias, P.H. Schmidt: Science *175*, 884 (1972)
1.30 R.A. Hein, J.W. Gibson, R. Mazelsky, R.C. Miller, J.K. Hulm: Phys. Rev. Lett. *12*, 320 (1964)
1.31 T.H. Geballe: In *Superconductivity in d- and f-Band Metals*, ed. by D.H. Douglass (American Institute of Physics, New York 1972) p.237
1.32 M. Kobayashi, I. Tsujikawa: J. Phys. Soc. Jpn. *46*, 1945 (1979)
1.33 L.N. Bulaevskii: Usp. Fiz. Nauk *116*, 449 (1975)

1.34 M. Boudeulle, P. Michel: Acta Crystallogr., Sect. A*28*, S193 (1972)
1.35 J. Bardeen, L.N. Cooper, J.R. Schrieffer: Phys. Rev. *106*, 162 (1957); *108*, 1175 (1957)
1.36 L.N. Cooper: Phys. Rev. *104*, 1189 (1956)
1.37 N.N. Bogolyubov, V.V. Tolmatchev, D.V. Shirkov: In *Novyj mjetod v tjeorii svjerkhprovodimosti*, ed. by K.P. Gurov (Izv. Akad. Nauk SSSR, Moscow 1958)
1.38 G.M. Eliashberg: JETP *38*, 966 (1960); *39*, 1437 (1960)
1.39 W.L. McMillan: Phys. Rev. *167*, 331 (1968)
1.40 A.B. Migdal: JETP *34*, 1438 (1958)
1.41 J.R. Schrieffer: *Theory of Superconductivity* (Benjamin, New York 1964)
1.42 B.T. Geilikman, V.Z. Kresin: Fiz. Tverd. Tela Leningrad *7*, 3294 (1965)
1.43 A.M. Clogston: Phys. Rev. *136*, 112 (1965)
1.44 W.J. Tomasch: Phys. Rev. Lett. *15*, 672 (1965); *16*, 16 (1966)
1.45 D.J. Scalapino, J.R. Schrieffer, J.W. Wilkins: Phys. Rev. *148*, 263 (1966)
1.46 G. Gladstone, M.A. Jensen, J.R. Schrieffer: In *Superconductivity*, ed. by R.D. Parks (Dekker, New York 1969)
1.47 E.G. Brovman, Yu.M. Kagan: Usp. Fiz. Nauk *112*, 369 (1974)
1.48 Yu.A. Izyumov, Yu.N. Skryabin: Phys. Status Solidi B*61*, 9 (1974)
1.49 S.V. Vonsovsky, Yu.A. Izyumov: Usp. Fiz. Nauk *77*, 377 (1962); *78*, 3 (1962)
1.50 A.A. Abrikosov, L.P. Gor'kov: JETP *39*, 1781 (1960); *42*, 1088 (1962)
1.51 S.V. Vonsovsky, M.S. Svirsky: JETP *46*, 1619 (1964)
1.52 N.F. Berk, J.R. Schrieffer: Phys. Rev. Lett. *17*, 1171 (1966)

Chapter 2

2.1 L.P. Gor'kov: JETP *34*, 735 (1958)
2.2 Y. Nambu: Phys. Rev. *117*, 648 (1960)
2.3 N.N. Bogolyubov: In *Kvazisrjednije v zadatchakh statistitcheskoj mjekhaniki*, ed. by E.L. Orlik (Naukova Dumka, Kiev 1971) Vol.3, p.174
2.4 A.A. Abrikosov, L.P. Gor'kov, I.E. Dzyaloshinsky: In *Mjetody kvantovoj tjeorii polja v statistitchjeskoj fizikje*, ed. by K.P. Gurov (Fizmatgiz, Moscow 1962)
2.5 J.R. Schrieffer: *Theory of Superconductivity* (Benjamin, New York 1964)
2.6 A.B. Migdal: JETP *34*, 1438 (1958)
2.7 G.M. Eliashberg: JETP *38*, 966 (1960); *39*, 1437 (1960)
2.8 P. Morel, P.W. Anderson: Phys. Rev. *125*, 1263 (1962)
2.9 D.J. Scalapino, Y. Wada, J.C. Swihart: Phys. Rev. Lett. *14*, 102 (1965)
2.10 D.J. Scalapino: In *Superconductivity*, ed. by R.D. Parks (Dekker, New York 1969)
2.11 V. Ambegaokar, L. Tewordt: Phys. Rev. *134*, 805 (1964)
2.12 N.N. Bogolyubov, V.V. Tolmatchev, D.V. Shirkov: In *Novyj mjetod v tjeorii svjerkhprovodimosti*, ed. by K.P. Gurov (Izv. Akad. Nauk SSSR, Moscow 1958)
2.13 A.E. Karakozov, E.G. Maksimov, S.A. Mashkov: JETP *68*, 1937 (1975)
2.14 D.A. Kirjnitz, E.G. Maksimov, D.I. Homskii: J. Low Temp. Phys. *10*, 79 (1973)
2.15 B.T. Geilikman, N.F. Mosharov: J. Low Temp. Phys. *6*, 137 (1972)
2.16 W. Kessel: Z. Naturforsch. A*29*, 445 (1974)
2.17 P. Hertel: Z. Phys. *248*, 272 (1971)
2.18 W.L. McMillan: Phys. Rev. *167*, 331 (1968)
2.19 M.V. Medvedev, E.A. Pashitskii, Yu.S. Pyatiletov: JETP *65*, 1186 (1973); J. Appel: Phys. Rev. Lett. *21*, 1164 (1968); C.R. Leavens, J.R. Carbotte: Can. J. Phys. *49*, 724 (1971); R.C. Dynes: Solid State Commun. *10*, 615 (1972)

2.20 A.E. Karakozov, E.G. Maksimov, S.A. Mashkov: Tezisy doklada na 18 Vsjes. sov. po Fiz. Nizk. Temp. (Institute of Physics of the Ukrainian SSR Academy of Sciences, Kiev 1974) p.83
2.21 Yu.A. Izyumov, Yu.N. Skryabin: Fiz. Met. Metalloved. *38*, 912 (1974)
2.22 C. Bergman, D. Rainer: Z. Phys. *263*, 59 (1973)
2.23 P.W. Anderson: J. Phys. Chem. Solids *11*, 26 (1959)
2.24 D.N. Zubarev: Usp. Fiz. Nauk *71*, 71 (1960)
2.25 E.G. Maksimov: JETP *57*, 1660 (1969)
2.26 D.W. Taylor, P. Vashishta: Phys. Rev. B*11*, 4410 (1972)
2.27 J.R. Gommersall, B.L. Gyorffy: J. Phys. Paris F*4*, 1204 (1974)
2.28 J.J. Hopfield: Phys. Rev. *186*, 443 (1969)
2.29 L.D. Landau, E.M. Lifshitz: *Kvantovaja mjekhanika* (Nauka, Moscow 1974)
2.30 P.B. Allen, R.C. Dynes: Phys. Rev. B*12*, 905 (1975)
2.31 P.B. Allen: In *Modern Trends in the Theory of Condensed Matter*, ed. by A. Pekalski, J. Przystawa, Lecture Notes in Phys., Vol.115 (Springer, Berlin, Heidelberg, New York 1980) p.388
2.32 P.B. Allen: Phys. Rev. B*13*, 1416 (1976)
2.33 W.H. Butler, P.B. Allen: In *Superconductivity in d- and f-Band Metals*, ed. by D.H. Douglass (Plenum, New York 1976) p.73

Chapter 3

3.1 S.V. Vonsovsky: JETP *16*, 980 (1946)
3.2 K. Maki: In *Superconductivity*, ed. by R.D. Parks (Dekker, New York 1969) p.1036
3.3 Yu.A. Izyumov, Yu.N. Skryabin: Fiz. Met. Metalloved. *37*, 921 (1974)
3.4 Yu.A. Izyumov, Yu.N. Skryabin, F.A. Kassan-Ogly: *Poljevyje mjetody v tjeorii fjerromagnjetizma* (Nauka, Moscow 1974)
3.5 Yu.A. Izyumov, Yu.N. Skryabin: Fiz. Tverd. Tela Leningrad *21*, 2045 (1979)
3.6 T.K. Lee, Yu.A. Izyumov, J.L. Birman: Phys. Rev. B*20*, 4494 (1979)
3.7 G. Sarma: J. Phys. Chem. Solids *24*, 1029 (1963)
3.8 B.S. Chandrasekhar: Appl. Phys. Lett. *1*, 7 (1962)
3.9 A.M. Clogston: Phys. Rev. Lett. *9*, 266 (1962)
3.10 S.V. Vonsovsky, M.S. Svirsky: Dokl. Akad. Nauk SSSR *182*, 204 (1958)
3.11 S.V. Vonsovsky, M.S. Svirsky: Fiz. Met. Metalloved. *15*, 318 (1963)
3.12 S.V. Vonsovsky, M.S. Svirsky: JETP *40*, 1676 (1961); *46*, 1619 (1964)
3.13 S.V. Vonsovsky, M.S. Svirsky: Izv. Akad. Nauk SSSR Ser. Fiz. *28*, 418 (1964)
3.14 S.V. Vonsovsky, M.S. Svirsky: Phys. Status Solidi *9*, 267 (1965)
3.15 Yu.A. Izyumov, Yu.N. Skrjabin: Phys. Status Solidi B*61*, 9 (1974)
3.16 T. Kasuya: Prog. Theor. Phys. *20*, 980 (1958)
3.17 A.I. Akhiezer, I.I. Pomerantchuk: JETP *36*, 859 (1959)
3.18 A.I. Akhiezer, I.A. Akhiezer: JETP *43*, 2208 (1962)
3.19 B.V. Karpenko: Fiz. Met. Metalloved. *9*, 794 (1962)
3.20 I.A. Privorotskii: JETP *43*, 2255 (1962)
3.21 A.A. Abrikosov, L.P. Gor'kov: JETP *39*, 1781 (1960); *42*, 1088 (1962)
3.22 B.T. Matthias, E. Corenzwit, H. Suhl: Phys. Rev. Lett. *1*, 92 (1958)
3.23 R.A. Hein, R.L. Falge, B.T. Matthias, E. Corenzwit: Phys. Rev. Lett. *2*, 500 (1959)
3.24 N.E. Phillips, B.T. Matthias: Phys. Rev. *121*, 105 (1962)
3.25 M. Peter, P. Donze, Ø. Fischer, A. Junod, J. Ortelli, A. Treyvaud, E. Walker, M. Wilhelm, B. Hillenbrand: Helv. Phys. Acta *44*, 345 (1971)
3.26 D.J. Erickson, C.E. Olsen, R.D. Taylor: *Proceedings of the Eighth Mössbauer Symposium* (Plenum, New York 1973)
3.27 P. Steiner, D. Gumprecht, S. Hufner: Phys. Rev. Lett. *30*, 1132 (1973)
3.28 S. Roth, K. Ibel, W. Just: Report, Int. Conf. of Magnetism, Moscow 1973
3.29 L.P. Gor'kov, A.I. Rusinov: JETP *46*, 1363 (1964)

3.30 A.A. Abrikosov, L.P. Gor'kov: JETP *39*, 1781 (1960)
3.31 A.I. Larkin, D.E. Khmelnitskii: JETP *58*, 1789 (1970)
3.32 G. Heber: Appl. Phys. *10*, 101 (1976)
3.33 V. Cannella, J.A. Mudosh: Phys. Rev. *6*, 4220 (1972)
3.34 P.W. Anderson: Mater. Res. Bull. *5*, 549 (1970)
3.35 Yu.A. Izyumov, Yu.N. Skryabin: Fiz. Met. Metalloved. *37*, 1127 (1974)
3.36 P. Fulde, K. Maki: Phys. Rev. *141*, 275 (1966)
3.37 K.H. Bennemann: Phys. Rev. Lett. *17*, 438 (1966)
3.38 K.H. Bennemann, S. Nakajima: Phys. Rev. Lett. *16*, 243 (1966)
3.39 K.H. Bennemann, F.M. Müller: Phys. Rev. *176*, 546 (1968)
3.40 K.H. Bennemann: In *Superconductivity*, ed. by P. Wallace (Gordon and Breach, New York 1969) p.1
3.41 A.I. Larkin, V.I. Malnikov, D.E. Khmelnitskii: JETP *60*, 846 (1971)
3.42 D. Rainer: Solid State Commun. *6*, 111 (1968)
3.43 D. Rainer: Z. Phys. *252*, 174 (1972)
3.44 M.A. Jensen, H. Suhl: In *Magnetism*, Vol.2B, ed. by G. Rado, H. Suhl (Academic, New York 1966) p.183
3.45 Ø. Fischer, M. Peter: In *Magnetism*, Vol.5, ed. by G. Rado, H. Suhl (Academic, New York 1973) p.327
3.46 S. Roth: Appl. Phys. *15*, 1 (1978)
3.47 C.A. Balseiro, L.M. Falikov: Phys. Rev. B*19*, 2548 (1979)
3.48 P.W. Anderson, H. Suhl: Phys. Rev. *116*, 898 (1959)
3.49 L.N. Bulaevski, A.I. Rusinov, M. Kulic: Solid State Commun. *30*, 59 (1979)
3.50 H.H. Miva: Proc. Phys. Soc. London *85*, 1197 (1965)
3.51 Yu.A. Izyumov, Yu.N. Skryabin: Fiz. Met. Metalloved. *49*, 903 (1980)
3.52 Ø. Fischer, A. Treyvaud, R. Chevrel, M. Sergent: Solid State Commun. *17*, 721 (1975)
3.53 R.N. Shelton, R.W. McCallum, H. Adrian: Phys. Lett. A*56*, 213 (1976)
3.54 M. Marezio, P.D. Dernier, J.P. Remeika, E. Corenzwit, B.T. Matthias: Mater. Res. Bull. *8*, 657 (1973)
3.55 M.B. Maple, L.E. De Long, W.A. Fertig, D.C. Johnston, R.W. McCallum, R.N. Shelton: In *Valence Instabilities and Related Narrow Band Phenomena*, ed. by R.D. Parks (Plenum, New York 1977)
3.56 M.B. Maple: In *Rare Earth in Modern Science and Technology*, ed. by G.J. McCarthy, J.J. Phyne (Plenum, New York 1978) p.381
3.57 R.W. McCallum, D.C. Johnston, R.N. Shelton, M.B. Maple: *Proceedings Second Rochester Conf. Superconductivity on d- and f-Band Metals*, ed. by D.H. Douglass (Plenum, New York 1976) p.625
3.58 R.W. McCallum, D.C. Johnston, R.N. Shelton, M.B. Maple: Solid State Commun. *24*, 391 (1977)
3.59 L.J. Azevedo, W.C. Clark, C. Murayama, R.W. McCallum, D.C. Johnston, M.B. Maple, R.N. Shelton: J. Phys. Paris *39*, C6-365 (1978)
3.60 R.W. McCallum, D.C. Johnston, R.N. Shelton, W.A. Fertig, M.B. Maple: Solid State Commun. *24*, 501 (1977)
3.61 M. Ishikawa, Ø. Fischer: Solid State Commun. *24*, 747 (1977)
3.62 J.W. Lynn, D.C. Moncton, G. Shirane, W. Thomlinson, J.J. Eckart: Appl. Phys. *49*, 1389 (1978)
3.63 N.E. Alekseevskii, G. Wolf, N.M. Dobrovolskii, C. Hohefeld: Phys. Status Solidi B*44*, K79 (1977)
3.64 D.E. Moncton, G. Shirane, W. Thomlinson, M. Ishikawa, Ø. Fischer: Phys. Rev. Lett. *41*, 1133 (1978)
3.65 W. Thomlinson, G. Shirane, D.E. Moncton, M. Ishikawa, Ø. Fischer: J. Appl. Phys. *50*, 1981 (1979)
3.66 B.T. Matthias, E. Corenzwit, J.M. Vandenberg, H.E. Barz: Proc. Natl. Acad. Sci. U.S.A. *74*, 1334 (1977)
3.67 M. Ishikawa, Ø. Fischer: Solid State Commun. *23*, 37 (1977)
3.68 J.M. Vandenberg, B.T. Matthias: Proc. Natl. Acad. Sci. U.S.A. *74*, 1336 (1977)

3.69 T. Jarlborg, A.J. Freeman, T.J. Watson-Yang: Phys. Rev. Lett. *36*, 1032 (1977)
3.70 W.A. Fertig, D.C. Johnston, L.E. De Long, R.W. McCallum, M.B. Maple, B.T. Matthias: Phys. Rev. Lett. *38*, 987 (1977)
3.71 L.D. Woolf, D.C. Johnston, H.B. MacKay, R.W. McCallum, M.B. Maple: J. Low Temp. Phys. *35*, 651 (1979)
3.72 M.B. Maple, H.C. Hamaker, D.C. Johnston, H.B. McKay, L.D. Woolf: J. Less Common Met. *62*, 251 (1978)
3.73 M.B. Maple: Bull. Am. Phys. Soc. *23*, 347 (1978)
3.74 D.C. Moncton, D.B. McWhan, J. Eckert, G. Shirane, W. Thomlinson: Phys. Rev. Lett. *39*, 1164 (1977)
3.75 G.K. Shenoy, B.D. Dunlap, F.Y. Fradin, C. Kimball: Bull. Am. Phys. Soc. *23*, 306 (1978)
3.76 H.R. Ott, W.A. Fertig, D.C. Johnston, M.B. Maple, B.T. Matthias: J. Low Temp. Phys. *33*, 159 (1978)
3.77 H.B. MacKay, L.D. Woolf, M.B. Maple, D.C. Johnston: Phys. Rev. Lett. *42*, 918 (1979)
3.78 L.D. Woolf, M. Tovar, H.C. Hamaker, M.B. Maple: Phys. Lett. A*72*, 481 (1979)
3.79 D.E. Moncton: J. Appl. Phys. *50*, 1880 (1979)
3.80 H.C. Hamaker, L.D. Woolf, H.B. MacKay, Z. Fisk, M.B. Maple: Bull. Am. Phys. Soc. *24*, 390 (1979)
3.81 J.M. Lynn, D.E. Moncton, W. Thomlinson, G. Shirane, R.N. Shelton: Solid State Commun. *26*, 493 (1978)
3.82 K.S. Liu, M.E. Fisher: J. Low Temp. Phys. *10*, 655 (1973)
3.83 I.F. Luksyutov, V.L. Pokrovskii, D.E. Khmelnitskii: JETP *69*, 1817 (1975)
3.84 W. Baltensperger: Helv. Phys. Acta *32*, 197 (1959)
3.85 E. Müller-Hartmann, J. Zittartz: Z. Phys. *234*, 58 (1970)
3.86 E.I. Blount, C.M. Varma: Phys. Rev. Lett. *42*, 1079 (1979)
3.87 H. Suhl: "Domain Walls", in *Magnetic Superconductors*, San Diego (1979) Preprint
3.88 H. Matsumoto, M. Tachiki, H. Umezawa: Solid State Commun. *31*, 157 (1979)
3.89 B.I. Halperin, T.C. Lubenski, S.K. Ma: Phys. Rev. Lett. *32*, 292 (1974)
3.90 M.V. Jaric, M. Belic: Phys. Rev. Lett. *42*, 1015 (1979)
3.91 R.A. Ferrell, J.K. Bhattacharjee, A. Bagchi: Phys. Rev. Lett. *43*, 154 (1979)
3.92 H. Suhl: J. Less. Common Met. *62*, 225 (1978)
3.93 S. Maekawa, M. Tachiki: Phys. Rev. B*18*, 4688 (1978)
3.94 A. Sakurai: Solid State Commun. *25*, 867 (1978)
3.95 H. Matsumoto, G. Simenoff, R. Teshima, T. Tachiki: Preprint, University of Alberta, Canada and University of Tohoku, Japan (1979)
3.96 T.K. Lee: Solid State Commun. *34*, 9 (1980)
3.97 R.M. Hornreich, H.G. Schuster: Phys. Lett. A*70*, 143 (1979)
3.98 D.R. Nelson, J.M. Kosterlitz, M.E. Fisher: Phys. Rev. Lett. *33*, 813 (1974); Phys. Rev. B*13*, 412 (1975)
3.99 N. Greve, B. Schuh: Z. Phys. B*36*, 89 (1979)
3.100 P.G. de Gennes: Solid State Commun. *1*, 132 (1963)
3.101 J. Keller, P. Fulde: J. Low Temp. Phys. *4*, 289 (1971); *12*, 63 (1973)
3.102 E.K. Bucher, J.P. Andres, J.P. Maita, G.W. Hull: Helv. Phys. Acta *41*, 723 (1968)
3.103 P. Fulde, L. Hirst, A. Luther: Z. Phys. *230*, 155 (1970)
3.104 P. Fulde: In *Handbook on the Physics and Chemistry of Rare Earths*, ed. by K.A. Gschneider, L. Eyring (North Holland, Amsterdam 1978) Chap.17
3.105 J. Kondo: Prog. Theor. Phys. *32*, 37 (1964)
3.106 Y. Nagaoka: Phys. Rev. *138*, 1112 (1965); Prog. Theor. Phys. *37*, 13 (1967)
3.107 D.R. Hamann: Phys. Rev. *158*, 570 (1967)
3.108 E. Müller-Hartmann, J. Zittartz: Z. Phys. *212*, 380 (1968)
3.109 E. Müller-Hartmann, J. Zittartz: Z. Phys. *232*, 11 (1970)

3.110 J. Zittartz: Z. Phys. *237*, 419 (1970)
3.111 E. Müller-Hartmann, J. Zittartz: Phys. Rev. Lett. *26*, 427 (1971)
3.112 H. Shiba: Prog. Theor. Phys. *40*, 435 (1968)
3.113 E. Müller-Hartmann: In *Magnetism*, Vol.5, ed. by G. Rado, H. Suhl (Academic, New York 1973) p.353
3.114 M.B. Maple. W.A. Fertig, A.C. Mota, L.E. De Long. D. Wohlleben, R. Fitzgerald: Solid State Commun. *11*, 829 (1972)
3.115 M.B. Maple, K.S. Kim: Phys. Rev. Lett. *23*, 118 (1969)
3.116 C.F. Ratto, A. Blandin: Phys. Rev. *156*, 513 (1967)
3.117 M.B. Maple, J.G. Hubor, B.R. Coles, H.C. Lawson: J. Low Temp. Phys. *3*, 137 (1970)
3.118 M.B. Maple: In *Magnetism*, Vol.5, ed. by G. Rado, H. Suhl (Academic, New York 1973) p.289
3.119 M.B. Maple, L.E. De Long, W.A. Fertig, D.C. Johnston, R.W. McCallum, R.N. Shelton: In *Valence Instabilities and Related Narrow Band Phenomena*, ed. by R.D. Parks (Plenum, New York 1977) p.17
3.120 H. Suhl: Phys. Rev. *141*, 483 (1966)
3.121 G. Yuval, P.W. Anderson: Phys. Rev. B*1*, 1522 (1970)
3.122 P. Schlottmen: Solid State Commun. *16*, 1297 (1975); J. Low Temp. Phys. *20*, 123 (1975)
3.123 K.G. Wilson: Nobel Symp. *24*, 68 (1973)
3.124 M.B. Maple: In *Superconductivity in d- and f-Band Metals*, ed. by D.H. Douglass (American Institute of Physics, New York 1972) p.175
3.125 P.W. Anderson: Phys. Rev. *124*, 41 (1961)
3.126 M.J. Zuckermann: Phys. Rev. *140*, 899 (1965)
3.127 M. Kiwi, M.J. Zuckermann: Phys. Rev. *164*, 548 (1967)
3.128 A.B. Kaiser: J. Phys. Paris C*3*, 410 (1970)
3.129 J.R. Schrieffer, D.C. Mattis: Phys. Rev. *140*, 1412 (1965)
3.130 D. Markowitz, L.P. Kadanov: Phys. Rev. *131*, 563 (1963)
3.131 R. Aoki, T. Ohtsuka: J. Phys. Soc. Jpn. *23*, 955 (1967); *26*, 651 (1969)
3.132 J.G. Huber, M.B. Maple: J. Low Temp. Phys. *3*, 537 (1970)
3.133 B. Coqblin, A. Blandin: Adv. Phys. *17*, 281 (1968)
3.134 A.D. Caplin, C. Rizzuto: Phys. Rev. Lett. *21*, 746 (1968)
3.135 B. Coqblin, M.B. Maple, G. Toulouse: Int. J. Magn. *1*, 333 (1971)
3.136 J.R. Schrieffer, P.A. Wolff: Phys. Rev. *149*, 491 (1966)
3.137 G. Toulouse, B. Coqblin: Solid State Commun. *7*, 853 (1969)
3.138 M.B. Maple, J. Wittig: Solid State Commun. *9*, 1611 (1971)
3.139 M.J. Zuckermann: J. Phys. Paris C*3*, 2130 (1970)
3.140 A. Theumann: Phys. Rev. B*5*, 4382 (1972)
3.141 S.Q. Wang, W.E. Evenson, J.R. Schrieffer: Phys. Lett. *23*, 92 (1969)
3.142 N. Rivier: J. Phys. Paris F*1*, L48 (1971)
3.143 N. Rivier, M.J. Zuckermann: Phys. Rev. Lett. *21*, 904 (1968)
3.144 C. Hargitai, C. Corradi: Solid State Commun. *7*, 1535 (1969)
3.145 M. Kiwi, Y. Roessler: Solid State Commun. *15*, 1581 (1974)
3.146 M.T. Beal-Monod, D.L. Mills: Phys. Rev. Lett. *24*, 225 (1970)
3.147 H.R. Krishna-Murthy, K.G. Wilson, J.R. Wilkins: Phys. Rev. Lett. *35*, 1101 (1975)
3.148 G. Riblet: Phys. Rev. B*3*, 91 (1971)
3.149 G. Riblet, M.A. Jensen: Physica *55*, 622 (1971)
3.150 S. Doniach: Phys. Rev. Lett. *18*, 554 (1967)
3.151 N.F. Berk, J.R. Schrieffer: Phys. Rev. Lett. *17*, 433 (1966)
3.152 J.R. Schrieffer: J. Phys. Paris Suppl. *32*, C1-19 (1971)
3.153 J. Lindhard: K. Dan. Videns. Selsk. Mat. Fys. Medd. *28*, 8 (1954)
3.154 M.A. Jensen, K. Andres: Phys. Rev. *165*, 545 (1968)
3.155 E. Bucher, W.F. Brinkman, J.P. Maita, H.J. Williams: Phys. Rev. Lett. *18*, 1125 (1967)
3.156 B.L. Gyorffy, A. Pindor, W.M. Temmerman: Phys. Rev. Lett. *43*, 1343 (1979)

3.157 F.J. Pinski, P.B. Allen, W.H. Butler: Phys. Rev. Lett. *41*, 431 (1978)
3.158 H. Rietschel, H. Winter: Phys. Rev. Lett. *43*, 1256 (1979)
3.159 W.H. Butler, H.G. Smith, N. Wakabayashi: Phys. Rev. Lett. *39*, 1004 (1977)
3.160 H. Rietschel: Z. Phys. B*30*, 271 (1978)
3.161 J.F. Janak: Phys. Rev. B*16*, 255 (1977)

Chapter 4

4.1 B.W. Roberts: Nat. Bur. Stand. U.S. Tech. Note *983* (1978) Suppl.
4.2 E. Bucher, J. Müller, J.L. Olsen, C. Palmy: Phys. Lett. *15*, 303 (1965)
4.3 B.T. Matthias, T.H. Geballe, E. Corenzwit, G.W. Hull: Phys. Rev. *129*, 1025 (1963)
4.4 E. Bucher, C. Palmy: Phys. Lett. A*24*, 340 (1967)
4.5 E. Maxwell: Rev. Mod. Phys. *36*, 144 (1964)
4.6 T.H. Geballe, B.T. Matthias, G.W. Hull, E. Corenzwit: Phys. Rev. Lett. *6*, 275 (1961)
4.7 D.K. Finnemore, D.E. Mapother: Phys. Rev. Lett. *9*, 288 (1962)
4.8 J.W. Gibson, R.A. Hein: Phys. Rev. *141*, 407 (1966)
4.9 T.H. Geballe, B.T. Matthias: IBM J. Res. Dev. *6*, 250 (1967)
4.10 R.A. Hein, J.W. Gibson: Phys. Rev. *131*, 1105 (1963)
4.11 R.E. Fassnacht, J.R. Dillinger: Phys. Rev. Lett. *17*, 255 (1966)
4.12 I. Giaever: Proc. Eighth Int. Conf. Low Temp. Phys., ed. by R.O. Davies, (Butterworths, London 1963) p.171
4.13 P. Townsend, J. Sutton: Proc. Phys. Soc. London *78*, 309 (1961)
4.14 M.D. Sherill, H.H. Edwards: Phys. Rev. Lett. *6*, 460 (1961)
4.15 I. Dieterich: Z. Naturforsch. A*17*, 94 (1962)
4.16 S.I. Ochia, M.L.A. MacVicar, R.M. Rose: Phys. Rev. B*4*, 2988 (1971)
4.17 H. Bohm, N. Horwitz: Proc. Eighth Int. Conf. Low Temp. Phys., ed. by R.O. Davies (Butterworths, London 1963) p.171
4.18 J.L. Brewster, M. Levy, I. Rudnick: Phys. Rev. *132*, 1062 (1963)
4.19 M. Levy, R. Kagiwada, I. Rudnick: Phys. Rev. *132*, 2039 (1963)
4.20 E.R. Dobbs, J.M. Perz: Rev. Mod. Phys. *36*, 257 (1964)
4.21 M. Levy, I. Rudnick: Phys. Rev. *132*, 1073 (1963)
4.22 N. Horowitz, H. Bohm: Phys. Rev. Lett. *9*, 313 (1962)
4.23 C.K. Jones, J.A. Rayne: Phys. Lett. *21*, 510 (1966)
4.24 J.M. Perz, E.R. Dobbs: Proc. R. Soc. London A*296*, 113 (1967)
4.25 N.V. Zavaritskii: JETP *45*, 1839 (1963); *48*, 837 (1965)
4.26 M.L.A. MacVicar, R.M. Rose: J. Appl. Phys. *39*, 1721 (1968)
4.27 A.J. Bennett: Phys. Rev. *140*, 1902 (1965)
4.28 D.H. Gualtieri: J. Appl. Phys. *45*, 1880 (1974)
4.29 B.T. Matthias: Phys. Rev. *92*, 874 (1953)
4.30 E.V. Galoshina: Usp. Fiz. Nauk *113*, 105 (1974)
4.31a J. Linde, O. Rapp: Phys. Lett. A*70*, 147 (1979)
4.31b I.M. Chapnik: Phys. Lett. A*72*, 255 (1979)
4.31c B.T. Matthias: Helv. Phys. Acta *41*, 1030 (1968)
4.32 H. Stern: Phys. Rev. B*8*, 5109 (1973); B*12*, 951 (1975)
4.33 L.N. Bulaevskii, V.L. Ginzburg, A.I. Golovashkin, G.F. Zharkov, D.A. Kirzhnitz, E.G. Maksimov, G.P. Motulevitch, D.I. Khomskii: Preprint FIAN SSSR, No.45, (Moscow 1974) p.40
4.34 J. Wittig: Phys. Rev. Lett. *24*, 812 (1970)
4.35 N.B. Brandt, N.I. Ginzburg: Usp. Fiz. Nauk *85*, 485 (1965)
4.36 T.F. Smith, J. Phys. Paris F*2*, 946 (1972)
4.37 T.F. Smith: In *Superconductivity in d- and f-Band Metals*, ed. by D.H. Douglass (American Institute of Physics, New York 1972) p.293
4.38 N.B. Brandt, O.A. Zarubina: Fiz. Tverd. Tela Leningrad *15*, 3423 (1973)

4.39 I.V. Berman, N.B. Brandt, O.A. Zarubina, A.L. Karuzskii: Fiz. Tverd. Tela Leningrad *15*, 1070 (1973)
4.40 M.M. Collver, R.H. Hammond: Phys. Rev. Lett. *30*, 92 (1973)
4.41 P.H. Schmidt, R.N. Castellano, H. Barz, A.S. Cooper, E.G. Spencer: J. Appl. Phys. *44*, 1833 (1973)
4.42 B. Stritzker, J.D. Meyer: Bull. Am. Phys. Soc. *25*, MG 14-433 (1980)
4.43 K. Gamo, H. Goshi, M. Takai, M. Iwaki, K. Masuda, S. Namba: Jpn. J. Appl. Phys. *16*, 1853 (1977)
4.44 Y.Y. Shy, L.E. Toth, R. Sonasundaram: J. Appl. Phys. *44*, 5539 (1973)
4.45 D.D. Koelling, F.M. Müller, A.J. Arko, J.B. Ketterson: Phys. Rev. B*10*, 4889 (1974)
4.46 L.F. Mattheiss: Phys. Rev. B*1*, 373 (1970)
4.47 L.F. Mattheiss: Phys. Rev. *139*, 1893 (1965)
4.48 T.M. Hattox, J.B. Conklin, J.C. Slater, S.B. Trickey: J. Phys. Chem. Solids *34*, 1627 (1973)
4.49 M. Yasui, E. Hayashi, M. Shimizu: J. Phys. Soc. Jpn. *29*, 1446 (1970)
4.50 S. Wakoh, J. Yamashita: J. Phys. Soc. Jpn. *35*, 1394 (1973)
4.51a D.A. Papaconstantopoulos, J.R. Anderson, J.W. McCaffrey: Phys. Rev. B*5*, 1214 (1972)
4.51b I.I. Geguzin, I.Ya. Nikiforov, G.I. Alperovitch: Fiz. Tverd. Tela Leningrad *15*, 931 (1973)
4.52a J.R. Anderson, D.A. Papaconstantopoulos, J.W. McCaffrey, J.E. Schirber: Phys. Rev. B*7*, 5115 (1973)
4.52b V.V. Nemoshkalenko, V.N. Antonov, V.G. Aleshin: Preprint IMF AN USSR *75*, 14 (1975)
4.53 V.A. Fomichev, T.M. Zimkina, A.V. Rudnev: In *Band Spectroscopy of Metals and Alloys*, ed. by D.J. Fabian, L.M. Watson (Academic, London 1973) p.259
4.54 D.W. Fischer: J. Appl. Phys. *40*, 4151 (1969)
4.55 J. Fuggle, D.J. Fabian: In *X-Ray Spectra and Electronic Structure of Matter*, ed. by A. Faessler, G. Wiech (Fotodruck Frank, München 1973) Vol.2, p.164
4.56 D.E. Eastman: Solid State Commun. *7*, 1697 (1969)
4.57 S.A. Nemnonov, E.Z. Kurmaev, V.I. Minin: In *Collected Papers: Rjentgjenovskije spjektry i eletronnaya struktura vjeschjestva*, IMF AN USSR (Kiev 1969) Vol.1, p.87
4.58 V.P. Belash, E.Z. Kurmaev, S.A. Nemnonov: Fiz. Met. Metalloved. *37*, 659 (1974)
4.59 N.F. Mott: Proc. Phys. Soc. London *47*, 571 (1935)
4.60 L.P. Bouckart, R. Smoluchowsky, E. Wigner: Phys. Rev. *50*, 58 (1936)
4.61 B.K. Vainshtein: *Modern Crystallography I*, Springer Ser. Solid-State Sci., Vol.15 (Springer, Berlin, Heidelberg, New York 1981)
4.62 J. Friedel: In *Collected Papers: The Physics of Metals, 1. Electrons*, ed. by J.M. Ziman (University Press, Cambridge 1969)
4.63 O. Jepsen: Phys. Rev. B*12*, 2988 (1975)
4.64 O. Jepsen, O.K. Andersen, A.R. Mackintosh: Phys. Rev. B*12*, 3084 (1975)
4.65 R.J. Cotton, J.T.J. Huang, J.W. Rabalais: Chem. Phys. Lett. *34*, 337 (1975)
4.66 S. Altmann, C.J. Bradley: Proc. Phys. Soc. London *92*, 764 (1967)
4.67 V. Bhokare, M. Yussouff: Nuovo Cimento B*19*, 149 (1974)
4.68 T.L. Loucks: Phys. Rev. *159*, 544 (1967)
4.69 L.F. Mattheiss: Phys. Rev. *151*, 450 (1966)
4.70 R. Asokamani, K. Iyakutti, V. Devancthan: Solid State Commun. *30*, 385 (1970)
4.71 N.E. Christinsen, B. Tenerbacher: Phys. Rev. B*10*, 2349 (1974)
4.72 J.B. Kettersen, D.D. Koelling, J.C. Shaw, L.R.W. Windmiller: Phys. Rev. B*11*, 1447 (1975)
4.73 O.K. Anderson, A.R. Mackintosh: Solid State Commun. *6*, 285 (1968); Phys. Rev. B*2*, 883 (1970)

4.74 U. Harrison: *Pseudopotentials in the Theory of Metals* (Benjamin, New York 1963)
4.75 G.S. Fleming, S.H. Liu, T.L. Loucks: Phys. Rev. Lett. *21*, 1524 (1968)
4.76 A.P. Cracknell, K.C. Wong: *The Fermi Surface* (Clarendon, Oxford 1973)
4.77 S.C. Keeton, T.L. Loucks: Phys. Rev. *168*, 672 (1968)
4.78 W.E. Evenson, S.H. Liu: Phys. Rev. Lett. *21*, 432 (1968)
4.79 L.F. Mattheis: Phys. Rev. *134*, 970 (1964)
4.80 E.H. Hygh, R.M. Welch: Phys. Rev. B*1*, 2424 (1970)
4.81 R.M. Welch, E.H. Hygh: Phys. Rev. B*4*, 4261 (1971)
4.82 M.H. Halloran, J.H. Condon, J.E. Graebner, J.E. Kunzler, P.S.L. Hsu: Phys. Rev. B*1*, 366 (1970)
4.83 W.L. McMillan: Phys. Rev. *167*, 331 (1968)
4.84 L.L. Boyer, D.A. Papaconstantopoulos, B.M. Klein: Phys. Rev. B*15*, 3685 (1977)
4.85 B. Allen: Phys. Rev. B*13*, 1416 (1976)
4.86 V.G. Bar'yakhtar, I.I. Falko, V.I. Makarov: JETP *49*, 1858 (1965)
4.87 V.G. Bar'yakhtar, V.V. Gann, V.I. Makarov, T.A. Ignatyeva: JETP *62*, 1118 (1972)
4.88 V.G. Bar'yakhtar, B.G. Lazarev, V.I. Makarov: Fiz. Met. Metalloved. *24*, 829 (1967)
4.89 A.L. Simons, C.M. Varma, W. Weber: Phys. Rev. B*23*, 2431 (1981)
4.90 D.J. Fabian, L.M. Watson, C.A.W. Marshall: Rep. Prog. Phys. *34*, 601 (1971)
4.91 V.V. Nemoshkalenko: *Rentgenovskaya emissionnaya spektroskopiya metallov i splavov* (Naukova Dumka, Kiev 1972)
4.92 A. Meisel, G. Leonhardt, R. Szargan: *Röntgenspektren und chemische Bindung* (Akademische Verlagsgesellschaft Geest and Portig, Leipzig 1977)
4.93 V.V. Nemoshkalenko, F.B. Aleshin: *Elektronnaya Spektroskopiya Kristallov* (Naukova Dumka, Kiev 1976)
4.94 J.E. Holliday: In *Soft X-Ray Spectra and the Electronic Structure of Metals and Materials*, ed. by D.J. Fabian (Academic, London 1968) p.101
4.95 E.C. Snow, J.T. Waber: Acta Metall. *17*, 623 (1969)
4.96 F. Heiniger, E. Bucher, J. Müller: Phys. Kondens. Mater. *5*, 243 (1966)
4.97 K.A. Gschneidner: Solid State Phys. *16*, 275 (1964)
4.98 V.P. Shirokovskii: Ph. D. Thesis, IFM UNC AN SSSR, Sverdlovsk (1976)
4.99 W.S. Gorak, B.B. Goodman, C.B. Satterthwaite, A. Wexler: Phys. Rev. *102*, 656 (1956)
4.100 C.H. Cheng, K.P. Gupta, E.C. van Reuth, P.A. Beck: Phys. Rev. *126*, 2030 (1962)
4.101 P.H. Keesom, R. Radebaugh: Phys. Rev. Lett. *13*, 685 (1964)
4.102 A.V. Rudnev, V.A. Fomitchev: In *Collected Papers: Eljektronnaja struktura pjerjekhodnykh mjetallov, ikh splavov i sojedinjenij* (Naukova Dumka, Kiev 1974) p.50
4.103 M. Ishikawa, L.E. Toth: Phys. Rev. B*3*, 1856 (1971)
4.104 I. Petroff, C.R. Viswanathan: Phys. Rev. B*4*, 799 (1971)
4.105 D.C. Rorer, D.G. Onn, H. Meyer: Phys. Rev. *129*, 1115 (1963)
4.106 R.C. Weast (ed.): *Handbook of Chemistry and Physics* (Chemical Rubber, Cleveland 1967, 1968) Vol.48
4.107 R.R. Hultgren, R.L. Orr, P.D. Anderson, K.K. Kelley: In *Selected Values of Thermodynamic Properties of Metals and Alloys*, ed. by J.H. Hollomon, J.E. Burke, B. Chalmers, R.L. Sproull, A.V. Tobolsky (Wiley, New York 1963) pp.176,272,308
4.108 B.M. Kelin, D.A. Papaconstantopoulos: Phys. Rev. Lett. *32*, 1193 (1974)
4.109 A.T. Stewart, B.N. Brockhouse: Rev. Mod. Phys. *30*, 250 (1958)
4.110 C.M. Eisenhauer, I. Pelah, D.J. Hughes, H. Palevsky: Phys. Rev. *109*, 1046 (1958)
4.111 K.C. Turberfield, P.A. Egelstaff: Phys. Rev. *127*, 1017 (1962)

4.112 M.G. Zemlyanov, Yu.M. Kagan, N.A. Chernoplekov, A.G. Checherin: In *Inelastic Scattering of Neutrons in Solids and Liquids* (IAEA, Vienna 1963) Vol.2, p.125
4.113 R. Hass, W. Kley, K.H. Krobs, R. Rubin: ibid., p.145
4.114 W. Gaser, F. Carvalho, G. Ehret: ibid., p.99
4.115 D.I. Page: Proc. Phys. Soc. London *91*, 76 (1967)
4.116 N.A. Chernoplekov, M.G. Zemlyanov, A.G. Checherin, V.G. Lyashenko: In *Inelastic Scattering of Neutrons in Solids and Liquids* (IAEA, Vienna 1963) Vol.2, p.159
4.117 B. Mozer, K. Otnes, H. Palevsky: In *Lattice Dynamics*, ed. by R.F. Wallis (Pergamon, New York 1965) p.63
4.118 R. Colella, B.W. Batterman: Phys. Rev. B*1*, 3913 (1970)
4.119 Y. Nakagawa, A.D.B. Woods: Phys. Rev. Lett. *11*, 271 (1963)
4.120 S.H. Chen, B.N. Brockhouse: Solid State Commun. *2*, 73 (1964)
4.121 W.H. Shaw, L.D. Muhlestein: Phys. Rev. B*4*, 969 (1971)
4.122 B.M. Powell, P. Martel, A.D.B. Woods: Phys. Rev. *171*, 727 (1968)
4.123 C. Stassis, D. Arch, B.N. Harmon, N. Wakabayashi: Phys. Rev. B*19*, 181 (1979)
4.124 C. Stassis, J. Zarestky, D. Arch, O.D. McMasters, B.N. Harmon: Phys. Rev. B*18*, 2632 (1978)
4.125 H.F. Bezdek, R.E. Schmunk, L. Finegold: Phys. Status Solidi B*42*, 275 (1970)
4.126 U. Harrison: *Solid State Theory* (McGraw-Hill, New York 1970)
4.127 W. Kohn: Phys. Rev. Lett. *2*, 393 (1959)
4.128 D. Bohm, T. Staver: Phys. Rev. *84*, 836 (1952)
4.129 W.L. McMillan, J.M. Rowell: Phys. Rev. Lett. *14*, 108 (1965)
4.130a L.Y.L. Shen: Phys. Rev. Lett. *24*, 1104 (1970)
4.130b E.L. Wolf, D.M. Burnell, R.J. Noer: Phys. Lett. A*82*, 300 (1981)
4.131a L.Y.L. Shen: In *Superconductivity in d- and f-Band Metals*, ed. by D.H. Douglass (American Institute of Physics, New York 1972) p.31
4.131b D.B. Kimbui, T.H. Geballe: Phys. Rev. Lett. *45*, 1039 (1980)
4.132a B. Schoneich, D. Elefant, P. Otschik, J. Schumaun: Phys. Status Solidi B*91*, 99 (1979)
4.132b Yu.F. Revenko, A.I. Dyatchenko, V.M. Svistunov, B. Shonaikh: Fiz. Nizk. Temp. *6*, 1304 (1980)
4.133 A.D.B. Woods: In *Superconductivity in d- and f-Band Metals*, ed. by D.H. Douglass (American Institute of Physics, New York 1972) p.45
4.134 H.G. Smith, N. Wakabayashi, R.M. Nicklow, W. Michailovich: In *Low Temperature Physics LT-13*, ed. by K.D. Timmerhaus, W.J.O. Sullivan, E.F. Hammel (Plenum, New York 1974) Vol.3, p.615
4.135 H.G. Smith, N. Wakabayashi, M. Mostoller: In *Superconductivity in d- and f-Band Metals*, ed. by D.H. Douglass (Plenum, New York 1976) p.223
4.136 C. Stassis, D. Arch. J. Zarestky, O.D. McMasters, B.N. Harmon: Solid State Commun. *35*, 259 (1980)
4.137 M.M. Kirillova, B.A. Tcharikov: Opt. Spektrosk. Akad. Nauk SSSR Otd. Fiz. Mat. Nauk *17*, 254 (1964)
4.138 A.I. Golovashkin, D.I. Mash, G.P. Motulevitch: Kratk. Soobshch. Fiz. *9*, 51 (1970)
4.139 G.S. Knapp, R.W. Jones, B.A. Loomis: In *Low Temperature Physics LT-13*, ed. by K.D. Timmerhaus, W.J.O. Sullivan, E.F. Hammel (Plenum, New York 1974) Vol.3, p.615
4.140 L.V. Nomerovannaya: Ph. D. Thesis, IFM UNC AN SSSR Sverdlovsk (1973)
4.141 M.M. Kirillova, B.A. Tcharikov: Fiz. Met. Metalloved. *16*, 205 (1963)
4.142 B. Robinson, J.M. Rowell: In *Transition Metals*, ed. by H.J.G. Lee, J.M. Perz, E. Fawcett (Institute of Physics, Bristol 1978) p.666
4.143 B. Chakraborty, W.E. Pickett, P.B. Allen: Phys. Rev. B*14*, 3227 (1976)
4.144 A.I. Golovashkin, I.E. Leksina, G.P. Motulevitch, A.A. Shubin: JETP *56*, 51 (1969)

4.145 E.L. Wolf, J. Zasadzinski: In *Transition Metals*, ed. by H.J.G. Lee, J.M. Perz, E. Fawcett (Institute of Physics, Bristol 1978) p.666
4.146 A.P. Lenham, D.M. Treherne: In *Optical Properties and the Electronic Structure of Metals*, ed. by F. Abeles (North Holland, Amsterdam 1966) p.196
4.147 I. Tutto, L.M. Kahn, J. Ruvalds: Phys. Rev. B*20*, 952 (1979)
4.148 M.M. Kirillova, G.A. Bolotin, V.M. Maevskii: Fiz. Met. Metalloved. *24*, 95 (1967)
4.149 A.P. Lenham, D.M. Treherne: J. Opt. Soc. Am. *56*, 1076 (1966)
4.150 M.M. Kirillova, L.V. Nomerovannaya, M.M. Noskov, N.B. Gorina, V.P. Polyakova, E.M. Savitskii: Fiz. Tverd. Tela Leningrad *20*, 1717 (1978)
4.151 L.V. Nomerovannaya, M.M. Kirillova, M.M. Noskov, G.A. Bolotin: Fiz. Met. Metalloved. *36*, 1197 (1973)
4.152 M.M. Kirillova, L.V. Nomerovannaya, M.M. Noskov: Fiz. Met. Metalloved. *34*, 291 (1972)
4.153 J.J. Hopfield: Phys. Rev. *186*, 443 (1969)
4.154 S. Barisic, J. Labbe, J. Friedel: Phys. Rev. Lett. *25*, 919 (1970)
4.155 W.H. Butler, J.J. Olson, J.S. Faulkner, B.L. Gyorffy: Phys. Rev. B*14*, 3823 (1976)
4.156 W.H. Butler: Phys. Rev. B*15*, 5267 (1977)
4.157 C.M. Varma, R.C. Dynes: In *Superconductivity in d- and f-Band Metals*, ed. by D.H. Douglass (Plenum, New York 1976) p.507
4.158 D.G. Pettifor: J. Phys. Paris F*7*, 1009 (1977)
4.159 R.C. Dynes, C.M. Varma: J. Phys. Paris F*6*, L215 (1976)
4.160 C.M. Varma, E.I. Blount, P. Vashishta, W. Weber: Phys. Rev. B*19*, 6130 (1979)
4.161 E.I. Maksimov: JETP *57*, 1660 (1969)
4.162 J.J. Hopfield: Comments Solid State Phys. *3*, 48 (1970)
4.163 N. Wakabayashi: Comments Solid State Phys. *10*, 11 (1981)
4.164 J.R. Gommersal, B.L. Gyorffy: Phys. Rev. Lett. *33*, 1286 (1974)
4.165 A.B. Migdal: JETP *34*, 1438 (1958)
4.166 E. Brovman, Yu. Kagan: Usp. Fiz. Nauk *112*, 369 (1974)
4.167 E. Brovman, Yu. Kagan: JETP *52*, 557 (1967)
4.168 G.D. Gaspari, B.L. Gyorffy: Phys. Rev. Lett. *28*, 801 (1972)
4.169 R. Evans, G.D. Gaspari, B.L. Gyorffy: J. Phys. Paris F*3*, 39 (1973)
4.170 P. Lloyd, P.V. Smith: Adv. Phys. *21*, 69 (1972)
4.171 W. John: J. Phys. Paris F*3*, L231 (1973)
4.172 J.R. Gommersal, B.L. Gyorffy: J. Phys. Paris F*4*, 1204 (1974)
4.173 V. Heine: Phys. Rev. *153*, 673 (1967)
4.174 O.K. Anderson: Phys. Rev. B*2*, 883 (1970)
4.175 D.A. Papaconstantopoulos, L.L. Boyer, B.M. Klein, A.R. Williams, V.L. Moruzzi, J.F. Janak: Phys. Rev. B*15*, 4221 (1977)
4.176 American Institute of Physics Handbook, 3rd ed. (McGraw-Hill, New York 1972) pp.4,115,116
4.177 K.H. Bennemann, J.W. Garland: In *Superconductivity in d- and f-Band Metals*, ed. by D.H. Douglass (American Institute of Physics, New York 1972) p.103
4.178 J.J. Hopfield: Physica *55*, 41 (1971)
4.179 R. Evans, V.K. Ratti, B.L. Gyorffy: J. Phys. Paris F*3*, L199 (1973)
4.180 V.K. Ratti, A. Jain: J. Phys. Paris F*3*, L69 (1973)
4.181 V.K. Ratti, R. Evans, B.L. Gyorffy: J. Phys. Paris F*4*, 371 (1974)
4.182 J. Friedel: Philos. Mag. *43*, 153 (1952)
4.183 J.W. Garland, K.H. Bennemann: In *Superconductivity in d- and f-Band Metals*, ed. by D.H. Douglass (American Institute of Physics, New York 1972) p.255

Chapter 5

5.1 E.M. Savitskii, V.V. Baron, Yu.V. Efimov, M.I. Bytchkova, L.F. Myzenkova: *Mjetallovjedjenije svjerkhprovodjaschikh matjerialov* (Nauka, Moscow 1969)
5.2 R.P. Elliot: *Constitution of Binary Alloys* (McGraw-Hill, New York 1962) First Supplement
5.3 T.H. Geballe: Rev. Mod. Phys. *36*, 134 (1964)
5.4 J.K. Hulm, R.D. Blaugher: Phys. Rev. *123*, 1569 (1961)
5.5 V.B. Compton, E. Corenzwit, J.P. Maita, B.T. Matthias, F.J. Morin: Phys. Rev. *123*, 1567 (1961)
5.6 K. Andres, E. Bucher, J.P. Maita, R.S. Sherwood: Phys. Rev. *178*, 702 (1969)
5.7 T. Mamiya, K. Nomura, Y. Masuda: J. Phys. Soc. Jpn. *28*, 380 (1970)
5.8 C.W. Chu, W.L. McMillan, H.L. Luo: Phys. Rev. B*3*, 3757 (1971)
5.9 R.D. Blaugher, A. Taylor, J.K. Hulm: IBM J. Res. Dev. *6*, 116 (1962)
5.10 K. Andres, M.A. Jensen: Phys. Rev. *165*, 533 (1968)
5.11 O. Rapp: J. Less Common Met. *21*, 27 (1970)
5.12 A.L. Giorgi, E.G. Szklarz: J. Less Common Met. *20*, 173 (1970)
5.13 G. Dummer: Z. Phys. *186*, 249 (1965)
5.14 B.T. Matthias: Phys. Rev. *97*, 74 (1955)
5.15 S.H. Autler, J.K. Hulm, R.S. Kemper: Phys. Rev. *140*, 1177 (1965)
5.16 M.A. Jensen, J.P. Maita: Phys. Rev. *149*, 409 (1966)
5.17 R.R. Hake: Phys. Rev. *123*, 1986 (1961)
5.18 B.T. Matthias, T.H. Geballe, V.B. Compton, E. Corenzwit, G.W. Hull: Phys. Rev. *128*, 588 (1962)
5.19 M. Ishikawa, L.E. Toth: Phys. Rev. B*3*, 1856 (1971)
5.20 E.M. Savitskii, V.V. Baron, Yu.N. Efimov: Tr. Inst. Metall. Akad. Nauk SSSR *4*, 230 (1960)
5.21 L.F. Myzenkova, V.V. Baron, E.M. Savitskii: In *Collected Papers: Mjetallovjedjenije, fiziko-khimija i mjetallofizika svjerkhprovodnikov* (Nauka, Moscow 1967) p.37
5.22 M.I. Bytchkova, V.V. Baron, E.M. Savitskii: ibid., p.68
5.23 G. Riblet, M.A. Jensen: Physica *55*, 622 (1971)
5.24 G. Dummer, E. Oftedal: Z. Phys. *208*, 238 (1968)
5.25 H. Wuhl: Z. Phys. *197*, 276 (1966)
5.26 K. Yasohama, T. Ogasawara: J. Phys. Soc. Jpn. *36*, 1349 (1974)
5.27 G. Riblet: Phys. Rev. B*3*, 91 (1971)
5.28 A.F. Prekul, V.A. Rassokhin, N.V. Volkenshtein: JETP Lett. *17*, 354 (1973)
5.29 A.F. Prekul, V.A. Rassokhin, N.V. Volkenshtein: Fiz. Met. Metalloved. *37*, 671 (1974)
5.30 E.W. Collings, J.C. Ho, R.L. Jaffee: Phys. Rev. B*5*, 4435 (1972)
5.31 C.C. Koch, R.H. Kernohan, S.T. Sekula: J. Appl. Phys. *38*, 4359 (1967)
5.32 B.W. Roberts: Nat. Bur. Stand. U.S. Tech. Note *482* (1969)
5.33 B.W. Roberts: Nat. Bur. Stand. U.S. Tech. Note *724* (1972); *825* (1974); *983* (1978)
5.34 B.W. Roberts: In *Intermetallic Compounds*, ed. by J.H. Westbrook (Wiley, New York 1967) p.581
5.35 G. Gladstone, M. Jensen, J. Schrieffer: In *Superconductivity*, ed. by R.D. Parks (Dekker, New York 1969)
5.36 P.H. Bellin, V. Sadagopan, H.C. Gatos: J. Appl. Phys. *40*, 3982 (1969)
5.37 V. Sadagopan, H.C. Gatos, G. Olson: J. Appl. Phys. *41*, 1874 (1970)
5.38 N.E. Alekseevskii, O.S. Ivanov, I.I. Raevskii, N.V. Stepanov: Fiz. Met. Metalloved. *23*, 28 (1977)
5.39 L.F. Myzenkova, Yu.V. Efimov, V.V. Baron, E.M. Savitskii: In *Collected Papers: Mjetallovjedjenije, fiziko-khimija i mjetallofizika svjerkhprovodnikov* (Nauka, Moscow 1965) p.39

5.40 B.G. Lazarev, O.N. Ovtcharenko, A.A. Matsakova, V.G. Volotskaya: In *Collected Papers: Mjetallovjedjenije, fiziko-khimija i mjetallofizika svjerkhprovodnikov* (Nauka, Moscow 1967) pp.76,98
5.41 N.E. Alekseevskii, O.S. Ivanov, I.I. Raevskii, N.V. Stepanov: Dokl. Akad. Nauk SSSR *176*, 305 (1967)
5.42 T. Doi, M. Masao, Y. Tagasi: J. Jpn. Inst. Metals *30*, 133 (1966)
5.43 M. Suenaga, K.M. Ralls: J. Appl. Phys. *40*, 4457 (1969)
5.44 T.G. Berlincourt, R.R. Hake: Phys. Rev. *131*, 140 (1963)
5.45 Yu.F. Bytchkov, V.Yu. Uzlov, A.N. Rozanov, A.A. Mitrofanskii: In *Collected Papers: Struktura i svojstava svjerkhprovodjaschikh matjerialov* (Nauka, Moscow 1974) p.123
5.46 N.B. Brandt, N.I. Ginzburg: JETP *51*, 59 (1966)
5.47 E.S. Itskyevitch, M.A. Ilyina, V.A. Sukhorukov: JETP *45*, 1378 (1963)
5.48 W. Gey, D. Kohnlein: Z. Phys. *255*, 308 (1972)
5.49 C.W. Chu, T.F. Smith: Phys. Rev. Lett. *20*, 198 (1968)
5.50 W. Gey: Z. Phys. *229*, 85 (1969)
5.51 T.F. Smith, R.N. Shelton: J. Phys. Paris F*5*, 911 (1975)
5.52 R.N. Shelton, T.F. Smith, C.C. Koch, W.E. Gardner: J. Phys. Paris F*5*, 1916 (1975)
5.53 N.S. Afonikova, V.F. Degtyareva, Yu.A. Litvin, A.G. Rabin'kin, Yu.A. Skakov: Fiz. Tverd. Tela Leningrad *15*, 1096 (1973)
5.54 A.G. Rabin'kin, L.A. Klishanova, L.P. Pronina: In *Collected Papers: Probljemy svjerkhprovodjaschikh matjerialov* (Nauka, Moscow 1970) p.141
5.55 B.T. Matthias: Phys. Rev. *92*, 874 (1953); *97*, 74 (1955); Phys. Today *24(8)*, 23 (1971)
5.56 M.M. Collver, R.H. Hammond: Phys. Rev. Lett. *30*, 92 (1973)
5.57 J.D. Bernall: Annu. Rep. Prog. Chem. *30*, 287 (1933)
5.58 S. Tanigushi, R.S. Tebble, D.E.G. Williams: Proc. R. Soc. London A*265*, 502 (1962)
5.59 D.O. Van Ostenberg, D.J. Lam, M. Shimizu, A. Katsuki: J. Phys. Soc. Jpn. *18*, 1744 (1963)
5.60 D.J. Lam, D.O. Van Ostenberg, D.W. Pracht: J. Appl. Phys. *35*, 976 (1964)
5.61 M. Shimizu, T. Takahashi, A. Katsuki: J. Phys. Soc. Jpn. *18*, 1192 (1963)
5.62 D.W. Jones, A.D. McQuillan: J. Phys. Chem. Solids *23*, 1441 (1962)
5.63 J.G. Booth: Philos. Mag. *17*, 205 (1968)
5.64 F. Heiniger, E. Bucher, J. Müller: Phys. Kondens. Mater. *5*, 243 (1966)
5.65 S.A. Nemnonov: Fiz. Met. Metalloved. *19*, 550 (1965)
5.66 F.E. Hoare, J.C. Matthews, J.C. Walling: Proc. R. Soc. London A*216*, 502 (1953)
5.67 E. Bucher, F. Heiniger, J. Muheim, J. Müller: Rev. Mod. Phys. *36*, 146 (1964)
5.68 E.W. Collings: Phys. Rev. B*9*, 3989 (1974)
5.69 D.I. Bolef: J. Appl. Phys. *32*, 100 (1961)
5.70 F.H. Featherston, J.R. Neighbours: Phys. Rev. *130*, 1324 (1963)
5.71 D.I. Bolef, J. de Klerk: Phys. Rev. *129*, 1063 (1963)
5.72 D.L. Davidson, F.R. Brotzen: J. Appl. Phys. *39*, 5768 (1968)
5.73 E.S. Fisher, D. Dever: Acta Metall. *18*, 265 (1970)
5.74 S.L. Sass: J. Less Common Met. *28*, 157 (1972)
5.75 A.R. Miedema: J. Phys. Paris F*3*, 1803 (1973)
5.76 A.R. Miedema, M.H. Van Maaren: Physica *69*, 308 (1973)
5.77 A.R. Miedema: J. Phys. Paris F*4*, 120 (1974)
5.78 J.C. Phillips: Phys. Rev. Lett. *26*, 543 (1971); *29*, 1551 (1972); J. Appl. Phys. *43*, 3560 (1972)
5.79 P.W. Anderson: Proc. Seventh Conf. Low Temp. Phys., ed. by G.M. Graham, A.C.H. Hallett (North Holland, Amsterdam 1961) p.298
5.80 A. Weinkauf, J. Zittartz: Solid State Commun. *14*, 365 (1974)
5.81 A. Weinkauf, J. Zittartz: J. Low Temp. Phys. *18*, 229 (1975)
5.82 P. Soven: Phys. Rev. *156*, 809 (1967)

5.83 B. Velicky, S. Kirkpatrick, H. Ehrenreich: Phys. Rev. *175*, 747 (1968)
5.84 H. Lustfeld: J. Low Temp. Phys. *12*, 595 (1975)
5.85 G. Kerker, K.H. Benneman: Solid State Commun. *14*, 399 (1974)
5.86 A.L. Giorgi, B.T. Matthias: Phys. Rev. B*17*, 2160 (1978)
5.87 A.F. Prekul, N.V. Volkenshtein: Fiz. Nizk. Temp. *4*, 1514 (1978)
5.88 A.F. Prekul, A.S. SCherbakov, N.V. Volkenshtein: Fiz. Nizk. Temp. *2*, 1399 (1976)
5.89 E.W. Collings, J.J. White: In *Transition Metals*, ed. by M.J.G. Lee, J.M. Perz, E. Fawcett (Institute of Physics, Bristol 1978) p.645
5.90 A.F. Prekul: Phys. Lett. A*64*, 473 (1978)
5.91 N.E. Alekseevskii, O.A. Balakhovskii, I.V. Kirillova: Fiz. Met. Metalloved. *40*, 50 (1975)
5.92 R. Asokamani, K. Iyakutti, V. Devanathan: Solid State Commun. *30*, 391 (1979)

Chapter 6

6.1 L.R. Testardi, J.H. Wernick, W.A. Royer: Solid State Commun. *15*, 1 (1974)
6.2 N.N. Mikhailov, I.V. Voronova, O.A. Lavrova, E.V. Melnikov, M.N. Smirnova: JETP Lett. *19*, 510 (1974)
6.3 E.M. Savitskii, V.V. Baron, Yu.V. Efimov, M.I. Bytchkova, L.F. Myzenkova: *Mjetallovjedjenije svjerkhprovodjaschikh matjerialov* (Nauka, Moscow 1969)
6.4 Yu.V. Efimov: Izv. Akad. Nauk SSSR Neorg. Mater. *2*, 598 (1966)
6.5 K. Girgis: J. Less Common Met. *65*, 1 (1979)
6.6 R.M. Waterstrat, E.C. van Reuth: In *Ordered Alloys* (Claitor's Publishing Division, Baton Rouge 1970) p.95
6.7 B.W. Roberts: Nat. Bur. Stand. U.S. Tech. Note *482* (1969); *724* (1972); *825* (1974); *983* (1978)
6.8 A.I. Golovashkin, I.S. Levtchenko, G.P. Motulevitch: Tr. Fiz. Inst. Akad. Nauk SSSR *82*, 72 (1975)
6.9 L.A. Pendrys, D.H. Douglass: Solid State Commun. *18*, 177 (1976)
6.10 R.H. Willens, T.H. Geballe, A.C. Gossard, J.P. Maita, A. Menth, G.W. Hull, Jr., R.R. Soden: Solid State Commun. *7*, 837 (1969)
6.11 R.A. Hein, J.E. Cox, R.D. Blaugher, R.M. Waterstrat: Solid State Commun. *7*, 381 (1969)
6.12 A. Junod, R. Flukiger, J. Müller: J. Phys. Chem. Solids *37*, 27 (1976)
6.13 V.M. Pan, V.P. Alekseevskii, A.G. Popov, Yu.I. Beletskii, L.M. Yupko, V.V. Yarosh: JETP Lett. *21*, 494 (1975)
6.14 D.S. Easton, C.C. Koch, D.M. Kroeger, J.W. Cable: Philos. Mag. *30*, 1117 (1974)
6.15 H.-L. Luo, E. Vielhaber, E. Corenzwit: Z. Phys. *230*, 443 (1970); E. Vielhaber, H.-L. Luo: Solid State Commun. *5*, 221 (1967)
6.16 J.R. Gavaler, M.A. Janocko, C.K. Jones: Appl. Phys. Lett. *21*, 179 (1972)
6.17 R. Flukiger: Ph. D. Thesis 1570, Geneva (1972)
6.18 W.L. Johnson, S.J. Poon: J. Less Common Met. *42*, 355 (1975)
6.19 P. Spitzli: Phys. Kondens. Mater. *13*, 22 (1971)
6.20 D.J. Bardos, R.M. Waterstrat, T.J. Rowland, J.B. Darby, Jr.: J. Low Temp. Phys. *3*, 509 (1970)
6.21 G.W. Webb, L.J. Vieland, R.E. Miller, A. Wicklund: Solid State Commun. *9*, 1769 (1971)
6.22 J.F. Cannon, D.L. Robertson, H. Tracy Hall, A.C. Lawson: J. Phys. Chem. Solids *35*, 1181 (1975)
6.23 H.D. Wiesinger: Phys. Status Solidi A*36*, 61 (1976)
6.24 H.D. Wiesinger: Phys. Status Solidi A*41*, 465 (1977)
6.25 E.M. Savitskii, Yu.V. Efimov, T.M. Frolova: Dokl. Akad. Nauk SSSR *236*, 582 (1977)

6.26 E.M Savitskii, V.G. Petrenko, L.F. Vereschagin, V.V. Evdokimova, V.I. Novokshonov: In *Fiziko-khimitchjeskii analiz svjerkhprovodjachikh splavov*, ed. by E.M. Savitskii (Nauka, Moscow 1979) p.152
6.27 R.E. Somekh, J.E. Evetts: Solid State Commun. *24*, 733 (1977)
6.28 G.R. Stewart, A.L. Giorgi: Phys. Rev. B*17*, 7534 (1978)
6.29 A.L. Giorgi, B.T. Matthias, G.R. Stewart: Solid State Commun. *27*, 291 (1978)
6.30 D. Dew-Hughes, V. Linde: J. Appl. Phys. *50*, 3500 (1979)
6.31 J. Müller: Rep. Prog. Phys. *43*, 641 (1980)
6.32 D. Dew-Hughes, V.G. Rivlin: Nature *250*, 723 (1974)
6.33 R. Flukiger, A. Paoli, J. Müller: Solid State Commun. *14*, 443 (1974)
6.34 A. Guha, M.P. Sarachik, F.W. Smith, L.R. Testardi: Phys. Rev. B*18*, 9 (1978)
6.35 R. Flükiger: In *Superconductor Materials Science*, ed. by S. Foner, B. Schwartz (Plenum, New York 1981) p.511
6.36 A. Junod, J.L. Staudenmann, J. Müller, P. Spitzli: J. Low Temp. Phys. *5*, 25 (1971)
6.37 B.T. Matthias, T.N. Geballe, R.H. Willens, E. Corenzwit, G.W. Hull, Jr.: Phys. Rev. *139*, 1501 (1965)
6.38 E.C. Van Reuth, N.J. Poulis: Phys. Lett. A*25*, 390 (1967)
6.39 G.H. Bongi: J. Phys. Paris F*6*, 1535 (1976)
6.40 G.H. Bongi, Ø. Fischer, H. Jones: J. Phys. Paris F*4*, L259 (1974)
6.41 P. Coppens: In *Transition Metals*, ed. by M.J. Lee, J.M. Perz, E. Fawcett (Institute of Physics, Bristol 1978) p.630
6.42 J.L. Staudenmann: Solid State Commun. *23*, 121 (1977)
6.43 J.L. Staudenmann, P. Coppens, J. Müller: Solid State Commun. *19*, 29 (1976)
6.44 V.I. Anisimov, V.A. Gubanov, A.L. Ivanovskii, E.Z. Kurmaev, J. Weber, R. Lacroix: Solid State Commun. *29*, 185 (1979)
6.45 D.U. Gubser: Phys. Rev. B*19*, 199 (1979)
6.46 V.F. Shamray: Izv. Akad. Nauk SSSR Neorg. Mater. *13*, 1230 (1977)
6.47 N.E. Alekseevskii, N.V. Ageev: Fiz. Met. Metalloved. *43*, 503 (1977)
6.48 L.A. Pandrys, D.H. Douglass: J. Low Temp. Phys. *23*, 367 (1976)
6.49 R.E. Siemens, D.J. Griffiths: Solid State Commun. *18*, 1097 (1976)
6.50 V.F. Shamray, A.M. Postnikov: Dokl. Akad. Nauk SSSR *226*, 1129 (1976)
6.51 N.V. Ageev, N.E. Alekseevskii, V.F. Shamray: Phys. Status Solidi B*77*, K129 (1976)
6.52 S.T. Zegler: Phys. Rev. *137*, 1438 (1965)
6.53 D.C. Johnston: Solid State Commun. *11*, 175 (1972)
6.54 F.J. Cadieu, N. Chencinski, C.Z. Rosen: J. Appl. Phys. *48*, 686 (1977)
6.55 B.A. Hatt, J.K. Page, V.G. Rivlin: J. Low Temp. Phys. *10*, 285 (1973)
6.56 J.C. Phillips: Solid State Commun. *18*, 831 (1976)
6.57 V.F. Shamray: Ph. D. Thesis, CNII Tchermjet, Moscow (1979)
6.58 S. Paidassi: Phys. Lett. A*71*, 140 (1979)
6.59 S. Paidassi: J. Appl. Phys. *50*, 3556 (1979)
6.60 R.W. Cahn: Nature *269*, 198 (1977)
6.61 N.E. Alekseevskii, N.V. Ageev, V.F. Shamray: Izv. Akad. Nauk SSSR Neorg. Mater. *11*, 2156 (1966)
6.62 B.T. Matthias, T.H. Geballe, L.D. Longinotti, E. Corenzwit: Science *156*, 645 (1967)
6.63 N.V. Ageev, N.E. Alekseevskii, N.N. Mikhailov, V.F. Shamray: JETP Lett. *6*, 901 (1967)
6.64 N.E. Alekseevskii, N.V. Ageev, N.N. Mikhailov, V.F. Shamray: JETP Lett. *9*, 28 (1969)
6.65 G. Arrhenius, E. Corenzwit, R. Fitzgerald, G.W. Hull, Jr., H.-L. Luo, B.T. Matthias, W.H. Zachariasen: Proc. Natl. Acad. Sci. U.S.A. *61*, 621 (1968)
6.66 J. Ruzicka: Z. Phys. *237*, 432 (1970)
6.67 G. Otto: Z. Phys. *215*, 323 (1968)

6.68 U. Kawabe, M. Kudo, S. Fukase: J. Phys. Soc. Jpn. *35*, 108 (1973)
6.69 E.M. Savitskii, K. Klyaynshtuk, P. Paufler, G. Ferstering, T. Shtefan, L.F. Myzenkova, V.G. Petrenko: In *Fiziko-khimitchjeskii analiz svjerkh-provodjachikh splavov*, ed. by E.M. Savitskii (Nauka, Moscow 1979) p.19
6.70 G. Ilonca, T.W. Barbce: Phys. Status Solidi A*47*, 357 (1978)
6.71 W.A. Lanford, P.H. Schmidt, J.M. Rowell, J.M. Poate, R.C. Dynes, P.D. Dernier: Appl. Phys. Lett. *32*, 339 (1978)
6.72 D.A. Papaconstantopoulos, D.U. Gubser, B.M. Klein, L.L. Boyer: Phys. Rev. B*21*, 1326 (1980)
6.73 F.J. Cadieu: J. Low Temp. Phys. *3*, 393 (1970)
6.74 V.F. Shamray, A.M. Postnikov: Dokl. Akad. Nauk SSSR *226*, 1129 (1976)
6.75 R. Somosundaram, L.E. Toth, H. Spitzer: J. Appl. Phys. *47*, 4656 (1976)
6.76 G. Otto: Z. Phys. *218*, 52 (1969)
6.77 A. Junod, F. Heiniger, J. Müller, P. Spitzli: Helv. Phys. Acta *43*, 59 (1970)
6.78 W. Kunz, E. Saur: Z. Phys. *189*, 401 (1966)
6.79 V.I. Surikov, V.I. Pryadein, A.K. Shtolts, A.P. Stepanov, P.V. Geld, A.F. Prekul, V.L. Zagryazhskii: Fiz. Met. Metalloved. *32*, 1222 (1972)
6.80 J. Hallais: Ann. Chim. Paris *6*, 309 (1971)
6.81 T. Asada, T. Horiuchi, M. Uchida: Jpn. J. Appl. Phys. *8*, 958 (1969)
6.82 G. Bongi, Ø. Fischer, H. Jones, R. Flukiger, A. Treyvaud: Helv. Phys. Acta *45*, 13 (1972)
6.83 Ø. Fischer, A. Treyvaud, G. Bongi, H. Jones, R. Flukiger: J. Phys. Paris F*4*, 1187 (1974)
6.84 G. Bongi, R. Flukiger, A. Treyvaud, Ø. Fischer, H. Jones, D. Schneider: J. Low Temp. Phys. *17*, 223 (1974)
6.85 G. Bongi, R. Flukiger, A. Treyvaud, Ø. Fischer: J. Low Temp. Phys. *23*, 543 (1976);
G. Bongi, A. Treyvaud, R. Flukiger, Ø. Fischer: J. Low Temp. Phys. *23*, 599 (1976)
6.86 A.M. Clogston, B.T. Matthias, M. Peter, H.J. Williams, E. Corenzwit, R.C. Sherwood: Phys. Rev. *125*, 541 (1962)
6.87 S. Moehlecke, D.E. Cox, A.R. Sweedler: Solid State Commun. *23*, 703 (1977)
6.88 G.W. Webb, S. Moehlecke, A.R. Sweedler: Mater. Res. Bull. *12*, 657 (1977)
6.89 F.E. Wang: Solid State Commun. *23*, 803 (1977)
6.90 F.E. Wang: J. Phys. Chem. Solids *35*, 273 (1974)
6.91 B.T. Matthias: Phys. Today *24*, 23 (1971)
6.92 T.V. Kopetskii, A.V. Pavlyutchenko: Fiz. Met. Metalloved. *43*, 498 (1977)
6.93 T.V. Kopetskii, A.V. Pavlyutchenko: Fiz. Met. Metalloved. *42*, 296 (1976)
6.94 R. Flukiger, P. Spitzli, F. Heiniger, J. Müller: Phys. Lett. A*29*, 407 (1969)
6.95 W.E. Blumberg, J. Eisinger, V. Jaccarino, B.T. Matthias: Phys. Rev. Lett. *5*, 149 (1960)
6.96 R.G. Shulman, B.J. Wuluda, B.T. Matthias: Phys. Rev. Lett. *1*, 278 (1958)
6.97 A.M. Clogston, J. Jaccarino: Phys. Rev. *121*, 1357 (1961)
6.98 A.V. Pavlyutchenko: Fiz. Tverd. Tela Leningrad *19*, 1431 (1977)
6.99 E. Ehrenfreud, A.C. Gossard, J.J. Wernick: Phys. Rev. B*4*, 2906 (1971)
6.100 M. Bernasson, P. Descouts, R. Flukiger, A. Treyvaud: Solid State Commun. *8*, 837 (1970)
6.101 F.J. Fradin, C.L. Wiley, J.W. Downey: Solid State Commun. *30*, 737 (1979)
6.102 L.A.G.M. Wulffers, G.A.M. Fijters, H.B. Kok, T.O. Klaasen, N.J. Poulis: Physica B*84*, 177 (1976)
6.103 L.A.G.M. Wulffers, G.A.M. Frijters, C.A. van Shie, T.O. Klaasen, N.J. Poulis: Physica B*93*, 180 (1978)
6.104 F.Y. Fradin, G. Cinader: Phys. Rev. B*16*, 73 (1977)

6.105 E.Z. Kurmaev, S.A. Nemnonov, V.P. Belash, Yu.V. Efimov: Fiz. Met. Metalloved. *33*, 578 (1972);
E.Z. Kurmaev, V.P. Belash, S.A. Nemnonov, A.S. Shulakov: Phys. Status Solidi B*61*, 365 (1974)
6.106 S.A. Nemnonov, E.Z. Kurmaev, V.A. Fomitchev, B.K. Ishmukhametov, V.P. Belash, A.V. Rudnev: Izv. Akad. Nauk SSSR Ser. Fiz. *36*, 317 (1972)
6.107 E.Z. Kurmaev, F. Werfel, O. Brümmer, R. Flukiger: Solid State Commun. *21*, 239 (1977)
6.108 A.I. Golovashkin, G.P. Motulevitch: Fiz. Tverd. Tela Leningrad *13*, 1232 (1971)
6.109 G.P. Motulevitch, A.I. Golovashkin, A.A. Shubin: In *Collected Papers: Eljektronnaja struktura pjerjekhodnykh mjetallov, ikh splavov i sojedinjenij* (Naukova Dumka, Kiev 1974) p.311
6.110 Y. Goldstein: Rev. Mod. Phys. *36*, 213 (1964)
6.111 S.W. McKnight, B.L. Bean, S. Perkowitz: Phys. Rev. B*19*, 1437 (1979)
6.112 J. Schumaun, D. Elefant: Phys. Status Solidi B*95*, 91 (1979)
6.113 V. Hoffstein, R.W. Cody: Phys. Lett. A*29*, 603 (1969)
6.114 H.J. Levinstein, J.E. Kunzler: Phys. Lett. *20*, 581 (1966)
6.115 L.Y.L. Chen: Phys. Rev. Lett. *29*, 1082 (1972)
6.116 S.I. Vedeneev, A.I. Golovashkin, I.S. Levtchenko, G.P. Motulevitch: JETP *63*, 1010 (1973)
6.117 D.R. Bosoworth, G.W. Gullen: Phys. Rev. *160*, 346 (1967)
6.118 G.D. Cody, R.W. Cohen: Rev. Mod. Phys. *36*, 121 (1964)
6.119 L.M. Fraas, P.F. Williams, S.P.S. Porto: Solid State Commun. *8*, 2113 (1970)
6.120 L.J. Vieland, A.W. Wicklund: Phys. Rev. *166*, 424 (1968)
6.121 J.D. Axe, G. Shirane: Phys. Rev. Lett. *30*, 214 (1973)
6.122 J.J. Hauser, D.D. Bacon, W.H. Haemmerlie: Phys. Rev. *151*, 296 (1966)
6.123 D.B. Tanner, A.J. Sievers: Phys. Rev. B*8*, 1978 (1973)
6.124 S.I. Vedeneev, A.I. Golovashkin, G.P. Motulevitch: Tr. Fiz. Inst. Akad. Nauk SSSR *86*, 140 (1975)
6.125 L.R. Testardi: In *Physical Acoustics*, ed. by W.P. Mason, R.N. Thurston, (Academic, New York 1973) Vol.10, p.193
6.126 B.W. Batterman, C.S. Barrett: Phys. Rev. Lett. *13*, 390 (1964)
6.127 R. Mailfert, B.W. Batterman, J.J. Hanak: Phys. Lett. A*24*, 315 (1967)
6.128 J. Wanagel, B.W. Batterman: J. Appl. Phys. *41*, 3610 (1970)
6.129 E. Nembach, K. Tachikawa, S. Takano: Philos. Mag. *21*, 869 (1970)
6.130 B.N. Kodess: Ph. D. Thesis, Perm (1973)
6.131 J.S. Bowles, C.S. Barrett, L. Guttman: Trans. Metall. Soc. AIME *188*, 1478 (1950)
6.132 L.J. Vieland, R.W. Cohen, W. Rehwald: Phys. Rev. Lett. *26*, 373 (1971)
6.133 P.W. Anderson, E.I. Blount: Phys. Rev. Lett. *14*, 219 (1965)
6.134 J. Perel, B.W. Batterman, E.I. Blount: Phys. Rev. *166*, 616 (1968)
6.135 B.T. Matthias, E. Corenzwit, A.S. Cooper, L.D. Longinotti: Proc. Natl. Acad. Sci. U.S.A. *68*, 56 (1971)
6.136 A.G. Rabin'kin, V.N. Galev, V.N. Laukhin: JETP *64*, 1724 (1973)
6.137 V.N. Galev, A.G. Rabin'kin: JETP *65*, 1061 (1973)
6.138 E.M. Savitskii, V.V. Baron, A.A. Korostelin, A.R. Kadyrbaev, V.N. Sumarokov: Dokl. Acad. Nauk SSSR *227*, 840 (1976)
6.139 J.P. Maita, E. Bucher: Phys. Rev. Lett. *29*, 931 (1972)
6.140 W. Dieterich, P. Fulde: Z. Phys. *248*, 154 (1971)
6.141 C.J. Williamson, C.S. Ting, H.K. Fung: Phys. Rev. Lett. *32*, 9 (1974)
6.142 J.D.N. Cheeke, H. Mallie, S. Roth, B. Seeber: Solid State Commun. *13*, 1567 (1973)
6.143 L.J. Vieland: J. Phys. Chem. Solids *31*, 1449 (1970)
6.144 C.S. Ting, A.K. Ganguly, R. Zeyer, J.L. Birman: Phys. Rev. B*8*, 3663 (1973)

6.145 T. Takashima: Jpn. J. Appl. Phys. *12*, 1286 (1973)
6.146 Yu.A. Izyumov, E.Z. Kurmaev: Usp. Fiz. Nauk *113*, 193 (1974)
6.147 L.J. Vieland, A.W. Wicklund: Solid State Commun. *7*, 37 (1969)
6.148 A.C. Gossard: Phys. Rev. *149*, 246 (1966)
6.149 J.S. Shier, R.D. Taylor: Phys. Rev. *174*, 346 (1967)
6.150 M. Milewitz, S.J. Williamson: J. Phys. Paris, Colloq. *39*, C6-408 (1978)
6.151 T. Fukase, T. Kobayashi, M. Isino, N. Toyota: J. Phys. Paris, Colloq. *39*, C6-406 (1978)
6.152 A.V. Skripov, A.P. Stepanov, V.A. Martchenko, V.M. Pan, A.D. Shevtchenko: JETP *77*, 2313 (1979)
6.153 J.B. Hastings, G. Shirane, S.J. Williamson: Phys. Rev. Lett. *43*, 1249 (1979)
6.154 L.R. Testardi, T.B. Bateman, W.A. Reed, V.G. Chirba: Phys. Rev. Lett. *15*, 250 (1965)
6.155 L.R. Testardi: Phys. Rev. B*3*, 95 (1971)
6.156 L.R. Testardi, T.B. Bateman: Phys. Rev. *154*, 402 (1967)
6.157 M. Rosen, H. Klimker, M. Weger: Phys. Rev. *184*, 466 (1969)
6.158 W. Rehwald, M. Rayl, R.W. Cohen, G.D. Cody: Phys. Rev. *6*, 363 (1972)
6.159 Z.R. Chang, G.R. Barsch: Phys. Rev. B*22*, 3242 (1980)
6.160 L.R. Testardi: Phys. Rev. Lett. *31*, 37 (1973)
6.161 P.F. Carcia, G.R. Barsch: Phys. Status Solidi B*59*, 595 (1973)
6.162 R.E. Larsen, A.L. Ruoff: J. Appl. Phys. *44*, 1021 (1973)
6.163 S. Huang, C.W. Chu: Phys. Rev. B*10*, 4030 (1974)
6.164 C.W. Chu: Phys. Rev. Lett. *33*, 1283 (1974); *32*, 766 (1974)
6.165 T.F. Smith: Phys. Rev. Lett. *25*, 1483 (1970)
6.166 K.R. Keller, J.J. Hanak: Phys. Rev. *154*, 628 (1967)
6.167 B.P. Schweiss, B. Renker, E. Schneider, W. Reichardt: In *Superconductivity in d- and f-Band Metals*, ed. by D.H. Douglass (Plenum, New York 1976) p.189
6.168 B.N.N. Achar, G.R. Barsch: Solid State Commun. *29*, 563 (1979)
6.169 B.N.N. Achar, G.R. Barsch: Solid State Commun. *30*, 419 (1979)
6.170 S. Schicktanz, R. Kaiser, W. Spengler, B. Seeber: Solid State Commun. *28*, 935 (1978)
6.171 H. Wipf, M.V. Klein, B.S. Chandrasekhar, T.H. Geballe, J.H. Wernick: Phys. Rev. Lett. *41*, 1752 (1978)
6.172 T.F. Smith: In *Superconductivity in d- and f-Band Metals*, ed. by D.H. Douglass (American Institute of Physics, New York 1972) p.293
6.173 C.W. Chu, P.H. Schmidt: Bull. Am. Phys. Soc. *21*, 290 (1976)
6.174 C.W. Chu, L.R. Testardi, P.H. Schmidt: Solid State Commun. *23*, 841 (1977)
6.175 R.D. Blaugher: IEEE Trans. Magn. *13*, 821 (1977)
6.176 G. Fasol, J. Schilling, B. Seeber: J. Phys. Paris, Colloq. *39*, C6-410 (1978)
6.177 L.R. Testardi, J.E. Kunzler, H.J. Levinstein, J.P. Maita, J.H. Wernick: Phys. Rev. B*3*, 107 (1971)
6.178 L.R. Testardi: Phys. Rev. B*5*, 4342 (1972)
6.179 M. Weger, B.G. Silbernagel: Phys. Rev. Lett. *13*, 521 (1964)
6.180 L.R. Testardi, J.M. Poate, H.L. Levinstein: Phys. Rev. Lett. *37*, 637 (1976)
6.181 C.M. Varma, J.C. Phillips, S.T. Chui: Phys. Rev. Lett. *33*, 1233 (1974)
6.182 B.T. Matthias, T.H. Geballe, V.B. Compton: Rev. Mod. Phys. *35*, 1 (1963)
6.183 P. Sen, J.D. McGervey, C. Knox, C.W. Chu: Phys. Lett. A*63*, 393 (1977)
6.184 M. Weger: Rev. Mod. Phys. *36*, 175 (1964)
6.185 J. Labbé, J. Friedel: J. Phys. Radium *27*, 153 (1966)
6.186 L.F. Mattheiss: Phys. Rev. *138*, 112 (1965)
6.187 L.F. Mattheiss: Phys. Rev. B*12*, 2161 (1975)
6.188 J.C. Slater, G.F. Koster: Phys. Rev. *94*, 1498 (1954)

6.189 M. Weger, I.B. Goldberg: In *Solid State Physics*, Vol.28, ed. by H. Ehrenreich, F. Seitz, D. Turnbull (Academic, New York 1973)
6.190 F. Debanaye, F. Brouers, J.P. Gaspard, H.R. Khan: Solid State Commun. *31*, 509 (1979)
6.191 B.M. Klein, L.L. Boyer, D.A. Papaconstantopoulos, L.F. Mattheiss: Phys. Rev. B*18*, 6411 (1978)
6.192 B.M. Klein, L.L. Boyer, D.A. Papaconstantopoulos: Phys. Rev. Lett. *42*, 530 (1979)
6.193 B.M. Klein, L.L. Boyer, D.A. Papaconstantopoulos: J. Phys. Paris F*8*, 617 (1978)
6.194 B.M. Klein, D.A. Papaconstantopoulos, L.F. Mattheiss: Bull. Am. Phys. Soc. *20*, 297 (1975)
6.195 P.B. Allen, R.C. Dynes: Phys. Rev. B*12*, 905 (1975)
6.196 C.C. Tsuei, S. von Molnar, J.M. Coey: Phys. Rev. Lett. *41*, 664, 1194 (1978)
6.197 P.B. Allen, W.E. Pickett, K.M. Ho, M.L. Cohen: Phys. Rev. Lett. *40*, 1532 (1978);
W.E. Pickett, K.M. Ho, M.L. Cohen: Phys. Rev. B*19*, 1734, 1751 (1979)
6.198 P.K. Lam, M.L. Cohen: Phys. Rev. B*23*, 6371 (1981)
6.199 T. Jarlborg, G. Arbmann: J. Phys. Paris F*6*, 189 (1976); *9*, 1635 (1977)
6.200 T. Jarlborg: J. Phys. Paris F*9*, 283 (1979)
6.201 T. Jarlborg: Ph. D. Thesis, Chalmers University of Technology (1977)
6.202 I.B. Goldberg: J. Phys. Paris C*8*, 1159 (1975)
6.203 G. Barak, I.B. Goldberg, M. Weger: J. Phys. Chem. Solids *36*, 847 (1975)
6.204 A.T. van Kessel, H.W. Myron, F.M. Müller: Phys. Rev. Lett. *41*, 181 (1978); *41*, 520 (1978); Bull. Am. Phys. Soc. *23*, 416 (1978); J. Less Common Met. *62*, 49 (1978)
6.205 O.K. Andersen: Phys. Rev. B*12*, 3060 (1975)
6.206 O.K. Andersen, O. Jepsen: Physica B+C *91*, 317 (1977)
6.207 T. Jarlborg, P.O. Nilsson: J. Phys. Paris C*12*, 265 (1979)
6.208 E.Z. Kurmaev, V.P. Belash, S.A. Nemnonov, A.S. Shulakov: Phys. Status Solidi B*61*, 365 (1974)
6.209 E.Z. Kurmaev, S.A. Nemnonov, V.P. Belash, Yu.V. Efimov: Fiz. Met. Metalloved. *33*, 578 (1972)
6.210 E.Z. Kurmaev, V.P. Belash, S.A. Nemnonov, I.A. Brytov, N.S. Vorob'eva, V.I. Sokolov: Fiz. Met. Metalloved. *31*, 753 (1971)
6.211 E.Z. Kurmaev, F. Werfel, O. Brummer, R. Flukiger: Solid State Commun. *22*, 375 (1977)
6.212 J. Riley, J. Azoulay, L. Ley: Solid State Commun. *19*, 993 (1976)
6.213 P.O. Nilsson, I. Curelaru, T. Jarlborg: Phys. Status Solidi B*79*, 277 (1977)
6.214 R.A. Pollak, C.C. Tsuei, R.W. Johnston: Solid State Commun. *23*, 879 (1977)
6.215 H. Höchst, S. Hufner, A. Goldman: Solid State Commun. *19*, 899 (1976)
6.216 S.A. Nemnonov, E.Z. Kurmaev: Phys. Status Solidi *24*, K43 (1967);
E.Z. Kurmaev, S.A. Nemnonov: Phys. Status Solidi B*43*, K49 (1971)
6.217 S.A. Nemnonov, E.Z. Kurmaev, V.P. Belash: Phys. Status Solidi *39*, 39 (1970)
6.218 A.J. Arko, D.H. Lowndes, F.A. Müller, L.W. Roeland, J. Wolfrat, A.T. van Kessel, H.W. Myron, F.M. Müller, G. Webb: Phys. Rev. Lett. *40*, 1590, (1978)
6.219 A.J. Arko, Z. Fisk, F.M. Müller: Phys. Rev. B*16*, 1387 (1977)
6.220 W.S. Farmer, F. Sinclair, S. Berko, G.M. Baardsley: Solid State Commun. *31*, 481 (1979)
6.221 J. Labbé: Phys. Rev. *158*, 647 (1967)
6.222 J. Labbé: Phys. Rev. *158*, 655 (1967)
6.223 J. Labbé, S. Barisic, J. Friedel: Phys. Rev. Lett. *19*, 1039 (1967)

6.224 P.C. Hohenberg: Phys. Rev. *158*, 383 (1967)
6.225 R.A. Ferrell: Phys. Rev. Lett. *15*, 330 (1964)
6.226 I.E. Dzialoshinsky, E.I. Kats: JETP *55*, 338 (1968)
6.227 Yu.A. Bytchkov, L.P. Gor'kov, I.E. Dzialoshinsky: JETP *50*, 778 (1966)
6.228 E.I. Kats: JETP *55*, 584 (1968)
6.229 I.E. Dzialoshinsky, A.I. Larkin: JETP *61*, 792 (1971)
6.230 K.B. Efetov, A.I. Larkin: JETP *66*, 2290 (1974)
6.231 B. Horovitz, M. Weger, H. Gutfreund: Phys. Rev. B*9*, 1246 (1974)
6.232 J. Labbé: Phys. Rev. *172*, 451 (1968)
6.233 W. Dieterich, P. Fulde: Z. Phys. *248*, 154 (1971)
6.234 R.W. Cohen, C.D. Cody, J.J. Halloran: Phys. Rev. Lett. *19*, 840 (1967)
6.235 W. Rehwald: Phys. Lett. A*27*, 287 (1968)
6.236 G.D. Cody, R.W. Cohen, L.J. Vieland: In Proc. Eleventh Int. Conf. Low Temp. Physics, St. Andrews, Scotland (1966) p.1009
6.237 R.W. Cohen, G.D. Cody, L.J. Vieland: In *Electronic Density of States*, Natl. Bur. Stand. U.S. Spec. Publ. *323*, 767 (1971)
6.238 L.P. Gor'kov: JETP *65*, 1658 (1973)
6.239 G.L. Bir, G.E. Pikus: *Simmjetrija i djeformacionnyje effjekty v poluprovodnikakh* (Nauka, Moscow 1972)
6.240 Yu.A. Izyumov, V.E. Naish, V.N. Syromyatnikov: Fiz. Met. Metalloved. *39*, 455 (1975)
6.241 Chin-Sen Ting, J.L. Birman: Phys. Rev. B*12*, 1093 (1975)
6.242 W. Dieterich: Phys. Lett. A*37*, 409 (1971)
6.243 L.P. Gor'kov: JETP Lett. *20*, 571 (1974)
6.244 L.P. Gor'kov, O.N. Dorokhov: J. Low Temp. Phys. *22*, 1 (1976)
6.245 G. Shirane, J.D. Axe, R.J. Brigenau: Solid State Commun. *9*, 397 (1971)
6.246 G. Shirane, J.D. Axe: Phys. Rev. B*18*, 3742 (1978)
6.247 T.-K. Lee, J.L. Birman, S.J. Williamson: Phys. Rev. Lett. *39*, 839 (1977)
6.248 T.-K. Lee, J.L. Birman, S.J. Williamson: Phys. Lett. A*64*, 89 (1977)
6.249 T.-K. Lee, J.L. Birman: Phys. Rev. B*17*, 4931 (1978)
6.250 R.N. Bhatt, L.F. Mattheiss: Phys. Rev. B*20*, 2542 (1979)
6.251 B.M. Klein, L.L. Boyer, D.A. Papaconstantopoulos: J. Phys. Paris F*8*, 617 (1978)
6.252 T.-K. Lee, J.L. Birman, S.J. Williamson: Phys. Rev. B*20*, 2547 (1979)
6.253 R.N. Bhatt: Phys. Rev. B*16*, 2804 (1977)
6.254 R.N. Bhatt: Phys. Rev. B*17*, 2947 (1978)
6.255 W.L. McMillan: Phys. Rev. *167*, 331 (1968)
6.256 A. Junod: Ph. D. Thesis 1661, Geneva (1974)
6.257 F.Y. Fradin, D. Zamir: Phys. Rev. B*7*, 4861 (1973)
6.258 F.Y. Fradin, J.D. Williamson: Phys. Rev. B*10*, 2803 (1974)
6.259 S.V. Verkhovsky, A.P. Stepanov, A.K. Shtolts: Fiz. Met. Metalloved. *40*, 529 (1975)
6.260 S.V. Verkhovsky: Ph. D. Thesis, Sverdlovsk (1975)
6.261 G.S. Knapp: J. Less Common Met. *62*, 127 (1978)
6.262 B.Y. Yao, S.E. Schnatterly: to be published
6.263 A.I. Golovashkin, G.P. Motulevich: JETP Lett. *17*, 1146 (1973)
6.264 I. Tutto, J. Ruvalds: Phys. Rev. B*19*, 5641 (1979)
6.265 J.M. Rowell, W.L. McMillan: Phys. Rev. Lett. *16*, 453 (1966)
6.266 D.E. Moore, M.R. Beasley, J.M. Rowell: J. Phys. Paris *39*, Colloq. 6/3, 1390 (1978)
6.267 I. Tutto, L.M. Kahn, J. Ruvalds: Phys. Rev. B*20*, 952 (1979)
6.268 I.R. Gomersall, B.L. Gyorffy: In *Low Temperature Physics LT-14*, ed. by M. Krusius, M. Vuorio (North Holland, Amsterdam 1975) Vol.5, p.449
6.269 H. Fröhlich: J. Phys. Paris C*1*, 544 (1968)
6.270 E.A. Pashitskii, V.L. Makarov, S.D. Tereshenko: Fiz. Tverd. Tela Leningrad *16*, 276 (1974)
6.271 E.A. Pashitskii, V.M. Chernousenko: JETP *33*, 802 (1971)

6.272 D. Pines: Can. J. Phys. *34*, 1379 (1956)
6.273 L.M. Kahn, J. Ruvalds, R. Histings: Phys. Rev. B*17*, 4600 (1978)
6.274 G. Shirane, J.D. Axe: Phys. Rev. B*4*, 2957 (1971)
6.275 L.J. Sham: Phys. Rev. B*6*, 3584 (1972)
6.276 L.J. Sham: Phys. Rev. Lett. *27*, 1725 (1971)
6.277 L.J. Sham: Phys. Rev. B*6*, 3581 (1972)
6.278 M. Born, Khuan Kun: *Dynamical Theory of Crystal Lattices* (Clarendon, Oxford 1954)
6.279 H. Wipf, M.V. Klein, B.S. Chandrasekhar, T.H. Geballe, J.H. Wernick: Phys. Rev. Lett. *41*, 1752 (1978)
6.280 J. Noolandi, L.J. Sham: Phys. Rev. B*8*, 2468 (1973)
6.281 J.D. Axe, G. Shirane: Phys. Rev. B*8*, 1965 (1973)
6.282 P. Horsch: Solid State Commun. *18*, 27 (1976)
6.283 P. Pietrass: Phys. Status Solidi B*68*, 553 (1975)
6.284 N. Srabo: J. Phys. Paris C*8*, L397 (1975)
6.285 M.W. Finnis, V. Heine: J. Phys. Paris F*4*, 960 (1974)
6.286 J.P. Staley, M.E. Fischer: J. Phys. A*6*, 1310 (1973)
6.287 Yu.M. Gufan: Fiz. Tverd. Tela Leningrad *13*, 225 (1971)
6.288 W. Klose, H. Schuster: Z. Phys. *241*, 348 (1971)
6.289 H.G. Schuster: Solid State Commun. *13*, 1361 (1973)
6.290 R.N. Bhatt, W.L. McMillan: Phys. Rev. B*14*, 1007 (1976)
6.291 G. Bilbro, W.L. McMillan: Phys. Rev. B*14*, 1887 (1976)
6.292 J. Birman: Phys. Rev. Lett. *17*, 11 (1966)
6.293 Yu.V. Kopaev: Tr. Fiz. Inst. Akad. Nauk SSSR *86*, 3 (1975)
6.294 R.N. Bhatt, P.A. Lee: Phys. Rev. B*16*, 4288 (1977)
6.295 M.V. Jaric, J.L. Birman: Phys. Rev. B*16*, 2564 (1977)
6.296 K.G. Wilson, J. Kogut: Phys. Rep. *12*, 75 (1974)
6.297 M.V. Jaric, J.L. Birman: Phys. Rev. B*17*, 4368 (1978)
6.298 G.R. Johanson, D.H. Douglass: J. Low Temp. Phys. *14*, 576 (1974)
6.299 V.M. Pan: JETP Lett. *21*, 228 (1975)

Chapter 7

7.1 M.Yu. Teslyuk: *Metallitcheskije sojedinjenija so strukturami faz Lavesa* (Nauka, Moscow 1969)
7.2 B.W. Roberts: In *Intermetallic Compounds*, ed. by J.H. Westbrook (Wiley, New York 1966) p.581
7.3 O. Rapp: J. Less Common Met. *21*, 27 (1970)
7.4 K. Inoue, K. Tachikawa: Jpn. J. Appl. Phys. *12*, 161 (1973)
7.5 O. Rapp, J. Invarsson, T. Claeson: Phys. Lett. A*50*, 159 (1974)
7.6 H.E. Hoenig, K. Baberschke, W. Schrittenlacher: Solid State Commun. *16*, 341 (1975)
7.7 P.F. Schippnick, A.C. Lawson: Solid State Commun. *15*, 1643 (1974)
7.8 A.C. Lawson, U. Engel, K. Baberschke: Phys. Lett. A*49*, 373 (1974)
7.9 M.N. Medvedeva, T.O. Malakhova: *Collected Papers: Fiziko-khimija, mjetallovjedjenije i mjetallofizika svjerkhprovodnikov* (Nauka, Moscow 1963) p.28
7.10 E.M. Savitskii, Yu.V. Efimov, N.D. Kozlova, B.P. Mikhailov, Yu.A. Uspenskii: In *Svjerkhprovodjachije sojedinjenija pjerjekhodnykh mjetallov*, ed. by V.V. Baron (Nauka, Moscow 1976) p.51
7.11 A.C. Lawson, J.E. Cannon, D.Z. Robertson, M.T. Hall: J. Less Common Met. *32*, 173 (1973)
7.12 Yu.V. Efimov: *Collected Papers: Fiziko-khimija, mjetallovjedjenije i mjetallofizika svjerkhprovodnikov* (Nauka, Moscow 1967) p.114
7.13 P. Wenneström, S. Byström, H. Larson, T. Lindqvist, O. Rapp: Phys. Scr. *11*, 378 (1975)

7.14 E.M. Savitskii, V.V. Baron, Yu.V. Efimov, M.S. Model: *Collected Papers: Svjerkhprovodjachije splavy i sojedinjenija* (Nauka, Moscow 1972) p.78
7.15 K. Inoue, K. Tachikawa: J. Jpn. Inst. Metals *39*, 1266 (1975)
7.16 K. Inoue, K. Tachikawa: Appl. Phys. Lett. *25*, 94 (1975)
7.17 K. Inoue, T. Kuroda, K. Tachikawa: IEEE Trans. Magn. *15*, 635 (1979)
7.18 K. Inoue, K. Tachikawa: J. Jpn. Inst. Metals *39*, 1274 (1975)
7.19 C.W. Chu, E. Bucher, A.S. Cooper, J.P. Maita: Phys. Rev. B*4*, 320 (1971)
7.20 V.M. Pan, I.E. Bulakh, A.D. Shevtchenko, V.I. Latysheva: Fiz. Met. Metalloved. *46*, 741 (1978)
7.21 P. Duffer, S.G. Sanhar, V.U.S. Rao, R.L. Bergner, R. Obermyer: Phys. Status Solidi A*31*, 655 (1975)
7.22 P. Duffer, D.H. Gualtiere, V.U.S. Rao: Phys. Rev. Lett. *37*, 1410 (1976)
7.23 V.U.S. Rao, D.H. Gualtiere, S. Krishnamurthy, A. Patkin, P. Duffer: Phys. Lett. A*67*, 223 (1978)
7.24 V.G. Pushin, V.I. Kozhanov, I.I. Syutkina, R.R. Romanova, E.P. Romanov: Dokl. Akad. Nauk SSSR *254*, 100 (1980)
7.25 A.C. Lawson, W.H. Zachariasen: Phys. Lett. A*38*, 1 (1972)
7.26 D.E. Moncton: Solid State Commun. *13*, 1779 (1973)
7.27 M. Levinson, C. Zahradnie, R. Bergh, M.L. MacVicar, J. Bostok: Phys. Rev. Lett. *41*, 899 (1978)
7.28 V.N. Kozhanov, E.P. Romanov, S.V. Verchovsky, A.P. Stepanov: Fiz. Met. Metalloved. *48*, 1249 (1979)
7.29 E.E. Ivanov, V.A. Finkel, E.A. Pushkarev: Dokl. Akad. Nauk SSSR *228*, 119 (1976)
7.30 A.C. Lawson, K. Baberschke, U. Engel: Phys. Lett. A*48*, 107 (1974)
7.31 R. Viswanathan, A.C. Lawson, C.S. Pandi: J. Phys. Chem. Solids *37*, 341 (1976)
7.32 Y. Kimura: J. Phys. Soc. Jpn. *36*, 306 (1974)
7.33 J.E. Doherty, D.F. Gibbons: Phys. Status Solid B*44*, K5 (1971)
7.34 V.N. Kozhanov, S.V. Verchovsky, A.P. Stepanov, E.P. Romanov, E.V. Galoshina: Fiz. Met. Metalloved. *44*, 1014 (1977)
7.35 I.E. Bulakh: Ph. D. Thesis, Kiev (1979)
7.36 M.M. Akhverdli, V.V. Burnashova, E.I. Gladyshevskii, M.V. Raevsakja, E.M. Sokolovskaja: Metallofizika *52*, 111 (1974)
7.37 F. Pop, M. Goldae, V.U.S. Rao: Phys. Status Solidi A*49*, 207 (1978)
7.38 T. Jarlborg, A.J. Freeman: Phys. Rev. B*22*, 2332 (1980)
7.39 S.V. Verchovsky, V.N. Kozhanov, A.P. Stepanov, E.P. Romanov, E.V. Galosi E.V. Galoshina: Fiz. Met. Metalloved. *49*, 1234 (1980)
7.40 V.M. Pan, I.E. Bulakh, A.L. Kasatkin, A.D. Shevchenko: J. Less Common Met. *62*, 157 (1978)
7.41 V.M. Pan, I.E. Bulakh, A.L. Kasatkin, A.D. Shevchenko: *Collected Papers: Voprosy atommoj nauki i tjekhniki*, Ser. Obschaja i Jadernaja Fizika *3*, 10 (1979)
7.42 V.M. Pan, I.E. Bulakh, A.L. Kasatkin, A.D. Shevchenko, V.E. Yatchmenev: Materialy 20 Vsjesojuznogo Sovjechanija po fizikje nizkikh tjemperatur, NT-20 (Institut Khim. Phys. Tchernogolovka 1978) Vol.3, p.161
7.43 V.M. Pan, I.E. Bulakh, A.L. Kasatkin, A.D. Shevchenko: JETP Lett. *27*, 629 (1978)
7.44 V.M. Pan, I.E. Bulakh, A.L. Kasatkin: Fiz. Tverd. Tela Leningrad *20*, 2487 (1978)
7.45 V.M. Pan, I.E. Bulakh, A.L. Kasatkin, A.D. Shevchenko: J. Phys. Paris *39*, C6-1037 (1978)
7.46 V.M. Pan, I.E. Bulakh, A.D. Shevchenko: In *Fiziko-khimitchjeskii analiz svjerkhprovodjachikh splavov*, ed. by E.M. Savitskii (Nauka, Moscow 1979) p.59
7.47 T. Takashima, H. Hayashi: Jpn. J. Appl. Phys. *12*, 1659 (1973)
7.48 A.C. Lawson: Phys. Lett. A*38*, 379 (1972)
7.49 T. Takashima, H. Hayashi: Phys. Lett. A*47*, 209 (1974)

7.50 A.C. Lawson: Ph. D. Thesis, University of California, San Diego (1972)
7.51 J.W. Hafstrom, G.S. Knapp, A.T. Aldred: Phys. Rev. B*17*, 2892 (1978)
7.52 G.S. Knapp: J. Less Common Met. *62*, 127 (1978)
7.53 Y. Kimura: Phys. Status Solidi A*43*, K141 (1977)
7.54 S.V. Verchovksy, V.N. Kozhanov, A.P. Stepanov, E.P. Romanov: In *Magnetic Resonance and Relation Phenomena*, ed. by E. Kundla, E. Lippmaa, T. Saluvere (Springer, Berlin, Heidelberg, New York 1979) p.427
7.55 Y. Kimura: Jpn. J. Appl. Phys. *16*, 577 (1977)
7.56 T.F. Smith, R.N. Shelton, A.C. Lawson: J. Phys. Paris F*3*, 2157 (1973)
7.57 C.W. Chu, E. Bucher, A.S. Cooper, J.P. Maita: Phys. Rev. B*4*, 320 (1970)
7.58 B.T. Matthias, E. Corenzwit, J.M. Vandenberg, H. Barz, M.B. Maple, R.N. Shelton: J. Less Common Met. *46*, 339 (1976)
7.59 E. Raub, W. Fritzsche: Z. Metallkd. *54*, 21 (1963)
7.60 R. Flükiger, F. Heiniger, J. Müller: Proc. Eleventh Int. Conf. Low Temp. Physics, St. Andrews, Scotland 1969, p.1017
7.61 M. Marezio, P.D. Dernier, C.W. Chu: Phys. Rev. B*4*, 2825 (1971)
7.62 L.B. Welsh, C.W. Chu: Phys. Rev. B*8*, 1028 (1973)
7.63 A.C. Switendick: In *Transition Metals*, ed. by H.I.G. Lee, J.M. Perz, E. Fawcett (Institute of Physics, Bristol 1978) p.386
7.64 E.Z. Kurmaev, Yu.M. Yarmoshenko, V.E. Dolgih, S.A. Nemnonov, F. Werfel, O. Brümmer: Solid State Commun. *29*, 59 (1979)
7.65 L.R. Testardi: Phys. Lett. A*35*, 117 (1971)
7.66 T.F. Smith, R.N. Shelton, A.C. Lawson, C.W. Chu: J. Phys. Paris F*4*, 1423 (1974)
7.67 C.W. Chu, S. Huang, T.F. Smith, E. Corenzwit: Phys. Rev. B*11*, 1866 (1975)
7.68 T.F. Smith, C.W. Chu: Bull. Am. Phys. Soc. *17*, 311 (1972)
7.69 M.V. Nevitt: In *Electronic Structure and Alloy Chemistry of the Transition Metals*, ed. by P.A. Beck (Interscience, New York 1963) p.97
7.70 K. Schubert: *Kristallstrukturen zweikomponentiger Phasen* (Springer, Göttingen, Heidelberg 1964)
7.71 D.S. Kasper: In *Collected Papers: Tjeorija faz v splavakh*, ed. by Ya.S. Umanskii (Mjetallurgija, Moscow 1962) p.320
7.72 R.M. Waterstrat, J.S. Casper: Trans. Metall. Soc. AIME *209*, 872 (1957)
7.73 R.J. Brown, J.B. Forsyth: Acta Crystallogr. *14*, 362 (1964)
7.74 K.P. Gupta: Trans. Metall. Soc. AIME *221*, 1047 (1961)
7.75 A. Müller: Z. Naturforsch. A*28*, 472 (1973)
7.76 E. Bucher, F. Heiniger, J. Müller: Phys. Kondens. Mater. *2*, 210 (1964)
7.77 L.R. Testardi, J.J. Hauser, M.H. Read: Solid State Commun. *9*, 1829 (1971)
7.78 L. Sausores, R.E. Glover: Bull. Am. Phys. Soc. *18*, 704 (1973)
7.79 C.M. Knoedler, D.H. Douglass: J. Low Temp. Phys. *37*, 189 (1979)
7.80 G.F. Hurdy, J.K. Hulm: Phys. Rev. *93*, 1004 (1954)
7.81 W.B. Pearson: *The Crystal Chemistry and Physics of Metals and Alloys* (Wiley-Interscience, New York 1972)
7.82 D.A. Robins: Philos. Mag. *3*, 313 (1958)
7.83 M.C. Ohmer, J.J. Wollan, J.C. Ho: IEEE Trans. Magn. *11*, 159 (1975)
7.84 S.P. Denker: J. Less Common Met. *14*, 1 (1968)
7.85 A.D. Uedsli: In *Njestjekhiomjetritcheskie sojedinjenija*, ed. by L. Mandelkorn (Khimija, Moscow 1971) p.109
7.86 D. Watanabe, J.R. Castles, A. Jostsons, A.S. Malin: Nature *210*, 934 (1966)
7.87 C. Froidevaux, D. Rossier: J. Phys. Chem. Solids *28*, 1197 (1967)
7.88 J.D. Venables, D. Kahn, R.G. Lye: Philos. Mag. *18*, 177 (1968)
7.89 J. Billingham, P.S. Bell, M.H. Lewis: Philos. Mag. *25*, 661 (1972)
7.90 K. Hiraga: Philos. Mag. *27*, 1301 (1973)
7.91 C.H. DeNovion, R.G. Lorenzelli, P. Costa: C. R. Acad. Sci. *263*, 775 (1966)
7.92 A.F. Shevakin, I.A. Tomilin, V.I. Sarrak: Fiz. Met. Metalloved. *32*, 1204 (1971)

7.93 V.G. Zubkov, L.B. Dubrovskaya, P.V. Geld, V.A. Tskhai, Yu.A. Dorofeev: Dokl. Akad. Nauk SSSR *184*, 874 (1969)
7.94 P.S. Bell, M.H. Lewis: Phys. Status Solidi A*7*, 431 (1971)
7.95 U. Mutschke, P. Ness: Z. Anorg. Allg. Chem. *292*, 82 (1957)
7.96 D. Watanabe, J.R. Castles, A. Jostsons, A.S. Malin: Acta Crystallogr. *23*, 307 (1967)
7.97 M.D. Banus, T.B. Reed: In *The Chemistry of Extended Defects in Non-metallic Solids*, ed. by L. Eyring, M.O. Keefe (North Holland, Amsterdam 1970) p.488
7.98 R.H. Willens, E. Bucher: Appl. Phys. Lett. *7*, 25 (1965); R.H. Willens, E. Bucher, B.T. Matthias: Phys. Rev. *159*, 327 (1967)
7.99 B.T. Matthias, J.K. Hulm: Phys. Rev. *87*, 739 (1952)
7.100 B.T. Matthias, T.H. Geballe, V.B. Compton: Rev. Mod. Phys. *35*, 1 (1963)
7.101 J.J. Veyssie, D. Brochier, A. Nemoz, J. Blanc: Phys. Lett. *14*, 261 (1965)
7.102 D.W. Bloom, L. Finegold, R.G. Lye, R. Radebaugh, J.D. Siegwarth: Phys. Lett. A*33*, 137 (1970)
7.103 E. Bucher, A.C. Gossard, K. Andres, J.P. Maita, A.S. Cooper: Materials Eighth Rare-Earth Conf., Reno, Nevada, 1970
7.104 O.I. Shulishova: Ph. D. Thesis, IPM Akad. Nauk USSR, Kiev (1966)
7.105 F. Hullinger, G.W. Hull: Solid State Commun. *8*, 1379 (1970)
7.106 J.K. Hulm, M.S. Walker, N. Pessall: Physica *55*, 60 (1971)
7.107 J.K. Hulm, C.K. Jones, R.A. Hein, J.W. Gibson: J. Low Temp. Phys. *7*, 291 (1972)
7.108 D.C. Johnston, A. Moodenbaugh: Phys. Lett. A*41*, 447 (1972)
7.109 S.V. Popova, L.N. Fomicheva, L.G. Khvostintsev: JETP Lett. *16*, 609 (1972)
7.110 L.E. Toth: *Transition Metal Carbides and Nitrides* (Academic, New York 1971)
7.111 M.W. Williams, K.M. Ralls, M.R. Pickus: J. Phys. Chem. Solids *28*, 333 (1967)
7.112 B. Stritzker, W. Buckel: Z. Phys. *257*, 1 (1972)
7.113 L.B. Boiko, S.V. Popova: JETP Lett. *12*, 101 (1970)
7.114 G.M. Klimashin, V.S. Neshpor, V.P. Nikitin, V.I. Novikov, S.S. Shalyt: JETP Lett. *12*, 147 (1970)
7.115 G.N. Schrauzer, H. Prakash: Solid State Commun. *14*, 1259 (1974)
7.116 A.L. Giorgi, E.C. Szklarz, T.C. Wallace: Proc. Br. Ceram. Soc. *10*, 183 (1968)
7.117 V. Sadagopan, H.C. Gatos: J. Phys. Chem. Solids *27*, 235 (1966)
7.118 B.T. Matthias: Phys. Rev. *92*, 874 (1953)
7.119 N. Pessall, R.E. Gold, H.A. Johansen: J. Phys. Chem. Solids *29*, 19 (1968)
7.120 N. Pessall, C.K. Jones, H.A. Johansen, J.K. Hulm: Appl. Phys. Lett. *7*, 38 (1965)
7.121 E.K. Storms, A.L. Giorgi, E.G. Szklarz: J. Phys. Chem. Solids *36*, 689 (1975)
7.122 J. Geerk, K.G. Langguth, P. Linker, O. Heyer: IEEE Trans. Magn. *13*, 662 (1977)
7.123 J. Geerk, K.G. Langguth: Solid State Commun. *23*, 83 (1977)
7.124 T.H. Geballe, B.T. Matthias, J.R. Remeika, A.M. Clogston, V.B. Compton, J.P. Maita, H.J. Williams: Physica *2*, 293 (1966)
7.125 C.M. Yen, L.E. Toth, Y.M. Shy, D.E. Andersen, L.G. Rosner: J. Appl. Phys. *38*, 2268 (1967)
7.126 L.E. Toth, J. Zbasnik: Acta Metall. *16*, 1177 (1968)
7.127 A. Ubellode: Usp. Khim. *7*, 1692 (1938)
7.128 Ya.S. Umansky: Izv. Sekt. Fiz. Khim. Anal. Inst. Obshch. Neorg. Khim. Akad. Nauk SSSR *16*, 127 (1943); Dokl. Akad. Nauk SSSR *204*, 664 (1972)
7.129 E. Dempsey: Philos. Mag. *8*, 285 (1963)

7.130 R.E. Rundle: Acta Crystallogr. *1*, 180 (1948)
7.131 W. Hume-Rothery: Philos. Mag. *44*, 1154 (1953)
7.132 R.G. Lye, E.M. Logothetis: Phys. Rev. *147*, 622 (1966)
7.133 J.B. Conclin, D.J. Silversmith: Int. J. Quantum Chem. Symp. *2*, 243 (1968)
7.134 I.A. Brytov, E.Z. Kurmaev, S.A. Nemnonov: Fiz. Met. Metalloved. *26*, 366 (1968)
7.135 E.Z. Kurmaev, G.P. Shveikin, S.A. Nemnonov: Phys. Status Solidi B*60*, K65 (1973)
7.136 L. Rumqvist, K. Hamrin, G. Johnsson, A. Fahlman, C. Nordling: J. Phys. Chem. Solids *30*, 1835 (1969)
7.137 V.E. Dolgikh, E.Z. Kurmaev, V.M. Tcherkashenko, G.P. Shveikin: Fiz. Met. Metalloved. *40*, 664 (1975)
7.138 A. Neckel, P. Rastl, P. Weinberger, R. Mechtler: Theor. Chim. Acta *24*, 170 (1972)
7.139 R.W. Simpson: Preprint No. 170, University of Florida, U.S.A. (June 1969)
7.140 K. Schwarz: Monatsh. Chem. *102*, 1400 (1971)
7.141 A. Neckel, K. Schwarz, R. Eibler, P. Rastl, P. Weinberger: Report No.13, Institut für Technische Elektrochemie, Vienna (1974)
7.142 P. Weinberger, P. Podloucky, C.P. Hallett, A. Neckel: J. Phys. Paris C*12*, 801 (1979)
7.143 E. Clementi, A.D. McLean: Phys. Rev. *133*, 419 (1964)
7.144 F. Herman, S. Skilman: *Atomic Structure Calculations* (Prentice-Hall, Englewood Cliffs, NJ 1963)
7.145 P.V. Geld, V.A. Tskhai, A.S. Boruchovich, L.B. Dubrovskaya, I.I. Matveenko: Phys. Status Solidi B*42*, 85 (1971)
7.146 A.S. Boruchovich, P.V. Geld, V.A. Tskhai, L.B. Dubrovskaya, I.I. Matveenko: Phys. Status Solidi B*45*, 179 (1971)
7.147 H.G. Smith, W. Glaser: Phys. Rev. Lett. *25*, 1611 (1970)
7.148 H.G. Smith: In *Superconductivity in d- and f-Band Metals*, ed. by D.H. Douglass (American Institute of Physics, New York 1972) p.321
7.149 W. Kress, P. Roedhammer, H. Bilz, W.D. Teuchert, A.N. Christensen: Phys. Rev. B*17*, 111 (1978)
7.150 A.N. Christensen, O.W. Dieterich, W. Kress, W.D. Teuchert: Phys. Rev. B*19*, 5699 (1979)
7.151 W. Weber, P. Roedhammer, L. Pintschovius, W. Reichardt, F. Compf, A.N. Christensen: Phys. Rev. Lett. *43*, 868 (1979)
7.152 W. Weber: Phys. Rev. B*8*, 5082 (1973)
7.153 M. Mostoller: Phys. Rev. B*5*, 1260 (1972)
7.154 M.P. Verma, B.R.K. Gupta: Phys. Rev. B*12*, 1314 (1975)
7.155 S.K. Sinha, B.N. Harmon: Phys. Rev. Lett. *35*, 1515 (1975)
7.156 M. Gupta, A.J. Freeman: In *Superconductivity in d- and f-Band Metals*, ed. by D.H. Douglass (Plenum, New York 1976) p.313
7.157 M. Gupta, A.J. Freeman: Phys. Rev. Lett. *37*, 364 (1976)
7.158 M. Gupta, A.J. Freeman: Phys. Rev. B*14*, 5205 (1976)
7.159 H.R. Zeller: Phys. Rev. B*5*, 1813 (1972)
7.160 H. Ihara, Y. Kimura: Jpn. J. Appl. Phys. *17*, 281 (1978) Suppl.17-2, p.281
7.161 W. Spengler, R. Kaiser, H. Bilz: Solid State Commun. *17*, 19 (1975)
7.162 W. Spengler, R. Kaiser: Solid State Commun. *18*, 881 (1976)
7.163 R. Kaiser, W. Spengler, S. Schicktanz, C. Politus: Phys. Status Solidi B*87*, 565 (1978)
7.164 M.V. Klein, J.A. Holy. W.S. Williams: Phys. Rev. B*17*, 1546 (1978)
7.165 B.M. Klein, D.A. Papaconstantopoulos: Phys. Rev. Lett. *32*, 1193 (1974); Phys. Rev. B*22*, 1946 (1980)
7.166 U. Haufe, G. Kerker, K.H. Bennemann: Solid State Commun. *17*, 321 (1975)
7.167 K. Schwarz, N. Rosch: J. Phys. Paris C*9*, 1433 (1976)
7.168 T. Skowskiewicz: Phys. Status Solidi A*11*, K123 (1972)
7.169 B. Stritzker, W. Buckel: Z. Phys. *257*, 1 (1972)

7.170 J.M.E. Harper: Phys. Lett. A*47*, 69 (1974)
7.171 N.E. Alekseevskii, Yu.A. Samarskii, G. Wolf, V.I. Tserbo, V.M. Zakosarenko: JETP Lett. *19*, 676 (1974)
7.172 J.E. Schirber: Phys. Lett. A*45*, 141 (1973)
7.173 T. Skowskiewicz, A.W. Szafranski, W. Bujnowski, B. Baranowski: J. Phys. Paris C*7*, 2670 (1974)
7.174 J.E. Schirber, C.J.M. Northrup: Phys. Rev. B*10*, 3818 (1974)
7.175 R.J. Miller, C.B. Satterthwaite: Phys. Rev. Lett. *34*, 144 (1975)
7.176 V.B. Gindlyak, L.N. Zherikhina, A.V. Inyutin: Fiz. Nizk. Temp. *5*, 810 (1979)
7.177 E.Z. Kurmaev, N. Märtenson, R. Nyholm, Yu.M. Yarmoshenko: Solid State Commun. *37*, 647 (1981)
7.178 A.I. Morozov: Fiz. Nizk. Temp. *3*, 835 (1977)
7.179 D.S. McLachlan, T.B. Doyle, J.P. Burger: J. Low Temp. Phys. *26*, 589 (1977)
7.180 N.W. Aschroft: Phys. Rev. Lett. *23*, 48 (1966)
7.181 T. Schneider: Helv. Phys. Acta *342*, 957 (1969)
7.182 N.F. Berk, J.R. Schrieffer: Phys. Rev. Lett. *17*, 433 (1966)
7.183 K.H. Bennemann, J.W. Garland: Z. Phys. *260*, 367 (1974); K.H. Bennemann, J.W. Garland: In *Superconductivity in d- and f-Band Metals*, ed. by D.H. Douglass (American Institute of Physics, New York 1972) p.103
7.184 D.E. Eastman, J.K. Cashion, A.C. Switendick: Phys. Rev. Lett. *27*, 35 (1971); F. Antonangeli, A. Balzarotti, A. Bianconi, E. Burattini, P. Perfetti, N. Nistico: Phys. Lett. A*55*, 309 (1975)
7.185 A.C. Switendick: Solid State Commun. *8*, 1463 (1970)
7.186 A.C. Switendick: Ber. Bunsenges. Phys. Chem. *76*, 535 (1972)
7.187 E.G. Maksimov, O.A. Pankratov: Usp. Fiz. Nauk. *116*, 385 (1975)
7.188 S. Aulick: Lett. Nuovo Cimento *7*, 545 (1973)
7.189 B.N. Ganguly: Z. Phys. *265*, 433 (1973)
7.190 J.R. Gommersal, B.L. Gyorffy: J. Phys. Paris F*3*, L138 (1973)
7.191 P. Hertel: Z. Phys. *268*, 111 (1974)
7.192 D.A. Papaconstantopoulos, B.M. Klein: Phys. Rev. Lett. *35*, 110 (1975)
7.193 G.D. Gaspari, B.L. Gyorffy: Phys. Rev. Lett. *29*, 801 (1972)
7.194 B.M. Klein, D.A. Papaconstantopoulos: Phys. Rev. Lett. *32*, 1193 (1974)
7.195 A.P. Miller, B.N. Brockhouse: Can. J. Phys. *49*, 704 (1971)
7.196 J.M. Rowe, J.J. Rush, H.G. Smith, M. Mostoller, H.E. Flotow: Phys. Rev. Lett. *33*, 1297 (1974)
7.197 D.A. Papaconstantopoulos, B.M. Klein, J.S. Faulkner, L.L. Boyer: Phys. Rev. B*18*, 2784 (1978)
7.198 B.M. Klein, E.N. Economou, D.A. Papaconstantopoulos: Phys. Rev. Lett. *39*, 574 (1977)
7.199 D.A. Papaconstantopoulos, B.M. Klein, E.N. Economou, L.L. Boyer: Phys. Rev. B*17*, 141 (1978)
7.200 R.J. Miller, C.B. Satterthwaite: Phys. Rev. Lett. *34*, 144 (1975)
7.201 A. Maeland, T.B. Flanagan: J. Phys. Chem. *68*, 1419 (1964)
7.202 W. Buckel, A. Eichler, B. Stritzker: Z. Phys. *263*, 1 (1973)
7.203 J.E. Schirber, B. Morosin: Phys. Rev. B*12*, 117 (1975)
7.204 B. Stritzker, J. Becker: Phys. Lett. A*51*, 147 (1975)
7.205 G. Heim, B. Stritzker: Appl. Phys. *7*, 239 (1975)
7.206 W. Buckel, B. Stritzker: Phys. Lett. A*43*, 403 (1973)
7.207 B. Stritzker: Z. Phys. *268*, 261 (1974)
7.208 C.G. Robbins, J. Müller: J. Less Common Met. *42*, 19 (1975)
7.209 C.B. Friedberg, J.C. Pickel, A.F. Rex, J. Ruvalds: Solid State Commun. *29*, 407 (1979)
7.210 B. Stritzker, H.L. Luo: Solid State Commun. *29*, 811 (1979)

7.211 D.A. Papaconstantopoulos, E.N. Economou, B.M. Klein, L.L. Boyer: Phys. Rev. B*20*, 177 (1979)
7.212 S. Kirkpatrick, B. Velicky, H. Ehrenreich: Phys. Rev. B*1*, 3250 (1970)
7.213 G.M. Stocks, R.W. Williams, J.S. Faulkner: J. Phys. Paris F*3*, 1688 (1973)
7.214 J.C. Phillips: J. Appl. Phys. *43*, 3560 (1972)
7.215 P.B. Allen, R.C. Dynes: Phys. Rev. B*12*, 905 (1975)
7.216 B.W. Zachariasen: Acta Crystallogr. *5*, 17 (1952)
7.217 A.E. Austin: Acta Crystallogr. *12*, 159 (1959)
7.218 M.C. Krupka: J. Less Common Met. *20*, 135 (1970)
7.219 M.C. Krupka, A.L. Giorgi, N.H. Krikorian, E.G. Szklarz: J. Less Common Met. *17*, 91 (1969)
7.220 M.C. Krupka, A.L. Giorgi, N.H. Krikorian, E.G. Szklarz: J. Less Common Met. *19*, 113 (1969)
7.221 A.L. Giorgi, E.G. Szklarz, M.C. Krupka: In *Superconductivity in d- and f-Band Metals*, ed. by D.H. Douglass (American Institute of Physics, New York 1972) p.147
7.222 M.C. Krupka, A.L. Giorgi, E.G. Szklarz: J. Less Common Met. *30*, 217 (1973)
7.223 A.L. Giorgi, E.G. Szklarz, M.C. Krupka, N.H. Krikorian: J. Less Common Met. *17*, 121 (1969)
7.224 A.L. Giorgi, E.G. Szklarz, N.H. Krikorian, M.C. Krupka: J. Less Common Met. *22*, 131 (1970)
7.225 A.L. Giorgi, H.H. Hilli, E.G. Szklarz, R.W. White: In *Superconductivity in d- and f-Band Metals*, ed. by D.H. Douglass (Plenum, New York 1976) p.361
7.226 E. Gillam: Acta Crystallogr. *15*, 1183 (1962)
7.227 A.S.Cooper, E. Corenzwit, L.D. Longinotti, B.T. Matthias, W.H. Zachariasen: Proc. Natl. Acad. Sci. U.S.A. *67*,313 (1970)
7.228 B.T. Matthias, E. Corenzwit, J.M. Vandenberg, H.E. Barz: Proc. Natl. Acad. Sci. U.S.A. *74*, 1334 (1977)
7.229 J.M. Vandenberg, B.T. Matthias: Proc. Natl. Acad. Sci. U.S.A. *74*, 1336 (1977)
7.230 D.E. Moncton, D.B. McWhen, J. Eckert, G. Shirane, W. Thomlinson: Phys. Rev. Lett. *39*, 1164 (1977)
7.231 D.C. Johnston: Solid State Commun. *24*, 699 (1977)
7.232 R.H. Wang, R.J. Laskowski, C.Y. Huang, J.L. Smith, C.W. Chu: J. Appl. Phys. *49*, 1392 (1978)
7.233 W.A. Fertig, D.C. Johnston, L.E. DeLong, R.W. McCallum, M.B. Maple, B.T. Matthias: Phys. Rev. Lett. *38*, 987 (1977)
7.234 R.N. Shelton, D.S. Johnston: Int. Conf. High-Pressure and Low-Temperature Physics, Cleveland, Ohio, USA, 1977
7.235 T. Jarlborg, A.J. Freeman, T.J. Watson-Yang: Phys. Rev. Lett. *39*, 1032 (1977)
7.236 R.H. Wang, C.Y. Huang, J.L. Smith: J. Phys. Paris *39*, C6-373 (1978)
7.237 H.R. Ott, W.A. Fertig, D.C. Johnston, M.B. Maple, B.T. Matthias: J. Phys. Paris *39*, C6-375 (1978)
7.238 B.D. Dunlap, G.K. Skenoy, F.Y. Fradin, C.W. Kimball: J. Phys. Paris *39*, C6-379 (1978)
7.239 J.C. Ho, C.Y. Huang, J.L. Smith: J. Phys. Paris *39*, C6-381 (1978)
7.240 R.N. Shelton: J. Less Common Met. *62*, 191 (1978)
7.241 H.E. Barz, A.S. Cooper, E. Corenzwit, M. Marezio, B.T. Matthias, P.H. Schmidt: Science *175*, 884 (1972)
7.242 A.D. Wadsley: Acta Crystallogr. *10*, 715 (1957)
7.243 GBL: Phys. Today *25*, 18 (1972)
7.244 Ø. Fischer: Appl. Phys. *16*, 1 (1978)
7.245 N.E. Alekseevskii: Proc. Symp. Physical Properties of Solids in High Magnetic Fields, Wroclaw, 1978, pp.37-57

7.246 R. Chevrel, M. Sergent, J. Prizent: J. Solid State Chem. *3*, 515 (1971)
7.247 B.T. Matthias, M. Marezio, E. Corenzwit, A.S. Cooper, H.E. Barz: Science *175*, 1465 (1972)
7.248 N.E. Alekseevskii, N.M. Dobrovolskii, D. Ekkert, V.I. Tsebro: JETP *72*, 1145 (1977)
7.249 N.E. Alekseevskii, N.M. Dobrovolskii, V.I. Tsebro: JETP Lett. *23*, 694 (1976)
7.250 Ø. Fischer, R. Odermatt, G. Bongi, H. Jones, R. Chevrel, M. Sergent: Phys. Lett. A*45*, 87 (1973)
7.251 R. Odermatt, Ø. Fischer, H. Gones, G. Bongi: J. Phys. Paris C*7*, L13 (1974)
7.252 S. Foner, R.J. McNiff, Jr., E.J. Alexander: Phys. Lett. A*49*, 269 (1974)
7.253 R.N. McCallum, L.D. Woolf, R.N. Shelton, M.B. Maple: J. Phys. Paris *39*, C6-353 (1978)
7.254 M. Marezio, P.D. Dernier, J.P. Remeika, E. Corenzwit, B.T. Matthias: Mater. Res. Bull. *8*, 657 (1973)
7.255 J. Guillevic, O. Bars, D. Grandjean: Acta Crystallogr. B*32*, 1338 (1976)
7.256 O. Bars, J. Guillevic, D. Grandjean: J. Solid State Chem. *6*, 335 (1973)
7.257 J. Guillevic, O. Bars, D. Grandjean: J. Solid State Chem. *7*, 158 (1973)
7.258 K. Yvon: Solid State Commun. *25*, 327 (1978)
7.259 B. Seeber, M. Decroux, Ø. Fischer, R. Chevrel, M. Sergent, A. Grüttner: Solid State Commun. *29*, 419 (1979)
7.260 A.A. Opalowski, V.E. Fedorov: Izv. Akad. Nauk SSSR Neorg. Mater. *2-3*, 443 (1966)
7.261 M. Spiesse, C. Marckal, J. Rouxel: C. R. Acad. Sci. *266*, 1583 (1968)
7.262 Ø. Fischer, B. Seeber: Phys. Bl. *35*, 655 (1979)
7.263 M. Sergent, Ø. Fischer, M. Decroux: J. Solid State Chem. *22*, 87 (1977)
7.264 C. Perrin, N. Sergent, J. Prigent: C. R. Acad. Sci. C*277*, 465 (1973)
7.265 F.S. Delk, M.J. Sienko: Solid State Commun. *31*, 699 (1979)
7.266 R. Chevrel: Ph. D. Thesis, University of Rennes (1974)
7.267 R. Chevrel, M. Sergent, J. Prigent: Mater. Res. Bull. *9*, 1487 (1974)
7.268 J. Guillevic, H. Lestrat, D. Grandjean: Acta Crystallogr. B*32*, 1342 (1976)
7.269 K. Yvon, A. Paoli, R. Flükiger, R. Chevrel: Acta Crystallogr. B*33*, 3066 (1977)
7.270 K. Yvon, A. Paoli: Solid State Commun. *24*, 41 (1977)
7.271 Ø. Fischer: Proc. Fourteenth Int. Conf. Low Temp. Physics, Vol.V, ed. by M. Krusius, H. Vuorio, Otaniemi, 1975
7.272 M. Sergent, R. Chevrel, C. Rossel, Ø. Fischer: J. Less Common Met. *58*, 179 (1978)
7.273 R.N. Shelton, A.C. Lawson, D.C. Johnston: Mater. Res. Bull. *10*, 297 (1975)
7.274 N.E. Alekseevskii, V.M. Dobrovolskii, V.I. Nizhankovskii, V.I. Tsebro: JETP *69*, 662 (1975)
7.275 A.W. Webb, R.N. Shelton: J. Phys. Paris F*8*, 261 (1978)
7.276 C.W. Kimball, L. Weber, G. Van Landuyt, F.Y. Fradin, B.D. Dunlop, G.K. Shenoy: Phys. Rev. Lett. *36*, 412 (1976)
7.277 N.E. Alekseevskii, G. Wolf, S. Krants, V.I. Tsebro: J. Low Temp. Phys. *28*, 381 (1977)
7.278 O.K. Andersen, W. Klose, H. Nohl: Phys. Rev. B*17*, 1209 (1979)
7.279 L.F. Mattheiss, C.Y. Fong: Phys. Rev. B*15*, 1760 (1977)
7.280 D.W. Bullett: Phys. Rev. Lett. *39*, 664 (1977)
7.281 T. Jarlborg, A.J. Freeman: Bull. Am. Phys. Soc. *24*, 389 (1979); Phys. Rev. Lett. *44*, 178 (1980)
7.282 G.D. Gaspari, B.L. Gyorffy: Phys. Rev. Lett. *28*, 801 (1972)
7.283 W.L. McMillan: Phys. Rev. *167*, 331 (1968)
7.284 R. Viswanathan, A.C. Lawson: Science *177*, 267 (1972)

7.285 H.E. Alekseevskii, E.I. Leyarowskii, L.N. Leyarowska, V.I. Tsebro: J. Phys. Paris *C8*, L585 (1975)
7.286 S.D. Bader, G.S. Knapp, S.K. Sinha, P. Schweiss, B. Renker: Phys. Rev. Lett. *37*, 344 (1976)
7.287 B.P. Schweiss, B. Renker, E. Schneider, W. Reichardt: Proc. Second Rochester Conf. on Superconductivity in d- and f-Band Metals, ed. by D.H. Douglass (Plenum, New York 1976) p.189
7.288 S.D. Bader, S.K. Sinha, R.N. Shelton: Proc. Second Rochester Conf. on Superconductivity in d- and f-Band Metals, ed. by D.H. Douglass (Plenum, New York 1976) p.209
7.289 P. Schweiss, B. Renker, J.-B. Suck: J. Phys. Paris *39*, C6-356 (1978)
7.290 U. Poppe, H. Wühl: J. Phys. Paris *39*, C6-361 (1978)
7.291 S.D. Bader, S.K. Sinha: Phys. Rev. B*18*, 3082 (1978)
7.292 F.J. Culetto, F. Pobell: Phys. Rev. Lett. *40*, 1104 (1978)
7.293 A.C. Lawson, R.N. Shelton: Ferroelectrics *16*, 73 (1977)
7.294 R.N. Shelton: Ph. D. Thesis, University of California, San Diego (1975)
7.295 D.C. Johnston, R.N. Shelton, J.J. Bugai: Solid State Commun. *21*, 949 (1977)
7.296 R. Flükiger, A. Junod, R. Baillif, P. Spitzli, A. Treyvaud, A. Paoli, H. Devantay, J. Müller: Solid State Commun. *23*, 699 (1977)
7.297 R. Flükiger, H. Devantay, J.L. Jorda, J. Müller: IEEE Trans. Magn. *13*, 818 (1977)
7.298 A.C. Lawson: Mater. Res. Bull. *7*, 773 (1972)
7.299 O. Bars, J. Guillevic, D. Grandjean: J. Solid State Chem. *6*, 335 (1973)
7.300 N.E. Alekseevskii, M. Alinski, H.M. Dobrovolskii, V.I. Tserbo: JETP Lett. *23*, 412 (1976)
7.301 S.A. Alterovitz, J.A. Woolam, L. Kammerdiner, H.L. Luo: Appl. Phys. Lett. *31*, 233 (1977)
7.302 N.E. Alekseevskii, N.M. Dobrovolskii, D. Ekkert, V.I. Tsebro: JETP *72*, 1145 (1977)
7.303 N.E. Alekseevskii, C. Bazan, K.H. Berthel, V.A. Ermolov, R.R. Hesske, G.S. Zaitsev: Phys. Status Solidi A*31*, 283 (1975)
7.304 J.P. Remeika, T.H. Geballe, B.T. Matthias, A.S. Cooper, G.W. Hull, E.M. Kelly: Phys. Lett. A*24*, 565 (1967)
7.305 D.C. Johnston, H. Prakash, W.H. Zachariasen, R. Viswanathan: Mater. Res. Bull. *8*, 777 (1973)
7.306 D.C. Johnston: J. Low Temp. Phys. *25*, 145 (1976)
7.307 N.H. Van Maaren, G.M. Schaeffer, F.K. Lotgering: Phys. Lett. A*25*, 238 (1967)

Chapter 8

8.1 B.T. Matthias: In *Superconductivity in d- and f-Band Metals*, ed. by D.H. Douglass (American Institute of Physics, New York 1972) p.367
8.2 J.C. Phillips: J. Appl. Phys. *43*, 3560 (1960)
8.3 L.R. Testardi: In *Physical Acoustics*, ed. by W.P. Mason, R.N. Thurston, (Academic, New York 1973) Vol.10, p.193
8.4 L.R. Testardi, J.H. Wernick, W.A. Roger, D.D. Bacon, A.L. Storm: J. Appl. Phys. *45*, 446 (1974)
8.5 N.E. Alekseevksii: Usp. Fiz. Nauk *95*, 253 (1968)
8.6 E.W. Collings, J.C. Ho, R.J. Jaffee: Phys. Rev. B*5*, 4435 (1972)
8.7 A.C. Lawson: Ph. D. Thesis, University of California, San Diego (1972)
8.8 Yu.A. Izyumov, E.Z. Kurmaev: Usp. Fiz. Nauk *118*, 53 (1976)
8.9 W. Buckel, B. Stritzker: Phys. Lett. A*43*, 403 (1973)
8.10 B. Stritzker: Z. Phys. *268*, 261 (1974)
8.11 H.E. Barz, A.S. Cooper, E. Corenzwit, M. Marezio, B.T. Matthias, P.H. Schmidt: Science *175*, 884 (1972)

8.12 I.J. Shen, J.H. Wernick, T.H. Geballe, S. Mahajan, S. Nakahara: Appl. Phys. Lett. *33*, 103 (1978)
8.13 J.J. Gilman: Aust. J. Phys. *13*, 327 (1960)
8.14 J.L. Staudenmann, P. Coppens, J. Müller: Solid State Commun. *19*, 29 (1976)
8.15 J.L. Staudenmann: Solid State Commun. *23*, 121 (1977)
8.16 P. Coppens: In *Transition Metals*, ed. by M.J.G. Lee, J.M. Perz, E. Fawcett (Institute of Physics, Bristol 1978) p.630
8.17 G.V. Samsonov, V.S. Sinelnikov: Metallofizika *46*, 29 (1973)
8.18 J.M. Ziman: *The Calculation of Bloch Functions*, Solid State Physics: Advances in Research and Applications, Vol.26, ed. by H. Ehrenreich, F. Seitz, D. Turnbull (Academic, New York 1971)
8.19 A.A. Levin: *Vvjedjenije v kvantovuje khimiju tverdogo tjela* (Khimija, Moscow 1974)
8.20 V.A. Gubanov, E.Z. Kurmaev: Int. J. Quantum Chem. Symp. *9*, 297 (1975)
8.21 V.A. Gubanov, B.G. Kasimov, E.Z. Kurmaev: J. Phys. Chem. Solids *36*, 861 (1975)
8.22 L.S. Palatnik, I.I. Falko: Dokl. Akad. Nauk SSSR *217*, 573 (1974)
8.23 L.S. Palatnik, I.I. Falko: Dokl. Akad. Nauk SSSR *234*, 1319 (1977)
8.24 V.M. Pan, V.P. Alekseevskii, A.G. Popov, Yu.I. Beletskii, L.M. Yupko, V.V. Yarosh: JETP Lett. *21*, 497 (1975)
8.25 D. Dew-Hughes, V. Linde: J. Appl. Phys. *50*, 3500 (1979)
8.26 F. Wang: J. Phys. Chem. Solids *34*, 273 (1974)
8.27 B. Stritzker, W. Buckel: Z. Phys. *257*, 1 (1972)
8.28 B.T. Matthias: Phys. Rev. *97*, 74 (1955)
8.29 S.A. Nemnonov, E.Z. Kurmaev: In *Collected Papers: Probljemy svjerkhprovodjaschikh matjerialov* (Nauka, Moscow 1970) p.5
8.30 P.D. Dernier, E. Bucher, L.D. Longinotti: J. Solid State Chem. *15*, 203 (1975)
8.31 F. Galasso, J. Ryle: Acta Crystallogr. *16*, 228 (1963)
8.32 K. Schubert, A. Raman, W. Rostenscher: Naturwissenschaften *51*, 506 (1964)
8.33 G.R. Johnson, D.H. Douglass: J. Low Temp. Phys. *14*, 575 (1974)
8.34 D.U. Gubser, R.A. Hein, R.M. Waterstrat, A. Junod: Phys. Rev. B*14*, 3856 (1976)
8.35 R.H. Hammond, S. Hazra: In *Low Temperature Physics LT-13* (Plenum, New York 1963)
8.36 D.U. Gubser, R.A. Hein, R.M. Waterstrat, A. Junod: Phys. Rev. B*14*, 3856 (1976)
8.37 M.W. Williams, K.M. Ralls, M.R. Pickus: J. Phys. Chem. Solids *28*, 333 (1967)
8.38 K.S. Keskar, T. Yamashita, Y. Onodera: Jpn. J. Appl. Phys. *10*, 370 (1971)
8.39 G. Lautz, E. Schroder: Z. Naturforsch. A*11*, 517 (1956)
8.40 E. Schroder: Z. Naturforsch. A*12*, 247 (1957)
8.41 R.H. Willens, E. Bucher: Appl. Phys. Lett. *7*, 25 (1965)
8.42 L.E. Toth, J. Zbasnik: Acta Metall. *16*, 1177 (1968)
8.43 L.R. Testardi, J.J. Hauser, M.H. Read: Solid State Commun. *9*, 1829 (1971)
8.44 Yu.A. Izyumov, V.E. Naish, V.N. Syromyatnikov: Kristallografia *21*, 256 (1976)
8.45 Yu.A. Izyumov, V.E. Naish, V.N. Syromyatnikov: In *Collected Papers: Problemy physiki tverdogo tjela* (Trydy Institute Physiki Metallov, Sverdlovsk 1976) Vol.31, p.131
8.46 A.C. Lawson, W.H. Zachariasen: Phys. Lett. A*38*, 1 (1972)
8.47 D.E. Moncton: Solid State Commun. *13*, 1779 (1973)
8.48 J. Birman: Phys. Rev. Lett. *17*, 11 (1966)
8.49 J. Perel, B.W. Batterman, E.I. Blount: Phys. Rev. *166*, 616 (1968)
8.50 O.V. Kovalev: *Njeprivodimyje prjedstavljenija prostranstvjennykh grupp* (Izv. Akad. Nauk SSSR, Kiev 1961)

8.51 M.I. Pyetrashen, E.D. Trofimov: *Primjenjenije tjeorii grupp v kvantovoj mjekhanikje* (Nauka, Moscow 1967)
8.52 V.I. Bobrovskii: Thesis, Uralskij Politjekhn. Institut, Sverdlovsk (1971)
8.53 D.K. Faddeev: *Tablicy osnovnykh unitarnykh pjedstavljenij federovskikh grupp* (Izd. Akad. Nauk SSSR, Kiev 1961)
8.54 E. Ascher: Phys. Lett. *20*, 352 (1966)
8.55 L.A. Shuvalov: Ph. D. Thesis, Institut Kristallografii Akad. Nauk SSSR, Moscow (1971)
8.56 G.Y. Lubarskii: *Tjeorija grupp i jeje primjenjenije v fizikje* (Fizmatgiz, Moscow 1958)
8.57 P.W. Anderson, E.I. Blount: Phys. Rev. Lett. *14*, 9 (1965)
8.58 G. Shirane, J.D. Axe: Phys. Rev. B*4*, 2957 (1971)
8.59 K. Inoue, K. Tachikawa: Jpn. J. Appl. Phys. *12*, 161 (1973)
8.60 M. Marezio, P.D. Dernier, C.W. Chu: Phys. Rev. B*4*, 2825 (1971)
8.61 R.H. Willens, T.H. Geballe, A.C. Gossard, J.P. Maita, A. Menth, G.W. Hull, Jr., R.R. Soden: Solid State Commun. *7*, 837 (1969)
8.62 B.N. Kodess: Ph. D. Thesis, PGU, Perm (1973)
8.63 B.N. Kodess: Ukr. Fiz. Zh. *19*, 1570 (1974)
8.64 N.E. Alekseevskii, E.P. Krasnoperov: JETP Lett. *16*, 522 (1972)
8.65 R. Viswanathan: Mater. Res. Bull. *9*, 277 (1974)
8.66 A.A. Boiko, B.N. Kodess, B.S. Shekhtman: Kristallografia *17*, 688 (1972)
8.67 E. Fawcett: Phys. Rev. Lett. *26*, 829 (1971)
8.68 A.C. Lawson: Mater. Res. Bull. *7*, 773 (1972)
8.69 D.C. Johnston, R.N. Shelton, J.J. Bugai: Solid State Commun. *21*, 949 (1975)
8.70 R. Flukiger, A. Junod, R. Baillif, P. Spitzli, A. Treyvaud, A. Paoli, H. Devantay, J. Müller: Solid State Commun. *23*, 699 (1977)
8.71 C.M. Varma, E.I. Blount, P. Vashishta, W. Weber: Phys. Rev. B*19*, 6130 (1979)
8.72 C.M. Varma, W. Weber: Phys. Rev. B*19*, 6142 (1979)
8.73 P.B. Allen: "Phonons and the Superconducting Transition Temperature", in *Dynamical Properties of Solids*, Vol.3, ed. by G.K. Horton, A.A. Maradudin (North-Holland, Amsterdam 1979)
8.74 S.K. Sinha: Phonons in Transition Metals, Preprint, Argonne Laboratory (1979)
8.75 K.L. Ngai, T.L. Reicke: Phys. Rev. B*16*, 1077 (1970)
8.76 G.M. Vujicic, V.L. Aksenov, N.B. Plakida, S. Stamenkovic: Phys. Lett. A*73*, 439 (1979)

Chapter 9

9.1 P.S. Swartz, H.R. Hart, R.L. Fleischer: Appl. Phys. Lett. *4*, 71 (1964)
9.2 S.T. Sekula, R.H. Kernohan: J. Appl. Phys. *39*, 2516 (1964)
9.3 H.T. Coffey, F.L. Keller, A. Patterson, S.H. Autlier: Phys. Rev. *155*, 355 (1967)
9.4 J.P. McEvoy, R.F. Decell, R.L. Novak: Appl. Phys. Lett. *4*, 43 (1964)
9.5 O. Meyer, B. Seeber: Solid State Commun. *22*, 603 (1977)
9.6 B.S. Brown, J.W. Hafstrom, T.E. Klippert: J. Appl. Phys. *48*, 1759 (1977)
9.7 A.R. Sweedler, D.E. Cox, S. Moehlecke: J. Nucl. Mater. *72*, 50 (1978)
9.8 D.G. Schweitzer, D.H. Parkin: Appl. Phys. Lett. *24*, 333 (1974)
9.9 H. Bauer, E.J. Saur, D.G. Schweitzer: J. Low Temp. Phys. *19*, 189 (1975)
9.10 A.E. Karkin, V.E. Arkhipov, S.A. Davydov, V.I. Savin, S.P. Kurushin, B.N. Goshitskii, E.Z. Valiev: Fiz. Met. Metalloved. *45*, 445 (1978)
9.11 D. Dew-Hughes, R. Jones: Appl. Phys. Lett. *36*, 856 (1980)
9.12 A.R. Sweedler, D.G. Schweitzer, G.W. Webb: Phys. Rev. Lett. *33*, 168 (1974)

9.13 R. Bett.: Cryogenics *14*, 361 (1974)
9.14 M. Soll, K. Boning, H. Bauer: J. Low Temp. Phys. *24*, 631 (1976)
9.15 A.R. Sweedler, G.W. Webb, L. Newkirk, T.H. Geballe, R. Schwall, B.T. Matthias, E. Corenzwit: Proc. Int. Conf. on Radiation Effects and Tritium Tech. for Fusion Reactors, Gatlinburg, TN (National Technical Information Service, U.S. Dept. of Commerce, Springfield, VA 1975) Vol.2, p.422
9.16 M. Gurvitch, A.K. Gosh, B.L. Gyorffy, H. Lutz, O.F. Kammerer, J.S. Rosner, M. Strongin: Phys. Rev. Lett. *39*, 1102 (1977)
9.17 G.W. Gullen, R.L. Novak: J. Appl. Phys. *37*, 3348 (1966)
9.18 M. Soll, H. Bauer, K. Boning, R. Bett: Phys. Lett. A*51*, 83 (1975)
9.19 H. Wiesmann, M. Gurvitch, A.K. Gosh, H. Lutz, K.W. Jones, A.N. Goland, M. Strongin: J. Low Temp. Phys. *30*, 513 (1978)
9.20 S. Moehlecke, D.E. Cox, A.R. Sweedler: J. Less Common Met. *62*, 111 (1978)
9.21 A.R. Sweedler, D.E. Cox, S. Foner, E.J. McNiff: Bull. Am. Phys. Soc. *3*, 461 (1975)
9.22 M. Conach, J. Doulat, E. Bonjour: IEEE Trans. Magn. *13*, 655 (1977)
9.23 A.E. Karkin, V.E. Arkhipov, B.N. Goshchitskii, E.P. Romanov, S.K. Sidorov: Phys. Status Solidi A*38*, 433 (1976)
9.24 D.M. Parkin, D.G. Schweitzer: Nucl. Technol. *22*, 108 (1974)
9.25 D.M. Parkin, A.R. Sweedler: IEEE Trans. Magn. *11*, 166 (1976)
9.26 K. Wohlleben: J. Low Temp. Phys. *13*, 269 (1973)
9.27 V. Arkhipov, A.E. Karkin, V.I. Voronin, A.V. Mirmelstein: Phys. Status Solidi (in press)
9.28 S. Klaumunzer, G. Ischenko, P. Müller: Z. Phys. *268*, 189 (1974)
9.29 H. Berndt, F. Seretz: Phys. Lett. A*33*, 427 (1970)
9.30 H. Neumüller, M. Lehmann, G. Ischenko, S. Klaumunzer, H. Adrian: Phys. Lett. A*62*, 446 (1977)
9.31 B. Stritzker: Phys. Rev. Lett. *42*, 1769 (1979)
9.32 A.R. Sweedler, D.E. Cox, S. Moehlecke, R.H. Jones, L.R. Newkirk, F.A. Valencia: J. Low Temp. Phys. *24*, 645 (1976)
9.33 D. Dew-Hughes, S. Moehlecke, D.O. Welch: J. Nucl. Mater. *72*, 225 (1978)
9.34 T.L. Fancavilla, B.N. Das, D.N. Gubser, R.A. Meussner, S.T. Sekula: J. Nucl. Mater. *72*, 203 (1978)
9.35 G. Ischenko, H. Adrian, S. Klaumunzer, M. Lehmann, P. Müller, H. Neumüller, W. Szymczak: Phys. Rev. Lett. *39*, 43 (1977)
9.36 M. Gurvitch, A.K. Ghosh, C.L. Snead, M. Strongin: Phys. Rev. Lett. *39*, 1102 (1977)
9.37 A.R. Sweedler, D. Cox: Phys. Rev. B*12*, 147 (1975)
9.38 A.I. Skvortsov, Y.V. Shemeljov, V.E. Klepatski, B.M. Levitski: J. Nucl. Mater. *72*, 198 (1978)
9.39 J.M. Poate, R.C. Dynes, L.R. Testardi, R.H. Hammond: Phys. Rev. Lett. *37*, 1308 (1976)
9.40 J.M. Poate, L.R. Testardi, A.R. Storm, W.M. Augustyniak: Phys. Rev. Lett. *35*, 1290 (1975)
9.41 A.R. Sweedler, D.G. Schweitzer, G.W. Webb: Phys. Rev. Lett. *33*, 168 (1974)
9.42 D.E. Cox, J.A. Tarvin: Phys. Rev. B*18*, 22 (1978)
9.43 R. Flukiger: Ph. D. Thesis, Geneva (1972)
9.44 M. Krivoglaz: *The Theory of X-Ray and Thermal Neutrons Scattering with Real Crystals* (Plenum, New York 1969)
9.45 A. Guinier: *Theorie et Technique de la Radiocristallographie* (Dunod, Paris 1956)
9.46 G.W. Cullen, G.D. Cody: J. Appl. Phys. *44*, 2838 (1973)
9.47 C.S. Pande: Solid State Commun. *24*, 241 (1977)
9.48 C.S. Pande: Phys. Status Solidi A*52*, 687 (1979)
9.49 L.R. Testardi, J.M. Poate, W. Weber, W.M. Augustyniak, J.H. Barrett: Phys. Rev. Lett. *39*, 716 (1977)
9.50 R.D. Burbank, R.C. Dynes, J.M. Poate: J. Low Temp. Phys. *36*, 573 (1979)

9.51 L.R. Testardi, J.M. Poate, H.J. Levinstein: Phys. Rev. B*15*, 2570 (1977)
9.52 R.H. Herber, R. Kalish: Phys. Rev. B*16*, 1789 (1977)
9.53 J.C. Glass, F. Pobell, J.M. Soder, J. Bolz: J. Low Temp. Phys. *36*, 701 (1979)
9.54 B.N. Goshchitskii: Fiz. Met. Metalloved. *48*, 707 (1979)
9.55 V.G. Chudinov, N.V. Moseev, B.N. Goshchitskii, V.I. Protasov: Phys. Status Solidi A*55*, 109 (1979)
9.56 V.G. Chudinov, V.I. Protasov, M.T. Telichko: Fiz. Met. Metalloved. *43*, 692 (1977)
9.57 E.S. Greiner, H. Mason: J. Appl. Phys. *35*, 3058 (1964)
9.58 A. Guha, M.P. Sarachik, F.W. Smith, L.R. Testardi: Phys. Rev. B*18*, 9 (1978)
9.59 R. Viswanathan, R. Caton: Phys. Rev. B*18*, 15 (1978)
9.60 O. Anderson: In *Physical Acoustics*, ed. by W.P. Mason (Academic, New York 1965) Vol.3B, p.43
9.61 B.W. Batterman, C.S. Barrett: Phys. Rev. Lett. *13*, 390 (1964)
9.62 G. Shirane, J.D. Axe, R.J. Birgenau: Solid State Commun. *9*, 397 (1977)
9.63 L.R. Testardi, R.L. Meck, J.M. Poate, W.A. Royer, A.R. Storm, J.H. Wernick: Phys. Rev. B*11*, 4304 (1975)
9.64 N.B. Brandt, N.I. Ginzburg: Usp. Fiz. Nauk *85*, 485 (1965)
9.65 H. Wiesmann, M. Gurvitch, A.K. Ghosh, H. Lutz, O.A. Kammerer, M.M. Strongin: Phys. Rev. B*17*, 122 (1978)
9.66 A.K. Ghosh, M. Gurvitch, H. Wiesmann, M. Strongin: Phys. Rev. B*18*, 6116 (1978)
9.67 P.B. Allen, R.C. Dynes: Phys. Rev. B*12*, 905 (1975)
9.68 R.C. Dynes, C.M. Varma: J. Phys. Paris F*6*, L215 (1976)
9.69 T. Jarlborg: Ph. D. Thesis, Chalmers University of Technology (1977)
9.70 P.W. Anderson: J. Phys. Chem. Solids *11*, 26 (1959)
9.71 D. Markovitz, L.P. Kadanoff: Phys. Rev. *131*, 563 (1963)
9.72 D.E. Farrell, B.S. Chandrasekhar: J. Nucl. Mater. *72*, 88 (1978)
9.73 M. Fahnle: Phys. Status Solidi B*84*, 541 (1977)
9.74 L.V. Meisel, P.J. Cote: Phys. Rev. B*19*, 4514 (1979)
9.75 I. Nakayama, T. Tsuneto: Prog. Theor. Phys. *58*, 1060 (1977)
9.76 I. Nakayama, T. Tsuneto: Prog. Theor. Phys. *59*, 1418 (1978)
9.77 J. Appel: Phys. Rev. B*13*, 3203 (1976)
9.78 A.S. Aleksandrov, V.F. Elesin, M.P. Kazeko: Fiz. Tverd. Tela Leningrad *21*, 2062 (1979)
9.79 E.V. Kuznetzov, V.I. Anisimov, V.A. Gubanov, B.N. Goshchitskii, E.Z. Kurmaev: Solid State Commun. *38*, 507 (1981)
9.80 M. Fahnle: Phys. Status Solidi B*92*, K127 (1979)
9.81 L.R. Testardi, L.F. Mattheiss: Phys. Rev. Lett. *41*, 1612 (1978)
9.82 L.F. Mattheiss, L.R. Testardi: Phys. Rev. B*20*, 2196 (1979)
9.83 A.A. Abrikosov, L.P. Gor'kov, I.E. Dzyaloshinskii: *Mjetody kvantovoi tjeorii polja v statistichjeskoj physikje* (GIFML, Moscow 1962)
9.84 J. Labbé, E.C. van Reuth: Phys. Rev. Lett. *24*, 1232 (1970)
9.85 J. Ruvalds, C.M. Soukoulis: Phys. Rev. Lett. *43*, 1263 (1979)
9.86 B.M. Klein, L.L. Boyer, D.A. Papaconstantopoulos, L.F. Mattheiss: Phys. Rev. B*18*, 6411 (1978)
9.87 C.J. Ballhausen, H.B. Gray: *Molecular Orbital Theory* (Benjamin, New York 1964)
9.88 H. Basch, A. Viste, H.B. Gray: J. Chem. Phys. *44*, 10 (1966)
9.89 K.H. Johnson: Adv. Quantum Chem. *7*, 143 (1973)
9.90 D.E. Ellis, G.A. Benesh, E. Byron: Phys. Rev. B*20*, 1198 (1979)
9.91 R. Hoffmann: J. Chem. Phys. *39*, 1397 (1963); *40*, 2474, 2745 (1964)
9.92 M. Wolfsberg, L. Helmholtz: J. Chem. Phys. *20*, 837 (1952)
9.93 V.A. Gubanov, E.Z. Kurmaev, G.P. Shveikin: J. Phys. Chem. Solids *38*, 201 (1977)

9.94 A.L. Ivanovskii, V.A. Gubanov, E.Z. Kurmaev, A.L. Hagstrom, S.E. Karlsson, L.I. Johansson: J. Electron Spectrosc. Relat. Phenom. *16*, 415 (1979)
9.95 V.I. Anisimov, V.A. Gubanov, A.L. Ivanovskii, E.Z. Kurmaev, J. Weber, R. Lacroix: Solid State Commun. *29*, 185 (1979)
9.96 R.C. Dynes, L.R. Testardi, R.H. Hammond: In *Superconductivity in d- and f-Band Metals*, ed. by D.H. Douglass (Plenum, New York 1976) p.489

Subject Index

MIX
Papier aus verantwortungsvollen Quellen
Paper from responsible sources
FSC® C105338

If you have any concerns about our products,
you can contact us on
ProductSafety@springernature.com

In case Publisher is established outside the EU,
the EU authorized representative is:
Springer Nature Customer Service Center GmbH
Europaplatz 3, 69115 Heidelberg, Germany

Printed by Libri Plureos GmbH
in Hamburg, Germany